International Conference on Materials, Mechanics and Structures (ICMMS 2020)

IOP Conference Series: Materials Science and Engineering
Volume 936

Online
14-15 July 2020

Editors:

Sajith A.S.
Robin Davis P.

ISBN: 978-1-7138-4772-4
ISSN: 1757-8981

Printed from e-media with permission by:

Curran Associates, Inc.
57 Morehouse Lane
Red Hook, NY 12571

Some format issues inherent in the e-media version may also appear in this print version.

Institute of Physics
Dirac House, Temple Back
Bristol BS1 6BE UK

Phone: 44 1 17 929 7481
Fax: 44 1 17 920 0979

techtracking@iop.org

Additional copies of this publication are available from:

Curran Associates, Inc.
57 Morehouse Lane
Red Hook, NY 12571 USA
Phone: 845-758-0400
Fax: 845-758-2633
Email: curran@proceedings.com
Web: www.proceedings.com

TABLE OF CONTENTS

BUILDING MATERIALS COMPOSITE STRUCTURES

BRIDGES

CONTROL AND HEALTH MONITORING

CONCRETE AND MASONRY STRUCTURES

FRACTURE MECHANICS

FLUID-STRUCTURE INTERACTION

LIGHTWEIGHT STRUCTURES

PERFORMANCE-BASED DESIGN

REPAIR, RETROFITTING AND REHABILITATION

SMART MATERIALS AND STRUCTURES

SPECIAL STRUCTURES

SAFETY, RELIABILITY AND RISK ANALYSIS

STRUCTURAL MECHANICS

STRUCTURAL STABILITY

TALL BUILDINGS

IOP Conf. Series: Materials Science and Engineering **936** (2020) 011001 doi:10.1088/1757-899X/936/1/011001

PREFACE

Civil Engineers are the builders of a nation. They always have played a significant role in providing the infrastructure required for the society. Ever increasing population, rapid urbanization and industrialization, global warming and climate change, increasing incidents of natural and manmade hazards and environmental degradation have become major concerns of the society today and these concerns pose several challenges to the Civil Engineers. There has been a noticeable change in the environment within which civil engineers carry out their work. There is a growing concern on the adverse consequences of the development activities. There is a need for Civil Engineers to rise up to face these challenges and contribute to the development and betterment of the nation in all core areas. A multitude of threats confront us in our effort to achieve the same. Civil engineers have always been looking for innovative tools and techniques for solving the challenging problems that they face during planning, analysis, design and construction of civil engineering structures. It was in this context that the Department of Civil Engineering, National Institute of Technology Calicut, Kerala, India, has come out with the proposal of organizing an *International Conference on Materials, Mechanics and Structures 2020 (ICMMS 2020)* with major funding from Technical Education Quality Improvement Program-Phase III. This conference was aimed to provide a forum to all Civil Engineers to deliberate on the innovative tools and techniques that are needed to face the challenges in the fields of Materials, Mechanics and Structures.

The advisory and technical committees of the conference consisted of experts from varied fields of materials, mechanics and structural engineering. Most of the relevant areas were identified as themes of the conference. We received an overwhelming response to our call for submission of abstracts to this conference. We received more than 200 abstracts from across the country. It is heartening to note that several of the contributions that we received are from young researchers/ faculty. Subsequently, we received 77 full length papers from the accepted abstracts. Through our review process we ensured that only quality papers were selected for presentation during the conference. After the peer review 71 full length papers were accepted for the oral presentation during the conference. The review process was conducted through *EasyChair* by a review panel consisting of 31 experts from different educational and research institutes around the world.

The original schedule of the conference was 26 – 28, March 2020. The event could not be conducted due to the Covid 19 pandemic and was postponed following orders from Government of India. Many of the countries had imposed travel restrictions by that time. Subsequently, the conference was postponed with due intimation to all concerned. Later on the event was conducted in the online mode during 14 – 15, July 2020, in the best interest of the participants and the Institute. Google Meet was used as the online platform for the conduct of the event.

The conference was inaugurated by Dr Sivaji Chakravorti, Director, National Institute of Technology Calicut, on 14th July 2020. There were 4 plenary sessions which consisted of 12 keynote lectures by eminent academicians in the thrust areas of structural engineering. Out of them 3 were by foreign speakers and the remaining, by Indian speakers. There were a total of 14 technical sessions each consisting of about 5 paper presentations by authors. A total of 67 papers were presented in the conference. All the sessions were well planned and trial online sessions were conducted in advance to make each and every stake holder comfortable with the online platform - Google Meet. All the sessions had fruitful interactions also. Duration stipulated for author paper presentation was 10 minutes, followed by a 3 minutes Q/A session by chair. In the plenary session each keynote speaker was given a time of 35 minutes to speak, again, followed by 10 minutes for Q/A. There were product presentation sessions from three technology players. One of them was a company having international presence. Altogether around 150 people from India and abroad (Australia, Spain, Sri Lanka, UK and USA) attended the conference during the two days. The conference was concluded with a valedictory function in the afternoon of 15th July 2020. Several people including participants, keynote speakers, invitees gave positive feedbacks about the conference and its conduct.

The conference was well received by all and is judged to have provided a platform, especially for young budding researchers to share their ideas and vision, notwithstanding the fact that it was conducted in the online (virtual) mode. New and innovative technologies that are needed for Civil Engineers to face the challenges in the present world were the objectives of this conference. It is hoped that this conference has helped in building a network of researchers, faculty and practicing engineers who can work towards the goal of achieving the same.

Dr Sajith A. S. & Dr Robin Davis P.

Editors

IOP Conf. Series: Materials Science and Engineering **936** (2020) 011002 doi:10.1088/1757-899X/936/1/011002

ICMMS 2020

TEQIP-III

Technical Education Quality Improvement Programme

Peer review declaration

All papers published in this volume of *IOP Conference Series: Materials Science and Engineering* have been peer reviewed through processes administered by the Editors. Reviews were conducted by expert referees to the professional and scientific standards expected of a proceedings journal published by IOP Publishing.

- **Type of peer review:** Single-blind
- **Conference submission management system:** *EasyChair*
- **Number of submissions received:** 77
- **Number of submissions sent for review:** 77
- **Number of submissions accepted:** 55
- **Acceptance Rate (Number of Submissions Accepted / Number of Submissions Received X 100):** 71.43%
- **Average number of reviews per paper:** 2
- **Total number of reviewers involved:** 31
- **Any additional info on review process:**
    - In addition to the peer review conducted before the oral presentation, remarks from the chair and co-chair during the technical paper presentations were also given due weightage in the selection of papers for publication.
    - Plagiarism/Similarity Index check was performed using Turnitin Service at National Institute of Technology Calicut, Kerala, India. The Cut-off was kept 25% without considering the references. Two papers with plagiarism slightly above the cut-off have also been forwarded as the exceedance is due to the similarity from their own submitted thesis works.
- **Contact person for queries:** Dr Sajith A. S. (sajithas@nitc.ac.in)

IOP Conf. Series: Materials Science and Engineering **936** (2020) 012001 doi:10.1088/1757-899X/936/1/012001

A Review on Use of Crushed Brick Powder as a Supplementary Cementitious Material

Mani Mohan[1*], Aditya Apurva[2], Nishant Kumar[2] and Abhishek Ojha[2]

[1]Assistant Professor, Department of Civil and Environmental Engineering, Birla Institute of Technology, Ranchi, India

[2] PG Student, Department of Civil and Environmental Engineering, Birla Institute of Technology, Ranchi, India

*mani.mohan@bitmesra.ac.in

Abstract. In India, it is estimated that 250 billion bricks are produced annually in about 100,000 kilns present all over the country. Mainly due to inefficient production processes, most of the blocks manufactured are of inferior quality, which ultimately becomes a significant source of waste production. The characteristics of Construction and Demolition wastes (C&D) are difficult to predict, but a considerable component of this type of solid waste comprises of bricks in masonry work. As more and more civil structures are nearing the end of their expected lifespan, proper management and disposal of C&D wastes are necessary to obtain a sustainable environment. Ground waste clay brick is a prospective pozzolanic material due to dehydroxylation of clay minerals during its manufacturing process at temperatures between 450°C and 700°C, leading to the disintegration of crystalline phases and formation of reactive anhydrous amorphous phases. A review of the existing literature regarding the use of crushed brick powder (CBP) from waste bricks as well as demolition debris as a supplementary cementitious material has been conducted. It has been observed that the pozzolanic character of clay used for making brick comes from calcining it at an optimum temperature, which is different for different clays. Cement replacement by CBP in concrete mostly causes increased compressive, flexural, and split – tensile strength at 28 and 90 days, indicating good pozzolanic behaviour. 10 – 25% replacement by CBP gives acceptable values of Strength Activity Index (SAI) at 7 and 28 days. Pore refinement causing the formation of additional hydration products due to pozzolanic reaction as well as filler effect of CBP are the primary factors causing improved durability of concrete. Limited research has been done on the effect of the chemical composition of raw clay on the pozzolanic potential of CBP. Future scope in this area can be on enhancement of pozzolanicity in CBP, the study of high volume replacement of CBP in concrete, and the effect of CBP replacement on corrosion of rebar.

1. Introduction

The most widely used construction binding material cement can be categorised as non-hydraulic and hydraulic. In today's fast-paced construction, non-hydraulic cement is preferred less because of desired conditions requirement with delay in strength gain, whereas hydraulic cement set and harden even in the presence of water. Hydraulic cement remains a primary construction material for the buildout of infrastructure all over the world, and there is less possibility that any other materials would replace it in the coming future. Global production of cement in the year 2015 was about 3.5 billion metric tons, and it is expected to touch 4.83 billion metric tons by 2030. China stands tall with 2.35 billion metric ton cement production in the year 2015, followed by India and the United States with the production of 270 and 83.4 million metric tons respectively in the same year [1]. Manufacture of cement is an energy-intensive process with the emission of carbon dioxide and other greenhouse gases. Industrial waste and agricultural wastes such as fly ash, blast furnace slag, silica fume, rice husk ash, and bagasse have been used as partial replacement of ordinary Portland cement (OPC) as an environment-friendly solution without having to sacrifice the quality and strength. It has already been reported that these materials improve concrete durability and reduce the risk of thermal cracking in mass concrete. However, the availability of industrial waste locally and season dependency of agricultural wastes brings out limitations of its use. Among all waste products applicable to replace cement partially, waste fired clay bricks are one which is readily available locally. Bricks are generally treated as waste when some damaged bricks are manufactured in the production line or due to poor handling at the construction site or when demolishing any existing structure with fire clay bricks as one of its building blocks.

Globally, 1.5 trillion bricks are produced annually with Southeast Asia, South Asia, and China being the large brick manufacturing region in the world, producing about 87 percent of the total bricks produced globally. Around 67 percent of the global bricks production is produced by China alone, followed by India, producing about 13 percent of the total bricks [2-4]. These bricks are typically fired in small-scale traditional kilns. The quality of fired brick depends both on the quality of green bricks as well as on the firing process. The tunnel kiln is ranked best in terms of fired brick quality. The Fixed chimney bull trench kiln (FCBTK) lacks uniform distribution of temperature in the firing zone, thereby 60-70% bricks are properly fired, whereas high draught zig-zag has more uniform temperature distribution, and hence 80-90% of the bricks are appropriately fired. 95%, 80%, and 60% are the quality bricks produced by tunnel kiln, high draught zig-zag kiln, and FCBTK, respectively. The remaining production is either inferior (under-fired and over-burnt), and there are losses due to breakages too. In India, the estimated annual production of bricks by FCBTK alone is 185 billion bricks out of an estimated 250 billion bricks in a year [4], i.e., 74 billion bricks generated are of not good quality.

Construction and Demolition Waste (C&D) characteristics are difficult to predict; however, the primary components of this waste comprise of bricks from masonry work. Annually India produces 10-12 million tons of Construction, Demolition, and Excavation waste material. Figure 1 shows the typical composition of C&D waste in India[5]. As more and more civil structures are nearing the end of their awaited lifespan, management and disposal of C&D wastes is the need of the hour to attain a sustainable built environment. In a report by Cheng et al. [6], it is stated that at least 20 billion m^3 of clay bricks produced in China will turn into solid waste in the next five decades, accounting for about 50% CDW. This scenario is highly applicable in the Indian backdrop too.

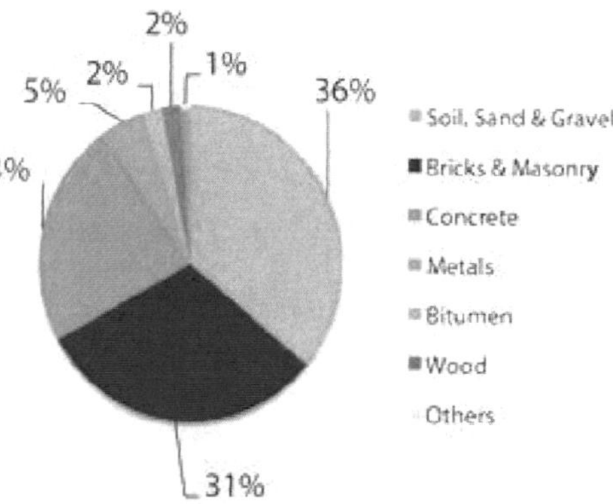

Figure 1. Typical composition of C&D waste in India

Grounded waste clay bricks are a prospective pozzolanic material due to the dehydroxylation of clay minerals during its manufacturing process at temperatures between 450°C and 700°C [7]. Hence, the reuse of waste brick in powdered form in cement-based composite materials can significantly bring social and economic benefits. In the present framework, there is an urgent requirement for sustainable use of waste bricks in the building and construction industry. Unless a systematic effort is made to correctly incorporate these types of waste materials into sustainable construction practices, dire wastage of finite natural resources will continue to occur. Keeping these views in context, a review of the existing literature regarding the use of crushed brick powder from waste bricks as well as demolition debris at this moment referred to as crushed brick powder (CBP) unless specified otherwise as a supplementary cementitious material has been conducted.

2. Pozzolanic nature of brick dust

Pozzolans' characteristics are supposed to be determined by their ability to react with lime and the ability to form insoluble products with binding materials. According to ASTM C125 [8], a pozzolan is defined as "a siliceous and aluminous material which, in itself, possesses little or no cementitious value but which will, in the finely divided form in the presence of moisture, react chemically with calcium hydroxide at ordinary temperature to form compounds possessing cementitious properties." Clay minerals like kaolinite, illite, montmorillonite, palygorskite, and others are present in clay deposits together with impurities of non-clay materials like calcite, quartz, feldspar, mica, sulfides, and anatase [9]. The primary raw material for preparing ordinary bricks are the clay deposits with less proportion of clay minerals. The obtained content is processed to get a workable mass to mould into a brick shape. These shaped bricks are then usually fired to a temperature above 800°C. Loss of structural water of clay minerals at temperatures between 450°C and 700°C leads to crystal lattice deterioration resulting in anhydrous compound with amorphous structure [7]. This transformation of the crystalline network to a disordered, unstable amorphous state as a result of thermal treatment is responsible for bringing pozzolanic characteristics to it. Baronio et al. [10] put forward the following three hypothetical chemical reactions between calcium hydrate, amorphous silica, and alumina in case of kaolinitic clays:

$$AS_2 + 6CH + 9H \rightarrow C_4AH_{13} + 2CSH \tag{1}$$

$$AS_2 + 5CH + 3H \rightarrow C_3AH_6 + 2CSH \tag{2}$$

$$AS_2 + 3CH + 6H \rightarrow C_2ASH_8 + CSH \tag{3}$$

AS_2- metakaolinite, CH-lime, H-water, C_4AH_{13}- tetra calcium aluminate hydrate, CSH-tobermorite, C_3AH_6- tricalcium aluminate hydrate, C_2ASH_8-hydrate gehlenite.

They studied both kaolinitic clay and common clay, which are used in the manufacture of bricks. When pozzolanic tests were carried out on both the types of clays, common clay did not show many positive results, which was attributed to its composition, whereas kaolinitic earth was shown to possess excellent pozzolanic characteristics. The authors concluded that modern bricks seldom possess pozzolanic characters as they are fired at higher temperatures and may not contain the proper content of clays. Firing temperature above 900°C was not recommended for producing pozzolanic materials. But this claim was disputed by Wild et al. [11]. In his study, eight different types of bricks were analysed. Based on the chemical tests for pozzolanic and strength development of mortars, it was concluded that all the ground brick types exhibit pozzolanic activity. The type of clay used in brick manufacture also influences the pozzolanic properties of CBP. Ambroise et al. [12] put the order of reactivity of thermally activated (TA) clays with lime and water-based on experimental observations as TA-kaolinite > TA-montmorillonite > TA- poorly-crystallized mica (illite) > TA-well crystallized micas. It can be concluded that depending on many factors like phase composition and range of firing temperature, fired clay building bricks may show a variety of pozzolanic activity; however, evaluation for the pozzolanic activity of CBP as supplementary cementitious material is necessary.

3. Chemical and physical characteristics of brick dust

As described by Akinshipe et al. (2017) [13], the clay brick firing process consists of 6 steps, with calcium sulphate ($CaSO_4$) being the most likely compound formed during the firing process. The first stage, evaporation, occurs at 20 – 150°C, during which the mechanical water is evaporated. At 149 – 650°C, i.e., dehydration stage, combustion of the combined water and carbonaceous matter takes place. At 300 – 982°C, oxidation of iron residues and remaining carbonaceous things takes place. The full strength of fired bricks is formed during the vitrification process (900 – 1316°C) through the sintering of clay particles and the melting of clay mass. The final stages are flashing (1150 – 1316°C), required to bring a proper colour to the brick and cooling (1316 - 20°C), after which the final product is obtained.

Crushed clay brick powder used by various researchers come from two primary sources: discarded bricks from brick kilns which do not meet the required standards due to improper heating regimes producing over burnt or under burnt bricks, presence of cracks, warping, asymmetrical dimensions and other imperfections [14-19] or demolished masonry from construction sites (construction and demolition waste) [20-23]. Both these materials fall under the category of solid wastes. Table 3 gives a comparison of the chemical constituents of bricks obtained from different sources. It can be observed that the variation of various chemical components is not very high. Manoharan et al. [24] studied the brick production process with the help of alluvial clay. The composition of alluvial clay before firing is included in Table 3 and can be referred for comparison with CBP. O' Farell et al. [16] had used heat-treated clay in addition to ground waste clay brick and reported that both were of similar composition, whereas the fineness was approximately the same as that of cement. Sintered red clay ceramic [25] classified under CDW also has similar properties as the other two sources, as can be seen from the table.

Table 1. Chemical constituents of CBP.

Component (wt %) →	SiO_2	Al_2O_3	Fe_2O_3	K_2O	MgO	Na_2O	CaO	P_2O_5	TiO_2	SO_3	LOI
Heat treated clay [16]	54.8	19.1	6	3.2	1.8	0.5	9.4	—	1	2.9	0.6
Ground crushed waste calcined clay brick [19]	63.89	25.49	7.73	0.95	0.04	traces	0.29	—	—	—	—
Red clay brick waste [25]	49.9	16.6	6.5	4.4	5.5	0.5	9.7	0.2	0.8	3.3	2.4
Clay brick demolished waste [20]	54.2	15.4	7.6	—	2.5	—	6.8	—	—	1.1	6.2
Alluvial Clay (TRY)	62.75	16.41	6.37	1.87	3.90	1.68	3.53	—	0.5	—	3.39
Alluvial Clay (KOM) [24]	64.23	15.7	5.52	3.62	2.35	0.8	3.42	—	0.92	—	4.74

Carbon content in cement must be carefully controlled as it is known to have detrimental effects on concrete properties. LOI can characterise carbon content, and it was experimentally proved [21] that LOI decreases as the amount of CBP in hybrid recycled concrete powder increases. The physical characteristics of CBP are summarised in Table 4.

Table 2. Physical characteristics of CBP.

Blaine specific surface area **[22]**	6485 m^2/kg
Density **[22]**	2660 kg/m^3
Specific gravity **[14]**	2.65
Mean particle size **[25]**	20.9 μm

Naceri et al. (2009) [26] found that specific weight decreased with the increasing quantity of CBP substituted in cement. Also, the incorporation of CBP in cement was found to reduce the grinding time (reduction in consumption of energy).

4. Effect of CBP on the hydration of cement-based on microstructural development

The study of microstructural properties of cement paste with the progress of hydration is essential because it is well established that the strength and durability properties of concrete originate from its internal microstructure and can be significantly enhanced by making suitable changes to the microstructure. Well-developed relations between microstructure and properties of concrete are not present, mainly because it is not an intrinsic characteristic of the material and subject to changes in morphology and formation of hydration products with the progress of time, humidity, and other ambient factors.

4.1. Normal consistency and setting time

Cement replacement by more percentage of brick dust, increased the water demand of cement paste, as reported by Naceri et al. (2009) [26]. A double effect was observed by the authors: the quantity of water required to have a paste of standard consistency increased along with a decrease in the set times. The reduction in the setting time was attributed to the increased gypsum/clinker ratio.

However, Lin et al. (2010) [27] reported that there was increase in setting time with the increase in amount of waste brick added to pastes. Both initial and final setting times were longer than that of corresponding OPC paste and varied as 5h 9m – 6h 30m (IST), 6h 49m – 8h 42m (FST). Kim et al. (2012) [28] also reported an increased initial and final setting times for cement pastes replaced with increased amount of waste concrete powder. Decrease of C_3A and C_3S responsible for accelerating the hydration process caused the delay in setting time values.

Increased setting times, although insignificant, was further reported by Bektas et al. [29]. The w/c ratio and IST were found to be increased as CBP was increased from 15 to 25%. FST remains the same as the control paste for a 15% sample, whereas a significant increase was not seen for a 25% sample.

Table 3. w/c ratio for normal consistency and setting times [26].

	% Waste brick replacement cement				
	0	5	10	15	20
w/c %	27.4	28.2	28.4	28.8	29.2
Increasing (+)%	0	+2.92	+3.65	+5.11	+6.57
IST (min)	162	156	135	129	124
Decreasing (-)%	0	-3.71	-16.67	-20.37	-23.46
FST (min)	274	254	250	248	243
Decreasing (-)%	0	-7.3	-8.76	-9.49	-11.32

4.2. Scanning electron microscopy (SEM)

Electron microscopy is one of the most powerful techniques for studying the microstructure of cementitious materials. Morphological studies of early age hydration showing fracture surfaces during the evolution of microstructure can be satisfactorily done with the help of SE images. However, it provides qualitative information only and is not of much use when the degree of reaction becomes significant. By using polished sections for BSE image analysis, quantitative information about the different phases formed at different hydration periods can be determined [30].

On conducting a microstructural analysis of paste specimens, Aliabdo et al. (2013) [20] concluded that CBP could be satisfactorily used as a filler material as the sample having the highest amount of CBP displayed the best pore structure. Densification of the matrix occurs as a result of the production of additional hydrates due to pozzolanic activity. SEM micrographs of cement paste prepared with varying percentages of red CBP was shown by Li et al. (2016) [31]. The presence of quartz sand and cement hydration products like CH and CSH were observed. The structure of blended paste was found to be denser than that of reference paste.

Quantitative evaluation of microstructure roughness of cement mortar replaced with 30% CBP using atomic force microscopy was carried out by Liu et al. (2014) [21]. The intrinsic roughness of cement paste is expressed by the root mean squared (RMS) roughness and lies within the range (115 – 492 nm). The RMS roughness of one of the regions containing sand and CBP reached a relatively high value of 595 nm, whereas the cement paste region had a value of 470 nm. Based on these results, it was concluded that the roughness of the cement paste is increased due to the presence of CBP but lies within an acceptable range. It was further determined that CSH gels formed a weaker bond with CBP than that with sand. Nanoindentation test results also revealed the presence of a weaker ITZ around hybrid CBP particles.

Heikal et al. (2012) [15] studied the SEM micrographs of cement paste substituted with 37.5% CBP hydrated for different periods with and without SCC admixture. The expected hydration products of CSH, CH, ettringite, etc. were observed. A denser microstructure was found in the presence of SCC superplasticiser with the formation of microcrystals of CSH as the main hydration product.

4.3. Thermal Analysis

Thermal analysis refers to any technique adopted to carry out measurements involving desired parameters by varying the temperature. Standard methods often used in conjunction with each other include thermogravimetric analysis and differential thermal analysis. In the thermogravimetric study, the change of weight of a substance is monitored while subjecting it to varying temperatures. In cases where the mass changes are tiny and cannot be determined by TGA, differential thermal analysis (DTA) is used. In this method, the temperature difference between the sample and an inert reference is determined such that the reaction can be classified as endothermic or exothermic.

Navrátilová et al. (2016) [7] studied the pozzolanic properties of brick dust on lime mortars. Three mortars MREF (reference), MFL (highest pozzolanic activity), and MOH (lowest pozzolanic activity) were studied for TG and DTG. MFL and MOH mortars were found to lose water in the temperature range of 100 - 400°C. These mortars were also found to have lower calcium hydroxide content than that of MREF mortar. TG curve of MFL containing brick dust with higher pozzolanic activity was found to decrease more than the TG curve of MOH mortar.

Aliabdo et al. (2013) [20] reported TGA results of CBP replaced cement paste. The highest amount of weight loss was seen for reference paste, indicating the highest amount of CSH and ettringite content. Reference paste was also determined to have the highest amount of calcium hydroxide, reaffirming the pozzolanic nature of CBP. Liu et al. (2014) [21] reported an exothermic peak at 340°C, corresponding to the burning of cellulose particles on using hybrid recycled powder obtained from demolition waste with cement. The rate of increase of CH from 7 to 28 days was found to be more significant for the control specimen than for the other samples. Further observations include the decrease of CH content from 28 to 91 days in all different samples except the control specimen, indicating ongoing pozzolanic reaction by Bektas et al. [29].

The degree of hydration is another parameter that can be calculated from the thermogravimetric analysis. Lin et al. (2009) [27] determined both the degree of hydration and gel/space ratio with the help of parameters obtained from TGA. Gel/space ratio is known to be significantly governing the concrete compressive strength, increasing with the progress of hydration. As the gel/space ratio increases, partial filling of pores takes place, causing an increase of concrete compressive strength. The authors reported a comparable gel/space ratio of 10% replacement with that of OPC at 60 days. But 50% replacement caused a significant decrease in the same. Increased degree of hydration with the increase of curing time indicating active pozzolanic reaction was also seen. A reduction of hydration for 50% replacement at 60 days was attributed to a low concentration of Ca^{2+} ions in the pore solution. Similar findings were reported by Bediako et al. (2018) [32] regarding the degree of hydration and pozzolanic reactivity of CBP. Filler effect of CBP particles was attributed to the promotion of early age hydration of cement.

Kishar et al. (2012) [33] studied the effect of calcium chloride on the outcome of CBP pastes and determined the amount of chemically combined water content in all the pastes studied with the help of ignition method. Combined water was found to increase for all the pastes up to 180 days as a result of the addition of $CaCl_2$. The formation of the inner product, CSH, is believed to have accelerated due to the acquisition of $CaCl_2$. Ge et al. (2015) [34] evaluated the amount of non-evaporable water (NEW) as the mass loss of ground powder specimen between 105 – 950°C. With an increase of CBP, the NEW values were found to decrease insignificantly but not in an equal manner. The phenomenon was attributed to enhanced cement hydration due to internal curing and pozzolanic activity of CBP.

The heat of hydration: Heat of hydration of binder pastes with the progress of time can be measured quantitatively with the help of a calorimeter. Bediako et al. (2018) [32] demarcated the heat evolution curve into four parts: initial stage, dormant part, accelerated peak point, and deceleration part with maximum heat flow occurring in the third portion. The peak point corresponding to maximum heat flow was found to be higher for control paste than CBP blended paste, proving that reduced heat liberation occurs in the early periods, which can be useful in the construction of mass concrete structures. Bektas et al. [29] reported similar findings on replacing 15 and 25% CBP in cement. They further noted a delay in time at which the peak temperature was recorded indicating the presence of CBP has a retarding effect on the hydration process. The presence of $CaCl_2$ in the blended paste containing CBP and cement had an accelerating effect on the hydration reaction as observed from the increased intensity of the endothermic peak of the DTA curve [33].

4.4. X-ray diffraction (XRD)

Qualitative information about the presence of different hydration products formed in cement paste, unhydrated cement minerals (allite, bellite, ferrite), and mineral phases (mullite, quartz) can be determined by XRD analysis. Refinement techniques provide quantitative information about the intensity of these phases. Afshinnia et al. (2015) [35] carried out an XRD analysis of mortar specimens with varying CBP percentages after one year. Portlandite peaks were found to have decreased or even disappeared in the samples replaced with CBP owing to its pozzolanic nature. Only at 10% replacement level, the peak was comparable with that of the reference sample. Further comparisons were carried out using brick as fine aggregate, and it was determined that CBP as SCM shows better pozzolanic activity as compared to CBP as fine aggregate owing to its higher specific surface area.

The presence of portlandite, ettringite, calcite, quartz, and CSH were detected by Aliabdo et al. (2013) [20] on CBP replaced cement pastes. The authors reported that the presence of CBP in cement paste does not have any significant effect on the mineral compositions of the matrix. XRD peaks of C_3S, CSH, CAH, CASH were reported by Lin et al. (2009) [27] with the conclusion that a dense and homogenous system is formed, which provides excellent long term strength. In a study on the utilization of red CBP for use as plaster [29], a reduced intensity of CH peak was reported for the samples replaced with CBP, indicating that the presence of red CBP has a strong influence on CH content. In the presence of $CaCl_2$, the intensity of CH peaks was found to be higher, indicating increased CH content, owing to the nature of $CaCl_2$ as an accelerating admixture [33].

5. Effect of CBP on properties of cement mortar

5.1. Workability of mortars

Schackow et al. (2015) [18] performed a flow test in which specimens with constant w/c ratio were cured for 28 and 90 days in $Ca(OH)_2$ saturated water at 23°C and 65 – 80% relative humidity. The flow index of the mixtures with an increasing percentage of replacement was found to decrease as finer CBP particles increased the water consumption. Hence, the negative impact on workability was observed, and the requirement of extra water or superplasticisers was seen. A replacement of 10% is recommended for satisfactory workability having a constant w/c ratio.

5.2. Strength activity index

Afshinnia et al. (2015) [35] determined the SAI of mortar cubes containing 0%, 10%, 25%, and 50% CBP at 7, 28 and 90 days according to ASTM C109. The 28 days SAI was reported as 92%, 86%, and 59%, respectively. The SAI values were found to increase gradually for mortar mixes with 25% CBP (79%, 86%, and 91% at 7, 28, and 90 days). However, this was not observed in mortar mixes with 10% CBP. 50% replacement by CBP did not show an acceptable level of pozzolanic activity. All kinds of CBP cannot be considered as pozzolanic material, especially if they contain large amounts of crystalline minerals like quartz and feldspar [20]. In this paper, the authors conducted the SAI test according to ASTM C618 requirements. SAI values were found to be 76.5% and 81.8% at 7 and 28 days, respectively, affirming the pozzolanic properties of CBP used. Two types of bricks: slab bricks (SB) and wall bricks (WB) were used in the experimentation process by [36]. The 75% limit proposed by ASTMC618 was met by both the materials used with SB showing better pozzolanic activity than WB. When tested with a 0.5 *w/b* ratio and 30% CBP replacement, the SAI was found to be 75% [21]. As concrete substituted with CBP requires a higher amount of water to maintain suitable workability, the compressive strength of mortar cubes was tested with a higher *w/b* ratio. The SAI value was found to be within limits (70%).

5.3. Compressive strength of mortar

Standard mortar mixes according to European standards containing cement replaced with CBP or heat-treated clay were prepared by [12]. The authors studied ten different mixtures. Strength at early ages was considerably low, but long term strength (90 days) was unaffected. Relative intensity (strength at particular age/strength of reference mortar at the same age) at different ages was found to decrease with the increase in the level of replacement. Heat-treated clay was found to give better early age strength, but ground brick gave better long term strength due to the presence of a higher quantity of glass.

Demir et al. (2011) [37] prepared CBP replaced cement mortars to calculate its compressive strength. Seven different types of mixtures were made and stored at 20°C for 24 hours after which they were cured under three different types of conditions: continuous curing under lime saturated tap water (TW), constant exposure to 5% sodium sulphate solution (SS) and a continual exposure to 5% ammonium nitrate solution (AN). On the simultaneous evaluation of all the different experimental conditions to which the cubes were exposed, it was determined that there was an improvement in 28 days strength values when replacement exceeded 5%. On observing long term strength of 90 and 180 days, 10 – 15% replacement gave promising results. O'Farell et al. (2001) [38] reported that increasing the amount of CBP in cement mortar caused decreasing compressive strength values relative to reference up to 28 days. However, after 90 days, the strength of some mortars with low levels of replacement was found to exceed the reference mortars. This effect was attributed to the formation of additional CSH gel due to the pozzolanic reaction, which causes the pores to fill up and hence refines the matrix. The connectivity of the capillary pores is significantly reduced as a result if this infilling effect. Reduced compressive strength on increasing percentages of CBP in cement mortar was also reported by [20]. 25% replacement caused a decrease of 25.5% in the 28-day compressive strength. The mortar was prepared in a mix proportion of 1:3:0.5 (binder: sand: water).

Replacement of waste glass, brick, and tile waste was done in percentages of 10%, 20%, 25%, 30%, and 40%, respectively [39]. 25% cement replacement independent of the waste type caused a reduction in compressive strength of 20%, whereas 40% replacement caused a decrease of 30%. The evolution of strength from 28 to 90 days was found to be higher for tile powder (25 – 30%) than for brick powder (15 – 20%). 25% addition of glass powder caused a reduction in mortar strength of 20 – 30%. At a maximum level of replacement, reduction in strength was found to be 37 – 45%. The compressive strength was also found to be improved with the increase in fineness of particle. The authors reported that out of all the various ceramic wastes studied, only tile waste powder provided sufficient pozzolanicity. The other materials being fired at a lower temperature did not possess pozzolanic properties. Cubic cement paste specimens of size 1 inch containing CBP and calcium chloride were prepared with a fixed water/ solid ratio of 0.25 [33]. The compressive strength of pastes incorporated with $CaCl_2$ was found to be higher than that of ordinary pastes. Mixtures with 10 – 30% CBP showed a steep increase in compressive strengths from 90 – 180 days due to the accelerating effect by $CaCl_2$ and decreased porosity of the paste. The use of 10% CBP was determined to be the optimum percentage for the improvement of compressive strength. Lin et al. [27] assessed the compressive strength of CBP blended cement pastes by an unconfined compressive strength test. Reference pastes had the highest compressive strength, whereas 50% CBP addition had the lowest strength due to an insufficient quantity of CH present to activate the CBP. 10 – 20% replacement gave strength values similar to OPC at 60 days.

5.4. Flexural strength of mortar

In the experiments conducted by Naceri et al. (2009) [26], flexural strength was found to decrease with an increasing amount of CBP up to 28 days. But at 90 days, 10% CBP replaced mortars attained strength comparable to that of reference mortars. Zheng et al. (2011) [40] showed that for varying levels of cement replacement and different average particle sizes, the flexural strength was not significantly affected. CBP obtained from bricks, and roofing tiles were used as cement replacement by Ioannou et al. (2009) [41]. From experimental data collected, it was concluded that long term strength (90 days) increased on using CBP, but on increasing *w/b* ratio, it decreased drastically. Due to the addition of surplus water, the porosity of the mortars increases, which adversely affects the compressive strength.

6. Effect of CBP on mechanical properties of concrete

Increasing values of the slump with the corresponding decrease in compacting factor with increased CBP replacement was stated by Wild (1996) [11]. Olofinnade et al. (2016) [17] put forward that for 10 – 20% cement replacement by CBP, the workability of concrete was possible to be kept constant. However, at constant *w/c* ratio, a decrease in a slump was observed for cement replacement up to 40%, resulting in reduced workability. Similar observations were reported by Ge et al. (2015) [34] regarding slump loss. Three different types of CBP having different particle sizes, water absorption values, and strength activity indices were used by the authors in this study (A – high, B – medium, C – low). The higher slump was observed when the replacement level was small, and the particles were coarser. Slump loss was acceptable up to 10% replacement for type A and B, whereas type C CBP showed acceptable slump loss up to 20% replacement. As established earlier, CBP has higher water demand, more angularity, and rough texture, all of which contribute towards loss of slump. High water demand is also responsible for decreasing the flow of the mix [23]. Typical properties of hardened concrete put forward by different authors are discussed in table 4.

Table 4. Comparison of mechanical properties of hardened concrete containing CBP.

Year	Author	Materials used	Conclusions	Remarks
2016	Olofinnade et. al. [17]	OPC (ASTM Type I), waste clay brick.	•A decrease in bulk density of hardened concrete. •Increase in 28 days compressive and tensile strength up to 10% replacement.	•CBP can be used as an SCM up to 15% replacement.
2013	Heikal et al. [15]	OPC, ground clay bricks, polycarboxylate SCC admixture, superplasticiser.	•In the absence of SCC, increasing GCB decreased compressive strength. •Increase of 28 days compressive strength in the presence of SCC admixture.	•GCB acts as a pozzolanic filler material reducing compressive strength. But it also causes the formation of denser matrix.
2015	Ge et al. [34]	OPC, waste clay brick, HRWR.	•A decrease in early age compressive strength but comparable compressive, flexural and split tensile strength at 90 days. •A decrease in static modulus of elasticity with an increase in replacement level.	•Replacement level influenced strength reduction more than the particle size. •The initial reduction in compressive strength is due to the dilution effect and low reactivity of CBP, but the final increase is due to internal curing and pore structure improvement.
2016	Zhu et al. [23]	42.5 OPC, powder obtained from clay bricks and cement solids, superplasticiser.	•A decrease in compressive and flexural strength on replacing silica fume in reactive powder concrete with recycled powder. •A reduction in flexural strength, a slight change in compressive strength on replacing cement in RPC with recycled powder. •RPC with recycled powder and GGBFS caused increased compressive and flexural strengths up to 10% replacement at 28 days.	•Silica fume and cement have higher pozzolanic activity than recycled powder. •GGBFS can replace RPC with silica fume and recycled powder up to 10%.

2017	Letelier et al. [42]	ASTM type P cement, CBP from demolition debris, natural and recycled aggregates.	•CBP replacement caused filler effect compensating the strength loss due to the use of recycled aggregates. •The combined impact of CBP and RA is more pronounced in flexural strength and static elasticity modulus.	•Optimum combination is determined as 30% RA and 5% CBP.
2012	Ge et al. [43]	42.5 OPC, CBP from recycled construction waste, HRWR.	•w/c ratio is the primary factor affecting the mechanical properties of concrete containing CBP. • Though the addition of CBP tends to reduce the compressive strength, it can be avoided by adopting proper mix design. •Flexural strength of CBP concrete is similar to that of reference.	•According to orthogonal experimental analysis, the optimal concrete mix consists of w/c ratio 0.26, 33% sand, 0.06 mm CBP size and 25% replacement.
1996	Wild [11]	OPC type I, waste Fletton brick, crushed flint.	•Increased compressive strength after 90 days with 10% replacement. •Comparable 90 days compressive strength with 20% replacement.	•CBP has potential pozzolanic properties making it a suitable SCM.
2007	Bektas et al. [29]	Type I Portland cement, ground clay brick, oolitic limestone, natural siliceous aggregate.	•The use of GCB reduces early age strength. •Ultimate compressive strength is comparable or higher than the control specimen. •Equal or higher split tensile, flexural strength and modulus of elasticity at 90 and 180 days.	•CBP from demolished masonry can be used as an SCM.

7. Effect of CBP on the durability of concrete

7.1. Shrinkage

The amount of water that can be held by cement paste is determined by its pore structure and amount of environmental humidity. Reduced relative humidity in the ambient atmosphere will cause concrete to become dimensionally unstable due to the loss of adsorbed water from CSH, resulting in the formation of shrinkage strain. Two types of shrinkage may occur; thermal shrinkage due to cooling effect and drying shrinkage due to loss of moisture [44]. It is observed that shrinkage increased with increasing time and percentage replacement from the variation of shrinkage strain with the increasing replacement of cement by CBP [26]. A significant reduction in autogenous shrinkage was reported by [34]. 10% replacement of cement by CBP was seen to decrease autogenous shrinkage by 35.9%. Reduction in the formation of hydration products at the early stages is a possible reason for this. Another probable mechanism is that the internal curing effect by CBP kept the capillary pores of the paste saturated, reducing internal stresses and associated strains.

Zhu et al. (2016) [23] also reported a reduced level of shrinkage with the increased replacement of cement by CBP in concrete due to smaller recycled powder particles filled up the pores in concrete, hindering water evaporation and hence reducing shrinkage. The durability of cement mortar and concrete replaced with different levels of CBP studied by various authors is summarised in table 5.

Table 5. Summary of durability properties

Parameters	Authors	Conclusions	Remarks
Sulphate Resistance	O'Farell et al. (2006) [16]	• Increasing replacement level gives increased resistance to expansion. • Performance of ground brick> heat treated clay> OPC. • Quenched clay provides better resistance than slowly cooled clay.	• QC mortars offer better performance than SC mortars due to better pozzolanic activity.
	Filho et al. (2007) [19]	• Reduction in tensile strength loss of control mortars exposed to 5% magnesium sulphate solution at 200 days.	• Reduction of CH, C_3A, and formation of additional hydrates refines the pore structure impeding the advancement of aggressive agents.
	Schackow et al. (2015) [18]	• Low sulphate resistance.	• The porous structure of control mortar is more efficient in accommodating the expansive reaction products formed during exposure.
	Wild et al. (1997) [45]	• CBP calcined at 1000 – 1100°C provides excellent sulphate resistance in a mortar. • Loss in sulphate resistance occurs when CBP is calcined at temperatures below 900°C.	• The poor performance of CBP calcined at low temperatures is due to the presence of crystalline activated clay and high initial levels of sulphate.
	O'Farell et. al. (1999) [46]	• High CaO/glass ratio increases sulphate expansion and vice versa.	• Chemical and mineralogical composition (especially CaO), type of CBP, glass content, oxides present, initial sulphate content are responsible for sulphate resistance.
	Binici et al. (2012) [47]	• Addition of red brick dust and basaltic pumice at 15% for each additive gives high sulphate resistance.	• Less ettringite formation in mortars exposed to sodium sulphate solution seen from SEM images.
Chloride Resistance	Filho et. al. (2007) [19]	• Reduction in chloride ion penetration.	• Pore refinement due to pozzolanic activity retarded ionic transport in the matrix.
	Schackow et al. (2015) [18]	• Low chloride absorption.	• Improvement in microstructure allows crystallisation/ dissolution of NaCl without causing stress to surround material.
	Gonçalves et al. (2009) [14]	• Reduction in chloride ion penetration more for metakaolin than for CBP.	• Pore structure refinement is responsible for chloride ion resistance.

	Ge et al. (2015) [34]	•30% replacement by CBP in concrete caused low chloride resistance, but excellent resistance was seen for low replacement levels.	•High CBP replacement caused high penetrability.
Resistance to synthetic seawater	O'Farell et al. (2000) [48]	•Reduced resistance is seen when ground brick contains high CaO, low glass, and low sulphate content. •Replacement by CBP minimises the formation of brucite and gypsum layer on the surface of the mortar. •30% replacement reduces strength loss on exposure to seawater.	•Chemical composition of CBP used is responsible for controlling the expansion of mortar.
	O'Farell et al. (2006) [16]	•Heat-treated clay is more effective in reducing expansion than CBP when exposed to synthetic seawater.	•Increased solubility of CH in seawater, the formation of brucite skin around mortar prism inhibits advancement of seawater solution.
Carbonation	Schackow et al. (2015) [18]	•A significant effect was not observed at 28 days on replacement by CBP. •At 90 days, 10 and 25% provided the lowest rate of carbonation.	•Refinement of pore structure due to the formation of additional CSH reduces carbonation rate.

7.2. Porosity and sorptivity

Water in its pure form is a cause of physical degradation in concrete, and as a transporter of aggressive ions acts as an agent of chemical degradation. The structural transformation of water trapped in the concrete matrix (like freezing) causes internal stress build-up. It is well known that during the process of hydration, the products of hydration formed occupy more space than that held initially by cement and water. On complete hydration of 1 cm^3 of cement, the space required for its total accommodation is approximately 2 cm^3. However, there remain some voids that remain unoccupied by either cement or any hydration product. These are called capillary voids. The porosity of concrete gives a picture of the volume of capillary voids present. Sorptivity, on the other hand, is described as the ability of a porous material to absorb or desorb any liquid by the action of capillarity.

Ortega et al. (2018) [22] reported that CBP caused increased pozzolanic reaction leading to pore refinement of the modified mortars. Filler effect of CBP (determined by impedance spectrometry) is also responsible for the improvement of pore structure. Pore size distribution and threshold radius are affected by the addition of CBP in mortar. At shorter times, increased intruded pore volume, reduction in microscopic pores, and the increase in threshold radius were observed. Positive effects were mostly found at longer curing ages. Pore refinement occurred on the addition of gypsum [38]. Compressive strength is seen to increase the reduction of the threshold radius.

Singh et al. (2006) [49] reported lower pore diameter and porosity at all curing periods, which contrasts with the behaviour of CBP, where lower porosity was observed at longer curing times only. Gonçalves et al. (2009) [14] reported a higher reduction in sorptivity for metakaolin than ground brick modified mortars. In the case of the use of CBP, total porosity was found to increase with a corresponding gradual decrease of sorptivity. Refinement of porous structure leading to smaller pore size in spite of higher total porosity is the reason for this observation. More replacement up to 40% caused an increase in both porosity and sorptivity. Sabir et al. (1998) [50] developed an automated testing system for the measurement of sorptivity in mortar and concrete. It was found that at short curing periods and higher replacement levels, the sorptivity of modified mortars increased as compared to that of control because of the presence of residual porosity, which had not yet been filled by hydration products of an ongoing slow pozzolanic reaction. However, sorptivity decreased as the curing time increased, with some mortars giving lower values than that of control mortar.

7.3. Alkali silicate reaction

Reactive siliceous minerals present in aggregate on coming in contact with free lime of concrete forms a gel. This gel on coming in contact with water absorbs a large amount of moisture by osmosis. As a result, concrete expands and generates internal tensile stresses in its matrix, which causes it to crack. Mineral additives like silica fume react with a sufficient amount of alkalis to reduce the volume of ASR products formed [44, 51].

Afshinnia et al. (2015) [35] carried out a study in which ground crushed brick was used as a cementitious material, and clay brick was used as an aggregate material. High replacement levels (25%) of CBP caused a significant reduction of ASR. On using crushed brick as a replacement for fine aggregate, no beneficial effect on ASR mitigation was seen. Thus, it was concluded that the fineness of CBP played an essential role in mitigating ASR. By conducting EDS analysis, it was further determined that the alkali content of ASR gel obtained from a control specimen was much higher than that obtained from CBP modified samples. ASR gel from CBP containing mortars also contained more elevated calcium, had higher viscosity, and caused considerable swelling.

A positive effect of CBP on ASR mitigation was also reported by Turanli et al. (2003) [52]. The use of CBP from both slabs and walls caused a reduction in expansion with an increase in replacement. 30% replacement provided optimum results with slab brick dust performing marginally better than wall brick dust. Incorporation of CBP caused a delay in the start of expansion due to filling up of pore space. But, the effectiveness of the material reduced at later ages. Observation of increased expansion rate at later stages led to this conclusion.

8. Conclusion

1. The principal chemical component of CBP is silica and alumina, along with other minor compounds in smaller amounts. SEM images show that CBP has angular morphology and is amorphous.
2. The pozzolanic character of clay used for making brick comes from calcining it at a particular optimum temperature, which is different for different clays. Hence, not all types of CBP show pozzolanic character. Future scope in this area remains in the investigation of parameters for the enhancement of pozzolanicity in CBP.
3. TG analysis of mortars and pastes containing CBP revealed that CH content at more extended curing periods is less as compared to standard OPC samples. CBP addition also results in an increased degree of hydration and decreased heat of hydration. A decrease of CH peaks was observed from XRD analysis.
4. 10 – 25% replacement by CBP gives acceptable values of Strength Activity Index at 7 and 28 days. The compressive strength of mortars containing CBP is lower than control at the beginning but increases at more extended curing periods (90 – 180 days). Similar results were observed for flexural strength tests.
5. Cement replacement by CBP in concrete mostly causes increased compressive, flexural, and split – tensile strength at 28 and 90 days, indicating good pozzolanic behaviour.
6. Pore refinement causing the formation of additional hydration products due to pozzolanic reaction as well as filler effect of CBP are the primary factors causing improved durability of concrete.
7. The future scope can be extended to the study of high volume replacement of CBP in concrete and the effect of CBP replacement on corrosion of rebar.

9. References

[1] Statista Major countries in worldwide cement production from 2012 to 2017 (million metric tons)
[2] Coalition C a C A Mitigating black carbon and other pollutants from brick production
[3] ICIMOD 2015 *Mitigating Air Pollution from Brick Production* (International center for integrated mountain development: Kathmandu, Nepal)

[4] Maithel S, Kumar S and Lalchandani D 2014 *Factsheets about Brick Kilns in South and South-East Asia* (Greentech Knowledge Solutions Pvt. Ltd.: New Delhi)
[5] Banerjee C C&D Waste Processing in India – Delhi Shows the Way (NBM&CW)
[6] Cheng H 2016 Reuse research progress on waste clay brick *Procedia Env. Sci.* **31** 218–26
[7] Navrátilová E and Rovnaníková P 2016 Pozzolanic properties of brick powders and their effect on the properties of modified lime mortars *Construction and Building Materials* **120** 530–9
[8] ASTM C 2005 125 *Standard Terminology Relating to Concrete and Concrete Aggregates*
[9] Habert G, Choupay N, Escadeillas G, Guillaume D and Montel J 2009 Clay content of argillites: Influence on cement based mortars *Applied Clay Science* **43** 322–30
[10] Baronio G and Binda L 1997 Study of the pozzolanicity of some bricks and clays *Construction and Building Materials* **11** 41–6
[11] Wild S 1996 Observations on the use of ground waste clay brick as a cement replacement material *Building Research & Information* **24** 35–40
[12] Ambroise J, Murat M and Péra J 1985 Hydration reaction and hardening of calcined clays and related minerals V. Extension of the research and general conclusions *Cement and Concrete Research* **15** 261–8
[13] Akinshipe O and Kornelius G 2017 Chemical and thermodynamic processes in clay brick firing technologies and associated atmospheric emissions metrics-a review
[14] Gonçalves J, Tavares L, Toledo Filho R D and Fairbairn E 2009 Performance evaluation of cement mortars modified with metakaolin or ground brick *Construction and Building Materials* **23** 1971–9
[15] Heikal M, Zohdy K and Abdelkreem M 2013 Mechanical, microstructure and rheological characteristics of high performance self-compacting cement pastes and concrete containing ground clay bricks *Construction and Building Materials* **38** 101–9
[16] O'Farrell M, Sabir B and Wild S 2006 Strength and chemical resistance of mortars containing brick manufacturing clays subjected to different treatments *Cement and Concrete Composites* **28** 790–9
[17] Olofinnade O M, Ede A N, Ndambuki J M and Bamigboye G O 2016 Structural properties of concrete containing ground waste clay brick powder as a partial substitute for cement *Materials Science Forum* 63–7
[18] Schackow A, Stringari D, Senff L, Correia S and Segadães A 2015 Influence of fired clay brick waste additions on the durability of mortars *Cement and Concrete Composites* **62** 82–9
[19] Toledo Filho R, Gonçalves J, Americano B and Fairbairn E 2007 Potential for the use of crushed waste calcined-clay brick as a supplementary cementitious material in Brazil *Cement and Concrete Research* **37** 1357–65
[20] Aliabdo A A, Abd-Elmoaty A-E M and Hassan H H 2014 Utilization of crushed clay brick in concrete industry *Alexandria Engineering Journal* **53** 151–68
[21] Liu Q, Tong T, Liu S, Yang D and Yu Q 2014 Investigation of using hybrid recycled powder from demolished concrete solids and clay bricks as a pozzolanic supplement for cement *Construction and Building Materials* **73** 754–63
[22] Ortega J M, Letelier V, Solas C, Moriconi G, Climent M Á and Sánchez I 2018 Long-term effects of waste brick powder addition in the microstructure and service properties of mortars *Construction and Building Materials* **182** 691–702
[23] Zhu P, Mao X, Qu W, Li Z and Ma Z J 2016 Investigation of using recycled powder from waste of clay bricks and cement solids in reactive powder concrete *Construction and Building Materials* **113** 246–54
[24] Manoharan C, Sutharsan P, Dhanapandian S, Venkatachalapathy R and Asanulla R M 2011 Analysis of temperature effect on ceramic brick production from alluvial deposits, Tamilnadu, India *Applied Clay Science* **54** 20–5
[25] Reig L, Tashima M M, Borrachero M, Monzó J, Cheeseman C and Payá J 2013 Properties and microstructure of alkali-activated red clay brick waste *Constrn. and Build. Mater.* **43** 98–106
[26] Naceri A and Hamina M C 2009 Use of waste brick as a partial replacement of cement in mortar *Waste management* **29** 2378–84

[27] Lin K-L, Chen B-Y, Chiou C-S and Cheng A 2010 Waste brick's potential for use as a pozzolan in blended Portland cement *Waste Management & Research* **28** 647–52
[28] Kim Y J and Choi Y W 2012 Utilization of waste concrete powder as a substitution material for cement *Construction and Building Materials* **30** 500–4
[29] Bektas F 2007 Use of ground clay brick as a supplementary cementitious material in concrete hydration characteristics, mechanical properties, and ASR durability
[30] Scrivener K, Snellings R and Lothenbach B 2018 *A Practical Guide to Microstructural Analysis of Cementitious Materials* (CRC Press)
[31] Li H, Dong L, Jiang Z, Yang X and Yang Z 2016 Study on utilization of red brick waste powder in the production of cement-based red decorative plaster for walls *Journal of Cleaner Production* **133** 1017–26
[32] Bediako M 2018 Pozzolanic potentials and hydration behavior of ground waste clay brick obtained from clamp-firing technology *Case Studies in Construction Materials* **8** 1–7
[33] Kishar E A, Ahmed D A, Mohammed M R and Noury R 2013 Effect of calcium chloride on the hydration characteristics of ground clay bricks cement pastes *Beni-Suef University Journal of Basic and Applied Sciences* **2** 20–30
[34] Ge Z, Wang Y, Sun R, Wu X and Guan Y 2015 Influence of ground waste clay brick on properties of fresh and hardened concrete *Construction and Building Materials* **98** 128–36
[35] Afshinnia K and Poursaee A 2015 The potential of ground clay brick to mitigate Alkali–Silica Reaction in mortar prepared with highly reactive aggregate *Construction and Building Materials* **95** 164–70
[36] Turanli L, Bektas F and Monteiro P 2003 Use of ground clay brick as a pozzolanic material to reduce the alkali–silica reaction *Cement and Concrete Research* **33** 1539–42
[37] Demir I, Yaprak H and Simsek O 2011 Performance of cement mortars replaced by ground waste brick in different aggressive conditions *Ceramics-Silikáty* **55** 268–75
[38] O'Farrell M, Wild S and Sabir B 2001 Pore size distribution and compressive strength of waste clay brick mortar *Cement and Concrete Composites* **23** 81–91
[39] Pereira-de-Oliveira L A, Castro-Gomes J P and Santos P M 2012 The potential pozzolanic activity of glass and red-clay ceramic waste as cement mortars components *Construction and Building Materials* **31** 197–203
[40] Zheng L, Ge Z, Yao Z and Gao Z 2011 *ICCTP 2011: Towards Sustainable Trans. Sys.* 3379–88
[41] Ioannou I, Ilia A and Philokyprou M 2009 Use of crushed fired clay ceramics in the production of mortars *WIT Transactions on Ecology and the Environment* **120** 257–64
[42] Letelier V, Tarela E and Moriconi G 2017 *Procedia Engineering* **171** 627–32
[43] Ge Z, Gao Z, Sun R and Zheng L 2012 Mix design of concrete with recycled clay-brick-powder using the orthogonal design method *Construction and Building Materials* **31** 289–93
[44] Mehta P K and Monteiro P J 2006 *Concrete: Microstructure, Properties and Materials*
[45] Wild S, Khatib J and O'Farrell M 1997 Sulphate resistance of mortar, containing ground brick clay calcined at different temperatures *Cement and Concrete Research* **27** 697–709
[46] O'Farrell M, Wild S and Sabir B 1999 Resistance to chemical attack of ground brick-PC mortar: Part I. Sodium sulphate solution *Cement and Concrete Research* **29** 1781–90
[47] Binici H, Kapur S, Arocena J and Kaplan H 2012 The sulphate resistance of cements containing red brick dust and ground basaltic pumice with sub-microscopic evidence of intra-pore gypsum and ettringite as strengtheners *Cement and Concrete Composites* **34** 279–87
[48] O'Farrell M, Wild S and Sabir B 2000 Resistance to chemical attack of ground brick–PC mortar: Part II. Synthetic seawater *Cement and Concrete Research* **30** 757–65
[49] Singh M and Garg M 2006 Reactive pozzolana from Indian clays—their use in cement mortars *Cement and Concrete Research* **36** 1903–7
[50] Sabir B, Wild S and O'farrell M 1998 A water sorptivity test for mortar and concrete *Materials and Structures* **31** 568
[51] Gambhir M L 2013 *Concrete Technology: Theory And Practice* (Tata McGraw-Hill Education)
[52] Tironi A, Trezza M A, Scion A N and Irassar E F 2013 Assessment of pozzolanic activity of different calcined clays *Cement and Concrete Composites* **37** 319–27

Study on Strength Properties of Concrete using Pumice Powder Pozzolan and Polypropylene Fibers

Ansu Lalu Abraham[1*] and Regi P Mohan[2]

[1] P.G. Student, Department of Civil Engineering, Sree Buddha College of Engineering, Alappuzha, Kerala, India

[2] Assistant Professor, Department of Civil Engineering, Sree Buddha College of Engineering, Alappuzha, Kerala, India

*ansu.j.lalu@gmail.com

Abstract. The construction industry is flourishing day by day, as a result, cement consumption also increases. Since cement production causes carbon footprint, partial replacement using waste materials from various industries is one of the solutions. Pumice powder, generated by the stone processing industry, is an inert by-product possessing excellent cementitious characteristics. Utilisation of pumice powder in concrete can reduce the cost of construction and at the same time address the waste disposal problem. The fiber dispersion into concrete is one of the techniques to further improve the ductile properties of concrete. This study is focused on the strength properties of pumice powder pozzolan concrete reinforced with polypropylene fibers. Mechanical strength characteristics were evaluated for normal and varying contents of pumice powder and polypropylene fiber added mix after 7 and 28 days of curing. The results indicate that the strength properties of the modified mix has improved considerably. The significance of this study is to promote the usage of by-product waste as a replacement for cement thereby reducing CO_2 emissions and thus resulting in the development of environmentally friendly concrete.

1. Introduction

To decrease the carbon footprint due to cement manufacturing and construction industries, one of the best ways is to reduce the cement used in concrete production. More than 4 billion tons of cement are produced annually for construction purposes [1]. This can be reduced by utilising alternative materials in concrete production which is called as supplementary cementitious materials (SCMs). Different types of wastes like industrial wastes, agricultural wastes, natural minerals, stone powders, and dust contributes to the production of an environment-friendly concrete [1]. The usage of SCMs in construction has the added advantage that it reduces excessive disposal of waste into the landfills.

This paper concentrates on the utilisation of one such industrial by-product of the stone processing industry for reducing CO_2 emission. Pumice powder (PP) is a powder ground form of natural lightweight-aggregate pumice which is formed by rapid cooling and solidification of volcanic lava. The notable characteristics which makes pumice a lightweight-aggregate are its reduced density, better thermal insulation and fire resistance, lower permeability and shrinkage, reasonable elasticity and porous structure [2, 3]. But when pumice as such is used without powdering in concrete, the strength improvement is found to be marginal. As the particle size of PP decreases, the rate of hydration, pozzolanic reaction and compressive strength increases [4]. Thus the pozzolanic activity of PP improves the mechanical characteristics of concrete [1]. Also, the utilisation of natural pumice pozzolan in concrete reduces global warming potential and improves the air quality index [5].

The fiber dispersion into concrete is one of the techniques to improve concrete's ductility properties like flexural capacity, crack control, etc. Polypropylene fibers are 3D fibers made from byproducts of petroleum. These fibers are nowadays well accepted in the construction industry due to the properties which they impart like high tensile strength, flexural strength, impact strength, toughness, etc. These fibers also improve the failure mode of concrete by helping in crack control [6].These fibers have a twisted geometry thereby improving mechanical anchorage with the cementitious matrix [6].They even show better strength and ductility properties when subjected to very high temperatures and exposure periods [7].

This study presents the results of the research conducted to assess the effect of partial replacement of cement with PP, along with a 3D fiber addition on mechanical strength characteristics of concrete. This work will provide a platform to investigate the performance of PP in the development of environment-friendly concrete thereby promoting its utilisation for the development of infrastructures.

2. Experimental program

2.1. Materials

The raw materials utilised in the study include cement, PP, fine aggregate, coarse aggregate, polypropylene fiber, superplasticiser, and water.

53 grade OPC (Ordinary Portland Cement) conforming to IS 12269-1987 was used throughout in this study. The properties of cement such as fineness, specific gravity, standard consistency, initial and final setting times were determined and are given in table 1.

Pumice powder which is an industrial byproduct of the stone processing industry i.e. pumice rock in powdered form is utilised as partial replacement of cement in this study. Color of pumice is normally found to be grey or white and ranges depending on the chemical composition of pumice. Properties of PP are shown in table 1.

Table 1. Properties of cement and PP used in the study.

Properties	Cement	PP
Fineness	7%	6.8%
Standard Consistency	28%	32.5%
Initial Setting time	75 minutes	50 minutes
Final Setting time	360 minutes	360 minutes
Specific gravity	3.1	2.84

Manufactured sand (M-sand) conforming to Zone II, passing 4.75 mm and retained on 150 μm sieve was used as fine aggregate in this study. The maximum size of the coarse aggregate used for the study is 20 mm. The physical properties of aggregates are shown in table 2.

Table 2. Physical Properties of aggregates used in the study.

Properties	Fine aggregate	Coarse aggregate
Fineness Modulus	4.708	5.22
Grade zone	II	-
Specific Gravity	2.645	2.9
Bulk density	1820 kg/m^3	1542 kg/m^3
Porosity	32%	44.3%
Void ratio	0.473	0.91

Polypropylene fibers are 3D fibers that provide reinforcement in x, y and z planes. Properties of PF are shown in table 3.

Table 3. Properties of polypropylene fiber (Source: Supplied by Manufacturer).

Properties	Value
Colour	White
Length	12mm
Diameter	24-44 micron
Aspect ratio approx.	500
Alkali resistance	100% alkali proof
Acid and salt resistance	High

Superplasticisers are chemical admixtures used to modify the workability property of concrete. In the present study, workability was improved by using a superplasticiser (SP) named Conplast SP430.The amount of SP dosage used in concrete is taken as 1.2% of the weight of the cement.

2.2. Mix proportion and specimen preparation

Test specimens were cast using optimum water-cement ratio for each mix. The optimum replacement of cement with pumice powder (PP) is determined by casting normal and PP mortar (PPM) cubes of size 70.6 mm ×70.6mm× 70.6 mm in 1:3 mix proportion as per IS:4031 (Part 6)-1988. Cement is replaced with PP at 0% (0PP), 5% (5PP), 10% (10PP), 15% (15PP) and 20% (20PP). The mix proportion of PPM cube specimens and PP-PF concrete specimens are given in table 4 and table 5 respectively.

Table 4. Mix proportion of PPM

Mix designation	Cement (kg)	PP (kg)	Sand (kg)	Water (kg)
0PP	1.2160	0	4.05	0.60
5PP	1.1552	0.0608	4.05	0.62
10PP	1.0944	0.1216	4.05	0.62
15PP	1.0336	0.1824	4.05	0.62
20PP	0.9728	0.2432	4.05	0.62

Table 5. Mix proportion of PP-PF concrete.

Mix designation	Cement (kg)	PP (kg)	PF (kg)	Sand (kg)	Coarse Aggregate (kg)	SP (kg)	Water (kg)
0 PP + 0 PF	23.29	0	0	38.04	63.67	0.279	10.2
10 PP + 0 PF	20.961	2.329	0	38.04	63.67	0.279	10.2
10 PP + 1 PF	20.961	2.329	0.484	38.04	63.67	0.279	10.2
10 PP + 1.25 PF	20.961	2.329	0.605	38.04	63.67	0.279	10.2
10 PP + 1.5 PF	20.961	2.329	0.726	38.04	63.67	0.279	10.2
10 PP + 1.75 PF	20.961	2.329	0.847	38.04	63.67	0.279	10.2

Along with optimum PP content, polypropylene fibers (PF) were added at 0% (10 PP + 0 PF), 1% (10 PP + 1 PF), 1.25% (10 PP + 1.25 PF), 1.5% (10 PP + 1.5 PF) and 1.75% (10 PP + 1.75 PF) of total volume of concrete to make pumice powder- polypropylene fiber (PP-PF) concrete. M30 grade of mix proportion 1: 1.47: 2.59 was adopted for concrete specimens. For each mix, the specimens cast are: 6 cube specimens (150 mm × 150 mm × 150 mm) each for compressive strength test after 7 days and 28 days of curing, 3 samples of cylinder (150 mm dia × 300 mm height) each for split tensile strength test after 28 days of curing and 3 beams (500mm × 100 mm×100 mm) each for flexural strength test after 28 days of curing. Specimens were removed from moulds after 24 hours and then cured under water until testing at 7 and 28 days.

2.3. Test methods

2.3.1. Compressive strength test on mortar and concrete specimens. Compressive strength test was conducted on PPM cubes as per IS 12269:2013. The optimum content of PP was found from samples which gave maximum average compressive strength value. Compressive strength test was also conducted on concrete cube specimensas per IS 516:1959. 3 cubes for 7 day and 3 cubes for 28 days were tested for each mix containing optimum content of PP and varying percentage of PF. Optimum PF content was obtained from the sample which gave maximum average compressive strength value.

2.3.2. Split tensile strength test and Flexural strength test on concrete specimens. Split tensile strength test was conducted after 28 days of curing on cylindrical specimens for determining the tensile strength of concrete as per IS 5816:1999. Three beam specimens were cast and tested after 28 days of curing to determine the flexural strength of concrete as per IS 516:1959 specification for all mixes.

3. Results and discussions

Tests were performed to investigate the effect of different PP and PF content on strength of PP-PF concrete.

3.1. Compressive strength

The variation in compressive strength of PPM and PP-PF concrete specimens are formulated in table 6 and table 7, respectively. The graphical representation of results in PPM and PP-PF concrete specimens is shown in figure 1 and figure 2, respectively.

Table 6. Variation in compressive strength with varying PP content.

Mix designation	7-day Compressive Strength (N/mm^2)	28-day Compressive Strength (N/mm^2)
0 PP	37.31	53.36
5 PP	40.53	59.10
10 PP	42.53	61.39
15 PP	39.72	57.82
20 PP	38.11	54.97

Table 7. Variation in compressive strength with varying PF content.

Mix designation	7-day Compressive Strength (N/mm^2)	28-day Compressive Strength (N/mm^2)
0 PP + 0 PF	24.44	39.28
10 PP + 0 PF	26.77	40.86
10 PP + 1 PF	27.24	42.20
10 PP + 1.25 PF	28.44	44.22
10 PP + 1.5 PF	29.96	46.67
10 PP + 1.75 PF	28.20	43.1

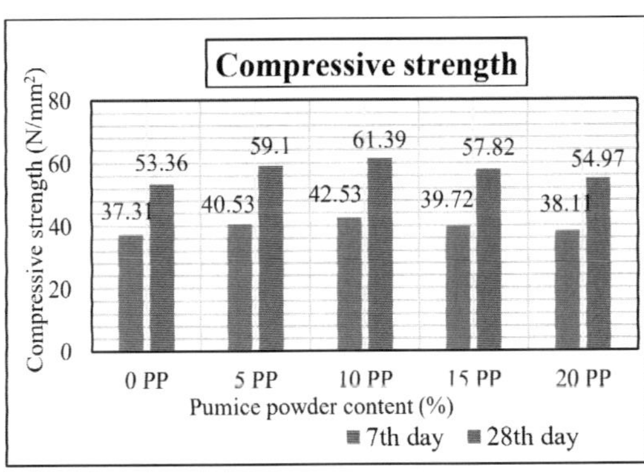

Figure1. Compressive strength of PPM.

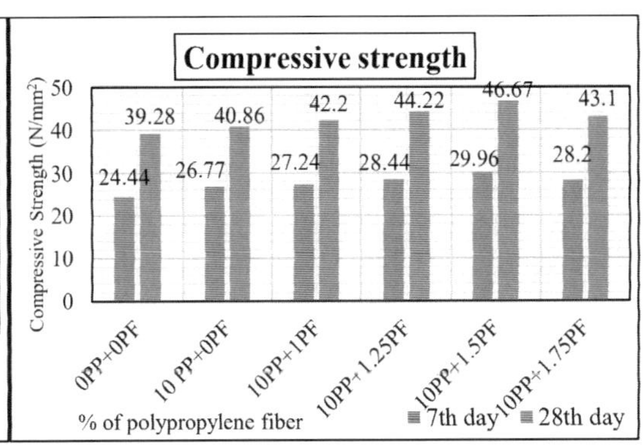

Figure 2. Compressive strength of PP-PF Concrete.

With the increase in PP content, compressive strength of mortar specimens increases up to 10% PP mix, which gave a maximum compressive strength of 42.53 N/mm^{2}and 61.39 N/mm^2 at 7 and 28 days respectively. Further, the strength is found to be slightly decreasing.

The compressive strength of concrete specimens is found to be increasing with the increase in PF content. 1.5% is the optimum content of PF which gave a maximum compressive strength of 29.96N/mm^2 and 46.67N/mm^2 at 7 and 28 days, respectively along with 10% PP content. Further, the strength is found to be slightly decreasing.

The silica in PP combines with calcium hydroxide of cement, in the presence of water to form the stable compound calcium silicate which is having cementitious properties [8]. This pozzolanic action of PP contributes to the enhancement of strength in PP added specimens. Hence addition of PP along with cement is beneficial.

3.2. *Split tensile strength and Flexural strength*

The results of the 28day split tensile strength and flexural strength test in PP-PF concrete specimens for various mixes are shown in table 8. Graphical representations of split tensile strength and flexural strength tests are shown in figure 3 and figure 4, respectively.

Table 8. Split tensile and flexural strength values of PP-PF concrete.

Mix designation	Split tensile strength (N/mm^2)	Flexural strength(N/mm^2)
0 PP + 0 PF	3.39	5.6
10 PP + 0 PF	3.80	5.8
10 PP + 1 PF	4.24	6.6
10 PP + 1.25 PF	4.81	8.0
10 PP + 1.5 PF	5.65	9.6
10 PP + 1.75 PF	4.52	7.2

Variation in tensile and flexural strength shows a trend similar to that of compressive strength. The 28 day split tensile strength and flexural strength increases with increase in PF content. Maximum split tensile strength was obtained as 5.65N/mm^2 with 10% PP and 1.5% PF content.A maximum flexural strength of 9.6 N/mm^2 was obtained for 10% PP and 1.5%PF content.

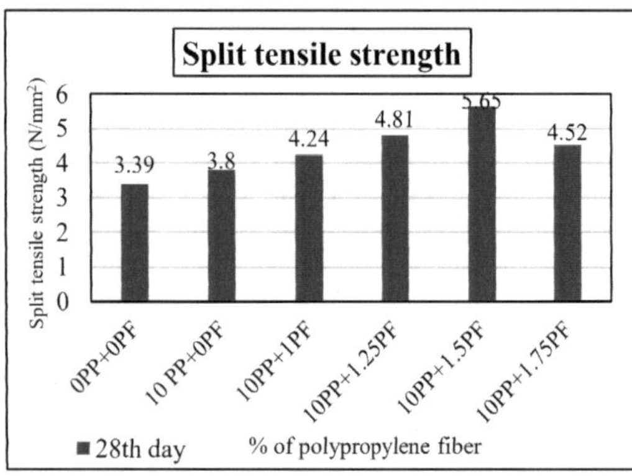

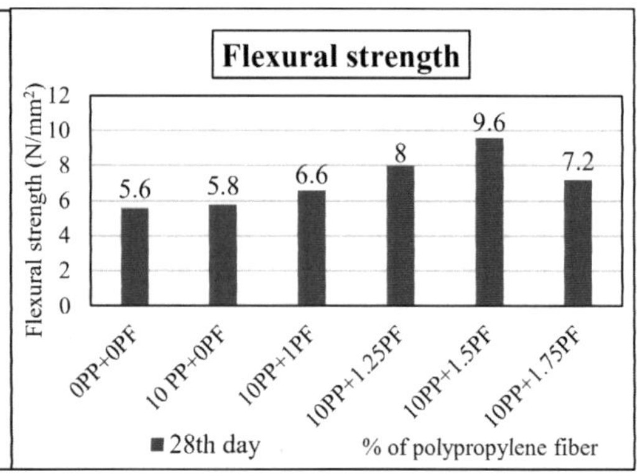

Figure 3. Split tensile strength of PP-PF concrete. **Figure 4.** Flexural strength of PP-PF concrete.

4. Conclusions

The study is focused on finding the combined effect of pumice powder and polypropylene fiber in concrete along with finding the optimum content for maximum strength development. The key findings of the study are as follows:

- Pumice powder exhibits pozzolanic characteristics similar to cement.
- With the addition of pumice powder, the compressive strength increased marginally and optimum replacement content of cement with PP was found to be 10%.
- The substitution of cement with cost-effective material like pumice powder improved the overall strength behavior of concrete although the strength improvement was marginal.

- With the addition of polypropylene fiber in concrete along with optimum content of pumice powder, mechanical strength characteristics increased and the optimum content of PF is found as 1.5%.
- The increase in compressive strength, split tensile strength and flexural strength for modified mix with optimum of PP and PF contents at 28 days of curing is 19%, 67%, and 72%, respectively.
- Fiber addition in concrete improved the tensile as well as the flexural properties of concrete.

Hence it can be concluded that strength properties of concrete incorporated with waste pumice powder and polypropylene fiber increased considerably as compared to normal concrete.

5. References

[1] Zeyad A M, Tayeh B A and Yusuf M O 2019 Strength and transport characteristics of volcanic pumice powder-based high strength concrete *Construction and Building Materials* **216** 314–24
[2] Ulusu H, Aruntas H Y and Gencel O 2016 Investigation on characteristics of blended cements containing pumice *Construction and Building Materials* **118** 11–19
[3] Kizilkanat A B, Oktay D, KabayN and Tufekci M 2016 Comparative Experimental Study of Mortars Incorporating Pumice Powder or Fly Ash *Journal of Materials in Civil Engineering* **28(2)** 040151191
[4] SaamiyaSeraj, Cano R, Ferron R D and Juenger M C G 2017 The role of particle size on the performance of pumice as a supplementary cementitious material *Cement and concrete Composites* **80** 135–42
[5] Hedayatinia F, Delnavaz M, Emamzadeh S S 2019 Rheological properties, compressive strength, and life cycle assessment of self-compacting concrete containing natural pumice Pozzolan *Construction and Building Materials* **206** 122–9
[6] Castoldi R S, Souza L M S and Silva F A 2019 Comparative study on the mechanical behavior and durability of polypropylene and sisal fiber reinforced concretes *Construction and Building Materials* **211** 617–28
[7] Rios J D, Cifuentes H, Leiva C and García C 2018 Behavior of high-strength polypropylene fiber-reinforced self-compacting concrete exposed to high temperatures *Journal of Materials in Civil Engineering* **30** 040182711
[8] Hossain K M A 2004 Development of volcanic pumice based cement and lightweight concrete *Magazine of Concrete Research* **56** 99–109
[9] IS 12269-1987: *Specification for 53 grade Ordinary Portland Cement, Bureau of Indian Standards* (New Delhi)
[10] IS 12269-2013: *Ordinary Portland Cement, 53 grade — Specification, Bureau of Indian Standards* (New Delhi)
[11] IS 516-1959: *Methods of tests for strength of concrete, Bureau of Indian Standards* (New Delhi)
[12] IS 5816-1999: *Splitting tensile strength of concrete - method of test, Bureau of Indian Standards* (New Delhi)

Acknowledgments

The authors thankfully acknowledge the financial support provided by Kerala State Council for Science, Technology, and Environment (KSCSTE) under the student project scheme [File No. 00736/ SPS64/ 2019/ KSCSTE], Kerala, India.

IOP Conf. Series: Materials Science and Engineering **936** (2020) 012003 doi:10.1088/1757-899X/936/1/012003

Design of Short Columns Reinforced with GFRP Bars Subjected to Axial Loading

Niyazuddin[1*], Trupti A Kinjawadekar[2], Praveen Nagarajan[1] and A P Shashikala[1]

[1] Department of Civil Engineering NIT, Calicut, India

[2] Manipal School of Architecture and Planning, MAHE, India

*niyazuddinansari999@gmail.com

Abstract. Due to the ongoing environmental changes, degradation of directly exposed material is seen very often. This results in issues related to aesthetics, function, and most importantly, the safety of the structures. Reinforced cement concrete (RCC) is the most commonly used material for construction where steel bars are used as reinforcement. Steel reinforcement in contact with the chemicals like chlorides and sulphates etc. which are very commonly present in urban and industrial areas and other aggressive environment, gets corroded, thus affecting the structural behaviour. Research studies propose replacing steel rebars with Glass Fibre Reinforced Polymer (GFRP) bars due to its non-corrosive properties. This paper presents the analysis and design of GFRP reinforced concrete columns. Present design methods mentioned in Indian Standard Code are not applicable to design GFRP reinforced RC columns directly. This study is an attempt to develop design provisions for GFRP reinforced concrete columns similar to those in practice for steel-reinforced columns. The spacing of ties influences the capacity of the column to carry load and failure type of compression member. So, this paper will also focus on the design of ties to provide proper confinement.

1. Introduction

Reinforced Concrete is a composite material that is widely used for construction. Various materials are being employed as reinforcing means for concrete which include bamboo, steel, wood, and some artificially produced composite materials [1]. Reinforcement in the form of steel reinforcing bar (rebar) is passively embedded in concrete before the concrete sets. However, rusting of steel rebar is identified as the foremost problem affecting structural stability [2]. In recent years, research has been carried out on Glass Fibre Reinforced Polymer (GFRP) bars as an alternative to steel reinforcement in concrete structures due to its non-corrosive nature [3]. The mechanical properties of the GFRP bar such as high tensile strength low density, lightweight, it is gaining prominence in the construction industry [3].

Castela et al. 2014 [4] investigated the effect of the corrosion rate of the steel bars wrapped with glass fibre-reinforced polymers (GFRP). They found that the corrosion rate and chloride ion penetration decrease due to the wrapping of GFRP. Experimental studies were conducted by Tu et al. [5], Mohamed et al. [6], Luca et al. [7], Hadi et al. [8] to find out the effect of compressive behaviour of GFRP bars in RC columns. From results, it is found that the columns having low longitudinal reinforcement ratio, fail in a brittle manner. The load-carrying capacity of GFRP reinforced concrete columns is 13-16% lower than the column reinforced with steel if provided longitudinal reinforcement ratio is the same[5]. Seismic analysis of GFRP reinforced concrete columns was conducted by Ali et al. Lateral cyclic quasi-static and constant axial loading applied during the testing. Experimental result shows that the drift capacity of GFRP-RC columns lies between 8.5 to 12.5% at failure. Due to well confined concrete core, columns shows insignificant strength degradation before failure[9]. Design equation of GFRP RC columns according to Indian standard is not available. Present study develops the design equation for GFRP reinforced concrete column considering provisions of Indian Standard code.

Figure 1. GFRP Bars.

2. Minimum longitudinal reinforcement

According to Indian standard (IS456: 2000) ratio of minimum reinforcement for compression member reinforced with steel is 0.8 percent of the gross cross-section area of the column[10]. Minimum reinforcement is necessary to cater to the bending moment in the column due to specified minimum eccentricity. Minimum reinforcement also prevents the passive yielding of the steel. From the literature survey, it is found that GFRP does not yield, so the same specification of the minimum reinforcement ratio can be applied to column reinforced with GFRP bars.

3. Design equation: ultimate load-carrying capacity

According to Indian standard (IS 456: 2000)[10] the capacity to carry the axial load by short column reinforced with steel bars is given by Eq. (1) (Without partial factor of safety) and column are designed according to Eq. (2).

$$P_u = 0.8\, f_{ck} A_c + f_y A_{sc} \tag{1}$$

$$P_u = 0.4\, f_{ck} A_c + 0.67 f_y A_{sc} \tag{2}$$

where P_u is the ultimate load-carrying capacity of the column, A_c is the area of cross-section of concrete, longitudinal reinforcement area is denoted by A_{sc}, f_y is yielding strength of the steel bar, f_{ck} is the characteristics strength in compression for the concrete.

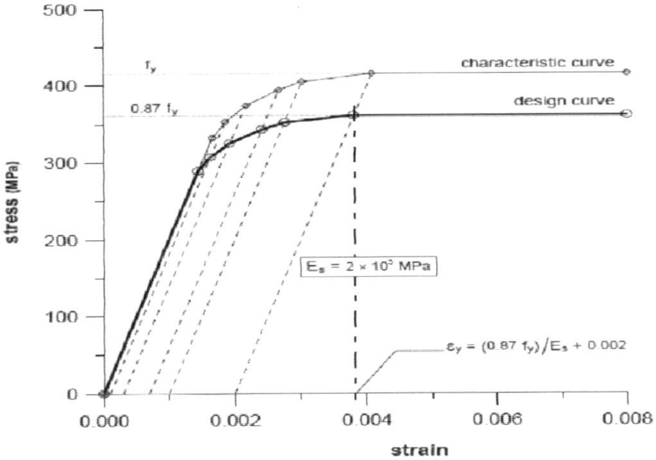

Figure 2. Cold work deformed bar: Stress vs Strain [11].

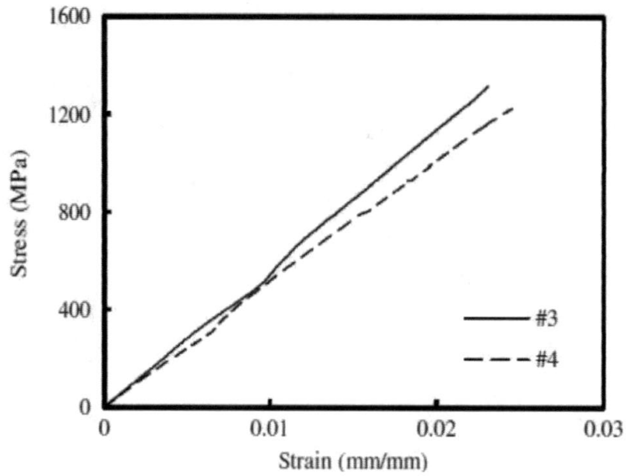

Figure 3. GFRP bar: Stress-Strain behavior[12].

GFRP bars show linear stress-strain behaviour. So, it is not easy to find out the yielding strength of these bars. Design equation of steel-reinforced concrete column given by IS456: 2000 is based on yielding strength of steel rebar because the yielding strength of these bars can be easily found out. Therefore, the design of the GFRP-RC column is not possible based on yield strength. It is assumed that elastic modulus in compression is comparable to the tension. The first part of the equation represents the load carried by concrete and the second part represents the load carried by GFRP bars. The experimental result shows that the ultimate stress compression for GFRP bars is 30 - 40 % of the ultimate stress in tension. Based on the experimental result, equation (3) is introduced to find out the load-carrying capacity of the GFRP reinforced concrete column. A new factor for strength reduction (β) is introduced. From literature (Mohamed et al. 2014 [6]), the value of the strength reduction factor is assumed 0.35.

$$P_u = 0.8\, f_{ck} A_c + \beta\, f_u A_{gf} \tag{3}$$

where f_u is the ultimate tensile strength of the GFRP bar. Based on equation (3) design equation of the column is given as

$$P_u = 0.4\, f_{ck} A_c + \beta\, f_u A_{gf} \tag{4}$$

4. Analytical results vs Experimental results

Experimental studies are carried out on square columns of size 150 ×150 × 850mm. GFRP bars are used as reinforcement for longitudinal bars and 6mm steel bars are used for stirrups. The grade of concrete used in studies is M30. The longitudinal reinforcement ratio is 1.39%. Four bars of 10mm are provided as longitudinal reinforcement. Table 1 shows the comparative analysis of theoretical and experimental results.

Table 1. Square columns reinforced with GFPR bars: Analytical vs Experimental results.

Specimen no.	Specimen ID[a]	Axial Compressive Strength (MPa)	Experimental load (kN)	Analytical load (kN)	% Difference in Experimental load and Eq.3
1	C-G75	42.35	1000.62	873.70	+12.6
2	C-G50	43.8	1177.2	899.44	+23.5

[a] C stands for control mix, G stands for GFRP bars and 75, 50 represents the spacing of stirrups.

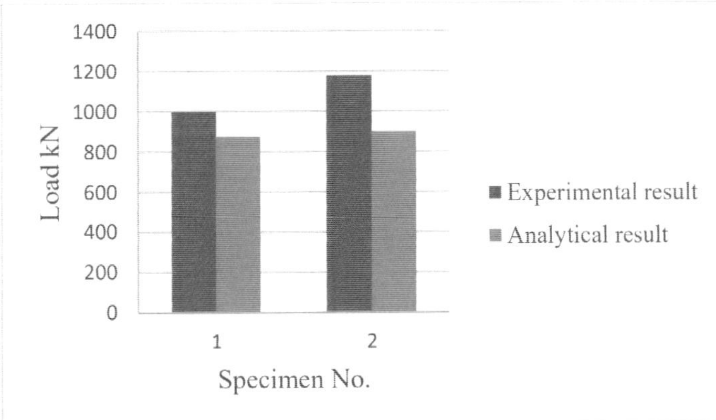

Figure 4. Experimental Analytical result comparison for square columns.

5. Validation of analytical equation

i. To validate the analytical equation for square columns experimental results are considered from literature (Tu et al. 2019)[5]. Eight GFRP reinforced concrete columns (200 × 200 ×600mm) were tested. The longitudinal reinforcement ratio is provided as 0.8, 1.1, and 1.5%. Four numbers of bars are provided as longitudinal reinforcement. All the columns are cast from the same concrete of having an axial compressive strength of 32.1 MPa. GFRP bars of 8 mm is used for stirrups. Table 2 shows the comparative analysis of theoretical and experimental results.

Table 2. Square columns reinforced with GFPR bars: Analytical vs Experimental results [5].

Specimen No.	Specimen ID[b]	Longitudinal reinforcement ratio %	Experimental load (kN)	Analytical Load (kN) Eq.3	% Difference Experimental load and Eq.3
1	G12-S30	1.1	970.0	1120.08	-15.37
2	G12-S50	1.1	951.6	1120.08	-17.70
3	G12-S80	1.1	937.7	1120.08	-19.45
4	G10-S50	0.8	936.8	1099.99	-17.42
5	G14-S50	1.5	981.7	1135.09	-15.62
6	G12-H30	1.1	954.0	1120.08	-17.41
7	G12-H50	1.1	943.2	1120.08	-18.75
8	G12-H80	1.1	927.7	1120.08	-20.74

[b] G represents GFRP bars, *S* stands for Spirals, and *H* represents hoop 10,12, 14 are the diameter of the longitudinal bars 30, 50, 80 are the spacing of the ties.

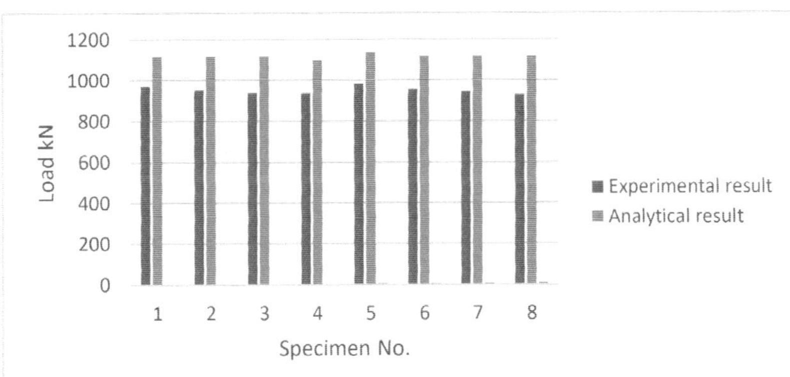

Figure 5. Experimental vs Analytical result comparison for square columns.

i. For circular columns, the experimental result of the load carried by the columns is taken from studies by Mohamed et al. 2014[6]. Concrete columns of diameter 300 mm. and height 1500 mm. reinforced with GFRP reinforced were tested. 8 numbers of GFRP bars of diameter 15.9 mm provided as longitudinal reinforcement. The longitudinal reinforcement ratio is provided at 2.2%. All the columns made from the same concrete of having an axial compressive strength of 42.9 MPa. GFRP bars are used for stirrups.

Table 3. Circular columns reinforced with GFPR bars: Analytical vs Experimental results.

Specimen No	Specimen ID[c]	Experimental load (kN)	Analytical Load (kN) Eq. 3	% Difference in Experimental load and Eq.3
1	G2S	2857	2890.68	-1.17
2	G3S	2920	2890.68	+1.00
3	G4S	3019	2890.68	+4.25
4	G3H200	2840	2890.68	-1.70
5	G3H400	2871	2890.68	-0.68
6	G3H600	2935	2890.68	+1.50

[c] G represents GFRP, 2-3-4 represents number of horizontal bar, *S* stands for spiral, *H* represents hoops, 200, 400 and 600 stands for the lap splices length in mm for GFRP hoops.

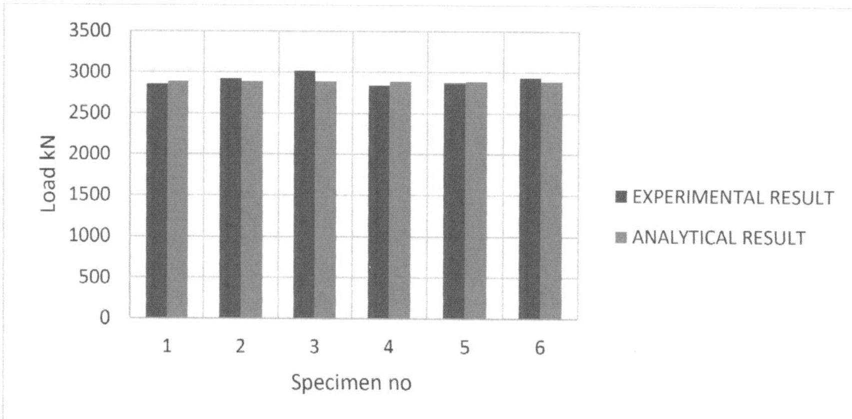

Figure 6. Experimental vs Analytical result comparison for circular columns.

6. Maximum spacing of transverse reinforcement

According to IS456:2000[10] for the steel-reinforced concrete column, spacing is limited to a minimum of

- Least lateral dimension of the column
- 16 times the longitudinal bar diameter
- 300 mm

The spacing between ties related to the longitudinal bar diameter based on the assumption that the bars are simply supported between the two-support provided only by the ties. For such member to avoid the buckling this condition must be satisfied [2].

$$\frac{\pi^2 EI}{S^2} \leq AE\epsilon \tag{5}$$

$$S_{max}^2 = \frac{\pi^2 I}{A\epsilon} \tag{6}$$

For a circular bar of diameter d_b this expression can be written as

$$S_{max} = \frac{\pi d_b}{4\sqrt{\epsilon}} \tag{7}$$

Where S is spacing of the ties, I is Moment of inertia of the bar, A is the cross-sectional area of the bar.

For steel bars to prevent buckling;$\epsilon = \epsilon_y = 0.002$. Substituting this value in Eq. (7)

$S_{max} = 17.5\ d_b$, which satisfies the code provision of IS456: 2000 ($S_{max} = 16\ d_b$).

For GFRP bars to prevent buckling before concrete crushing;$\epsilon = \epsilon_{cu} = 0.0035$, therefore

$S_{max} = 13.27\ d_b$ so considering all these points the spacing of the transverse reinforcement S_{max} is minimum of (1) least lateral dimension of column (2) 12 longitudinal bar diameter (3) 300 mm

7. The failure mode of GFRP columns

Failure of GFRP columns mainly depends on the spacing of stirrups. GFRP columns having large spacings failed in an explosive and brittle manner, and columns with small stirrups spacings are failed in a ductile manner. So, it is very important to provide proper confinement so that columns should be failed in a ductile manner.

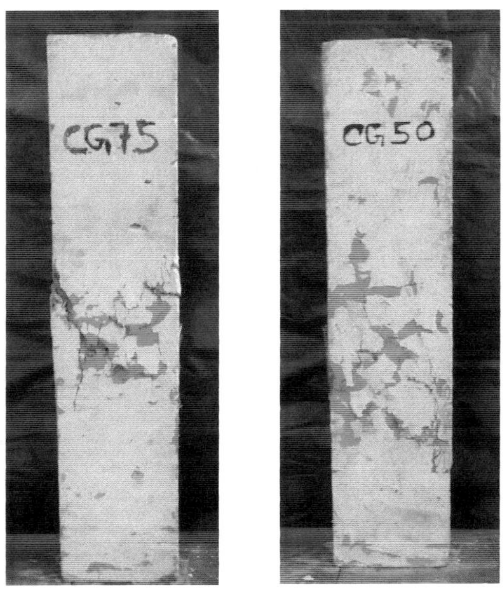

Figure 7. The failure mode of GFRP columns.

8. Conclusions

- Using Equation 3, the variation in experimental and theoretical results are observed in the range of 12.6 % to 23.5 % for square columns and circular columns, respectively. Difference between experimental and theoretical results are observed in the range of 0.68% to 4.25%.
- It is observed that the columns with large stirrups spacing failed in a brittle manner compared to small stirrups spacing. Experimental results indicate that change in stirrups spacing causes variation in the load-carrying capacity of the column.
- By changing the spacing of the ties, difference between experimental and analytical results also varies and it is seen that reduction in spacing results in a decrease in the variation.
- Due to the limited availability of the data for GFRP bars, the design provision given in this paper does not apply to the structures in the seismic zone.

IOP Conf. Series: Materials Science and Engineering **936** (2020) 012003 doi:10.1088/1757-899X/936/1/012003

9. References

[1] Lin H and Liao C 2004 Compressive strength of reinforced concrete column confined by composite material *Composite Structures* **65** 239–50

[2] Zadeh H J and Nanni A 2013 Design of RC columns using glass FRP reinforcement *Journal of Composites for Construction* **17(3)** 294–304

[3] Nanni A, De Luca A and Zadeh H J 2014 *Reinforced Concrete with FRP Bars: Mechanics and Design* (CRC Press)

[4] Da Fonseca B S, Castela A S, Silva M A, Duarte R G, Ferreira M G and Montemor M F 2015 Influence of GFRP confinement of reinforced concrete columns on the corrosion of reinforcing steel in a salt water environment *Journal of Materials in Civil Engineering* **27(1)** 04014107

[5] Tu J, Gao K, He L and Li X 2019 Experimental study on the axial compression performance of GFRP-reinforced concrete square columns *Advances in Structural Engineering* **22(7)** 1554–65

[6] Mohamed H M, Afifi M Z and Benmokrane B 2014 Performance evaluation of concrete columns reinforced longitudinally with FRP bars and confined with FRP hoops and spirals under axial load *Journal of Bridge Engineering* **19(7)** 04014020

[7] De Luca A, Nardone F, Matta F, Nanni A, Lignola G P and Prota A 2011 Structural evaluation of full-scale FRP-confined reinforced concrete columns *Journal of Composites for Construction* **15(1)** 112–23

[8] Hadi M N S 2009 Behaviour of eccentric loading of FRP confined fibre steel reinforced concrete columns *Constr. Build. Mater.* **23(2)** 1102–8

[9] Ali MA and El-Salakawy E 2016 Seismic performance of GFRP-reinforced concrete rectangular columns *Journal of Composites for Construction* **20(3)** 04015074

[10] IS 456 2000 *Plain And Reinforced Concrete - Code of Practice* (Bureau of Indian Standards: New Delhi)

[11] Easy Engineering 2017 *Reinforced Concrete Structures* (https://easyengineering.net/)

[12] Hadi M N, Karim H and Sheikh M N 2016 Experimental investigations on circular concrete columns reinforced with GFRP bars and helices under different loading conditions *Journal of Composites for Construction* **20(4)** 04016009

An Investigation on Pozzolanicity of Mechanically Activated Pond Ash

V Vidyadhara[1*], T Shamanth Gowda[1] and R V Ranganath[2]

[1] Research Scholar, Department of Civil Engineering, BMS College of Engineering, Bengaluru, India

[2] Professor, Department of Civil Engineering, BMS College of Engineering, Bengaluru, India

*vidyadhara92@outlook.com

Abstract. Pond ash (PA) is one of the major industrial solid wastes from thermal power plants. In the present study, PA of different storage durations, i.e. freshly ponded, two year ponded and Six-year ponded ash were collected from Bellary Thermal Plant (BTP). All three kinds of PA were mechanically activated by grinding in Industrial Hammer Mill. Unprocessed Pond Ash (UPA) and Ground Pond Ash (GPA) were characterised for Particle Size Distribution (PSD) by Laser-diffraction analyser. Mineralogical and Morphological characteristics were analysed through X-ray Diffraction and Electron Microscopic studies. Chemical composition of PA was analysed. The pozzolanicity of PA and other Supplementary Cementitious Materials (SCM) was determined and compared by conducting Lime reactivity test. XRD patterns and Chemical analysis clearly indicate the presence of relatively more amorphous phases in a fresh batch of PA compared to the one which is ponded for a longer duration. Lime reactivity strength (LRS) of UPA was far behind the LRS of commonly used SCM's. Mechanical activation of PA improved the LRS from 0.57 MPa to 2.41 MPa. The results suggest the possibility of enhancing the reactivity of PA with early use and more efficient grinding process, thus add value to PA for its use in concrete.

1. Introduction

Concrete is the man-made material which is most widely used by mankind next to water in terms of substance consumption. Cement is the largely used binding material in concrete. One ton of CO_2 is released to the environment in the production of a unit ton of cement [1]. Utilisation of supplementary binding materials in concrete, especially waste materials from different industries is the current trend in concrete technology. Fly Ash, Bottom Ash, Silica Fume, Blast Furnace Slag, Paper Sludge and calcium carbide are some of the industrial waste materials used in concrete by earlier researchers [1-6]. Usage of these materials in concrete results in the reduction of its carbon footprint.

Coal Fuel ash from Thermal Power Plants has been used in concrete as admixtures to replace cement as well as aggregates. Coal ashes are of different types which include Fly ash, Bottom Ash, Ponded fly ash and Pond Ash. Fly ash is the most reactive material among all ashes, because of the presence of fine, spherical particles with a more amorphous phase. Fly ash, because of its high pozzolanic property, is used as a replacement for cement in conventional concrete [7]. In addition, due to the presence of amorphous silica and alumina, it is used as a binder material in geopolymer concrete [2]. Bottom ash is the coarser ash that is collected from the bottom of the boiler. After milling it is transported for open-disposal by simple water- wash [8]. It is less reactive compared to fly ash, as it contains coarser, porous, irregular shaped particles with fewer amorphous phases. Bottom ash is used as fine aggregate in concrete, as it comprises of well-graded, angular, rough-textured and light particles [9]. Presence of porous particles in Bottom ash makes it a better alternative to be used as coarse aggregate in pervious concrete [10]. Grinding of bottom ash results in the decrease of porosity and increase in the surface area thereby increasing its reactivity [11]. Ground bottom ash is used in cement mortar as partial replacement to cement [12]. The additional gain in the reactive components of bottom ash after grinding has made researchers to use it as aluminosilicate precursor in geopolymer [13]. Countries like USA and UK, dispose unused fly ash into ponds, which is termed as Ponded fly ash. Ponding fly ash in such a manner will result in increased size of particles and reduction of amorphous phase compared to virgin fly ash [14]. Ponded fly ash is used as a replacement to cement in concrete and the strengths achieved are relatively less compared to concrete made using virgin fly ash [15].

Fly ash from electrostatic precipitator which will be unused and bottom ash from boiler after milling is conveyed to ponds in the form of water mixed slurry. This stored slurry in ponds results in the generation of Pond ash [16] which is most commonly seen in Indian Thermal Power Plants. Pond ash reactivity lies in between fly ash and bottom ash as it includes reactive fine fly ash particles and non-reactive or less reactive large bottom ash particles [17]. Similar to Bottom ash, Pond ash is also used as fine aggregate in concrete since it comprises of coarser bottom ash particles. Pond ash, when used as fine aggregate, has resulted in a decrease in the workability of concrete [18]. Reduction in strength is also noticed when Pond ash is used as a partial replacement to cement in concrete [16]. However, there is a lack of research on the pozzolanic behaviour of fine ground or mechanically activated Pond ash, unlike Bottom ash. Pond ash is more reactive than bottom ash, as it contains some traces of reactive fly ash particles. There is a need for an investigation on the pozzolanicty of ground or mechanically activated pond ash so that processed Pond ash could be assessed for its potentiality to be used as a binder in conventional concrete and as aluminosilicate precursor in geopolymer concrete. The current research is an attempt to investigate the possible improvement in the pozzolanic nature of pond ash by means of simple mechanical activation technique.

2. Experimental program

2.1. Materials

Pond ash (PA) was procured from Bellary thermal power plant, Karnataka. PA of three kinds, Fresh ponded, two-year-old and Six-year-old was taken from the plant-based on their age of ponding. In addition, one more type of sample i.e. mixed sample was prepared by blending above three samples in equal proportions. All the samples of PA were oven-dried at 110±5 °C. A portion of PA from all respective samples was subjected to grinding in Industrial Hammermill. Further, Fly Ash (FA), Ultrafine Fly Ash (UFA), Ground Granulated Blast Furnace Slag (GGBS) and Ultrafine Slag (US) were purchased from commercial sources for their comparative analysis with PA. Lime was locally procured for use in Lime reactivity test. Standard sand (Ennor) [19] was used as fine Aggregate.

The Fresh Unprocessed Pond ash (FUPA), Two-year-old Unprocessed Pond ash (2UPA), Six-year-old Unprocessed Pond ash (6UPA), Blend Unprocessed Pond ash (BUPA), Fresh Ground Pond ash (FGPA), Two-year-old Ground Pond ash (2GPA). Six-year-old Ground Pond ash (6GPA) and Blend Ground Pond ash (BGPA) were checked for their pozzolanicity.

2.2. Physical properties of pond ash

The Particle Size Distribution (PSD) of all the PA samples were analysed using Laser Diffraction Analyser and PSD curves are shown in figure 1. Specific gravity, Surface area, weight retained on sieve no. 325 and mean diameter of PA particles are represented in table 1. The specific gravity of GPA is recorded more compared to UPA, as grinding eliminates the pores of PA thereby increasing mass to volume ratio. Since UPA was coarser, Surface area by Blaine's test couldn't be found out [3]. Blaine Surface area of GPA varies in the range of 222-288 m^2/kg. Weight retained on No.325 sieve and median particle size of GPA is greatly reduced compared to UPA. Out of all unprocessed samples, 6UPA comprised of higher coarser particles. There is an overall increment in fineness of PA after subjecting it to the selected grinding process. This can be noticed from the PSD graph and from the median particle size of samples.

Table 1. Physical properties of different PA.

Samples	Specific Gravity	Surface area (m^2/kg)	% by weight Retained on a No. 325 sieve	Median Particle Size, d50 (mm)
FUPA	1.96	-	90	160
2UPA	1.79	-	87	127
6UPA	2.09	-	92	188
BUPA	1.86	-	93	156
FGPA	2.32	230	62	59
2GPA	2.19	222	62	59
6GPA	2.26	288	47	40
BGPA	2.25	240	55	49

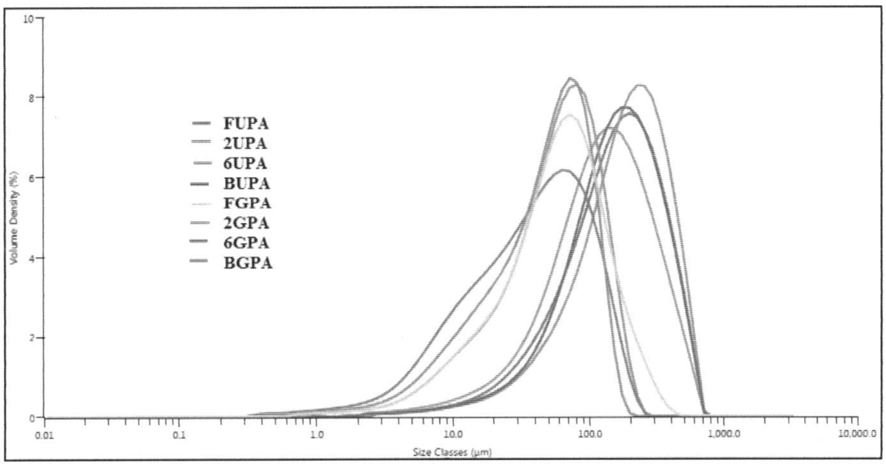

Figure 1. Particle size distribution curves of different PA.

2.3. XRD, SEM and EDAX studies of pond ash

Before grinding, PA is observed to have porous, irregular, agglomerated -larger particles and grinding process resulted in a reduction of its size and porosity [1, 3, 12] which is clearly indicated in SEM images shown in figure 2 a) and b). XRD analysis was carried out for finding the mineralogical composition of UPA. Quartz was the main mineral phase present in PA as shown in figure 3a. Scanning was concentrated between 20° and 30° to compare the amorphous phase of different UPA. Hump formation can be noticed at 26.5° in case of FUPA and 2UPA indicative of the presence of the amorphous phase. Micrograph shown in figure 3 b) was obtained by Energy-dispersive X-ray Analysis (EDAX) for the analysis of elements in PA. The presence of Si, Al, O and Fe were confirmed as the main elements in PA as per the analysis [11].

Figure 2. SEM images of a) Unprocessed pond ash b) Ground pond ash.

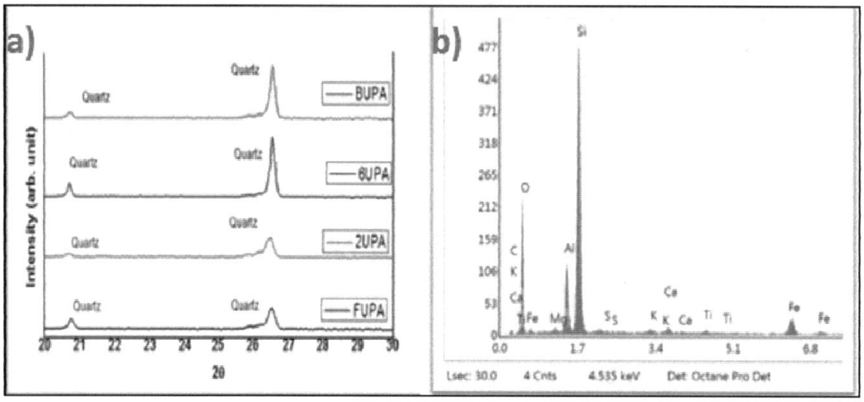

Figure 3. a) XRD patterns of unprocessed pond ash b) Micrograph of pond ash obtained by EDAX.

2.4. Chemical composition of pond ash

Chemical composition of Unprocessed PA samples is given in table 2. As per ASTM C618 [20], Specification for Coal Fly Ash and Raw or Calcined Natural Pozzolan for use in Concrete, SiO_2+ Al_2O_3 + Fe_2O_3 minimum limit should be 70% and LOI maximum limit should be 6%. Chemical composition of all the PA samples are well within the aforementioned limits and are classified as ASTM Class F ash [21]. Reactive silica content represents the percentage of amorphous silica in total silica content. FUPA has a good amount of amorphous silica and 6UPA has the lowest value of the same. This depicts, increase in ponding duration reduces the reactive silica in PA.

Table 2. Chemical composition of UPA.

Samples	LOI	SiO_2	Al_2O_3	Fe_2O_3	CaO	MgO	Reactive SiO_2	SiO_2+ Al_2O_3 + Fe_2O_3
FUPA	5.30	58.38	25.34	5.68	2.69	0.98	20.02	89.80
2UPA	3.87	59.24	26.94	5.68	2.24	1.16	18.74	91.86
6UPA	4.47	55.90	28.76	5.56	2.86	1.28	16.02	90.22
BUPA	5.90	57.12	26.06	5.84	2.57	1.37	17.86	88.98

2.5. Pozzolanicity of pond ash

Pozzolanic reactivity of supplementary cementitious material (SCM) can be assessed from various tests and Indian standard IS: 1727 – 1967[22] recommends for Pozzolanic reactivity of SCM by Lime reactivity test and same was adopted in the current study. SCM, Standard sand (Ennor) [20] and Lime will be mixed with water in standard proportions. Cubes of size 50mm are cast and demoulded after 2 days. Demoulded samples are kept in desiccators for curing in an oven at standard temperature and relative humidity conditions for 8 days. Later cubes are tested for compressive strength. This test can assess the potentiality of SCM to react with Calcium Hydroxide ($Ca(OH)_2$) to form Calcium Silicate Hydrate (CSH) and Calcium Aluminate Hydrate (CAH) gels.

3. Results and discussion

3.1. Water requirement

As per lime reactivity test method IS: 1727 – 1967[22], the amount of water added to the mix should produce a flow of 170±5 mm on flow table. Water required by different PA samples to achieve the prescribed flow is given in table 3. From table 3, it is evident that between UPA samples and GPA samples, UPA samples consumed more water compared to GPA samples to meet the flow requirement. Since UPA samples contain porous and dry bottom ash particles, more water is absorbed demanding extra water for lubrication as seen in earlier studies [3, 8, 23]. Pores are eliminated after grinding, thereby reducing the water requirement in case of ground samples to maintain the same flow [12].

Table 3. Lime reactivity test results of pond ash.

Samples	Lime reactivity 8 days (MPa)	Lime reactivity 28 days (MPa)	Water added(ml) to achieve flow of 170±5mm
FUPA	0.57	0.98	390
2UPA	0.60	0.96	390
6UPA	0.40	0.71	400
BUPA	0.46	0.96	400
FGPA	2.41	3.35	310
2GPA	2.22	3.01	310
6GPA	2.14	3.05	320
BGPA	2.38	3.13	320

3.2. Lime reactivity strength (LRS)

LRS as observed from table 3 records low values for UPA samples compared to that of GPA. The strength is evaluated after 8 and 28 days of desiccator curing and the results are graphically represented in figure 4. Even though UPA was rich in Silica and alumina, due to more coarser and porous particles with less specific surface area, it has resulted in a lesser degree of reaction of silica and alumina with $Ca(OH)_2$. In contrast, grinding resulted in eliminating pores, reducing size, increasing both surface area and reactive silica content in PA particles, as observed in studies on mechanical activation of red mud [24]. Thus the degree of reaction of silica and alumina present in GPA with $Ca(OH)_2$ was relatively more contributing for higher strength both at 8th and 28th day of testing. In addition to increased reactivity, non-reactive GPA particles contribute strength by filling the voids in the matrix.

Among UPA samples 6UPA samples got lesser LRS, because Six years of Ponding resulted in an increase in the size of their particles as well as a decrease in reactive silica content which can be observed from PSD graph (Figure 1) and chemical composition (Table 2). Among the GPA samples, 6GPA had finer particles compared to other samples, in spite of this LRS of 6GPA was less compared to FGPA which is having higher reactive silica content. It clearly indicates in addition to size reduction, the presence of silica in amorphous phase (Reactive silica) is obligatory for a higher degree of pozzolanic reaction [12].

SEM images shown in figure 5 a, further verifies the reason for Lower LRS due to presence of more unreactive coarser PA particles in case of unprocessed PA [17]. Whereas for ground PA, a relatively dense matrix with more reacted phase can be noticed from figure 5 b.

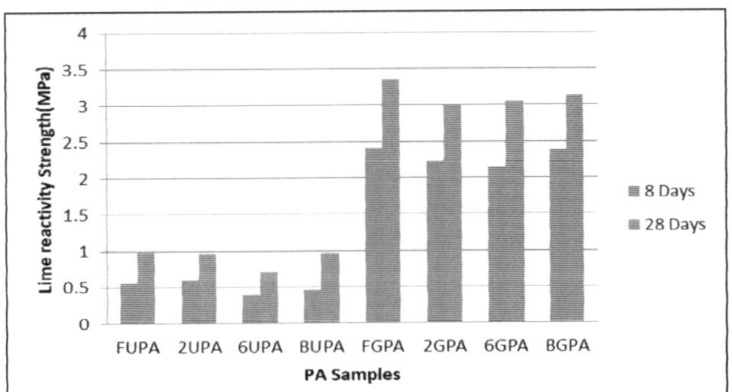

Figure 4. Graph showing lime reactivity strength results of different PA.

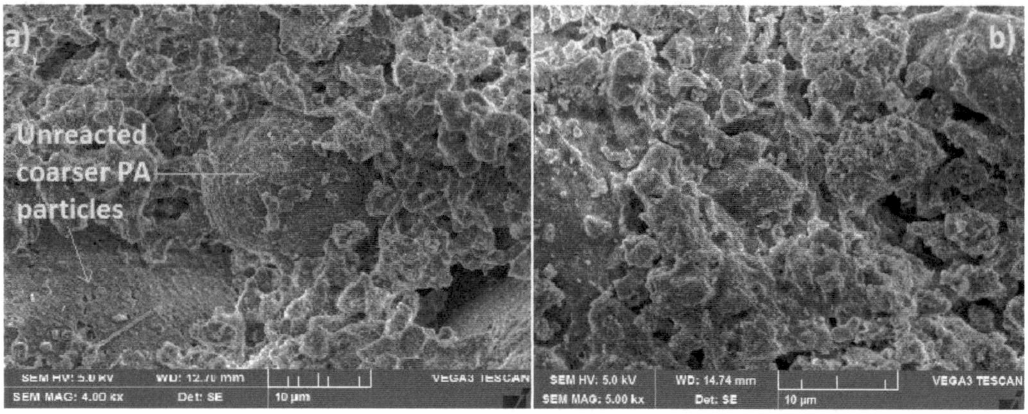

Figure 5. SEM images of lime reactivity tested specimens of a) UPA b) GPA.

3.3. Comparison of lime reactivity strength

It can be observed from the LRS of SCMs shown in table 4, the reactivity of popular SCMs is more prominent compared to PA. However, after certain mechanical activation, the reactivity of PA is improved over that of UPA. The reduction in particle size as recorded from figure 2 may have created more reactive phases of silica and alumina after processing [24]. Yet the comparative reactivity is less than the commercial mineral admixtures. This indicates that with a better grinding process and a further reduction in Particle size of raw PA, the pozzolanic behaviour may be made equivalent to that of FA which is noticed in earlier studies on pozzolanicity of bottom ash [12].

Considering the microstructure of LR specimen, it is clear that commercial SCMs such as FA, UFA, GGBS and UFS have a denser matrix (Figure 6) compared to PA samples (Figure 5).

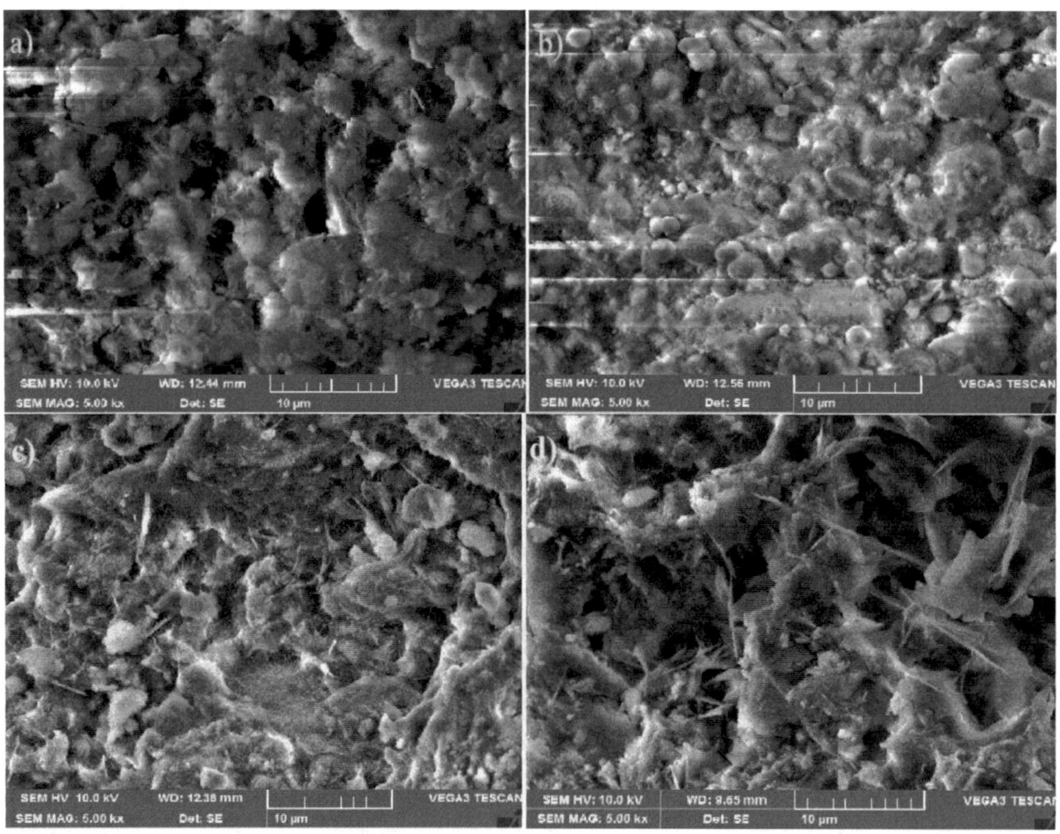

Figure 6. SEM images of lime reactivity tested specimens of a) FA b) UFA c) GGBS d) UFS.

Table 4. Lime reactivity strength of SCMs and PA.

Samples	Lime reactivity 8 days (MPa)
Ultrafine Slag	13.46
Ground Granulated Blast Furnace Slag	7.21
Ultrafine Fly Ash	6.70
Fly Ash	5.33
UPA	0.46
GPA	2.38

4. Conclusions

The brief characterisation of abundantly available PA along with LR test results has indicated the possibility of utilising the same in cement composites. Simple mechanical activation procedure adopted in the current study has shown promising results towards enhancing the reactivity of PA for its large-scale utilisation.

- PSD curves and SEM analysis of both UPA and GPA indicate reduced particle size after processing that can induce reactivity and better filling ability of PA particles when used in heterogeneous material like concrete.
- XRD and Chemical analysis of PA indicates the reduction in its reactivity with longer storage period in ponds and hence it is of benefit to use PA as early as possible from the period of discharge. Based on Chemical analysis result PA may be classified as ASTM Class F ash.
- Results of LRS of GPA have confirmed increased pozzolanic reactivity compared to UPA. The increment in strength is recorded to be around 5 times that of UPA as on 8th day of incubation of test specimens. The strength gain has continued even up to 28th day of incubation for GPA samples which are 3 to 4 times higher than that of UPA samples.
- The comparative pozzolanic reactivity of PA even after grinding is recorded lesser than that of common SCMs. However, the study suggests the possibility of enhancing the reactivity of PA by early use after disposal along with the efficient grinding process, thus adding value to PA for its use in concrete.

5. References

[1] Penpichcha Khongpermgoso, Akkadath Abdulmatin, Weerachart Tangchirapat and Chai Jaturapitakkul 2019 Evaluation of compressive strength and resistance of chloride ingress of concrete using a novel binder from ground coal bottom ash and ground Calcium carbide residue *Construction and Building Materials* **214** 631–40

[2] Prinya Chindaprasirt, Chai Jaturapitakkul, Wichian Chalee and Ubolluk Rattanasak 2009 Comparative study on the characteristics of fly ash and bottom ash geopolymers *Waste Management* **29** 539–43

[3] Chai Jaturapitakku and Raungrut Cheerarot 2003 Development of Bottom Ash as Pozzolanic Material *Journal of Materials in Civil Engineering* **15(1)** 48–53

[4] Hisham M Khater 2013 Effect of silica fume on the characterisation of the geopolymer materials *International Journal of Advanced Structural Engineering* **5** 12

[5] Zhuguo li and Sha li 2018 Carbonation resistance of fly ash and blast furnace slag based geopolymer concrete *Construction and Building Materials* **163** 668–80

[6] Rozineide A Antunes Boca Santa , Adriano Michael Bernardin , Humberto Gracher Riella and Nivaldo Cabral Kuhnen 2013 Geopolymer synthetised from bottom coal ash and calcined paper Sludge *Journal of Cleaner Production* **57** 302–7

[7] Manomi N, Sathyan D, Anand K B 2018 Coupled effect of superplasticiser dosage and fly ash content on strength and durability of concrete *Materials Today: Proceedings* vol 5 pp 24033–42

[8] Ng H J, Al Bakri Abdullah M M, Tan S J, Sandu A V and Hussin K 2018 Characterisation and understanding of Portland cement mortar with different sizes of bottom ash *Advances in Cement Research* **30(2)** 66–74

[9] Iman M Nikbin , Saman Rahimi R , Hamed Allahyari and Mohammad Damadi 2016 A comprehensive analytical study on the mechanical properties of concrete containing waste bottom ash as natural aggregate replacement *Construction and Building Materials* **121** 746–59

[10] Yuwadee Zaetang , Ampol Wongsa , Vanchai Sata and Prinya Chindaprasirt 2015 Use of coal ash as geopolymer binder and coarse aggregate in pervious concrete *Construction and Building Materials* **96** 289–95

[11] Rozineide A, Antunes Boca Santa , Cíntia Soares and Humberto Gracher Riella 2017 Geopolymers obtained from bottom ash as source of aluminosilicate cured at room temperature *Construction and Building Materials* **157** 459–66

[12] Akkadath Abdulmatin, Weerachart Tangchirapat and Chai Jaturapitakkul 2018 An investigation of bottom ash as a pozzolanic material *Construction and Building Materials* **186** 155–62

[13] Vanchai Sata, Apha Sathonsaowaphak and Prinya Chindaprasirt 2012 Resistance of lignite bottom ash geopolymer mortar to sulfate and sulfuric acid attack *Cement & Concrete Composites* **34** 700–8

[14] Tarunjit Singh Butalia 2019 Beneficial use of ponded fly ash in structural concrete *UKIERI Concrete Congress Key Note Address* (Dr B R Ambedkar National Institute of Technology Jalandhar) https://ukiericoncretecongress.com/Home/files/Proceedings/pdf/1B-T%20Butalia-Keynote-Conf-1.pdf

[15] M J McCarthy , L Zheng, R K Dhir and G Tella 2018 Dry-processing of long-term wet-stored fly ash for use as an addition in concrete *Cement and Concrete Composites* **92** 205–15

[16] A Sofi and B R Phanikumar 2015 An experimental investigation on flexural behaviour of fibre-reinforced pond ash-modified concrete *Ain Shams Engineering Journal* **6** 1133–42

[17] R V Ranganath, B Bhattacharjee, and S Krishnamoorthy 1998 Influence of size fraction of ponded ash on its pozzolanic activity *Cement and Concrete Research* **28(5)** 749–61

[18] Ranganath R V 1995 A study on the charecterisation and use of Ponded Fly Ash as Fine Aggregate in Mortar and Concrete *Ph.D Thesis Report* (IITD– Delhi)

[19] Bureau of Indian Standards 2002 IS 650: Specification for Standard Sand for Testing of Cement (New Delhi India)

[20] American Society for Testing and Materials 2015 ASTM C618 Standard specification for coal fly ash and raw or calcined natural pozzolan for use in concrete (ASTM International, West Conshohocken, PA)

[21] M Cheriaf, J Cavalcante Rocha and J Péra 1999 Pozzolanic properties of pulverised coal combustion bottom ash *Cement and Concrete Research* **29** 1387–91

[22] Bureau of Indian Standards 1996 IS :1727-1967 Methods of test for pozzolanic materials (New Delhi India)

[23] Apha Sathonsaowaphaka, Prinya Chindaprasirta and Kedsarin Pimraksab 2009 Workability and strength of lignite bottom ash geopolymer mortar *Journal of Hazardous Materials* **168** 44–50

[24] Smita Singh , M U Aswath and R V Ranganath 2018 Effect of mechanical activation of red mud on the strength of geopolymer binder *Construction and Building Materials* **177** 91–101

A Study on Tensile Properties of Bamboo Textile Reinforced Composites

I S Jazeel[1*] and Mini Remanan[2]

[1] Research scholar, National Institute of Technology Calicut, Kerala, India

[2] Assistant professor, National Institute of Technology Calicut, Kerala, India

*jazeelis@gmail.com

Abstract. A composite material is a material, which combines complementary nature and features of two or more constituents to result in a new material with unique and outstanding characteristics when compared with its original constituents. Conventional fibres such as glass and carbon can be replaced by environment-friendly bamboo fibres having good mechanical properties. Also being fastest-growing plants in the world (growing at a rate of 3 cm per hour), bamboo is abundantly available. Better fibre-matrix interaction will result in good interfacial adhesion between fibre and matrix and thereby reduces voids in the composite. This paper describes the tensile properties of bamboo textile reinforced composites. In this study, two different types of bamboo textiles i.e. knitted and weaved are used. Hand layup technique was incorporated in the production of composites and after that, air-cooling is adopted. All the tensile testing specimens are prepared according to ASTM D3039. Variation of the tensile properties of composite with different layers of bamboo textile and age are studied. The number of layers varies from 1 to 10. Tensile properties of these two bidirectional bamboo textile reinforced composites are tested for 7 days and 28 days. Keeping, the fibre volume fraction constant, increasing the number of layers enhances the ultimate tensile strength. Further increase in the number of layers shows no considerable variation in its ultimate tensile strength. Modulus of elasticity and percentage elongation of bamboo textile reinforced composites are also investigated.

1. Introduction

Composite materials or composites are engineering materials made up of two or more constituent materials that remain distinct and separate on a macroscopic level while forming a single component. For the past few decades a significant increase in the demand for the materials that are stiffer and stronger, but lighter, have been observed especially in the field of energy, aerospace and civil infrastructure [2]. As a result, manufacturers started producing composite materials that exactly fit the required properties for a particular purpose by selecting an appropriate combination of reinforcement and matrix material. Composite material systems offer great advantages of flexible design and results in a performance that are unreachable by individual constituents. For the past several years, natural fibres have been commonly used as reinforcement material across the world [3]. More recently, it is introduced in combination with plastics. Several types of natural fibres have been employed for use in plastics which include hemp, flax, straw, jute, wood fibre, barley, wheat, rye, oats, cane (sugar and bamboo), ramie, grass reeds, coir, sisal, rice husks, pineapple leaf fibre, pennywort, banana fibre, papyrus etc. [4]. For the last few years, there has been a heightened interest in natural fibre as a replacement for conventional FRP materials such as carbon and glass fibres. The advantages of using sustainable natural fibre or textile are weight saving, lower raw material cost, thermal recycling etc. Among the various natural fibres, bamboo fibre reinforced polymer composites, coir fibre reinforced polymer composites

IOP Conf. Series: Materials Science and Engineering **936** (2020) 012005 doi:10.1088/1757-899X/936/1/012005

and silk fibre-reinforced composites are of great interest, as these composites have high impact strength besides having moderate flexural and tensile properties. However, currently, several advanced fibre treatments have been developed which improve their tensile properties considerably [1]. There are many works carried out in composites using natural and conventional fibre reinforced polymer composites. Not much works have been carried out using textile reinforced polymer composites.

2. Experimental procedure

Experimental procedure includes the preparation of composite materials and testing the tensile properties of the composite specimen using Universal Testing Machine. Two types of bamboo textiles are collected which differ in their manufacturing process i.e. knitted and weaved bamboo textile. Knitting is the process in which yarn or thread is made into cloth and other crafts. It consists of loops or stitches of the material consecutively run together. Weaving is the process in which two types of threads or yarn are interlaced together to form a cloth or fabric. The two types of threads or yarns used are warp or weft thread, which is running in a different direction, in which warp threads are running in lengthwise and weft thread in crosswise or horizontal direction. Both knitted and weaved fibres are bidirectional. In the present work, a total of twelve specimens are prepared for each layer, out of which six specimens are prepared using weaved type fabric and the remaining six using knitted type fabric. Out of six specimens in each type, three specimens are used for finding 7-day ultimate tensile strength and remaining three for testing 28-day ultimate tensile strength. Number of layers in each type varies from 1 to 10. Altogether, 120 specimens are considered in the present study.

Composite consists of two parts, which are reinforcement and matrix. Bamboo textile is used as reinforcement and corn-starch based polyvinyl epoxy is used as matrix. For epoxy, EPOFINE-230 and for hardener, FINEHARD-950 are used. The property of epoxy and hardener is shown in table 1. Acetone is introduced in the epoxy to reduce the viscosity of the matrix system. Being a natural fibre it absorbs more amount of matrix as compared to other fibres, therefore to reduce the viscosity of the matrix, 5% of the volume of matrix acetone has been used. It mainly allows the matrix to ease the flow over the fabric. It should be kept in mind that the acetone amount must not be applied more than 10% of the matrix volume as it affects the strength of composite badly. The property of acetone is shown in table 2. The experiment consists of casting the composite and finding its tensile properties by conducting tension test using Universal Testing Machine.

Table 1. Properties of Epoxy and hardener system.

Characteristics	Finehard 951	Epofine 230	Test method	Unit
Viscosity at 25^{0}C	1200 – 1900	<20	ASTM – D 445	mPa s
Epoxy Content	4.00 – 4.50	-	ISO – 3001	Eq/kg
Density at 25^{0}C	1.12 – 1.16	0.97 – 0.99	ASTM – D 4052	g/cc
Flash Point	190 – 200	110 – 120	ASTM – D 93	^{0}C
Storage life	3	1		years

The advantages that make hardener and epoxy ideal for industrial use are (i) simple mixing ratio (Resin: Hardener = 100: 10 by weight), (ii) tolerant mixing ratio (2%), (iii) low viscosity resin/hardener mix, (iv) proper flow into the mould, (v) very good water resistance, (vi) excellent chemical resistance and (vii) electrical insulation.

Table 2. Properties of Acetone.

Property	Value	Standard
Colour	Colourless	HSDB 1992
Physical state	Liquid	HSDB 1992
Melting point	−95.35^{0}C	HSDB 1992
Boiling point	56.2^{0}C @ 1 atm	HSDB 1992
Density	0.78 g/ml	HSDB 1992

Molecular weight	58.08	-

2.1. Fabrication of composite

The fabrication of composite is done by hand layup technique. The fibre volume fraction is fixed as 35%. It is found that many of the literatures are showing high value for the tensile strength, modulus of elasticity and percentage elongation at 35% fibre volume fraction. After the application of each layer of fibre, the roller has been applied to remove all the bubbles present inside the composite. Initially composite is kept in ambient temperature for 24 hours and further, it is cured in an oven for 6 hours @ 80^0C and 2 hours @ 120^0C. The specimens are prepared according to ASTM D7565M. Initially, composite sheets of size 30 cm × 10 cm are prepared. From each sheet three tensile specimens of size 27 cm×2.5 cm are cut according to ASTM D3039/D3039M. Figure 1 shows a sample coupon used for testing the tensile strength. The average thickness of each composite was found using Vernier callipers. Tensile load is applied to each specimen using a universal testing machine. Ultimate tensile strength, percentage elongation, modulus of elasticity, breaking load and stress-strain behaviour are found. Figure 2 shows one of the specimens after breaking.

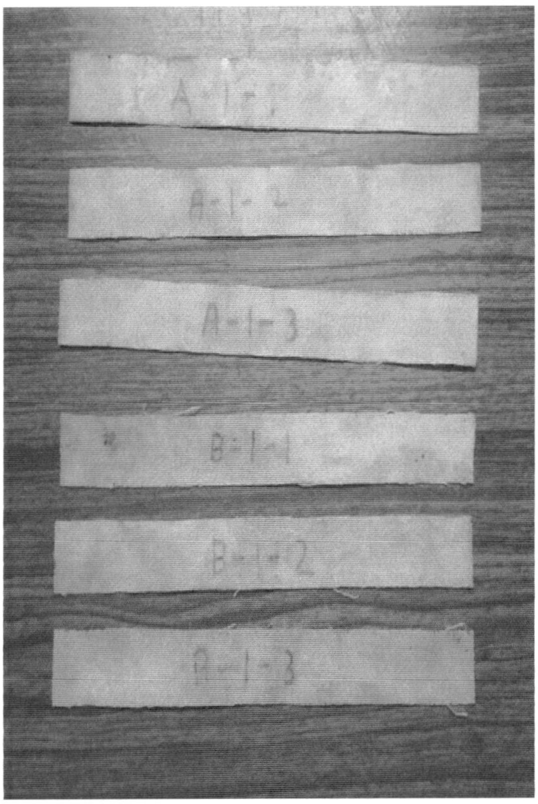

Figure 1. Tensile test specimen.

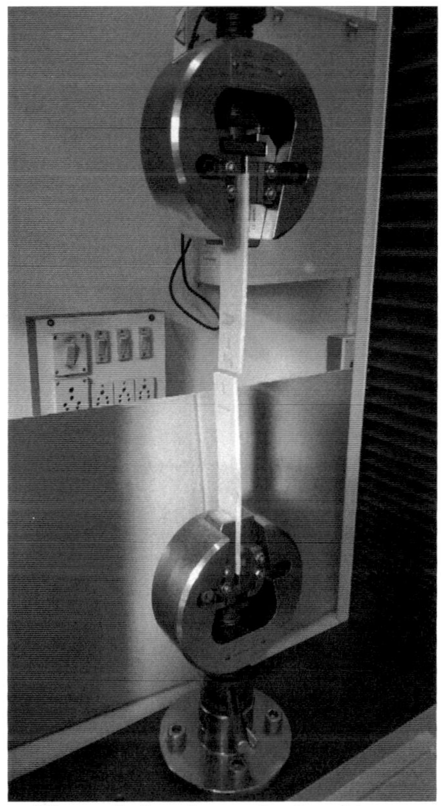

Figure 2. Specimen after breaking.

3. Results and discussions

Ultimate tensile strength of Knitted Bamboo Textile Reinforced Polymer (KBTRP) composite and Weaved Bamboo Textile Reinforced Polymer (WBTRP) composite with one layer to ten layers are found at 7 days and 28 days and is shown in figure 3. For all layers from one to ten, WBTRP composites have higher ultimate tensile strength than KBTRP composite. Ultimate tensile strength of WBTRP increased to 200% on varying the number of layers from one to five. Further, increasing the number of layers from six to ten showed no considerable variation in its ultimate tensile strength. In the case of KBTRP, ultimate tensile strength increased by 190% on varying number of layers from one to five. Further, increasing the number of layers from six to ten showed no considerable variation. For WBTRP composite and KBTRP composite, it is noted that 28-day ultimate tensile strength is more than 7-day ultimate tensile strength by 3 to 4 % and 2 to 3% respectively.

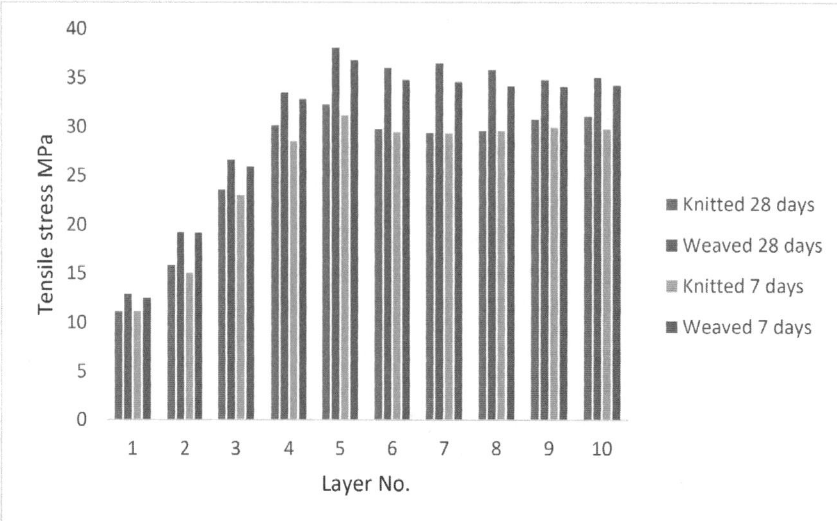

Figure 3. Maximum tensile stress for different layers of weaved and knitted bamboo textile reinforced composites.

Increase in the thickness on increasing the number of layers is shown in figure 4. The thickness of composite for number of layers from one to four is found to be almost same for KBTRP composite and WBTRP composite. As the number of layers increases from four to ten it is observed that KBTRP composite has more thickness when compared to WBTRP composite. It is found that the thickness of composite goes on increasing and maximum thickness is observed for 10 layer knitted bamboo textile reinforced polymer composite. Ultimate tensile load carrying capacity of KBTRP composite and WBTRP composite is also found for number of layers from one to ten and is shown in figure 5. It is observed that the ultimate tensile load keeps on increasing as the number of layer increases.

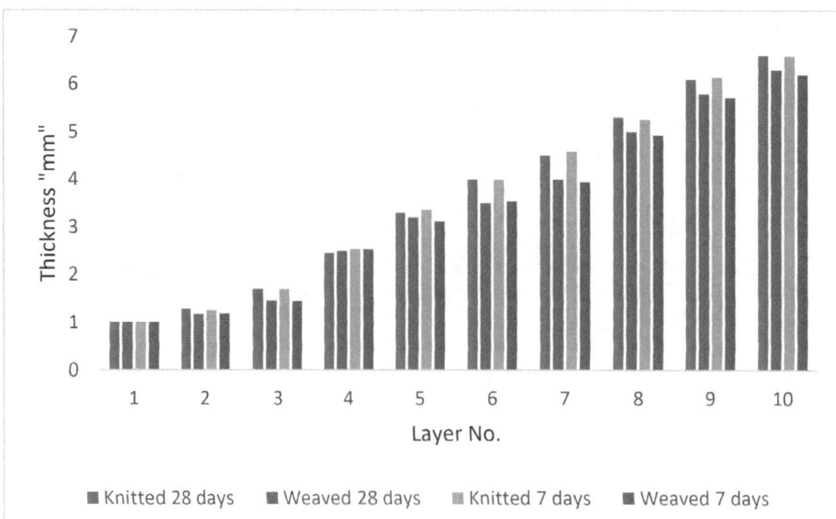

Figure 4. Thickness for different layers of weaved and knitted bamboo textile reinforced composites.

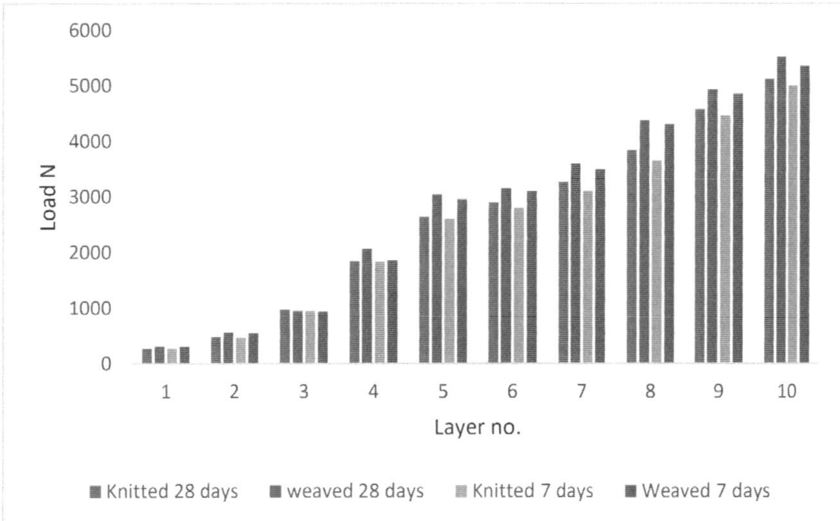

Figure 5. Loads for different layers of weaved and knitted bamboo textile reinforced composites.

From the results, it is found that the ultimate tensile strength of WBTRP composites is more than KBTRP composites. So, the stress-strain characteristics of WBTRP composites are shown for number of layers from one to ten at 28 days in figure 6. Modulus of elasticity and percentage elongation is found for each layer of WBTRP composite and is shown in Table 3. It is found that percentage elongation is maximum with 7 layers of WBTRP composite and it is 3.35%. From table 3 and figure 6 it can be concluded that WBTRP composite with 7 layers has more ductility. From figure 6 it is observed that the slope of the curve does not vary much as the number of layers change, i.e. modulus of elasticity has not much variation. Modulus of elasticity of WBTRP composite is found to be varying from 1.17 GPa to 1.92 GPa, and the maximum is obtained for WBTRP composite with five layers. It can also be observed that ultimate tensile strength is maximum for WBTRP composite with 5 layers which is having a higher modulus of elasticity. All the tensile specimens are found to break at an angle 90^0 with the specimen axis, hence it is a brittle failure.

Table 3. Modulus of elasticity and % elongation for different layers of weaved bamboo textile reinforced composites.

Number of layers	Modulus of elasticity GPa	% elongation
1	1.17	1.10
2	1.37	1.49
3	1.42	2.18
4	1.75	2.62
5	1.92	2.87
6	1.81	3.02
7	1.67	3.35
8	1.62	3.05
9	1.52	2.65
10	1.48	2.51

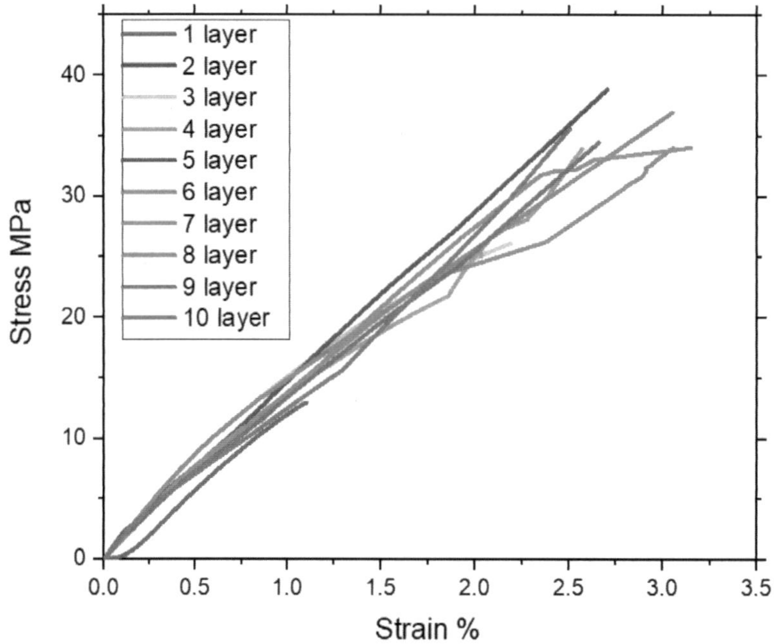

Figure 6. Stress-strain curves for different layers of weaved bamboo textile reinforced composites at 28 days.

4. Conclusions

From the study, it can be concluded that the ultimate tensile strength for WBTRP composites is greater than KBTRP composites for all layers of the composite specimen. Ductility, as well as percentage elongation, is maximum for WBTRP composite with seven layers. Ultimate tensile strength and modulus of elasticity are maximum for WBTRP with five layers. Modulus of elasticity of WBTRP composite for different layers ranges from 1.17 GPa to 1.92 GPa. WBTRP composite with more than three layers has failure strain in the range 2.25 to 3 percentage.

5. References

[1] Kai Z, Fangxin W, Wenyan L, Zhenqing W, Zhiwei D and Bin Y 2018 Thermal and mechanical properties of bamboo fiber reinforced epoxy Composites *Polymers* **10** 608

[2] Mohammad Z R K, Sunil K S and Gupta M K 2018 Tensile and flexural properties of natural fiber reinforced polymer composites: A review *Journal of Reinforced Plastics and Composites* **37(24)** 1435–55

[3] Ain U, Mohammed T H S, Mohammed J, Francisco C and Abu T 2016 A review on the tensile properties of bamboo fiber reinforced polymer composites *Bioresources* **11(4)** 10654–76

[4] Saira T, Munawar A M and Shafiullah K 2007 Review Natural fibre-reinforced polymer composites *Proc. of the Academy* of *Sciences* vol 44 (Pakistan) pp 129–44

IOP Conf. Series: Materials Science and Engineering **936** (2020) 012006 doi:10.1088/1757-899X/936/1/012006

Sustainability of Construction with Textile Reinforced Concrete- A State of the Art

Anjana Elsa Alexander[1*] and A P Shashikala[2]

[1] Research Scholar, Department of Civil Engineering, National Institute of Technology Calicut, India

[2] Professor, Department of Civil Engineering, National Institute of Technology Calicut, India

*anjanaelsaalexander@gmail.com

Abstract. Steel reinforcement in reinforced concrete structure is prone to corrosion. It has been found that a sustainable building material that can be used to replace steel from the construction industry is textile reinforcement. Textiles of high strength and fine-grained mortar are used to make a composite material termed as Textile Reinforced Concrete (TRC). This review paper presents an evaluation of different textiles that can be used as reinforcement in TRC, their bonding behaviour, durability characteristics and applications. A comprehensive overview on TRC reveals that TRC has excellent mechanical properties and durability characteristics. TRC is applicable for constructing lightweight, thin structural elements and for strengthening or repair of damaged structural elements.

1. Introduction

Steel has been used in concrete as reinforcement for over 100 years. But steel reinforcement is prone to corrosion reducing the effective cross-sectional area of bar resulting in spalling of the concrete leading to structural failure in extreme cases. Gries T et al., 2015 [1] have shown that "durability and reliability of structures can be increased by replacing steel reinforcement by textile reinforcement". Textile Reinforced Concrete (TRC) consist of fine-grained mortar and high strength textile fabrics. Textiles are multifilament yarns having high tensile strength usually made of carbon (Figure 1 a), basalt, alkali-resistant (AR) glass or polymer materials. Maximum size of aggregate used in TRC depends on the yarn distance and dimensions of the structural elements. Usually, aggregate used in TRC has less than 2mm size. High tensile strength, pseudo ductile behaviour and corrosion resistance are the major advantages of TRC. Also, TRC has excellent mechanical properties and durability characteristics which make this composite applicable for the construction of lightweight structural elements (Figure 1 b)) and for repair and rehabilitation of old structural elements [2].

Figure 1. (a) Carbon textile (b) Formwork element made of TRC [3].

2. Sustainability aspects of construction with TRC

In the modern era, new technologies to provide sustainability are becoming a major driving force for innovation in the construction industry. The sustainability qualities offered by TRC spans over a wide range:

a) Potential for making components with considerably smaller amount of material.
b) Longer service life than conventional concrete.
c) To extend the life-span of existing structures that are undergoing deterioration or need upgrading of their mechanical performance to withstand higher static and dynamic loads.

3. Fibre reinforcements

Fibres such as carbon, AR glass, basalt or aramid are commonly used in the manufacturing of textile materials. These fibres have very high mechanical properties compared to metals. Properties of those selected fibres are given in table 1 and described below [4]. In general, textile reinforcements are available in woven, nonwoven or knitted forms [5]. Dimensional characteristics of these textile reinforcements can be given in the form of planar (2D) or spacer (3D). Spacer reinforcements withstand loads within a volume.

Table 1. Fibre properties [4].

Fibre	Tensile Strength in MPa	Modulus of elasticity in GPa	Ultimate Strain (%)	Density (g/cm^3)
Glass (AR)	2500	70	3.6	2.78
Carbon	3500-6000	230-600	1.5-2.0	1.60-1.95
Aramid	3000	60-130	2.1-4.0	1.4
Basalt	3000-4840	79.3-93.1	3.1	2.7
Steel	1200	200	3-4	7.85

3.1. Glass

The basic ingredient is silica (SiO_2), and other oxides are added to modify the three- dimensional network structure. Glass fibres are amorphous and isotropic. The most economical and widely used glass fibres are E-glass (high electrical resistivity). However, in an alkaline environment, the sensitivity of E-glass fibres is quite higher. Alkali-resistance to glass fibre can be provided by adding 15 percentage of zirconia (ZrO_2). Presently, AR glass filaments are widely used in TRC applications because of their good adhesion properties in cement matrices and are economical in nature [4].

3.2. Carbon

Carbon fibres are commercially available as continuous tow and chopped (6–50 mm long) fibres. Continuous carbon yarns contain numerous (~10,000) filaments with diameters of 7–15 μm. These filaments exhibit high resistance to acid and alkaline environments and to organic solvents while their adhesion to cement-based material is not as good as that of AR glass [4].

3.3. Aramid

Aramid fibres are highly crystalline aromatic polyamide fibres. In contrast to glass and carbon, aramid filaments fracture in a ductile manner with considerable necking and fibrillation. This property is considered beneficial for impact or dynamic loading applications [6]. Aramid fibres are unaffected by temperature up to 160°C, but the filament is likely to lose most of its strength above 300°C. Aramid fibres absorb moisture and exhibit internal cracking and longitudinal splitting at increased moisture content, but the effect of moisture on fibre tensile properties are minimum. Also, aramid fibres are sensitive to ultraviolet light. Thousands of filaments bundled together comprise an aramid yarn, each 10–15 μm in diameter.

3.4. Basalt
Basalt fibres exhibit increased modulus of elasticity and strength, high temperature (1,100°C–1,200°C) and corrosion resistance [7][8]. Basalt fibres have strength higher than that of glass fibre and cost almost similar to glass fibre. It has good resistance to the alkaline environment (pH 13 or 14) but relatively less stable in strong acid.

4. Matrix composition
Considering the necessities of workability, durability and adequate bonding, a cement-based matrix for the TRC production have been developed. High binder content (40–50% by volume) of the matrix helps to achieve sufficient bonding with the filaments of the textile reinforcement. The maximum aggregate size is determined based on yarn distance, the spacing between textile layers and the dimensions of the structural elements. Typically, the aggregate size used ranges between 1 and 2 mm [9]. The advantage of using cement-based matrix in TRC is that it prevents a catastrophic failure because the cement matrix cracks much before the full potential of fabric is utilised [5].

Protecting RC structures against corrosion are achieved to an extent by the alkaline matrix of concrete, but in case of fibres, it will result in a very aggressive environment [10][11][12][13]. Hence, to develop a new matrix with improved energy, lower CO_2 emission, high durability and optimal bonding to reinforcement are essential for sustainability. This can be achieved either by using

a) Mineral Additives - lower energy consumption and CO_2 emission.
b) Alternative Cements- evaluating the use of alternative cements like High-alumina cements, Calcium aluminate cement, Inorganic phosphate cements or Geopolymers.
c) Polymer-modified cement-based systems- adding polymers help in modifying the properties of the fine-grained concrete [14].

5. Bonding in TRC
Strength, ductility and toughness of the composite are largely influenced by the textile-matrix bond [15]. Strong bonding results in a composite with high strength and low ductility, which may lead to brittle failure; whereas weak bonding shows a more ductile behaviour causing fibre pull-out. The bond between the cement matrix and fibre is strongly influenced by the elastic nature of fibre [5][16]. By increasing modulus of elasticity of fibre the clamping stresses that develop around the fibre increases, which in turn increases the bond strength of fibre [17] as shown in figure 2. Fabric geometries and anchoring of fabrics play an important role in increasing bonding with the cement matrix in TRC composites [18].

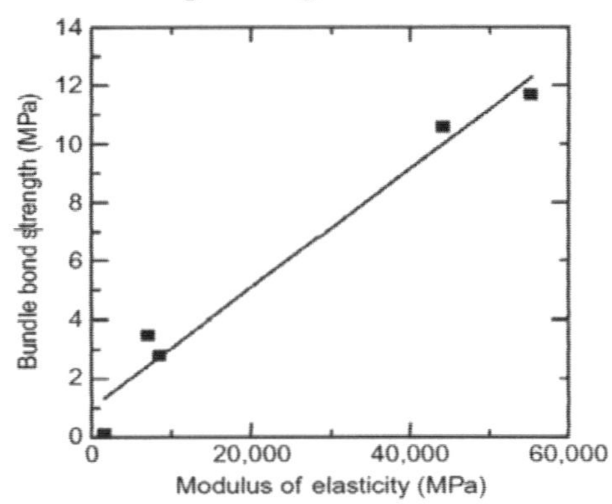

Figure 2. Graph showing variation of modulus of elasticity on the bond strength of yarn.

5.1. Bonding mechanism: Telescopic pull-out
In TRC, hundreds or thousands of multifilament yarns (rovings) bundle together to form a fabric. But using this fabric in the cementitious matrix is challenging, as large cement particles of about 5-70 μm cannot completely penetrate the gaps between the internal filaments of fabric (<5 μm). As a result, external filaments of yarns have direct contact with the cementitious matrix whereas the inner filaments

(core) have no contact with the hydration product resulting in a unique microstructure as shown in figure 3 (a & b). Thus, external filaments have better bonding with matrix [15][19][20][21][22].

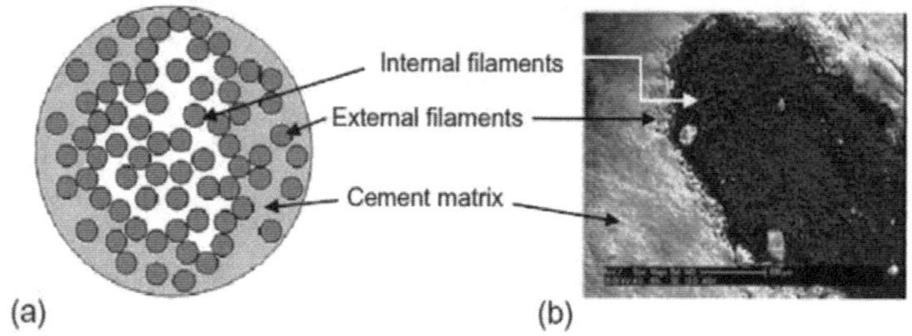

Figure 3. (a) Cross-section of cementitious matrix comprising of yarn (b) its actual view [15][22].

During loading, these external filaments become fractured. After they fail, the internal filaments slip against the external filaments resulting in a telescopic pull-out form as in figure 4 [21]. So, these core filaments are considered as wasted. When the tensile behaviour of carbon fabric in the cement matrix and fabric itself is compared, effective use of the internal filament is not achieved due to low cement penetration as depicted in figure 5 [22].

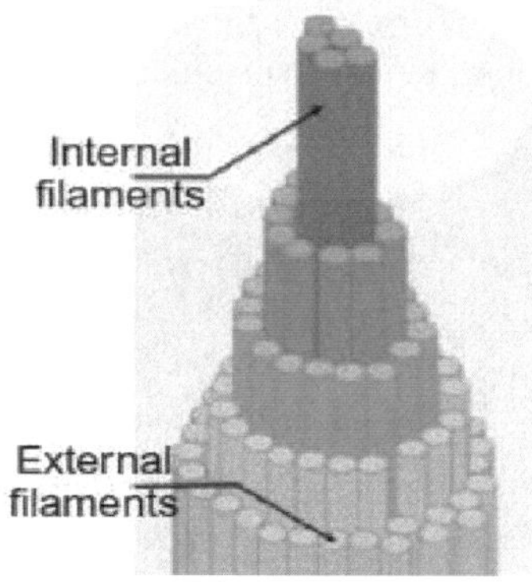

Figure 4. Pull-out behaviour of a fabric yarn [21].

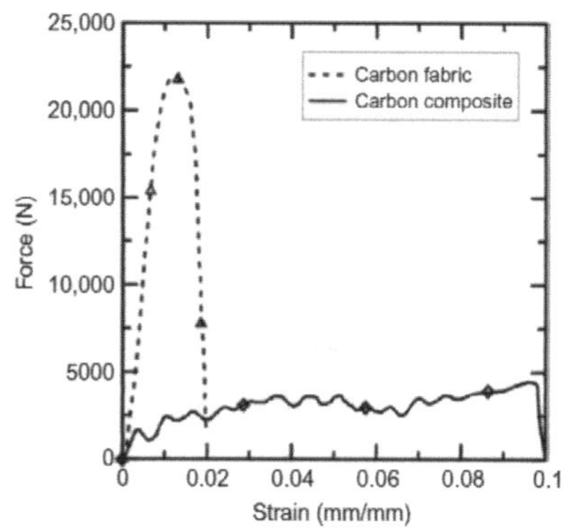

Figure 5. Tensile behaviour of carbon fabric in cement matrix and fabric itself [22].

5.2. Bond improvement

Filling the spaces between bundles inner filament can improve the stress transfer between them. Several researchers have examined the effect of polymer coating and bundle gaps filling. Xu et al. (2004) [23] observed that fabrics coated with epoxy exhibit enhanced bond strength and stiffness when compared to uncoated bundle filaments. This was observed for AR glass, carbon and aramid textiles and also for epoxy impregnated prestressed yarns. Good mechanical performance and penetrability were seen for polymer fillers in carbon fabrics. But they are affected at high temperature. To overcome these limitations, the use of nanoparticles to fill these gaps have been studied by few researchers. Bentur et al., 2010 [4] and Cohen and Peled, 2010 [21] compared mineral nanoparticles and polymer-based coating to evaluate the performance of AR glass and carbon fabric. Improved tensile strength, strong bonding and good filling were shown by mineral filler composites.

6. Durability

6.1. Durability of matrix

Durability of the TRC matrix may get affected by frost action, chemical attack or by abrasion. Deterioration of the composite can be hindered to some extent by adding short fibres in TRC [24].

6.2. Durability of textile fibre

In the manufacture of TRC, carbon, AR glass and basalt textiles are commonly used. Carbon fibre has high resistance to chemical attack. Compared to conventional glass fibre, AR glass fibre shows increased resistance to alkaline environments. But with increasing temperature and pH value of matrix, corrosion damage of AR-glass fibre occurs (Yilmaz and Glasser, 1991). But to a large extent such damages can be delayed by applying organic polymer size to the glass filament surfaces during its production. Butler et al. (2009) [12] concluded that if the organic polymer size is removed largely at some areas of filament then due to stress concentration, this nanoscopic defect becomes an important cause for damage. However, this failure of individual filament will be compensated in case of multifilament yarns in TRC due to redistribution of the load to the neighbouring filaments. Basalt fibres have low cost, temperature resistance and are available locally but it has reduced load-bearing capacity due to corrosive nature of fibre in an alkaline environment [25][10]. Generally, basalt fibre has low resistance to the alkaline environment compared to AR-glass fibre, however much better compared to E-glass fibre. Moreover, by adding zirconia to basalt fibre, its alkali resistance can be improved [26].

7. Fire resistance

Tanano et al., 1999 [27] stated that "Material used for building construction should have resistance to initial fire (flammability) and smoke and gas-generating properties".

7.1. Fibre response

At high temperature, the thermomechanical properties of commonly used fibres differ. Above 400–500°C temperature, carbon fibres start to oxidise while aramid fibres tend to oxidise above 150°C, limiting their use. In the case of glass fibres, oxidation will not happen but when reaching 600-800°C they start softening [4]. Carbon-fibre textiles, compared to AR-glass textiles shows superior performance at high temperature [28][29][30].

7.2. Fire resistance of TRC

Antons et al. (2012) [31] conducted flexural tests on heated prisms loaded with 30, 50 and 80 percent of its ultimate flexural capacity. They concluded that below 300°C, the temperature has no clear impact on the load levels considered; between 300°C and 600°C, as load is increased there is a negative influence of temperature; and beyond 620°C for any load level, melting of AR glass fibre was observed causing sudden failure.

8. Applications of TRC

Highly-finished surface, minimised product thickness, possibilities of freeform architecture, increased mechanical and durability characteristics are the advantages of TRC technology. TRC finds applications in making precast structural elements and in repair and rehabilitation of reinforced structures.

8.1. TRC in precast industry

Prefabricated TRC products are used for both architectural and structural elements due to less concrete consumption; minimised production, transportation and erection costs; and less wastage. Hegger et al., 2015 [32] reported the use of textiles like AR glass or carbon for facade construction. Engberts, 2006 [33] also reported the usage of a combination of continuous AR-glass yarns and fibre meshes embedded in a self-compacting fibre-reinforced micro concrete for the construction of larger ventilated facade elements (4–7 m^2). TRC was used in precast pedestrian bridges. The bridge built in Oschatz, Germany was the first TRC Bridge having a span of 8.6 m. AR glass yarns woven into mats was used as reinforcement inserted in four layers in fine concrete. The textile reinforcement helped to reduce the crack width as it served as minimum reinforcement. Then, a 17-m long TRC bridge in Kempten, Germany, was built. Vehicles weighing 3.5 t were allowed on it. To account for this load to the superstructure, additional crossbeams have been added. In 2010, the longest TRC Bridge in the world was constructed in Albstadt, Germany with 97 m length. Alkali resistant glass textile was used as reinforcement [34]. Prefabricated garages composed of a ceiling of TRC was widely used as transportation costs for the lighter TRC parts are very low. AR glass fibres impregnated with epoxy resin have been used as textile reinforcement. Freeform structural members are becoming the most remarkable trends in modern architecture. Pavilion building is an example for freeform construction, RWTH Aachen University [35]. Hybrid pipes consisting of an outer TRC ring enveloping an inner polymer tube have been developed to meet the water supply and sewage disposal needs [36]. Another application of TRC is in noise barrier parapets production. Further TRC can be used in the manufacture of handmade furniture, sculptures, ornaments and small boats.

8.2. TRC in repair and rehabilitation of reinforced concrete structures
Textile reinforced mortar (TRM) was used for flexural strengthening and shear strengthening of damaged RC beams and slabs[37][38]. TRM jackets also help in bridging shear cracks. When TRM jackets are provided by fully wrapping the cross-section of RC members, they give maximum shear enhancement (Figure 6 a). Open TRM systems are also effective but proper anchorage should be provided (Figure 6 b).

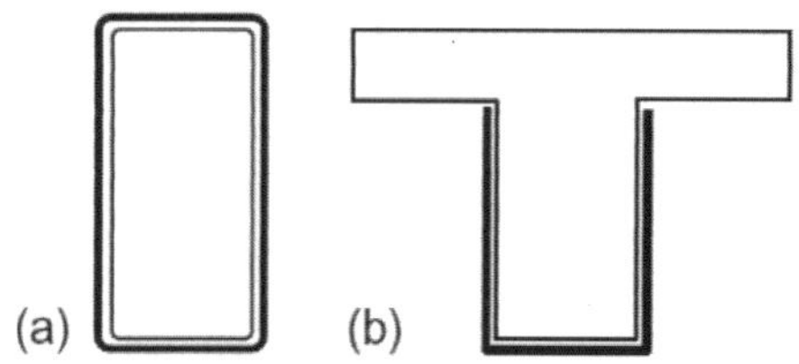

Figure 6. Shear strengthening of RC elements (a) Fully wrapped, (b) three-side wrapped [4].

9. Further scope of the research
Future research directions could be related to the following

- Bonding of various types of textile fibres at high temperature to the cementitious matrix are not studied.
- Application of TRC in thin structural elements is mentioned in most of the literature. But its behaviour in large structural elements is not presented.
- The behaviour of TRC in various types of high strength inorganic matrix such as ultra-high performance concrete, self-compacting concrete are not studied.

10.Conclusions
In this review paper, fibres such as glass, carbon, aramid, and basalt are examined for their applicability in TRC. Carbon fibres are inert and compared to other fibres has superior characteristics in terms of tensile strength and modulus of elasticity. But carbon has high cost. Alkali Resistant glass is found to be the most cost-effective and gives satisfactory results in terms of tensile strength. Improved tensile properties, good filling in between bundles and strong bonding are observed for matrix with mineral filler composite.

As TRC has advantages like thin size, good load-bearing capacity, resistance to corrosion, excellent ductility, no magnetic disturbances and lightweight of components, it is now widely used for various applications including precast constructions, repair, rehabilitation and structural strengthening of existing structures. With TRC it possible to produce lightweight members which save up to 80 % of concrete thereby reducing 80% of transportation costs and hence a sustainable material for the construction industry.

11.References
[1] Stolyarov O, Quadflieg T and Gries T 2015 Effects of fabric structures on the tensile properties of warp-knitted fabrics used as concrete reinforcements *Textile research journal* **85** 1934–45
[2] Lieboldt M, Butler M and Mechtcherine V 2008 Application of textile reinforced concrete (TRC) in prefabrication *In: Gettu, R. (Ed.) 7th Int. RILEM Symp. on Fibre Reinforced Concrete: Design and Applications. RILEM Publications S.A.R.L* pp 253–62

[3] Hegger J, Aldea C, Brameshuber W and Curbach M 2006 Applications of textile reinforced concrete *Brameshuber, W.(edited) State-of-the-Art Report of Rilem Technical Committee* pp 237–66
[4] Peled A, Bentur A and Mobasher B 2017 *Textile reinforced concrete* (CRC Press)
[5] Peled A and Bentur A 2003 Fabric structure and its reinforcing efficiency in textile reinforced cement composites *Composites Part A: Applied Science and Manufacturing* **34(2)** 107–18
[6] Mallick P K 2007 *Fiber-reinforced composites: materials, manufacturing, and design* (CRC press)
[7] Artemenko S E 2003 Polymer composite materials made from carbon, basalt and glass fibers - structure and properties *Fiber Chemistry* **35(3)** 226–9
[8] Militky J K and Kovacic V 1996 Ultimate mechanical properties of basalt filaments *Textile Research Journal* **66**(4) 225–9
[9] Triantafillou T 2016 *Textile fibre composites in civil engineering* (Woodhead Publishing)
[10] Scheffler C, Forster T, Mader E, Heinrich G, Hempel S and Mechtcherine V 2009 Aging of alkali-resistant glass and basalt fibers in alkaline solutions: Evaluation of the failure stress by Weibull distribution function *Journal of Non-Crystalline Solids* **355** 2588–95
[11] Scheffler C, Gao SL, Plonka R, Mader E, Hempel S, Butler M and Mechtcherine V 2009 Interphase modification of alkali-resistant glass fibres and carbon fibres for textile reinforced concrete I: Fibre properties and durability *Composites Science and Technology* **69** 531–8
[12] Butler M, Mechtcherine V and Hempel S 2009 Experimental investigations on the durability of fibre–matrix interfaces in textile-reinforced concrete. *Cement and Concrete Composites* **31(4)** 221–31
[13] Hempel S, Butler M and Mechtcherine V 2015 Bond behaviour and durability of basalt fibres in cementitious matrices *In3rd ICTR Int. Conf. on Textile Reinforced Concrete, ed Brameshuber W* (Aachen, Germany) pp 225–33
[14] Mechtcherine V and Lieboldt M 2011 Permeation of water and gases through cracked textile reinforced concrete *Cement and Concrete Composites* **33(7)** 725–34.
[15] Brameshuber W 2006 *Textile Reinforced Concrete: State of the Art Report (RILEM TC-201-TRC, Report 36)* (Aachen/ RILEM Publications)
[16] Peled A and Bentur A 2000 Geometrical characteristics and efficiency of textile fabrics for reinforcing cement composites *Cement and concrete research* **30(5)** 781–90
[17] Stang H 1996 Significance of shrinkage-induced clamping pressure in fiber-matrix bonding in cementitious composite materials *Advanced Cement Based Materials* **4** 106–15
[18] Peled A and Bentur A 1998 Reinforcement of cementitious matrices by warp knitted fabrics *Materials and structures* **31(8)** 543–50
[19] Bartos P 1980 Analysis of pull-out tests on fibres embedded in brittle matrices *Journal of Materials Science* **15(12)** 3122–8
[20] Zhu D, Mobasher B and Peled A 2013 Experimental study of dynamic behaviour of cement-based composites *Journal of Sustainable Cement-Based Materials* **2(1)** 1–2
[21] Cohen Z and Peled A 2010 Controlled telescopic reinforcement system of fabric cement composites durability concerns *Cement and Concrete Research* **40(10)** 1495–506.
[22] Zhu D, Peled A and Mobasher B 2011 Dynamic tensile testing of fabric cement composites *Construction and Building Materials* **25(1)** 385–95
[23] Xu S, Krüger M, Reinhardt H W and Ozbolt J 2004 Bond characteristics of carbon, alkali resistant glass, and aramid textiles in mortar *Journal of materials in civil engineering* **16(4)** 356–64
[24] Barhum R and Mechtcherine V 2013 Influence of short dispersed and short integral glass fibres on the mechanical behaviour of textile-reinforced concrete *Materials and structures* **46(4)** 557–72
[25] Sim J and Park C 2005 Characteristics of basalt fiber as a strengthening material for concrete structures *Composites Part B: Engineering* **36** 504–12
[26] Rybin V A, Utkin A V and Baklanova N I 2013 Alkali resistance, microstructural and mechanical performance of zirconia-coated basalt fibers *Cement and Concrete Research* **53** 1–8

[27] Tanano H, Masuda Y and Tomosawa F 1999 Characteristics and performance evaluation methods of continuous fiber bars-State of the art studies on fire properties and durability of continuous fiber reinforced concrete in Japan *American Concrete Institute* **188** 523–34

[28] Kruger M and Reinhardt H W 2006 Composite materials-6.4, Fire resistance *Textile reinforced concrete—State-of-the-art report of RILEM TC 201*-TRC (RILEM Publications S.A.R.L. France) pp 211–9

[29] Buttner T, Orlowsky J and Raupach M 2007 Fire resistance tests of textile reinforced concrete under static loading-results and future developments *In Fifth Int. Workshop on High Performance Fiber Reinforced Cement Composites (HPFRCC 5)* (Mainz, Germany) pp 10-13

[30] Reinhardt HW, Kruger M, Raupach M and Orlowsky J 2008 Behavior of textile-reinforced concrete in fire *American Concrete Institute* **250** 99–110

[31] Antons U, Hegger J, Kulas C and Raupach M 2012 High-temperature tests on concrete specimens reinforced with alkali-resistant glass rovings under bending loads *In Proc. of the 6th Int. Conf. on FRP Composites in Civil Engineering* (Rome, Italy) pp 13–5

[32] Hegger J, Will N, Chudoba R and Scholzen A 2015 Classification of approaches to dimensioning and assessment of lightweight TRC structures *In FERRO 11–11th Int. Symp. on Ferrocement and 3rd ICTRC-Int. Conf. on Textile Reinforced Concrete* pp 45–50

[33] Engberts E *2006* Large-size facade Elements of textile reinforced concrete *In ICTRC'2006-1st Int. RILEM Conf. on Textile Reinforced Concrete 2006 (RILEM Publications/SARL)* pp 309–318

[34] Raupach M, Orlowsky J, Büttner T, Dilthey U, Schleser M and Hegger J 2006 Epoxy-impregnated textiles in concrete-load bearing capacity and durability *In ICTRC'2006 1st Int. RILEM Conf. on Textile Reinforced Concrete (RILEM Publications/SARL)* pp 77–88

[35] Scholzen A, Chudoba R and Hegger J 2015 Thin-walled shell structures made of textile-reinforced concrete: part I: structural design and construction *Structural Concrete* **16(1)** 106–14

[36] Lieboldt M, Helbig U and Engler T 2006 Textile reinforced concrete multilayer composite pipes *Proceedings of the ICTRC* pp 369–78

[37] Amir SL, Raphael C, Emmanuel F and Patrice H 2011 Flexural strengthening of reinforced concrete beams with textile reinforced concrete (TRC) *In Advances in FRP Composites in Civil Engineering* (Springer, Berlin, Heidelberg) pp 665–7

[38] Bruckner A, Ortlepp R and Curbach M 2008 Anchoring of shear strengthening for T-beams made of textile reinforced concrete (TRC) *Materials and Structures* **41(2)** 407–18

IOP Conf. Series: Materials Science and Engineering **936** (2020) 012007 doi:10.1088/1757-899X/936/1/012007

Biochar Amended Concrete for Carbon Sequestration

Jyothi J Nair[1*], S Shika[2] and Vandana Sreedharan[3]

[1] P G student, Vimal Jyothi Engineering College, Kannur, India

[2] Associate Professor, Vimal Jyothi Engineering College, Kannur, India

[3] Associate Professor, Government Engineering College, Kannur, India

*jyothijnair009@gmail.com

Abstract. With the rapid increase in built-up areas around the world, and the emergence of new cities, there is a potential of sequestering carbon in civil infrastructures. Biochar is the by-product of the pyrolysis of biomass, which can be introduced to the cement mortar and the cement concrete, as a replacement of cement as well as an additive. Biochar is effective in the carbon sequestration activity and results in the high internal curing action due to the high-water absorptive capacity. As the percentage of biochar increases the amount of carbon dioxide adsorbed is also observed to increase. However, it is imperative that biochar addition should not compromise the strength characteristics of concrete. As a well-accepted fact, the pozzolanic activity plays large importance on the durability and the strength gain of concrete. The present study is aimed at investigating the pozzolanic activity of the biochar. Biochar specimens were subjected to different pre-treatments, and the corresponding changes in the pozzolanic activity were studied. The biomass used for the study was locally available coconut shells and the pre-treatment was done with the 0.1N HCl. The pozzolanic activity was checked by calculating the amount of $Ca(OH)_2$ fixation during the titration. It was observed that treatments at higher temperatures have no significant effect on the pozzolanic activity. And this suggests that pre-treatment of biochar at room temperature is enough for enhanced pozzolanic activity. The study has reaffirmed the potential use of biochar in making sustainable concrete fit for carbon sequestration.

1. Introduction

Cement is the most broadly perceived material used in the construction industry. It has critical importance among the construction materials. In the production process of cement, the amount of carbon dioxide emission is more and thus it leads to many of the environmental impacts. India lies second in the production of cement for satisfying the needs of the construction work. Thus, there is a requirement of replacement of cement with a suitable material which will go with the different characteristics of cement. Various research papers were addressed on the characteristics of concrete after replacement of the cement with different industrial by-products such as silica fume [6][12], metakaolin [1][14], fly ash [6][7], bottom ash[13], rice husk ash[1], palm oil fuel ash[3][15]. Biochar is formed after the pyrolysis of biomass. The biomass may be any material such as food waste, agricultural waste, poultry waste [19] etc.

The importance of biochar is mainly on carbon sequestrating activity. Now in the present time, the biochar is successfully used in the soil for enhancing the soil, and it also holds the carbon in soil and thus makes the soil more fertile. This paper focuses on the effective use of the coconut shell biochar in the replacement of the cement. The studies reveal that the biochar showed better carbon sequestration characteristics in concrete without compromising the strength requirements. As the emission of carbon dioxide is increasing in the present world, it has become a pertinent problem that requires special attention. So that the material that is used for the building construction should have a capacity of the carbon sequestration. This situation leads to the effective use of the biochar in the concrete as the replacement of cement. Biochar is a material that can hold water and carbon which in turn results in the internal curing action and carbon sequestration.

When a material is used for the partial replacement of cement it may not always assure the strength properties, therefore some sort of treatment is to be done prior to replacement. It was clear that the strength property depends on the pozzolanic activity of the material, which can be enhanced by the pre-treatment. The effective treatment is done by using the 0.1 N HCl [19], and higher normality does not influence the pozzolanic activity of the biochar. The increased pozzolanic activity leads to better strength characteristics. The possible use of biochar in concrete can cater to the growing need of disposal of biomass in addition to producing concrete with enhanced carbon sequestration property.

2. Materials and methods

2.1. Biochar production

Coconut shell, which is commonly available in Kerala, was used for this study. Though the coconut shells are used for different purposes such as fuel, craft and for purifying the water, the amount of coconut shells that are effectively utilised is minimal. The coconut shells were collected, cleaned, and converted to biochar by the pyrolysis process at controlled temperature. The temperature range was fixed at 600°C to 650°C. The prepared biochar was then ground into fine particles, having a size less than 180 microns.

2.2. Pre-treatment

In order to study the pozzolanic activity, biochar was pre-treated with 0.1N HCl [19]. The biomass ie., the coconut shells were immersed in the HCl solution at varying temperatures viz., at room temperature, at 100°C, 500°C and 800°C for different time intervals of 30 minute and 1 hour at each of these temperatures. After this treatment, the sample of treated biomass was washed with distilled water and it was then oven dried. Then it was turned into biochar using a small pyrolysis unit. Since it was evident from earlier studies that the normality of HCl had little effect on the enhancement of the pozzolanic activity [19] it was decided to carry out the pre-treatment with 0.1N HCl. The biochar was ground using mechanical grinders.

2.3. Pozzolanic activity measurement

Chapelle test is a French standard (NF P18-513) used for finding pozzolanic activity.25 ml of deionised water is used as a solvent, and in that 1g biochar and 2 g of calcium oxide powder were added. The solution was then kept for continuous stirring for a period of 16 hours at a temperature of 90^0C by keeping it in a hot stirring plate. Then, after stirring it for 16 hr, the slurry was cooled and then 250 ml of 0.7 M sucrose was added to it, accompanied by stirring for 15 minute and followed by filtration of suspension. A few drops of 0.1% phenolphthalein were added to the 25 ml of the filtrate and followed by filtration with 0.1 N HCl. The pozzolanic activity is determined [19] using the titration results, as shown in equation (1):

$$Ca\,(OH)_2 \text{ fixed in mg} = 2\,\frac{(V_b - V_m)}{V_b}\frac{74}{56}*1000 \tag{1}$$

The terms V_m and V_b denote the volume of 0.1 N HCl solution required for blank samples with and without biochar, respectively. The calcium hydroxide formed during titration in mg gives the measurement of the pozzolanic activity.

2.4. Test for identifying carbon Sequestration

Biochar amended mortar cubes with different percentage of biochar were placed inside an airtight chamber. The setup is as shown in figure 1.

Burning candles were kept inside the containers to produce carbon dioxide within the containers. Mortar cubes prepared with different percentages of biochar were stacked in the container in order to check the possible sorption of carbon dioxide produced by the burning candles. MQ135 carbon dioxide sensors were fixed inside the container to monitor concentration of carbon dioxide. Containers were properly sealed to ensure the measurement of carbon dioxide within the container. Mortar cubes with 1%, 2%, 5% amendments were used in the study. The studies were done with single and double burning candles. Table 1 and table 2 give the properties of cement and the biochar that were used for the experiment.

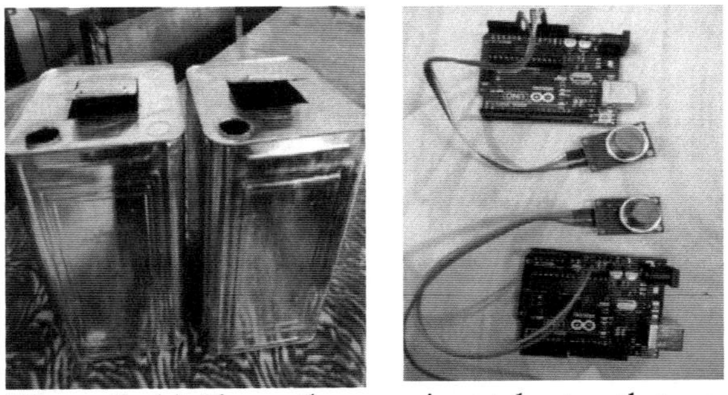

Figure 1. (a) Shows the experimental setup that was used for the identification of the carbon sequestration property (b) MQ135 carbon dioxide sensors.

Table 1. Properties of Cement.

Particulars	Values
Grade	OPC 43
Specific gravity	3.16
Standard consistency	32%
Initial setting time	116 min
Fineness	5%

Table 2. Test result for biochar.

Particulars	Values
Size	150 microns down-size
pH	9
Specific gravity	2.64

3. Results and discussion

3.1. Pozzolanic activity

Pozzolan is a broad class of siliceous or siliceous and aluminous materials which, in themselves, possess little or no cementitious value but which will, in finely divided form and in the presence of water, react chemically with calcium hydroxide at ordinary temperature to form compounds possessing cementitious properties. The pozzolanic activity has a great importance in the strength properties.

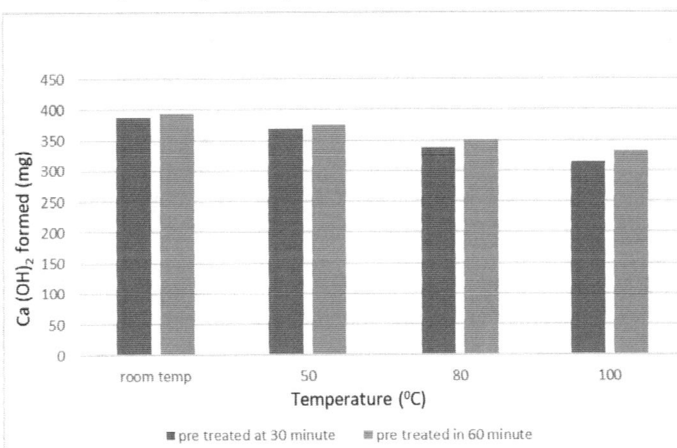

Figure 2. Variation of pozzolanic activity at different temperature.

Figure 2 shows the effect of pre-treatment at different temperatures and time conditions. The pre-treatment of the biomass followed by the pyrolysis process is shown to enhance the pozzolanic activity of the biochar produced.

It was observed that higher temperatures do not enhance the pozzolanic activity significantly. The untreated biochar shows $Ca(OH)_2$ formation of 208.7 mg whereas the pre-treated biochar exhibited almost double this value. This confirms that pozzolanic activity can be increased considerably with the use of biochar produced from pre-treated biomass. As mentioned, the pozzolanic activity has considerable importance in the strength properties; this points out that the strength characteristics can be ameliorated by the treatment of the biomass before pyrolysis and thereby carryout the pyrolysis under controlled temperature conditions. The increase in pozzolanic activity is due to the increase in the surface area and porosity of the biochar produced. Better pozzolanic activity was achieved by the pre-treatment of the biomass even at room temperature. However, it was found that the pozzolanic activity reduced with the increase in the pre-treatment temperature. Moreover, with the increase in the time of immersion there was no appreciable increase in the pozzolanic activity. Hence it can be stated that pre-treatment with 0.1N HCl acid at room temperature for 60 minutes was well enough to enhance the pozzolanic activity of biochar.

3.2. Carbon sequestration

Results of the possible sorption of carbon dioxide by biochar amended mortar specimens are tabulated in table 5. Then the percentage of sorption was calculated for each biochar percentage. The results are tabulated in table 3.

Table 3. Percentage sorption of carbon dioxide in single candle, single specimen setup.

	Initial concentration of CO_2 (ppm)	Peak concentration of CO_2 (ppm)	Final concentration of CO_2 (ppm)	Percentage sorption of CO_2 (%)
Control specimen (no biochar)	400	561	539	13.66
Cement + 1% biochar	404	543	516	19.42
Cement + 2% biochar	471	579	530	45.37
Cement + 5% biochar	548	674	579	75.30

It is evident from the table that there was an increased percentage of sorption of carbon with the increasing biochar content in the cement mortar. The experiments with double candle (Table 4) also showed similar trend.

Table 4. Percentage sorption of carbon dioxide in double candle, single specimen setup.

	Initial concentration of CO_2 (ppm)	Peak concentration of CO_2 (ppm)	Final concentration of CO_2 (ppm)	Percentage sorption of CO_2 (%)
Control specimen (No biochar)	408	570	557	8.02
Cement + 1% biochar	400	543	521	15.38
Cement + 2% biochar	440	548	516	21.29
Cement + 5% biochar	408	570	530	24.69

This points out that the biochar amended is increasing the carbon sequestration ability of cement mortar. This opens up the possible use of biochar in concrete and cement mortar for sequestering carbon and mitigating climate change. Further studies with biochar amended concrete have clearly brought up the fact that concrete prepared with pre-treated biochar had higher strength. The table 5 shows the compressive strength of concrete amended with biochar. M25 mix design was used, mix design was done with reference to IS 10262:2009.

Table 5. Compressive strength test results using treated biochar amended concrete.

Specimen specification	7 days compressive strength (N/mm^2)	28 days compressive strength (N/mm^2)
Control specimen (No biochar)	37.2	44.7
Cement + 5% biochar	40.4	53.5
Cement +10% biochar	38.6	47.8
Cement + 15% biochar	33.6	40.6

These results categorically points that addition of biochar into concrete was not compromising on the strength of concrete.

4. Conclusions

From the experimental studies it is evident that the amendment of biochar prepared from pre-treated biomass produces mortar with better carbon sequestration ability which can be extensively used for improving the indoor air quality. The percentage of sorption increase with the increase of biochar. The strength of the biochar amended concrete is not compromised due to the pre-treatment which involves the simple procedure by treating with 0.1N HCl. The study has brought out clearly that the biochar produced from pre-treated biomass is a potential candidate for combating the pollution from cement industry and catering to the need to save the environment.

5. References

[1] Gill A S and Siddique R 2018 Durability properties of self-compacting concrete incorporating metakaolin and rice husk ash *Constr. Build. Mater.* **176** 323–32

[2] Akhtar A and Sarmah A K 2018 Novel Biochar-Concrete Composites: Manufacturing, Characterization and Evaluation of the Mechanical Properties *Science of the Total Environment* **616** 408–16

[3] Thomas B S, Kumar S and Arel H S 2017 Sustainable concrete containing palm oil fuel ash as a supplementary cementitious material – a review *Renew. Sustain. Energy Rev.* **80** 550–61

[4] Cha J S, Park S H, Jung S C, Ryu C, Jeon J K, Shin M C and Park Y K 2016 Production and utilization of biochar: A review *Journal of Industrial and Engineering Chemistry* **40** 1–15

[5] Cuthbertson D, Berardi U, Briens C and Berruti F 2018 Biochar from residual biomass as a concrete filler for improved thermal and acoustic properties *Biomass and Bioenergy* **120** 77–83

[6] Wang D, Zhou X, Meng Y and Chen Z 2017 Durability of concrete containing fly ash and silica fume against combined freezing-thawing and sulphate attack *Constr. Build. Mater.* **147** 398–406

[7] Bentz D P and Ferraris C F 2010 Rheology and setting of high-volume fly ash mixtures *Cem. Concr. Compos.* **32** 265–70

[8] Gupta S and Kua H W 2017 Factors determining the potential of biochar as a carbon capturing and sequestering construction material: critical review *Journals of Materials in Civil Engineering, ASCE* **29(9)** 04017086

[9] Gupta S, Kua H W and Koh H J 2018 Applications of biochar from wood and wood waste as green admixture for cement mortar *Science of the Total Environment* **619** 419–35

[10] Gupta S, Kua H W and Low C Y 2018 Use of biochar as carbon sequestering additive in cement mortar *Cement and Concrete Composites* **87** 110–29

[11] Gupta S, Kua H W and Pang S D 2018 Biochar-mortar composite: Manufacturing, evaluation of physical properties and economic viability *Construction and Building Material* **167** 874–89

[12] Rossen J E, Lothenbach B and Scrivener K L 2015 Composition of C–S–H in pastes with increasing levels of silica fume addition *Cem. Concr. Res.* **75** 14–22

[13] Minane J R, Becquart F, Abriak N E and Deboffe C 2017 Upgraded mineral sand fraction from MSWI bottom ash: an alternative solution for the substitution of natural aggregates in concrete applications *Procedia Eng.* **180** 1213–20

[14] Coopamootoo K and Masoero E 2018 Cement pastes with UV-irradiated polypropylene: fracture energy and the benefit of adding metakaolin *Constr. Build. Mater.* **165** 303–9

[15] Jumaat M Z, Alengaram U J, Ahmmad R, Bahri S, Islam A B M S 2015 Characteristics of palm oil clinker as replacement for oil palm shell in lightweight concrete subjected to elevated temperature *Constr. Build. Mater.* **101** 942–51

[16] Parthiban Kathirvel, George Amal Anik, Saravana Raja Mohan 2019 Effect of partial replacement of cement with Prosopis juliflora ash on the strength and the microstructural characteristics of cement concrete *Construction and Building Materials* **225** 273–82

[17] Rayane Mrad and Ghassan Chehab 2019 Mechanical and Microstructure Properties of Biochar-Based Mortar: An Internal Curing Agent for PCC *Sustainability* **11(9)** 2491

[18] Biskri Y, Achoura D, Chelghoum N and Mouret M 2017 Mechanical and durability characteristics of High-Performance Concrete containing steel slag and crystalised slag as aggregates *Constr. Build. Mater.* **150** 167–78

[19] Zeidabadi Z A, Bhakhtiari S, Abbaslou H and Ghanizadeh A R 2018 Synthesis characterization and evaluation of biochar from agricultural waste biomass for use in building materials *Construction and Building Material* **181** 301–8

IOP Conf. Series: Materials Science and Engineering **936** (2020) 012008 doi:10.1088/1757-899X/936/1/012008

Impact Resistance of GGBS-Dolomite Geopolymer Concrete

P Saranya*, Praveen Nagarajan and A P Shashikala

Department of Civil Engineering, National Institute of Technology Calicut, Kerala, India

*saranyasankar81@gmail.com

Abstract. Geopolymers are new alternative binders for cement in which polymerization gives strength to concrete. The present study highlights the development of geopolymer concrete from industrial by-products such as Ground Granulated Blast furnace Slag (GGBS) and dolomite. They are obtained from steel industries and rock crushing plants respectively. Standard mix design guidelines are not available for geopolymer concrete. The trial and error method was chosen for attaining a target compressive strength of 60 MPa. Impact strength plays an important role in structures like bridges, harbours, and barriers. The present paper investigates the impact resistance of GGBS dolomite geopolymer concrete using drop weight method. Procedures of ACI Committee 544 were used, and results were compared with that of cement concrete. Cylindrical specimens having 150 mm diameter and 50mm thickness were used. Maximum impact resistance was observed when GGBS and dolomite become 80:20, which is 42% more than GGBS geopolymer concrete without dolomite.

1. Introduction

Geopolymers are a promising alternative to ordinary portland cement concrete from the sustainability point of view. Production of geopolymer concrete (GPC) decreases the release of 80% greenhouse gases [1] and 60% energy required [2] compared to cement concrete. The main component of geopolymer concrete is alumina-silica based source material [3]. Fly ash [4], Ground Granulated Blast furnace Slag (GGBS) [5], Metakaolin [6], Rice husk ash [7], etc. are normally used binder materials for the development of geopolymer concrete. Alkaline solutions such as sodium-potassium hydroxide and silicates are essential for geopolymerisation, which gives strength to geopolymer concrete [8]. Geopolymer concrete has advanced engineering and durability properties, but they are brittle in nature [9]. Geopolymer concrete has better bond strength [10] and impact resistance [11] compared to cement concrete. Numerous studies were reported on the mechanical properties of GPC. Studies on the impact strength of geopolymer concrete were found to be scanty. Several methods are available for finding the impact resistance of concrete. Drop weight method (ACI (American Concrete Institution) committee 544) [16] is the simplest method used for finding impact resistance of concrete.

Recently [12], Geopolymer concrete was developed from GGBS and dolomite, which have advanced engineering and durability properties. Structures such as buildings in the hurricane region, bridges, harbours, etc. are subjected to impact loading [13]. In these structures impact resistance plays a main role than other strength properties. The present study investigates the impact resistance of GGBS-dolomite GPC and compares with that of OPC concrete.

2. Materials used

2.1. GGBS and Dolomite

The main source material used in this study was GGBS, which is obtained from steel industries as a by-product. Effect of rock crushing plant by-product (dolomite) on GGBS geopolymer concrete also has been studied. The specific gravity of GGBS and dolomite are 2.9 and 2.85 respectively. Blaine's fineness was observed as 4032 and 3500 cm^2/gm for GGBS and dolomite, respectively.

2.2. Aggregate

River sand and coarse aggregate (size 12 mm) were used as aggregate [15].

2.3. Alkaline solution

The alkaline medium was prepared from sodium hydroxide and sodium silicate solutions. The required quantity of sodium hydroxide (NaOH) pellets (97% purity) was dissolved in water to prepare a particular molarity NaOH solution. Sodium silicate solution (SiO_2 (27.2%), Na_2O (8.9%), and H_2O (63.9%)) is mixed with sodium hydroxide solution, 30 minutes before casting.

2.4. Superplasticiser

Sulphonated naphthalene based superplasticisers with a specific gravity of 1.2 was added to improve the workability [12].

3. Mixing procedure for geopolymer concrete

Proper mix proportioning guidelines are not available for slag based geopolymer concrete when subjected to ambient curing. GPC having compressive strength 60 MPa were achieved through trial mix proportioning process. Quantity of materials for developing $1m^3$ geopolymer concrete and cement concrete are given in table 1 [9].

Table 1. Mix proportioning of geopolymer and cement concrete.

Materials	Quantity (kg/m^3)	
	Geopolymer concrete	Cement Concrete
Binder content/cement	400	556
Alkaline solution/water	240	158
Fine aggregate	646	698
Coarse aggregate	1021.5	1067
Superplasticiser	6	6

GGBS was replaced with 10%, 20%, 30%, 40% and 50% by dolomite and referred as G90:D10, G80:D20, G70:D30, G60:D40 and G50:D50 respectively. Binder material and aggregates were mixed in a pan mixer one minute. Alkaline solution followed by a superplasticiser was added and mixed for three minutes.

4. Compressive strength test

Compressive strength test was performed on concrete cube specimens (size 150 mm) as per the guidelines given in IS 516:1959. All specimens were demoulded within 24 hours and cured under ambient conditions (25-28°C). Compressive strength test was conducted in a load controlled compression-testing machine at a rate of 140 kg/sq cm/min. Table 2 shows the compressive strength at the age of 28 days of geopolymer concrete.

5. Impact resistance test

5.1. Specimen details

Cylindrical specimens having 150 mm diameter and 50 mm height were used for studying the impact resistance. Specimens are shown in figure 1.

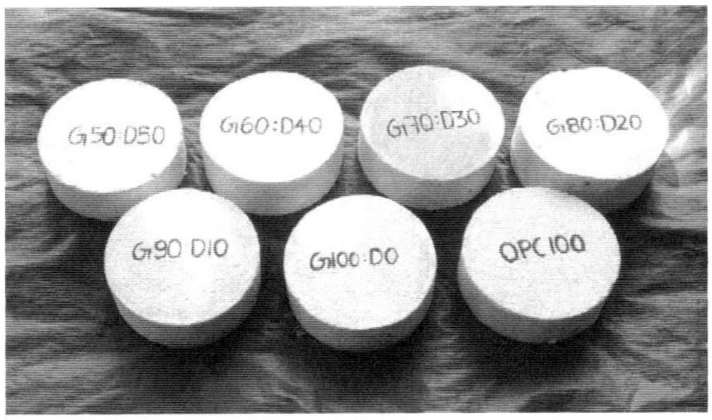

Figure 1. Specimens for impact test.

5.2. Test setup
Impact resistance test was carried out by dropping weight method, in which a steel ball of weight 3 kg was dropped freely from a height of 457 mm. Number of blows that causes visible first crack are denoted as first crack blows, and failure blows were measured when the complete spreading of cracks occurred [14]. The test setup used for finding impact resistance is shown in figure 2.

Figure 2. Test setup for impact resistance.

5.3. Impact strength
Impact resistance of GGBS geopolymer concrete with different proportions of dolomite were found out by using dropping weight method. Number of blows that cause first crack (First crack blows) and failure crack are tabulated in table 2. Impact energy was calculated from the equation $E = N \times m \times g \times h$, where N is the number of blows which cause the failure of specimen, m is the weight of steel ball, h is the freely falling height of ball and g is the acceleration due to gravity. Impact energy for different specimens are also shown in table 2.

Table 2. Impact strength of geopolymer and cement concrete.

Mix	Compressive strength(N/mm^2)	Impact resistance test		
		First crack blows	Failure crack blows	Impact Energy Nm
G100:D0	64.5	26	102	1371.85
G90:D10	68.4	30	130	1748.43
G80:D20	70	42	145	1950.17
G70:D30	72.5	35	140	1882.93
G60:D40	65.7	22	90	1210.45
G50:D50	60.4	20	63	847.31
OPC100	71.5	18	120	1613.94

The first crack blows of GGBS-dolomite geopolymer concrete were found as 2.3 times that of cement concrete. Maximum impact energy was observed for G80:D20, which is 21% more than cement concrete. G50:D50 less impact resistance compared to all other specimens. Variation of Impact energy along with the addition of dolomite, is shown in figure 3.

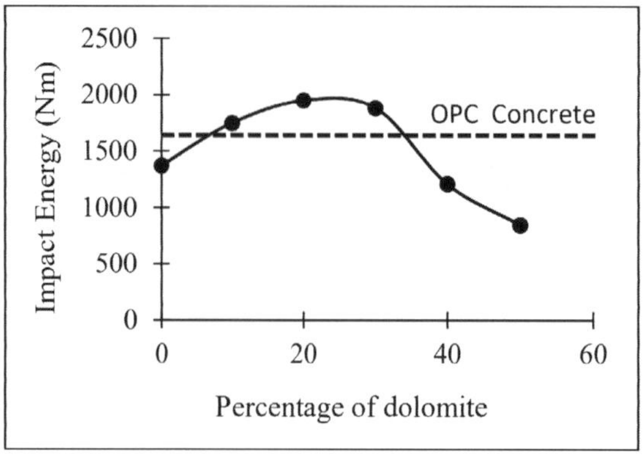

Figure 3. Variation of impact energy with dolomite.

Addition of dolomite (10-30%) improved impact resistance up to 28-42%. Microstructural (Scanning Electron Microscopy-SEM Analysis) studies were conducted on GGBS GPC and GGBS-dolomite GPC and are shown in figure 4.

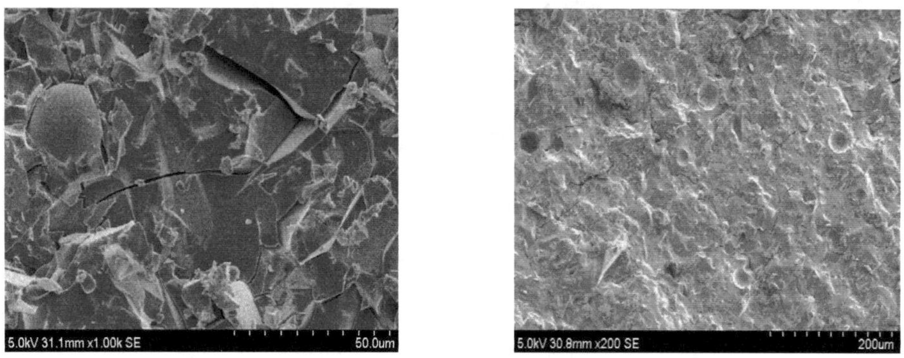

(i) GGBS GPC (ii) GGBS-dolomite GPC

Figure 4. SEM analysis of GPC.

Microstructural studies reveal that maximum denser microstructure was observed in GGBS-dolomite GPC than GGBS GPC [12]. This leads to more impact strength in GGBS-dolomite GPC.

5.4. Failure modes and crack patterns

Failure pattern for different mixes after impact testing is shown in figure 5.

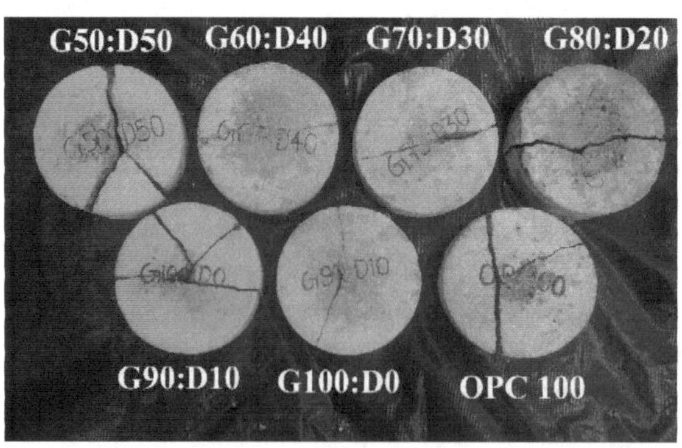

Figure 5. Failure pattern after impact test.

All the specimens were failed into two or three pieces. G60:D40, G70:D30 and G80:D20 split exactly into two pieces, which indicate the higher impact energy of these mixes.

6. Conclusions

Blended geopolymer concrete has been developed from GGBS and dolomite, which are industrial waste by-products. Inclusion of dolomite up to 30% improved the compressive strength of GGBS geopolymer concrete. GGBS-dolomite concrete having compressive strength 60 N/mm^2 was developed through trial and error process. Impact resistance was 21% more for GGBS-dolomite GPC than that of cement concrete due to its dense microstructure.

7. References

[1] Li Z Ding and Z Zhang Y 2004 Development of sustainable cementitious materials *Proc. Int. sustainable dev. Conc. Tech.* **1(1)** 55–76
[2] Duxson P, Provis J L, Lukey G C and Van Deventer J S 2007 The role of inorganic polymer technology in the development of 'green concrete *Cem. Concr. Res.* **37(12)** 1590–7
[3] Salihoglu N K and Salihoglu G 2018 Marble sludge recycling by using geopolymerization technology *J. Hazard. Toxic. Radioactive Waste* **22(4)** 04018019
[4] Paija N, Kolay P K, Mohanty M and Kumar S 2019 Ground Bottom Ash Application for Conventional Mortar and Geopolymer Paste *J. Hazard. Toxic. Radioactive Waste* **24(1)** 04019025
[5] Salih Moslih Amer, Nima Farzadnia, Abang Abdullah Abang Ali and Ramazan Demirboga 2015 Development of high strength alkali activated binder using palm oil fuel ash and GGBS at ambient temperature *Constr. Build. Mater.* **93** 289–300
[6] Marín-López C, J L Reyes Araiza, A Manzano-Ramírez, J C Rubio Avalos J, J Perez-Bueno, M S Muñiz-Villareal, E Ventura-Ramos and Yuri Vorobiev 2009 Synthesis and characterization of a concrete based on metakaolin geopolymer *Inorg. Mater.* **12** 1429
[7] Mehta A. and Siddique R 2018 Sustainable geopolymer concrete using ground granulated blast furnace slag and rice husk ash: strength and permeability properties *J. cleaner prod.* **205** 49–57
[8] Hardjito D, Wallah S E, Sumajouw D M and Rangan B V 2005 Fly ash-based geopolymer concrete *Aust. J. Struct. Eng.* **6** 77–86
[9] Saranya P, Praveen Nagarajan, A P Shashikala 2019 Performance evaluation of geopolymer concrete beams under monotonic loading *Struct* **20** 560–9
[10] Castel A and Foster S J 2015 Bond strength between blended slag and Class F fly ash geopolymer concrete with steel reinforcement *Cem. Concr. Res.* **72** 48–53
[11] Li W and Xu J 2009 Mechanical properties of basalt fiber reinforced geopolymeric concrete under impact loading *Mat. Sci. Eng. A* **505(1-2)** 178–86
[12] Saranya P, Nagarajan P and Shashikala A P 2019 Development of Ground-Granulated Blast-Furnace Slag-Dolomite Geopolymer Concrete *ACI Mater. J.* **116(6)** 235–43
[13] Khalil E, Abd-Elmohsen M and Anwar A M 2015 Impact resistance of rubberised self-compacting concrete *Water Sci.* **29(1)** 45–53
[14] Feng J, Sun W, Zhai H, Wang L, Dong H & Wu Q 2018 Experimental study on hybrid effect evaluation of fiber reinforced concrete subjected to drop weight impacts *Mater.* **11(12)** 2563
[15] IS 2386(Part III)-1963: Methods of test for aggregate for concrete Part 3-specific gravity, density, voids, absorption and bulking *Bureau of Indian standards* New Delhi pp 17
[16] ACI Committee 544. ACI 544.2R-89, 1996 Measurement of properties of fibre reinforced concrete. In: ACI Manual of Concrete Practice, Part 5: Masonry, Precast Concrete and Special Processes *American Concrete Institute*

Recycling of Domestic Plastic Waste Bags as Fine Aggregate in Concrete

Ramakrishna Bag[1*], Aman Agarwal[2] and Ranga Praneeth[2]

[1] Department of Civil and Environmental Engineering, Indian Institute of Technology Patna, India-801103

[2] Department of Civil Engineering, National Institute of Technology Rourkela, India-769008

*rkbag@iitp.ac.in

Abstract. Plastic has been used in many ways in modern life both in industry product as well as in domestic use. The waste generated by using plastic poses an enormous threat to the environment due to its non-degradable nature for a very long time. There has been a lot of effort to minimise plastic waste generation and reuse it. However, the plastic waste generation has been increased considerably in the past decade. This paper presents a potential use of plastic waste as fine aggregate in concrete. Experimental investigations were carried out to determine the strength characteristics of concrete by replacing fine aggregate with 10, 15, 20, and 25% of processed plastic waste. Tests were conducted on concrete specimens under tensile, compressive, and flexural loading conditions to understand its behaviour. Although there was a notable reduction in strength characteristics of the concrete specimens, the optimum strength was obtained at 15% sand replacement. The concrete mix prepared using plastic waste can be used for mass concreting and low load-bearing structures.

1. Introduction

A modern lifestyle, alongside the advancement of technology, has led to an increase in the amount of waste being generated. The various types of waste generated lead to disposal crisis. Humankind had generated 6.3 billion metric tons of plastic waste until the previous decade. Out of this, only 9% was recycled and 12% incinerated. The vast majority of 79% was thrown away. It is expected to generate four times more plastic waste over the next 30 years. If these trends continue 26 billion metric tons of plastic waste will be produced by 2050. Almost half of which will be dumped in landfills. Due to non-degradable nature of plastic, huge amount of waste would be accumulated in near future. Plastic takes more than 400 years to degrade, so most of it still exists in some form. Therefore, the management of such waste is one of the major environmental concerns worldwide [1].

In the past, several researchers have studied the possibility of recycling the plastic waste in various forms. The reuse of plastic waste helps in minimizing the environmental problem. With the scarcity of space for landfilling and cost associated, waste utilization has become an alternative to disposal [1]. Use of plastic waste as aggregate in concrete is considered as one of the possible options. The development of new construction material using recycled plastics is important to both the construction and the plastics recycling industries [2]. The use of plastic waste in concrete is economical and environment friendly. Several studies conducted to evaluate the properties of cement-composites containing various types of plastic waste as aggregate, filler or fiber [3-5]. Some researchers used glass, plastics, and demolished concrete to recycle in concrete [6]. Ground plastics and glass were used to replace the fine aggregates. The crushed concrete was used to replace coarse aggregates in concrete. The recycled materials partially substituted sand and coarse aggregates in concrete mixture. Studies were conducted to recycle waste polymeric materials as aggregate in concrete [7]. The efficiency of reusing plastic waste and plastic bags in concrete also investigated [8-13]. This paper presents an optimum content of plastic waste as a fine aggregate in concrete. The recycled plastic bags in concrete would reduce environmental pollution. One series of tests were conducted using sand and coarse aggregates for the concrete mix design of M25. Other tests were conducted by replacing sand with 10, 15, 20, and 25 % of plastic waste by volume. The results were compared with the concrete prepared without adding any plastic waste as fine aggregate.

2. Materials and method

2.1. Cement
Cement is a crystalline compound of calcium silicates and other calcium compounds. Cement is used as a binder material, a substance that sets, hardens and binds other elements together. In the current investigation portland slag cement was used. The specific gravity of the cement used for this study was determined using Le Chatelier flask found to be 2.99. The specific gravity tests were repeated three times, and the average value is reported.

2.2. Coarse aggregate
Coarse aggregates are particles of sizes greater than 4.75 mm. Usually 9.5 to 37.5 mm diameter coarse aggregates are used in concrete mix design. Coarse aggregates can be obtained from primary, secondary, or recycled sources. In this study, crushed stones of size below 20 mm were used as coarse aggregates. The water absorption test conducted by weighing the aggregates soaked in water for 24 hours was found to be 1.2%. The specific gravity of the coarse aggregate was found out by weighing the certain quantity of coarse aggregate within a wire basket inside and outside the water. The specific gravity of coarse aggregates was found to be 2.84.

2.3. Fine aggregates
Fine aggregates consist of natural sand or crushed stone of particle sizes lower than 4.75 mm. Locally available river bed sand was used as primary fine aggregate in this study. The water absorption and specific gravity of the sand were found to be 1.5% and 2.71, respectively. Zone II sand was used in this study. One series of tests were conducted using sand as a fine aggregate. For all other tests sand was replaced with the plastic waste of 10, 15, 20 and 25% by volume.

2.3.1. Waste plastic as fine aggregate. Processed plastic waste bags were used in the concrete mix design. The specific gravity of waste plastic was determined using kerosene as it is lighter than water. The specific gravity of kerosene is 0.82. The specific gravity of plastic waste used in this study was found to be 0.95.

3. Process of plastic waste recycling
The various stages of processing to recycling plastic bags used in the current investigation are shown in figure 1. Plastic waste was collected from various sources and separated into several categories. The low-density plastic wastes were shredded into tiny pieces or flakes. The shredded plastic flakes were cleaned and compressed by applying heat to densify. The densified plastic wastes were formed into granules. The granules formed in this process were used in the current investigation.

4. Details of tests
Series of tests were conducted to determine compressive strength, tensile strength, flexural strength, bond stress and ultrasonic pulse velocity of concrete. Concrete mix were prepared replacing 10, 15, 20 and 25% by volume of fine aggregates by plastic waste. The mix design ratio of concrete used for the study was 1:1.4:2.5. The mix design was prepared for M25 grade of concrete. One series of tests was conducted without using plastic waste as fine aggregate. Compressive strength, ultrasonic pulse velocity, and pull out tests were carried out on cube specimens of 150 mm size each. The diameter of the reinforcement used during the pullout test was 16 mm. Tensile splitting tests were carried out on cylindrical specimens of height 300 mm and diameter 150 mm. Flexural strength tests were carried out on specimens of size 500 mm × 100 mm × 100 mm. In total, 54 tests were conducted. Three identical

specimens were cast at each condition and tests were conducted. The average values of the three tests are reported.

(a) Shredded plastic waste (b) Compressed plastic waste

(c) Final granules/threads used in concrete

Figure 1. Various stages of plastic waste processing for recycle.

4.1. Compressive strength test

Compressive strength is the capacity of material or structure to withstand the load applied to it. It is measured using compressive strength testing machine. Cube or cylindrical samples are usually tested to determine the compressive strength of concrete. In this study, concrete cubes were cast using the design mix as per IS code guidelines [14]. For each mix, six cubes of dimension 150 mm × 150 mm × 150 mm were cast. After seven days of curing inside water bath, three cubes were tested to determine compressive strength using a universal compression testing machine. Similarly, another three cubes were tested after 28days to determine its compressive load-carrying capacity.

4.2. Tensile splitting test

Concrete is very weak in tension due to its brittle nature. The tensile strength is one of the important properties of concrete. The tensile splitting test is usually carried out on cylindrical concrete specimens. The test was conducted using compressive strength testing machine. For each mix, six cylindrical specimens of 150 mm diameter and 300 mm length were prepared. After seven days of curing underwater, three specimens were tested, and the average strength was noted [15]. Other specimens were tested after 28days of curing.

4.3. Pull out test

Pull out test measures the force required to pull out the reinforcement bar inserted into the concrete specimen. For each mix, three cubes of 150 mm × 150 mm × 150 mm were cast. For pullout test, 16 mm diameter reinforcement bars were used inside the concrete. The reinforcement were placed inside the cubes up to about 10 mm from bottom. The bond stress was measured after 28 days of curing underwater [16].

4.4. *Ultrasonic pulse velocity test*

Ultrasonic pulse velocity (UPV) test is a non-destructive test to analyse the quality of concrete. The velocity of an ultrasonic pulse traveling through the concrete can indicate the strength and quality of concrete. An ultrasonic wave pulse was sent through concrete specimens and the time required to travel was noted. Higher pulse velocities indicate good quality and continuity of the material. Whereas, lower velocities may indicate concrete with voids or lesser density. In the current investigation, UPV tests were conducted after 28 days of curing [17].

4.5. *Flexural strength test*

Flexural strength test was conducted on hardened concrete. Flexural strength is one measure of the tensile strength of concrete. It is a measure of an unreinforced concrete beam or slab to resist failure in bending. Flexural strength tests were carried out [18] after seven days and 28 days of curing the specimens under water.

5. Results and discussion

The various tests conducted with and without plastic wastes were compared.

5.1. *Effect of plastic waste as fine aggregate on compressive strength of concrete*

Figure 2 presents the effect of plastic content on compressive strength of concrete after 7 and 28 days curing. The compressive strength was found to be constant after seven days curing for the sand replacement up to 15%. Beyond 15% replacement of sand by plastic waste, the compressive strength of the concrete was found to decrease. However, for the tests conducted after 28 days of curing, the compressive strength was found to decrease gradually. A change in the compressive strength of 30 and 35% was noted for 25% sand replacements after curing for 7 and 28 days, respectively. The plastic waste granules used in this investigation were found to be smoother and slippery. Therefore, the bonding between granules was weaker, resulting in lower compressive strength as compared to aggregates. Therefore, the concrete with a lower percentage of plastic as a fine aggregate can be used in large construction projects. High percentage of waste plastic can be used to replace sand for lighter structures where the total amount of load is lesser.

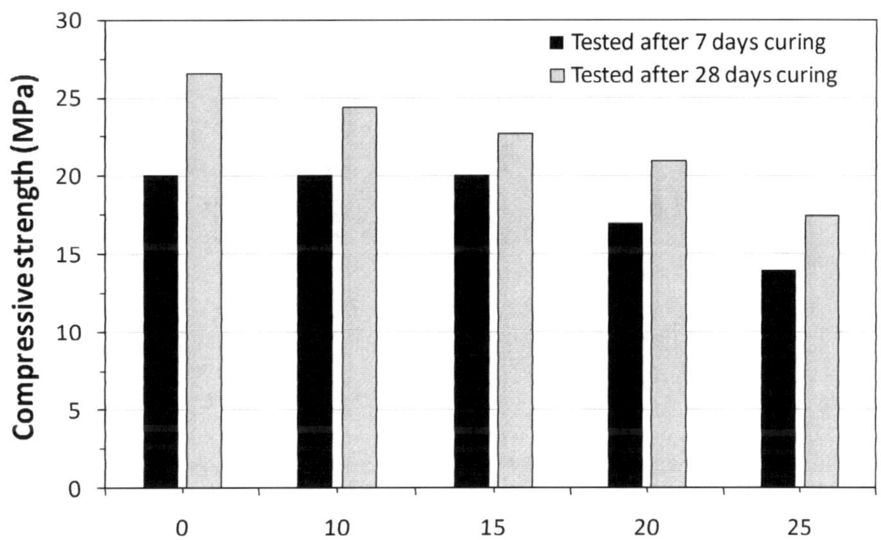

Figure 2. Change in compressive strength due to addition of plastic as fine aggregate

5.2. Effect of plastic waste as fine aggregate on tensile strength of concrete

Figure 3 presents the change in tensile strength with various percentages of plastic waste used in concrete after 7 and 28 days curing. The crack propagation during the tensile test on the cylindrical concrete specimen is shown in figure 4. The variation in tensile splitting strength for various percentages of plastic waste used in concrete was not significant after 7 days curing. Any particular trend of change in flexural strength was not observed. However, the tensile splitting strength determined after 28 days curing was found to be constant up to 15% sand replacement. For a further increase in plastic waste content, the tensile splitting strength was found to decrease by about 18%. Hence, the concrete with up to 15 % of plastic replacement can be used for various construction purposes.

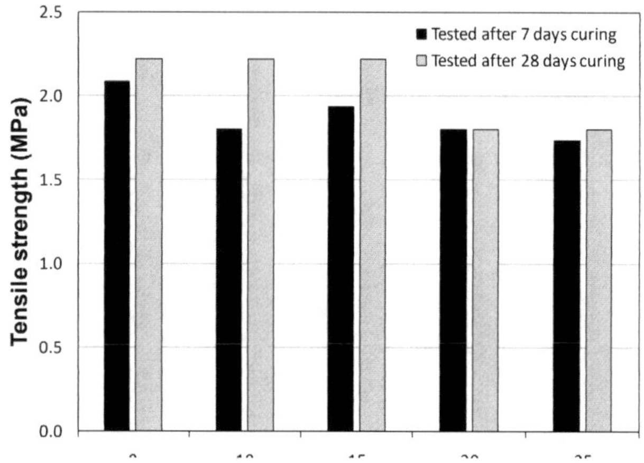

Figure 3. Variation in tensile strength due to addition of plastic waste as fine aggregate.

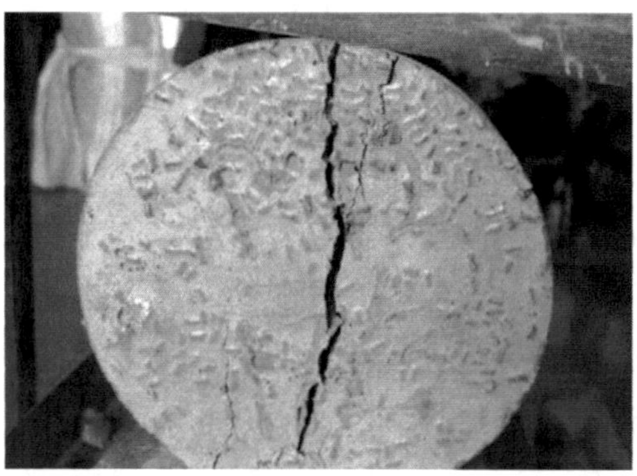

Figure 4. Crack propagation during tensile splitting test in concrete cylinders.

5.3. Effect of plastic waste as fine aggregate on flexural strength of concrete

Flexural strength is one of the measures of tensile strength of concrete. The force required to resist the failure of an unreinforced concrete beam or slab by bending is determined. Figure 5 shows the crack propagation of the specimen after failure. The effect of plastic waste on flexural strength of concrete is shown in figure 6. After 7 days curing, the flexural strength decreased for 10% addition of plastic waste as compared to concrete without any plastic waste. Upon further increase in plastic content, the flexural strength was found to be constant. However, for 28 days curing the flexural strength was found to increase by 10% addition of plastic waste. At higher percentage of plastic content similar behaviour as observed for 7 days curing was found.

5.4. Effect of plastic waste as fine aggregate on pull out strength of concrete

Figure 7 presents the bond stresses measured from pull out tests with various percentages of plastic waste. There was no particular trend obtained from pull out test results. The maximum bond stress was obtained at 15% sand replacement by plastic waste.

5.5. Effect of plastic waste as fine aggregate on ultrasonic pulse velocity of concrete

The UPV is a non-destructive test to determine the strength of concrete. Figure 8 presents the variation of ultrasonic pulse velocity with different percentages of plastic waste. The UPV was found to decrease continuously with an increase in plastic content. Therefore, the strength of the concrete tested for the investigation would be decreased with higher percentages of plastic content. The results indicate that the rigidity of the material reduces.

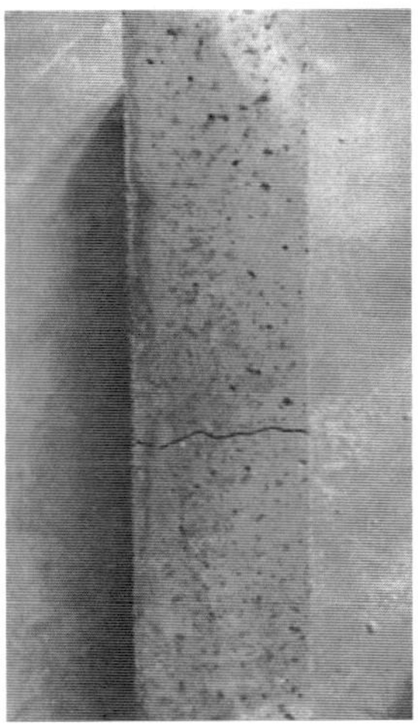

Figure 5. Crack propagation during flexural strength test.

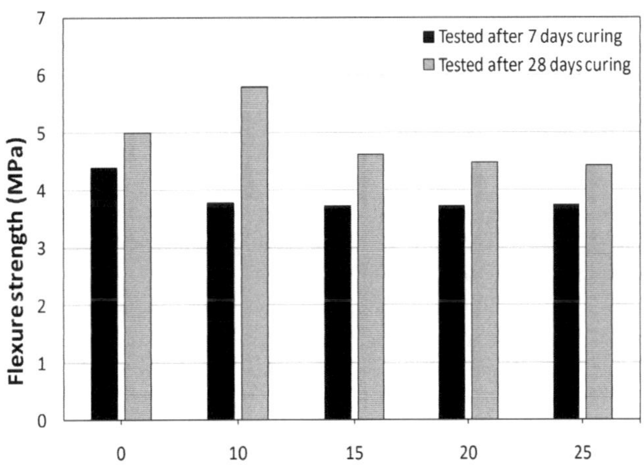

Figure 6. Change in flexural strength due to addition of plastic as fine aggregate.

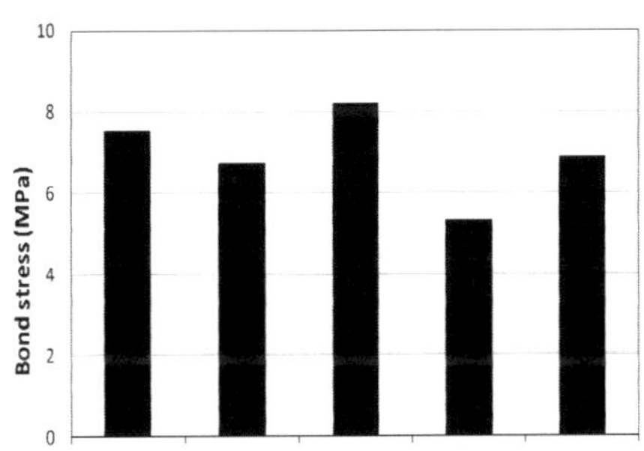

Figure 7. Change in bond stress due to addition of plastic as fine aggregate.

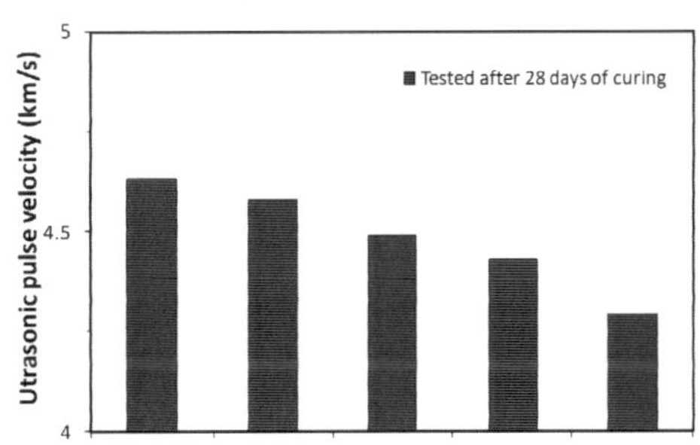

Figure 8. Change in Ultrasonic pulse velocity due to addition of plastic as fine aggregate.

6. Conclusions

Series of tests were conducted to determine the suitability of plastic waste bags as fine aggregates in concrete. Tests were conducted after curing the concrete specimens for 7 and 28 days. Based on the results obtained, following conclusions are drawn:

- The compressive strength of recycled plastic concrete was found to decrease with increase in plastic content. The surface of the plastic is smoother than sand, and hence the bonding between elements in the concrete mix is weaker.

- The tensile strength was found to be constant up to 15% sand replacement after 28 days curing under water.
- The bond stress was observed to be maximum at 15% replacement of sand replacement. The strength of the modified concrete mix with up to 15% addition of plastic waste as fine aggregate was noted to be within the permissible limit.
- Therefore, 15% of sand can be replaced by plastic waste in concrete for light structures. The use of plastic waste in concrete would be helpful in reducing the environmental pollution. The concrete with plastic waste would be lighter as well as economically viable.

7. References

[1] Siddique R, Khatib J and Kaur I 2008 Use of recycled plastic in concrete: A review *Waste Mgmt.* **28** 1835–52
[2] Rebeiz K S and Craft A P 1995 Plastic waste management in construction: technological and institutional issues *Resour. Conserv. and Recyc.* **15** 245–57
[3] Elzafraney M, Soroushian P and Deru M 2005 Development of energy-efficient concrete buildings using recycled plastic aggregates *ASCE J. Arch. Eng.* **11(4)** 122–30
[4] Saikia N and De Brito J 2012 Use of plastic waste as aggregate in cement mortar and concrete preparation: A review *Cons. and Build. Mat.* **34** 385–401
[5] Jibrael M A and Peter F 2016 Strength and behaviour of concrete contains waste plastic *J. Ecos. & Ecograph.* **6** 186
[6] Batayneh M, Marie I and Asi I 2007 Use of selected waste materials in concrete mixes *Waste Mgmt.* **27(12)** 1870–6
[7] Rahman M A, Islam M A, Ahmed M and Salam M A 2012 Recycled polymer materials as aggregates for concrete and blocks *J. Chem. Eng.* **27(1)** 53–7
[8] Ismail Z Z and Al-Hasmi E A 2008 Use of waste plastic in concrete mixture as aggregate replacement *Waste Mgmt.* **28** 2041–7
[9] Panyakapo P and Panyakapo M 2008 Reuse of thermosetting plastic waste for lightweight concrete *Waste Mgmt.* **28(9)** 1581–8
[10] Choi Y W, Moon D J, Kim Y J and Lachemi M 2009 Characteristics of mortar and concrete containing fine aggregate manufactured from recycled waste polythelne terephthalate bottles *Const. and Build. Mat.* **23** 2829–35
[11] Al-Maaded M, madi N K, Kahraman R, Hodzic A and Ozerkan N G 2012 An overview of solid waste management and plastic recycling in Qatar *J. Polymer. Env.* **20** 186–94
[12] Safi B, Saidi M, Aboutaleb D and Maallem M 2013 The use of plastic waste as fine aggregate in the self-compacting mortars: Effect on physical and mechanical properties *Const. and Build. Mat.* **43** 436–42
[13] Kumar K S and Baskar K 2015 Recycling of E-plastic waste as a construction material in developing countries *J. Mat. Cyc. and Waste Mgmt.* **17** 718–24
[14] IS 10262 2009 *Concrete Mix Design* (Bureau of Indian standard: New Delhi)
[15] IS 5816 1999 *Splitting Tensile Strength of Concrete - Method of Test* (Bureau of Indian standard: New Delhi)
[16] IS 2770-1 1976 *Methods of Testing Bonds in Reinforced Concrete* (Bureau of Indian standard: New Delhi)
[17] IS 13311-1 1992 *Method of Non-Destructive Testing of Concrete* (Bureau of Indian standard: New Delhi)
[18] IS 516 1959 *Method of Tests for Strength of Concrete* (Bureau of Indian standard: New Delhi)

IOP Conf. Series: Materials Science and Engineering **936** (2020) 012010 doi:10.1088/1757-899X/936/1/012010

Precast Bamboo Reinforced Furniture Elements using Self Compacting Concrete

Jagadish Vengala[1*], K B Namratha[2] and S Anusha[3]

[1] Associate Professor, Department of Civil Engineering, PVP Siddhartha Institute of Technology, Andhra Pradesh, India

[2] Graduate Student, Purdue University, 320 Brown Street, WL IL 47906, USA

[3] Former Student, Civil Engineering Department, BMSIT&M, Bangalore, Karnataka, India

*jdvengala@gmail.com

Abstract. Public furniture in India are designed based on ergonomics standards derived from the international standards proving a significant hurdle in comfort for the users. We propose to tackle this by conducting ergonomic studies of true users. This study aims to check the viability of using an eco-friendly composite as street furniture, i.e., Park Bench. Additionally, owing to the increasing negative impact steel manufacturing has on the environment, the furniture is designed using a Bamboo-SCC composite. The objective of the study is to use bamboo as reinforcement in SCC to make furniture. The furniture is made using self-compacting concrete to ensure greater stability of the bamboo reinforcement and further to increase the durability of the benches. The final product is analysed based on deflection tests, cost-benefit analysis, and embodied energy savings. It was concluded that in areas where bamboo is readily available, this project can be implemented as a more energy-efficient and low-cost option.

1. Introduction

Street-furnishing at public places creates a sense of identity when properly designed and integrated. They are generally looked at as places where one can comfortably sit, eat, rest, and meet people. Street-furniture is important for the elderly who have very minimal mobility and people with kids. The benches and tables at parks and squares attract people. Properly designed and placed furniture is socially significant as they are a crowd-puller at public spots. In order to tackle the challenge of human comfort in this project, ergonomic design is used.

Concrete has suitable building properties, such as its capability to take large compressive loads. It is a widely used ingredient in the foundation of infrastructure, readily available, and cheap. The only drawback is that it has low tensile strength. This limits the use of concrete. Overcoming this limitation, we use one of the most popular reinforcing bar (rebar) that is steel. Developing countries that are in need for steel as reinforcement but lack proper means to meet the demand of steel production can opt for a more efficient, maintainable, and extremely resilient building material, i.e., Bamboo. Instead of being dependent on the developed countries which control the global market, bamboo has the potential to replace steel as a low-cost alternative in places it can't be produced.

The tensile strength of bamboo is relatively high and can reach 370 MPa [1]. This makes bamboo an attractive alternative to steel in tensile loading applications. This is due to the fact that the ratio of tensile strength to the specific weight of bamboo is six times greater than that of steel [1].

Studies conducted by Gavami et al., [2] established the durability of bamboo as a reinforcement in concrete, besides improving the bonding of bamboo reinforcing bars. The results of the investigations show that bamboo can substitute steel satisfactorily. The structural elements developed and studied could be used in many building constructions.

Self-Compacting Concrete (SCC) was developed initially in Japan in the 1980s when contractors were experiencing a severe shortage of skilled manpower. SCC does not require any external vibration; it flows under its own weight and completely fills the forms without segregation or bleeding. For SCC, it is generally necessary to use superplasticisers to obtain mobility and viscosity modifying admixtures or large volume of powder material to eliminate possible segregation. The powdered materials are fly ash, Ground Granulated Blast furnace Slag (GGBS), silica fume, limestone powder [3].

The objective of the study is to use bamboo as reinforcement in SCC to make furniture. Ergonomic studies, Cost-benefit analysis and lab tests were conducted to substantiate the same. The furniture is made using SCC to ensure greater stability of the bamboo reinforcements and further to increase the durability of the benches. The benches were precast as 3 separate slabs, one top slab, and two bottom supports and later assembled. Deflection tests were conducted to ascertain the amount of load that the bench could take, and the benches tested satisfies the design load criteria.

2. Experimental program

The Experimental program was conducted in three phases, mix design of SCC, ergonomic studies, and finally the assembly of the bench.

2.1. Materials

Ordinary Portland cement of 53 grade, conforming to the requirements of IS-12269[4], was used in the study. GGBS used was obtained from M/s JSW, Karnataka, and satisfies the requirements of IS 16714:2018[5].The specific gravity values of the cement and GGBS were 3.14 and 2.91, respectively.

Crushed stone sand was used as fine aggregate, and crushed stone with a maximum size of 10mm was used as coarse aggregate. Commercially available polycarboxylic ether-based Superplasticisers (Glenium sky 8233) was used in this study. Bamboo splits of 9mm width were used as reinforcement. Bamboo splits were treated using dip diffusion method. Commercially available boric and borax solution was used in dip diffusion method[6].

2.2. Mix design of SCC

Guidelines from the standard Japanese method and method proposed by Jagadish Vengala was used to arrive at the mix proportions [7]. Marsh cone test was conducted to arrive at the optimum super plasticiser dosage. Once the SCC mix is designed, the fresh mix is tested for its flowability and passing ability by conducting slump flow, L-box, and V-funnel tests. The test results were compared with the EFNARC guidelines [8].

SCC trial mixes were cast out into 100mm cubes. The specimens were cured and tested at 7, 14, and 28 days. The specimen is removed from the water after specified curing time, and excess water from the surface is wiped out. In order to finalise the design mix of concrete, target strength is a desired input which can also be achieved by resorting to accelerated curing. In the present study, accelerated curing tests were conducted as per IS 9013[9] to arrive at the target strength of concrete.

2.3. Ergonomic studies

First, the potential users of the bench were identified, so that the precast benches could be designed based on the anthropometric dimensions of the potential users. The following flow diagram illustrates the steps followed in arriving at the final dimensions of the bench (Figure 1).

First, the target users were decided based on the location of placing the bench- near the college canteen, thereby making the potential users- teachers, students, workers, gardeners, and outside visitors. The age range of the people using the bench is between 18-65 yrs. A random crowd of 50 women and 50 men were selected from college; this group included all- teachers, students, workers, gardeners, and outside visitors. Dimensions of their shin and their femur were tabulated based on gender. The 25th, 50th, and 75th percentile of the data were ascertained in a spread sheet.

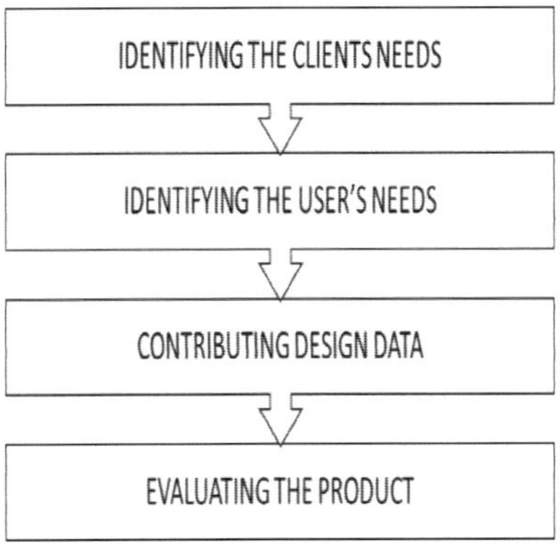

Figure 1. Flow diagram for conducting ergonomic studies.

2.4. Preparation of precast park bench

The next step was to prepare the reinforcements. After the treatment, three slabs were cast out of which one was the top slab, and the other two were bottom support slabs. Bamboo strips were placed at 1.5 inches from the concrete surface, and the strips were staggered to ensure the uniform cross-section of the slab. Binding wire was used to tie the bamboo reinforcement.

SCC is produced and directly poured to the location of concrete placement. Sufficient carpenters, along with the supervisor inspected the behaviour of supports below the slab during the casting. The curing was started immediately after 24 hrs. Curing was done using membrane curing for a minimum period of 7 days. One main slab and two supports were cast. A one-inch deep socket was made along the breadth of the main slab to fix the supports. Two supports were placed, and the main slab was placed above it, gently fixed by the socket provided. The voids were filled in with cement mortar.

2.5. Testing of precast park bench

Once the slab was assembled, its performance was tested. The center line of the slab was located, and a span length of 80cm was taken by taking 40 cm on either side. From the center, an offset of 10cm was taken, and a dial gauge was placed at the bottom of the slab. Loading was done along the span length using uniform hollow blocks of approximately 20kg and sand bags of 40 kg. Deflection was measured, and the behaviour of the bench under loading such as crack formations was observed.

3. Results and discussion

3.1. Mix design and tests on Self Compacting Concrete

Self compacting concrete was designed based on the Japanese method of mix design. Initial trials were made to arrive at final mix proportions. Figure 2 shows the marsh cone test. Table 1 gives the results of the Marsh cone test, which gives the optimum dosage of superplasticiser.

Figure 2. Marsh cone test.

IOP Conf. Series: Materials Science and Engineering **936** (2020) 012010 doi:10.1088/1757-899X/936/1/012010

Table 1. Results of marsh cone test.

Sl. No.	Cement	Admixture	W/P ratio	Water (ml)	Optimum Dosage (%)	Avg. Marsh cone time,s
1	OPC 53 grade cement	Masterglenium sky 8233	0.50	1170	0.1	17.40
2					0.2	15.20
3					0.3	11.77
4					0.4	11.74
5					0.5	11.71

Trials have been carried out to arrive at the optimum dosage of superplasticiser using Marsh Cone test for the SCC mix of M25 grade. The details of the mix proportions of the SCC mix used for the present study is shown in table 2.

Table 2. Mix proportions of the SCC mix adopted in the study.

Materials	Quantities
Coarse aggregates (kg/m^3)	800
Fine aggregates (kg/m^3)	1200
Cement (kg/m^3)	250
GGBS (kg/m^3)	300
Water (litres)	275
W/P ratio	0.5
superplasticiser dosage	0.3%
Slump (mm)	650

Table 3. Results of fresh Properties of SCC mix.

Fresh Properties	Range EFNARC[3]	Obtained values
V -funnel	6 - 12	8sec
V_{5min}	[6-12] + 3sec	11 sec
Slump	500 – 700 mm	650 mm
L – box = H_2/H_1	>0.8	0.9

(a) Slump flow test **(b)** L-Box test **(c)** V- funnel test

Figure3. Tests conducted for SCC to ascertain fresh properties.

Tests have been conducted to ascertain the fresh properties and check the performance of SCC mix. Figure 3 represents tests conducted for SCC to ascertain fresh properties. The results of the fresh properties are shown in table 3. From table 3 it can be inferred that the filling ability of the SCC mix, which was checked by conducting V funnel and slump flow test satisfies the requirement and the values lie within the range recommended. The passing ability test conducted using the L Box satisfies the requirement, and the values lie within the range recommended.

After casting the concrete specimens, tests have been conducted for specimens after 7, 14 and 28 days. All the tests were conducted as per IS: 516[10]. Table 4 gives the accelerated curing test results and compressive strength results of the mix used in the study.

Table 4. Accelerated curing test results and compressive strength results of mix used in the study.

Avg. accelerated curing strength R_a(MPa)	Average R_{28} (MPa)	Cube compressive strength (MPa)		
		7 days	14 days	28 days
18.6	38.6	27.8	39.4	43.5

3.2. Design of precast park bench based on ergonomic data

The shin (knee) length and the femur length of target users are collected based on gender. A random crowd of 100 men and women of age group of 18-65 years have been taken. The 25th, 50th, and 75th percentile of the data were ascertained in the excel sheet. Based on the data collected the different percentile ranking and the average is tabulated as below in table 5.

Table 5. Different percentile and average values of anthropometric data in cm.

Percentile	50th	75th	Average of 50th and 75th percentile	
Female (femur)	51.0	54.0	52.5	
Male (femur)	51.6	53.3	52.45	
Average	51.3	53.65	52.5	Breadth
Female (knee)	40.0	42.0	41.0	
Male (knee)	47.2	50.5	48.8	
Average	51.3	53.65	44.9	Height

The bench dimensions were decided by taking the average of the 50th and the 75th percentile of both males and females. The final dimensions of the bench were set as follows. Table 6 gives the dimensions of the precast park bench.

Table 6. Dimensions of the precast park bench in cm.

Length	152.4
Breadth	52.5
Height of the supports	45.7
Thickness of the slab	7.6
Spacing of reinforcement	7.6

3.3. Preparation of precast park bench

Treated bamboo reinforcement was used to make the reinforcement cage (figure 4) and the same was placed in the prepared mould. Bamboo strips were staggered to ensure the uniform cross-section of the slab. SCC is directly poured into the mould as shown in figure 5. Curing was done using membrane curing for a minimum period of 7 days. Figure 6 shows the assembled precast park bench.

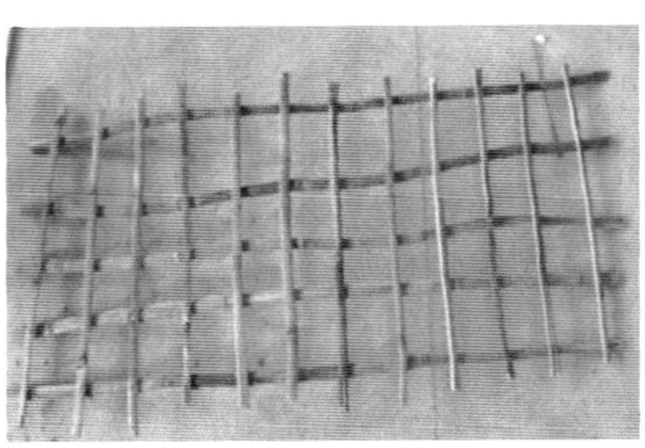

Figure 4. Treated bamboo reinforcement cage.

Figure 5. Precast mould while pouring SCC.

Figure 6. Assembled precast park bench.

3.4. Test results of precast park bench

The loads were applied to the precast park bench in increments of load upto a maximum of two and a half times of the design load. Figure 7 shows the precast park bench while testing. Table 7 gives the load increment and corresponding deflection values of the bench.

Table 7. Deflection values of the precast park bench at various loads.

Load(kg)	Deflection(mm)
0	0
80	0.06
160	0.15
240	0.21
320	0.30
400	0.57
480	1.41
560	3.44
640	3.53
720	4.17
720@24 hrs	4.38

Figure 7. Precast park bench while testing.

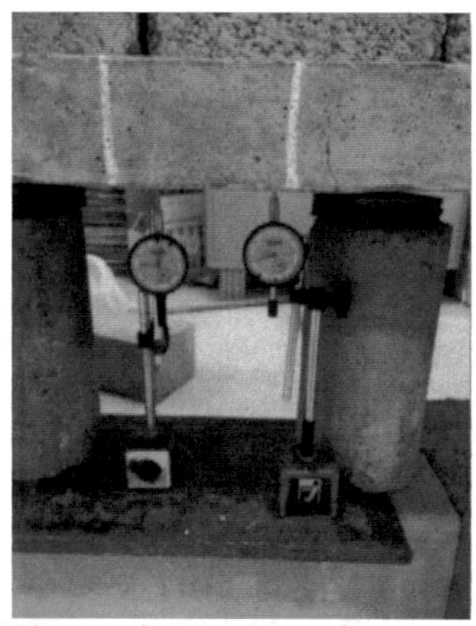

Figure 8. Testing of precast park bench under progress.

Taking design load as 300kgf, considering 3 people of each 100kgs approximate weight, it can be seen that the load-carrying capacity is almost two and a half times of the design load. Moreover, the load of 750 kgf was left to rest on the bench for one day, and small cracks were seen but deflection remained less than 5mm (Figure 8).

3.5. Embodied Energy

Based on the cost analysis, the precast concrete bench prepared using bamboo reinforcement is 18% low-cost than the bench made using steel reinforcement. In addition to the cost-benefit, the embodied energy of steel is 30MJ/kg, whereas bamboo's embodied energy is 0.5MJ/kg. The carbon emission from steel is 1.85 tonnes / tonnage of steel, whereas bamboo intakes CO_2 which intern reduces pollution; the greater the CO_2 level faster the growth of bamboo. Also, steel is non-renewable resources after a few decades, the availability of steel decreases, whereas bamboos have faster growth and cost-efficient when compared to steel.

4. Concluding remarks

In this study, the furniture is made using self-compacting concrete to ensure greater stability of the bamboo reinforcements and further to increase the durability of the benches. It is important to design the benches based on ergonomics to maintain a correct posture and to increase the comfort level of potential users. Based on the study conducted, it can be seen that concrete furniture elements can be made using bamboo as a reinforcement, and SCC ensures the stability of the reinforcement as there is no vibration required to consolidate the concrete. The permeability of SCC is low, and GGBS as part replacement of cement provides more resistance against chemical attacks moreover, treated bamboo ensures longer life. Precast slabs ensure consistency in quality and reduce errors in design. The performance study conducted on the bench indicates that it can take a load much greater than the design load without significant deflection. The load-carrying capacity of the bench is almost two and a half times of the design load. Based on the cost analysis, the precast concrete bench prepared using bamboo reinforcement is 18% low-cost than the bench made using steel reinforcement.

5. References

[1] Ghavami K 1995 Ultimate load behaviour of bamboo reinforced lightweight concrete beams *Cement and Concrete Composites* **17** 281–8

[2] Khosrow Ghavami 2005 Bamboo as reinforcement in structural concrete elements *Cement and Concrete Composites* **27** 637–49

[3] Vengala J, Sudarshan M S and Ranganath R V 2003 Experimental study for obtaining self-compacting concrete *Indian concrete journal* **77(8)** 1261–6

[4] IS 12269 2013 *Specificationfor 53 grade Ordinary Portland cement* (Bureau of Indian Standards: New Delhi)

[5] IS 16714: 2018 *Specification for Ground Granulated Blast Furnace Slag for Use in Cement, Mortar & Concrete* (Bureau of Indian Standards: New Delhi)

[6] Vengala J, Jagadeesh H N and Pandey C N 2008 Development of bamboo structures in India *Modern Bamboo Structures-Xiao et al.(eds)* 51–63

[7] Vengala J and Ranganath R V 2004 Mixture proportioning procedures for self-compacting concrete *Indian concrete journal* **78(8)** 13–21

[8] Self-Compacting Concrete European Project Group 2005 The European guidelines for self-compacting concrete: Specification, production and use *International Bureau for Precast Concrete (BIBM)*

[9] IS 9013 1978 *Method of Making, Curing and Determining Compressive Strength of Accelerated Cured Concrete Test Specimens* (Bureau of Indian Standards: New Delhi)

[10] IS 516 1959 *Method of Tests for Strength of Concrete* (Bureau of Indian Standards: New Delhi)

Acknowledgments
We acknowledge the support from Dr. Rudresh H B from M/s Vijay Metal Finishers for his support during the project.

IOP Conf. Series: Materials Science and Engineering **936** (2020) 012011 doi:10.1088/1757-899X/936/1/012011

Feasible use of Recycled Foam Concrete in Cement Mortar

Evangeline Lalrinmawii[1], Santosini Sahu[2*], Pradip Sarkar[3] and Robin Davis[4]

[1] M.Tech. Scholar, National Institute of Technology Rourkela, India-769 008

[2] Ph.D. Scholar, National Institute of Technology Rourkela, India-769 008

[3] Associate Professor, National Institute of Technology Rourkela, India-769 008

[4] Assistant Professor, National Institute of Technology Calicut, India-673 601

*santosinisahu62@gmail.com

Abstract. Lightweight foam concrete blocks have gained popularity over the last decade, particularly in the construction of interior infill walls. Consequently, the construction and demolition waste generated by such foam concrete blocks are increasing regularly. Recycling of these wastes and using them in cement mortar or concrete as partial replacement of conventional material can be an essential step towards the management of construction and demolition waste as well as conservation of natural resources leading to sustainability. The present study aims to assess the feasibility of recycled Autoclaved Aerated Concrete (AAC) and Cellular Lightweight Concrete (CLC) blocks as a possible replacement of sand and cement in the manufacturing of mortar. This study investigates various physical and mechanical properties of mortar in which sand and cement are partially replaced with material from recycled AAC and CLC block. Chemical analysis of demolished AAC and CLC block samples determined using X-Ray Fluorescence (XRF) analysis to find out the differences in composition between the conventional materials and replaced material. This study concludes that materials from recycled AAC and CLC block can be used in cement mortar as partial replacement of either cement or sand up to 15-20% without significant reduction of strength.

1. Introduction

Continuous urbanization and infrastructural development worldwide lead to a huge amount of construction and demolition waste generation. Growing usage of lightweight foam concrete blocks in recent times results in a considerable portion of the construction and demolition waste. Disposal of these wastes requires a huge open area for landfilling and creates other environmental issues. The use of waste foam concrete blocks in cement concrete or mortar provides a reasonable solution for problems related to waste management. In the present study, an attempt is made to partially replace sand and cement by materials recycled from demolished wastes of two popular forms of foam concrete blocks: AAC and CLC.

The quantity of natural sand in the cement mortar and concrete varies from 20 to 80%. Hence, the demand for natural sand is quite high, and the excessive use of this causes a severe threat to the environment and society. To encounter the present requisite, accessibility of good quality of sand is decreasing day-by-day [1]. The shortage of such natural resources has introduced the concept of the use of various substitute materials to replace the sand partially. Alternative materials such as fly ash, crushed brick, recycled alumina, pond ash, quarry dust, marble dust or limestone, waste foundry sand, waste plastic, etc. are previously used in concrete and mortar mixtures as a partial or full replacement of normal sand to check the feasibility [2-6].

Cement occupies 15 to 20% of the volume and is the most expensive component in concrete and mortar production [7]. The cement manufacturing industry is a major source of greenhouse gas emissions, causing environmental pollution. At this stage, the use of supplementary cementitious materials to replace a portion of the cement in concrete or mortar can be a promising idea. Many previous studies are carried out to investigate possible replacement materials for Portland cement (OPC). Some substitute materials have already been used as a replacement of OPC such as fly ash, rice husk ash, sugarcane ash, silica fume, limestone, and siliceous stone powder, etc. in past literature [8-13].

Recycling of the AAC and CLC blocks in construction work is not studied yet. Replacement of cement by demolished CLC and AAC block dust can be promising for both reducing CO_2 emission and problems of waste management. Similarly, the replacement of sand by demolished CLC and AAC block dust can help to conserve natural resources. Therefore, the prime focus of the present study is identified as to explore the prospect of utilizing the demolished CLC and AAC dust as an alternative material for partial replacement of sand and OPC in cement mortar.

2. Material and experiments

2.1. Materials

As a part of experimental work, cement, sand, recycled AAC, and CLC powder are used to prepare the test specimens. Ordinary Portland cement of 43-grade and normal tap water are used throughout the study. The fine aggregate used in the experiment is natural river sand conforming to ASTM C144 [14]. The CLC and AAC powder used in this study for sand and cement replacement is recycled from waste CLC and AAC blocks. AAC aggregate (AAC-A) and CLC aggregate (CLC-A) finer than 4.75 mm is used as the replacement material for natural sand in the test specimen. Specific gravity and water absorption of sand, AAC-A, and CLC-A are evaluated experimentally as per relevant ASTM C128 [15] and reported in table 1. It can be observed from table 1 that the specific gravity values of AAC-A and CLC-A are lesser than that of sand. This is because AAC-A and CLC-A are obtained from lightweight foam concrete blocks, which contain 65-70% of fly ash. The more porous structure of AAC-A and CLC-A than sand is indicated by their higher water absorption rates. Table 1 also presents the fineness modulus, coefficient of curvature, and coefficient of uniformity for sand, AAC-A, and CLC-A obtained from sieve analysis. The fineness modulus data show that sand is coarser than AAC-A and CLC-A. The coefficient of uniformity and coefficient of curvature values show that the grading of all three samples is of similar nature. The particle size distribution curve for all three samples, including the ASTM C-144 [14] limits are shown in figure 1. It can be seen from the figure that AAC-A and CLC-A are within the permissible range to qualify as fine aggregate in cement mortar.

Table 1. Physical properties of fine aggregate used in the study.

Properties	Sand	AAC-A	CLC-A
Specific gravity	2.6	1.7	1.6
Water absorption (%)	0.8	8.7	13.2
Fineness modulus	3.26	2.98	3.17
Coefficient of uniformity	3.25	2.94	3.53
Coefficient of curvature	0.78	0.92	0.88

AAC binder (AAC-B) and CLC binder (CLC-B) finer than 90μ is used as the replacement material for cement in the test specimen. Physical properties of AAC-B, CLC-B, and OPC are tested in the laboratory as per relevant ASTM standards [16-17], and reported in table 2. AAC and CLC block has fly ash as a chief constituent material, which results in the specific gravity of AAC-B and CLC-B much lower than that of OPC. In addition, these materials require more water than OPC to form a consistent paste because of the presence of high absorptive fly ash particles.

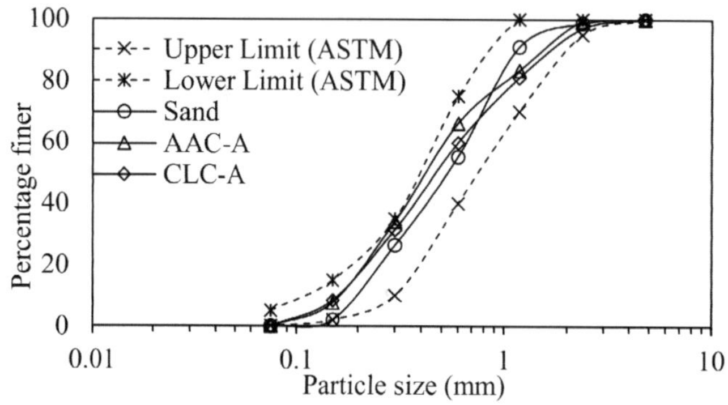

Figure 1. Particle size distribution of Sand, AAC-A, and CLC-A.

Table 2. Physical properties of OPC, AAC-B, and CLC-B.

Properties	AAC-B	CLC-B	OPC
Specific gravity	2.10	2.18	3.15
Consistency (%)	45	53	31

The microstructural analysis of AAC-B and CLC-B is carried out by X-Ray Fluorescence (XRF) analysis to evaluate their chemical composition and applicability as cementitious materials. The major oxide compositions of OPC, AAC-B, and CLC-B obtained with Zetium PANalytical XRF spectrometer are reported in table 3. It tends to be seen from table 3 that both AAC-B and CLC-B contain all the major oxide elements of OPC. However, the percentage of these oxides in AAC-B and CLC-B varies from OPC. The major difference is observed to be with the percentage of CaO, SiO_2, Al_2O_3, and MgO. Cao and MgO present in higher amounts while SiO_2 and Al_2O_3 in a lesser amount in OPC as compared to that of AAC-B and CLC-B. The XRF analysis result fortifies the feasible use of AAC-B and CLC-B as a partial cementitious material in cement mortar or concrete.

Table 3. Chemical composition (%) of OPC, AAC-B, and CLC-B.

Oxide	OPC	AAC-B	CLC-B
CaO	56.422	19.432	16.473
SiO_2	20.495	40.016	44.531
Al_2O_3	5.736	15.830	18.018
MgO	4.526	0.737	0.760
Fe_2O_3	3.218	4.133	3.379

2.2. Replacement of fine aggregate

The mortar cubes of 50 mm size are prepared with a water-to-binder (*w/b*) ratio of 0.485 and sand-to-cement ratio 2.75 as per ASTM C109 [18]. Cement mortars cubes are prepared with natural sand replaced separately by AAC-A and CLC-A. Sand replacement ratio considered in this study are 0% (natural sand only), 5%, 10%, 15%, 20%, 25% and 30% by weight. It is to be noted that OPC is used uniformly as a binder in all the mortar specimens considered here. The compressive strength and water absorption test of the cube specimen from different categories are tested after 28 days of curing.

2.3. Replacement of cement

AAC-B and CLC-B are used to replace OPC in cement mortar by 0% (OPC only), 5%, 10%, 15%, 20%, 25%, and 30% of weight. The proportions of materials for the standard mortar are kept uniform, as mentioned in the previous section. Natural sand is used as fine aggregate uniformly for all the mortar specimens considered here. The mortar cubes of 50 mm size are prepared according to ASTM C-109

(2016) for the compressive strength test. The compressive strength and water absorption of mortar cubes are tested after 28 days of curing as per the procedure outlined in relevant ASTM Standard.

3. Result and discussion

3.1. Replacement of fine aggregate

The effect of replacement of sand by different percentages of AAC-A and CLC-A on the mean compressive strength of cement mortar cubes are presented in figure 2. The test results illustrate that the replacement causes a reduction in mortar compressive strength. Sand replacement up to 15% with AAC-A and CLC-A cause a reduction in compressive strength of mortar cubes up to 20% when compared with controlled mortar specimens. Further increase in the replacement of sand is found to reduce the strength significantly. This reduction may be due to the low specific gravity, high water absorption of AAC-A and CLC-A as compared to that of sand and the presence of the foam-forming agent. A high volume of AAC-A and CLC-A particles led to the creation of pores in the mixes. It can be suggested from the above result that replacement of sand maximum up to 15% with AAC-A and CLC-A can be considered for waste utilization without significant loss in compressive strength.

The water absorption test results of AAC-A and CLC-A replaced cube specimen are shown in figures 3-4. From the test results, it is found that the water absorption rate increases as the AAC-A and CLC-A replacement increases. The water absorption rate of the mortar specimens containing 15% of AAC-A and CLC-A found to be increased by 30 to 40 % as compared to the controlled mortar specimen. This high rate of absorption of replaced mortar specimen is because of the high water absorption of the AAC and CLC fines as compared to sand.

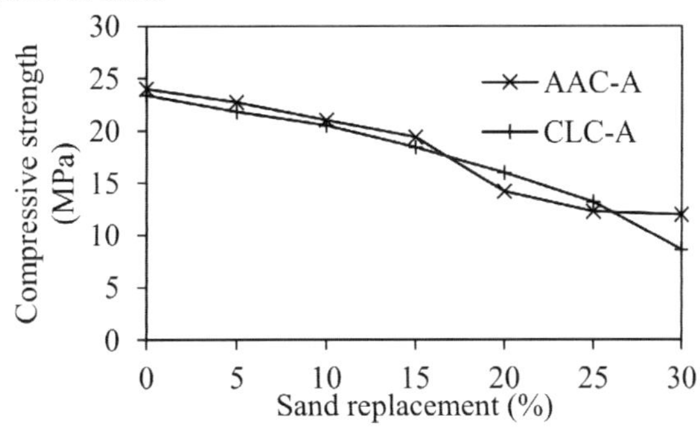

Figure 2. Variation of 28-day compressive strength of sand replaced samples.

3.2. Replacement of cement

The compressive strength of hardened mortar is presented in figure 5 for mortar cubes with cement replaced by AAC-B and CLC-B. It can be observed from figure 5 that the compressive strength of mortar cube decreases gradually as AAC-B replacement increases. However, this decrease is quite small (less than 10%) up to 15% replacement. Similarly, the compressive strength of mortar cube increases with the increase of CLC-B replacement up to 10%. Further increase of CLC-B replacement reduces the compressive strength. However, up to 20% of CLC replacement results in no reduction of compressive strength as compared to the control specimen.

The results of the water absorption test for AAC-B and CLC-B replaced mortar specimen are shown in figures 6-7. From the test results, it is found that the water absorption rate increases as the AAC-B and CLC-B replacement increases. An increase in water absorption rate in AAC-B replaced specimen is found to be much higher than that in CLC-B replaced specimen. This may be the reason behind the strength reduction in AAC-B replaced cube specimens. The water absorption rate is increased by 20% in 15% AAC-B replaced specimen, and about 6% in 15% CLC-B replaced mortar specimens.

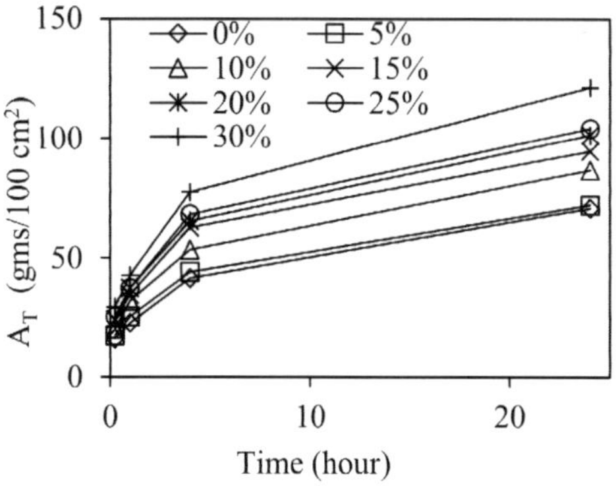

Figure 3. Water absorption rate (A_T) of AAC-A replaced cube specimen.

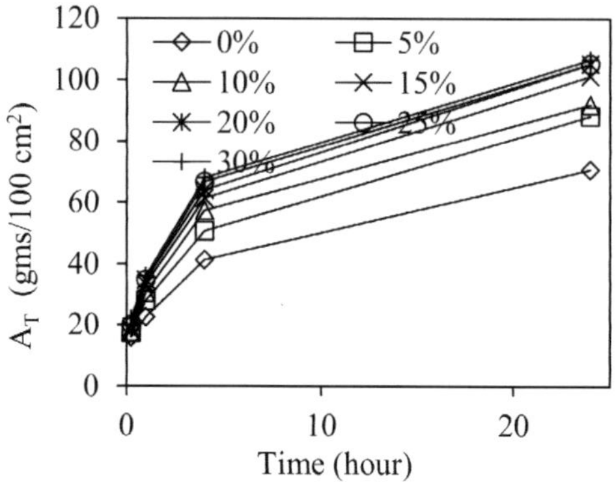

Figure 4. Water absorption rate (A_T) of CLC-A replaced cube specimen.

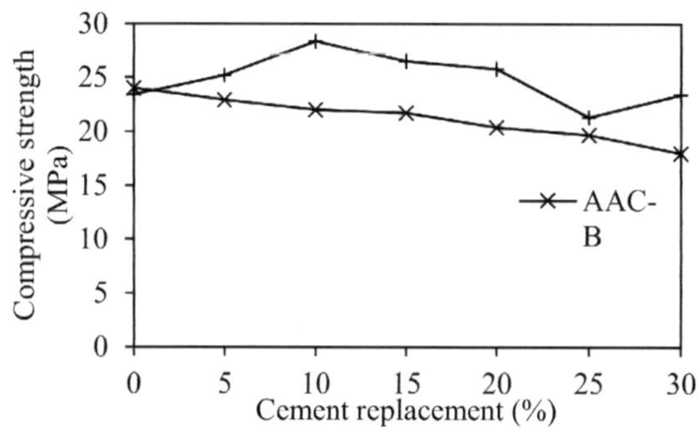

Figure 5. Variation of 28-day compressive strength of cement replaced samples.

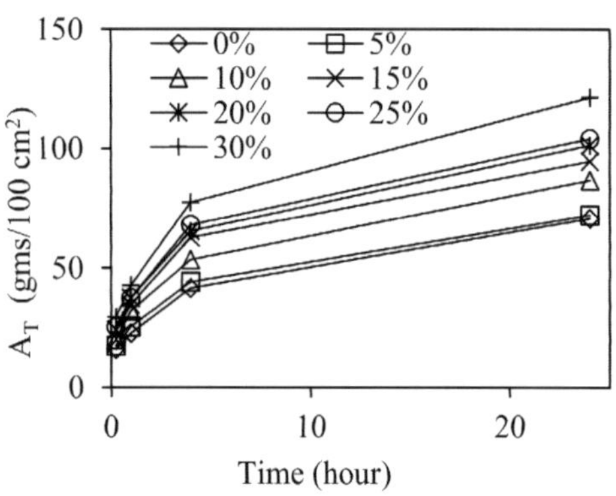

Figure 6. Water absorption rate (A_T) of AAC-B replaced cube specimen.

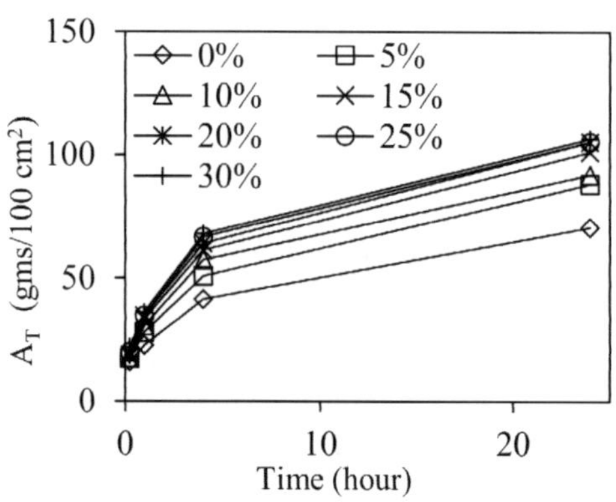

Figure 7. Water absorption rate (A_T) of CLC-B replaced cube specimen.

4. Conclusions

In order to find an alternative application, which has a higher economic and environmental value, the feasibility of demolished AAC and CLC fines as a possible sand and cement replacement in the mortar preparation is investigated. The experimental result shows that the compressive strength decreases, and the water absorption rate increases with the increase of sand replacement. However, from the strength reduction rate of mortar cubes, it can be suggested that AAC-A and CLC-A can be used as recycled material up to 15% replacement of sand in mortar. The microstructural analysis of AAC-B and CLC-B confirms that AAC-B and CLC-B can also be used as cementitious materials. The strength and durability tests on mortar specimens replaced with AAC-B and CLC-B indicate that these materials can effectively replace cement usage in mortar up to 15%. Replacement of cement with CLC-B can even increase the compressive strength to some extent. The experimental investigation conducted in this study makes it clear that recycled foam concrete waste can be used as partial replacement of cement and sand ensuring a sustainable solution to waste management with the conservation of natural resources

5. References

[1] Rashad A M 2013 A preliminary study on the effect of fine aggregate replacement with metakaolin on strength and abrasion resistance of concrete *Construction and Building Materials* **44** 487–95
[2] Gara E J I, Rivera, R X M and Gorokhovsky A 2009 Waste gypsum-blast furnace slag cement in mortars with granulated slag and silica sand as aggregates *Construction and Building Materials* **23(8)** 2851–5
[3] Sua-iam G and Makul N 2013 Use of recycled alumina as fine aggregate replacement in self-compacting concrete *Construction and Building Materials* **47** 701–10
[4] Lim S K, Tan C S, Li B, Ling T C, Hossain M U and Poon C S 2017 Utilizing high volumes quarry wastes in the production of lightweight-foamed concrete *Construction and Building Materials* **151** 441-8
[5] Heidari A, Hashempour M, Javdanian H and Karimian M 2018 Investigation of mechanical properties of mortar with mixed recycled aggregates *Asian Journal of Civil Engineering* **19(5)** 583–93
[6] Sarumathi K, Elavenil S, and Vinoth A S 2019 Use of waste foundry sand with multiscale modellingin concrete *Asian Journal of Civil Engineering* **20(2)** 163–70
[7] Mirza W H and Al-Noury S I 1998 Powdered lightweight aggregate as a cement replacement material *The International Journal of Cement Composites and Lightweight Concrete* **10(3)** 171–3
[8] Payá J, Borrachero MV, Monzó J, Mora E P and Bonilla M 2002 Long-term mechanical strength behaviour in fly ash/Portland cement mortar prepared using processed ashes *Journal of Chemical Technology and Biotechnology* **77** 336–44
[9] Ganesan K, Rajagopal Kand Thangavel K 2007 Evaluation of bagasse ash as supplementary cementitious material *Cement and Concrete Composites* **29(6)** 515–24
[10] Naceri A and Hamina M C 2009 Use of waste brick as a partial replacement of cement in mortar *Waste Management* **29(8)** 2378–84
[11] Uygunoglu T 2011 Reuse of hydrated mortar as partial replacement material in cement *Advances in cement research* **23(3)** 113–22
[12] Bentz D P, Ferraris C F, Jones S Z, Lootens D and Zunino F 2017 Limestone and silica powder replacements for cement: Early-age performance *Cement and Concrete Composites* **78** 43–56
[13] Kanthe V N, Deo S V and Murmu M 2018 Effect of fly ash and rice husk ash on strength and durability of binary and ternary blend cement mortar *Asian Journal of Civil Engineering* **19(8)** 963–70
[14] ASTM 2018 Standard specification for aggregate for masonry mortar C144-18 West Conshohocken PA: ASTM
[15] ASTM 2015 Standard Test Method for Relative Density (Specific Gravity) and Absorption of Fine Aggregate C128-15 West Conshohocken PA: ASTM
[16] ASTM 2016 Standard Test Method for Amount of Water Required for Normal Consistency of Hydraulic Cement Paste C187-16 West Conshohocken PA: ASTM
[17] ASTM 2017 Standard Test Method for Density of Hydraulic Cement C188-17 West Conshohocken PA: ASTM
[18] ASTM 2016 Standard Test Method for Compressive Strength of Hydraulic Cement Mortars C109-16 West Conshohocken PA: ASTM

A Comparative Study on FBD and DBD Methodology with Emphasis on Bridges with Seismic Irregularity

E T Abey[1*], T P Somasundaran[2] and A S Sajith[3]

[1] Assistant Professor, Civil Engineering Department, College of Engineering Aranmula, India

[2] Retd. Professor, Civil Engineering Department, NIT Calicut, India

[3] Associate Professor, Civil Engineering Department, NIT Calicut, India

[*]abeyethomas@gmail.com

Abstract. One of the major developments in seismic design over the past few decades is the increased emphasis for limit states design now generally termed as Performance Based Engineering. Performance Based Seismic Design (PBSD) uses Displacement Based Design (DBD) methodology where-in structures are designed for a target level of displacement rather than Force Based Design (FBD) methodology, where-in force or strength aspect is used. Indian codes still follow FBD methodology compared to other modern codes like CalTrans, which follow DBD methodology. Hence in the present study, a detailed review of the two most common design methodologies i.e. FBD and DBD will be discussed, along with its comparison on bridge models with and without 'seismic regularity'. Bridges with irregular column height are studied to discuss the inherent discrepancy associated with FBD while dealing with 'seismic irregularity'.

1. Introduction

Design for seismic resistance has been undergoing a critical reappraisal in recent years, with the emphasis changing from "strength" to "performance" [6]. Previously they were believed to be synonymous, i.e. an increase in strength will result in an increase in performance in terms of safety and damage reduction. Also, it is now a proven fact that, inelastic characteristics can be seldom described through elastic idealisation. But, still Indian Standards like IRC 112 (2011) follows FBD methodology in comparison to modern codes like CalTrans (2013), which follows DBD methodology. Among the various DBD procedures, the one established by Priestley et al. (2007) i.e., DDBD, is the most preferred among the researchers. Hence, in the present study a detailed review of both FBD and DDBD methodologies are discussed with design illustrations for better understanding the significance of the latter. Also, the study evaluates the performance level achieved by structures designed as per both design methodologies during seismic excitation.

Unequal column height is one of the main irregularities seen in bridges, particularly while negotiating steep valleys but, IRC 112 (2011) does not provide any specific design recommendations for these bridges. Thus, it is inevitable to assess the performance level achieved by the irregular bridges during seismic excitation as irregularities make bridges vulnerable to seismic damage [7].

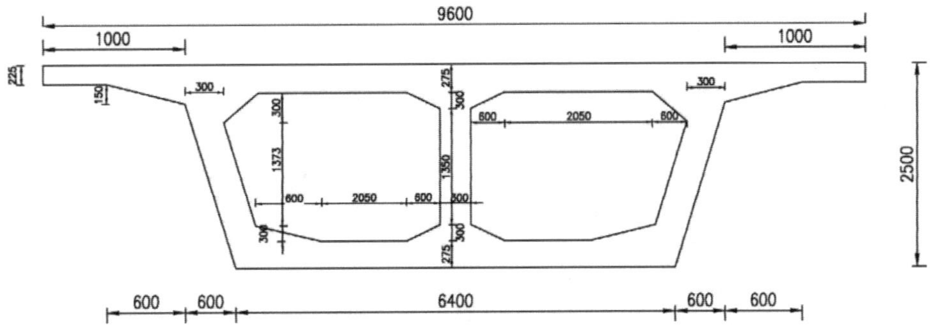

Figure 1. Geometry and dimension detail of the deck (all dimensions in mm).

2. Details of Modelling and Analysis

For bridges, it is generally assumed that, the deck or the superstructure portion behaves elastic and the substructure portion mainly columns, act as ductile members during seismic excitations. Thus, the lateral load resisting members are the columns and the performance of the entire structure depends on their performance. So, in the present study columns are designed for the two design methodologies while deck remains the same, which is designed as per FBD. Three dimensional (3D) finite element model of bridges are done in SAP2000 NL to assess its global seismic performance through pushover analysis. Three span PSC box-girder bridges with individual span length of 30 m are used in all the cases studied. The geometry and dimension detail of the twin celled box-girder deck used in the present study is shown in figure 1. The deck is cast monolithic with the columns at bents, and at abutments it is provided with sliders permitting translational motion along the longitudinal direction similar to that of a semi-integral bridge. Super Imposed Dead Load (SIDL) of 2 kN/m is provided for considering the mass of wearing course and other secondary elements which does not contribute additional stiffness to the structure. The bridge models considered are of single column bents and the columns at both bents are of the same cross-sectional dimension 1.5 m × 3 m. The height of column at Bent-1 (B-1) is varied relatively from 0.5, 0.65, 0.8 and 1 with respect to the column at Bent-2 (B-2), which is of 16 m height, as shown in figure 2. Hence, the column at B-1 will have heights of 8 m, 10.4 m, 12.8 m and 16 m, respectively for Relative Height (RH) ratio of 0.5, 0.65, 0.8 and 1. M40 concrete and Fe 415 steel is used in all the models considered.

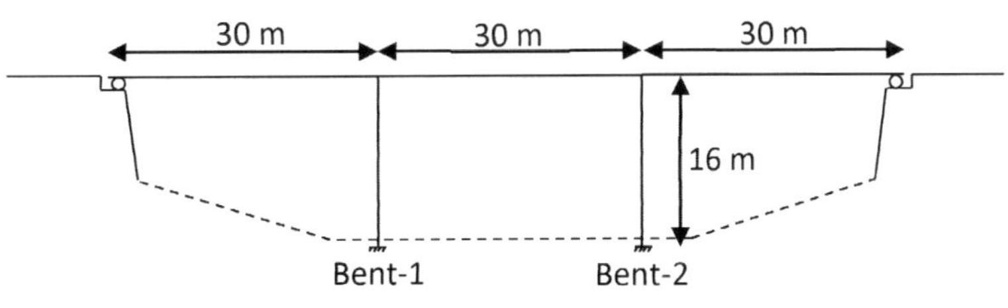

Figure 2. Bridge geometry used in the study of seismic irregularity.

3. Comparison of DDBD and FBD Methodology

Although the current force-based design method is considerably improved compared to the procedures used earlier, there still exist certain fundamental problems with this procedure when applied to reinforced concrete or reinforced masonry structures. After the Loma Prieta earthquake in 1989, extensive research has been conducted to develop improved seismic design criteria for bridges, emphasizing the use of displacements rather than forces as a measure of earthquake demand and damage in the structures [4][8][9]. Extensive work on the application of capacity design principles to assure ductile mechanisms and concentration of damage in specified regions has also been conducted.

Several DBD methodologies have been proposed, among them, the Direct Displacement-Based Design Method (Priestley, 1993) has proven to be effective for performance-based seismic design of bridges, buildings and other types of structures [10][11][12][13][14].

3.1. Discussion on FBD and DDBD procedure

The sequence of operation required for force-based design of a structure is summarised in the flow chart shown in figure 3.

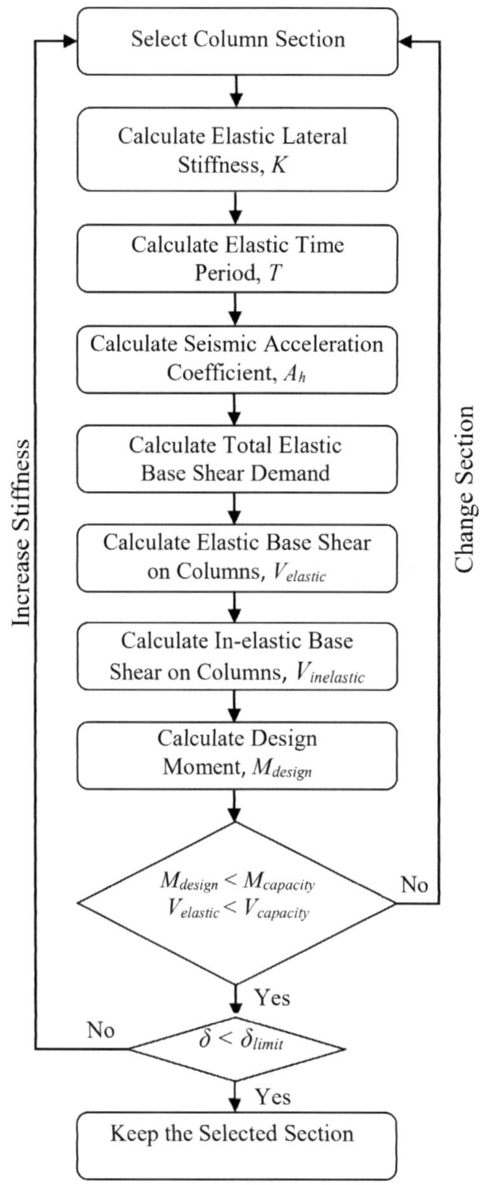

Figure 3. Flow chart of force-based design methodology.

In this method of design, elastic stiffness of the member is used in the estimation of total base shear demand in the structure. The total base shear demand, V_b, is calculated as, $V_b = A_h \times W$; where, A_h is the seismic acceleration coefficient calculated from the elastic acceleration response spectrum and W is the seismic weight of the structure. V_b is then distributed to the individual columns according to their initial stiffness. Hence, the distributed shear will be proportional to $1/H_i^3$, where, H_i is the height of the corresponding column. Considering the ductility, overstrength and redundancy of the column, the code recommends a Reduction factor (R), for the columns. Thus, the in-elastic base shear demand on columns $V_{inelastic}$ is calculated as $V_{elastic}/R$. This base shear value is used in the estimation of M_{design}. To check the safety condition, the moment capacity, $M_{capacity}$, is estimated through Moment-Curvature analysis and $V_{capacity}$ based on empirical relations given in the code. The section is considered safe, if both moment and shear demand values are less than their corresponding capacity values. Here, the elastic base shear estimated is used instead of $V_{inelastic}$. Finally, the deflection (δ) of the structure (column top displacement for bridges) is checked against the maximum permissible limits stipulated by the codes, to ensure whether it is serviceable under the selected lateral loading condition.

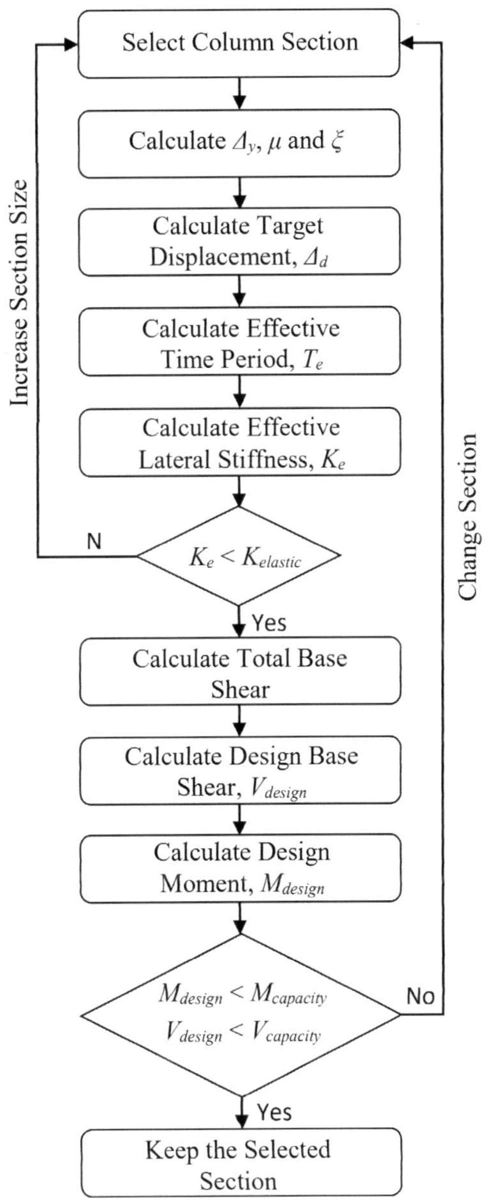

Figure 4. Flow chart of direct displacement-based design methodology.

Displacement-Based techniques gained momentum with the knowledge of the discrepancies associated with the Force-Based Design method. Consequently, a number of new design methods or improvements to the existing methods are evolved. The two main methodologies associated with Deformation-Specification Based Design are Equal Displacement-Based Design, which uses the pre-yield elastic stiffness as in conventional FBD and hence requires a number of iterations; and DDBD method which utilises the secant stiffness corresponding to the maximum displacement and hence need little or no iterations to design a structure corresponding to a specified displacement limit. Thus, DDBD is the most preferred technique among the several displacement-based methods owing to its simplicity in application and wider applicability to several structures.

DDBD is a design procedure for determining the required strength of different structural systems to ensure that, a given performance state defined by flexural strain or drift limits is achieved, under a specified level of seismic intensity. From this design strength, the required moment capacity at intended locations of plastic hinges or shear capacity of seismic isolation devices for seismic isolated structures can be determined.

The procedure for DDBD is portrayed in figure 4, where-in a target displacement and the effective stiffness of the structure have to be determined in order to calculate the base shear demand. The procedure described in Priestley *et al.* (2007) has been adopted in this study. Here, the structure is designed for a target displacement Δ_d ($\Delta_d = \Delta_y \times \mu$); where, Δ_y is the yield displacement which is based on the section size and μ is the displacement ductility. The target displacement value corresponding to the 'damage control' limit state is used in the design. Based on the ductility, equivalent viscous damping of the system, ξ is estimated. Using the displacement response spectra generated for the estimated level of damping and target displacement, effective time period (T_e) of the structure is calculated. The T_e obtained is used for the estimation of the effective stiffness (K_e) of the structure. If the effective stiffness estimated is less than the initial elastic stiffness of the structure ($K_{elastic}$), the procedure is continued else the section size is increased. The total base shear, V_d, is calculated as, $V_d = K_e \times \Delta_d$, which is then distributed to the individual columns in inverse proportion to their height. Finally, the column is checked for moment and shear values as portrayed in Figure 4. In this method, the design moment M_{design} for all the columns remain the same as the V_d is distributed in inverse proportion to the column height.

3.2. Discrepancies Associated with FBD

In FBD, constant member stiffness independent of the member strength is assumed, which implies that, the yield curvature is directly proportional to strength, but in reality, the yield curvature is independent of member strength and the stiffness is directly proportional to strength and hence, the assumption of constant member stiffness is no longer valid. As a consequence of this invalid assumption, successive iteration must be carried out before an adequate elastic characterization of the structure is obtained, which is rarely performed by the designers. Also, the member stiffness may be based on gross-section stiffness or sometimes a reduced stiffness to represent the influence of cracking. And, clearly the stiffness values assumed affects the design seismic forces selected from the elastic response spectrum. The basic assumption associated with FBD is that the elastic characteristic of the structure is the best indicator of its inelastic performance. But it is a well-known fact that, beyond yield the initial elastic stiffness is no longer valid due to; the crushing of concrete, Bauschinger softening of reinforcing steel and damage on crack surfaces. This gives the impression that, structural characteristics that represent the performance at maximum response might be better predictors of performance than initial values of stiffness and damping.

The force reduction factor calculated based on the ductility demand following an equal-displacement approximation involves many uncertainties. It is a known fact that, the equal -displacement approximation is inappropriate to very short period and very long period structures and its validity on medium period structures is also doubtful when the hysteretic character of the inelastic system deviates significantly from elasto-plastic behaviour [15]. Also, there exists a significant difference in the criteria by which yield and ultimate displacement values are chosen in different countries, leading to significantly varied values of force reduction factors used in the codes of those countries (e.g. America, Japan, India etc.). This gives an impression that the absolute value of the member strength is of relatively minor importance. A key tenet of FBD is the usage of unique ductility capacity and hence, unique force reduction factor for different structural systems. In the case of a bridge, the column/pier displacement ductility (μ_d) and hence, force reduction factor is found to be dependent on its height as evident from (1). Thus, using the same response reduction factor irrespective of the ductility of the structural system is absurd [15].

$$\mu_d = 1 + 3\frac{\varphi_P L_P}{\varphi_y H} \tag{1}$$

where, φ_P = Idealised plastic curvature capacity, φ_y = yield curvature, L_p= Plastic Hinge Length.

Following an equal displacement approximation in FBD, it is assumed that, as the strength of the structure increases by reducing the force-reduction factor, its safety increases. This is based on the assumption that, the members will have constant stiffness independent of strength, which is proven to be wrong in the discussion above. By conducting numerical experiments on bridge piers with varied percentage of longitudinal reinforcements Priestley et al. (2007) found that, the strength and stiffness of the member increases with increase in reinforcement ratio while its ductility capacity and displacement capacity reduces. But, as the member strength increases its stiffness also increases which leads to a reduction in its elastic period and hence, a reduced displacement demand from the displacement demand spectra. Thus, it can be summarised based on their study that, the reduction in displacement demand to capacity ratio with increase in member strength is negligibly low, which proves that the argument of safety enhancement with increase in member strength is invalid.

Another serious issue with FBD is its application on structures with dual load path, like that of a bridge, where its superstructure is designed to behave elastically while its columns/piers are designed to have in-elastic response upon seismic action. This is particularly important while conducting transverse seismic analysis of bridges. In this case, assuming force-reduction factor for piers/columns may adversely affect the elastic behaviour assumed in its superstructure design.

3.3. *Design of Bridges based on FBD and DDBD*

The bridge models considered for the comparison of DDBD and FBD methodologies include both regular and irregular bridges. Irregularity in seismic action is brought by providing bridge columns of unequal height along the length of the bridge. Here, the longitudinal and confinement reinforcement of bridge columns are provided in accordance to the following three conditions; (a) based on the actual moment and shear values obtained from the analysis, (b) satisfying the minimum reinforcement requirements as in IRC 112 and CalTrans (2013) for FBD and DDBD, respectively, and (c) modified minimum longitudinal reinforcement ratio for FBD in accordance with AASHTO (2012), while keeping all other structural characteristics the same as discussed in section 2. A brief discussion on the analysis and design of bridge with RH = 0.5 is provided below for both DDBD and FBD for better clarity.

3.3.1. Illustration of FBD. In FBD, the design base shear demand (V_d) is estimated based on the seismic acceleration coefficient (A_h) obtained from the response spectrum (see Figure 5) and the seismic weight (W) of the bridge. The bridge is assumed to be situated in Zone V and having an Importance factor (I) of 1, corresponding to 'Normal' type of bridges as per IRC 6 (2010). The elastic time period (T) of the bridge model considered is calculated based on modal analysis, which is found to be 0.406 s and the initial elastic stiffness can be estimated based on the expression given in (2), which is estimated as 409066.7 kN/m.

$$K_{elastic} = \frac{V_d}{\Delta_{max}} \tag{2}$$

where, $V_d = A_h \times W$ and $\Delta_{max,elastic} = \frac{T^2}{4\pi^2} S_a g$.

The elastic base shear demand ($V_{elastic}$) is found to be 15078.2 kN after using a load factor of 1.5 and the in-elastic base shear demand ($V_{inelsatic}$) after applying a reduction factor of 3 as per IRC 6 (Amendment: August 2014) is estimated to be 5026.1 kN. The total elastic base shear demand ($V_{elastic}$) is distributed to the individual columns based on their initial stiffness i.e. the base shear distribution is inversely proportional to the cube of their corresponding heights as shown in (3). This results in a base shear value of 13402.8 kN in the short column at B-1 and 1675.4 kN in the long column at B-2. The design moment (M_{design}) is calculated from inelastic base shear demand ($V_{inelastic}$) and is found to be 17870.4 kNm and 4468 kNm in the columns at B-1 and B-2, respectively. Here, the design moment will

IOP Conf. Series: Materials Science and Engineering **936** (2020) 012012 doi:10.1088/1757-899X/936/1/012012

be distributed to the individual columns as inverse proportion to the square of their column heights. Table 1 gives the value of similar design parameters for the bridge models with other RH ratios considered in the present study. The variation in design base shear and moment is portrayed in Figure 7 and 8, respectively, and the corresponding reinforcement is given in Table 3, 4 and 5, for all the three cases discussed above.

$$V_i = \frac{1/H_i^3}{\Sigma 1/H_i^3} V_{elastic} \qquad (3)$$

Table 1. Design parameters for FBD.

RH	T (s)	K_i (kN/m)	V_d (kN)	$V_{inelastic}$ (kN)
1	0.827	102551.0	10289.8	3429.9
0.8	0.686	146731.0	12168.8	4056.3
0.65	0.552	223942.2	15260.4	5086.8

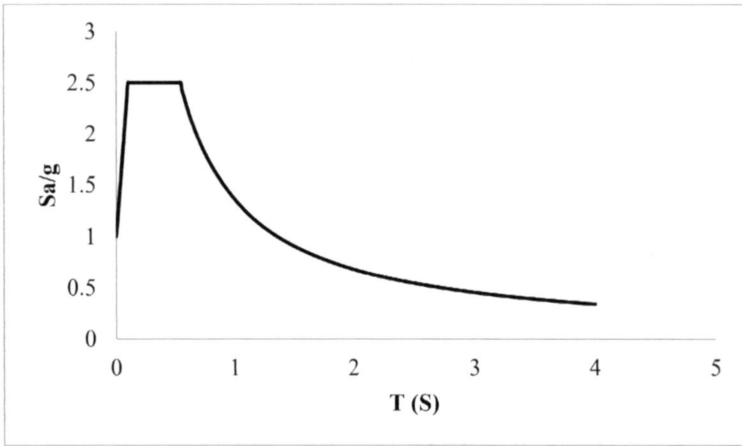

Figure 5. Response spectrum as per IRC 6 (2010).

3.3.2. Illustration of DDBD. In the selected bridge model with RH of 0.5, the column height at B-1 is 8 m while at B-2 it is 16 m. Since, column at B-1 is the shortest, the yield and target displacement will be governed by this column. Confinement reinforcement of 12 mm dia. @ 200 mm c/c connecting each layer of longitudinal bar is provided, so that, for a longitudinal reinforcement percentage of 1, the confinement reinforcement percentage is 0.565, which is within the minimum specified value as per CalTrans (2013), for a Peak Ground Acceleration (PGA) equivalent to Zone V of IRC 6 (2010). According to Montes and Aschleim (2003) the effective yield curvature for a member section with Fe 400 grade steel is as given in (4), as per moment-curvature analysis. And, Priestley et al. (2007) advocates a rotation (θ) value of 0.01 for achieving 'damage control' limit state. At this limiting state non-structural damage is least expected. The 'damage control' limit state can also be defined based on the compressive strain limit of confined concrete as given in (5). Based on (5) and using the expression for yield displacement as in (6), the yield and target displacement corresponding to 'damage control' limit state are found to be 36.57 mm and 120 mm, respectively. Hence, the ductility of columns at B-1 and B-2 are 3.28 and 0.82, respectively. Now, the equivalent viscous damping ratio of individual column can be estimated using (7), where the first part of the equation is for 5% material damping of concrete and the second part is for hysteresis damping, calculated from ductility [15]. The equivalent viscous damping of columns at B-1 and B-2 are found to be 0.148 and 0.05, respectively. The equivalent viscous damping for the whole system can be estimated using (8). In DDBD the base shear is distributed as inverse proportion to the column height and hence, the equivalent viscous damping of the whole system is found to be 11.53%. The spectral reduction factor corresponding to the equivalent system damping can be calculated from (9).

$$\varphi_y = 2.4 \times \frac{\varepsilon_{ys}}{d} \tag{4}$$

$$\varepsilon_{cm} = 0.004 + 1.4\rho_s f_{yh} \frac{\varepsilon_{suh}}{f_{cc}} \tag{5}$$

$$\Delta_y = \frac{\varphi_y H^2}{6} \tag{6}$$

$$\xi_{eq} = 0.05 + 0.444 \frac{(\mu - 1)}{\pi\mu} \tag{7}$$

$$\xi_{sys} = \frac{\Sigma m V_i \xi_i}{\Sigma m V_i} \tag{8}$$

$$R_\xi = \left(\frac{0.07}{0.02 + \xi_{sys}}\right)^{0.5} \tag{9}$$

where, φ_y = yield curvature; ε_{ys} = yield strain of longitudinal reinforcement; d = effective depth of the section; ε_{cm} = limiting compressive strain of confined concrete; ρ_s = volumetric ratio of confinement reinforcement; f_{yh} = yield strength of confinement; ε_{suh}= strain at maximum stress; f_{cc} = compressive strength of confined concrete [18]; Δ_y = yield displacement; H = effective height of column; ξ_{eq} = equivalent viscous damping ratio and μ = displacement ductility; V_i = base shear in each column.

The demand displacement response spectra based on the Indian code, IRC 6 (2010) is shown in figure 6. From the displacement response spectra the effective time period (T_e) of the structure can be found out for the corresponding target displacement calculated, the value of which is found to be 1.38 s. Knowing the effective seismic weight (W_e) of the bridge the effective stiffness (K_e) can be calculated using the expression given in (10). The value of W_e for the bridge model selected is 22338 kN and the corresponding value of K_e is found to be 47203.8 kN/m.

$$K_e = \frac{4\pi^2 W_e}{g T_e^2} \tag{10}$$

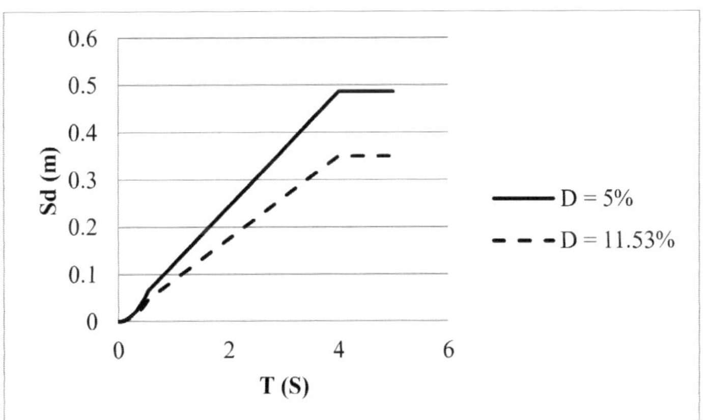

Figure 6. Displacement response spectra.

The total base shear demand V_d, which is the product of effective stiffness and target displacement is estimated to be 5664.5 kN. This base shear is distributed to each individual column in inverse proportion to their height. Hence, the base shear in each individual column (V_i) can be estimated using the expression given in (11) and is found to be 3776.3 kN and 1888.2 kN, respectively at B-1 and B-2. This method of base shear distribution in DDBD leads to equal bending moment, and hence equal longitudinal reinforcement in columns at both bents. Table 2 gives the value of similar design parameters for the bridge models with other RH ratios considered in the present study. The variation in design base shear and moment is portrayed in Figure 7 and 8, respectively, and the corresponding reinforcement is given in Table 3, 4 and 5 for all the three cases discussed.

$$V_i = \frac{1/H_i}{\sum 1/H_i} V_d \tag{11}$$

The base shear at B-1 is found to be increasing in Figure 7 as the irregularity in bridges increases, while at B-2 it gets decreased in the case of FBD. But, for the bridges designed as per DDBD methodology, the base shear remains comparatively similar and well below the corresponding value at RH = 1. This clearly indicates the relevance of DDBD over FBD methodology. As discussed earlier, the variation in base shear value points out that, in FBD short columns determine the failure criteria, whereas in DDBD a more regularity in seismic performance can be achieved.

Table 2. Design parameters for DDBD.

RH	Δ_y (mm)	Δ_d (mm)	μ_1	μ_2	ξ_1	ξ_2	ξ_{sys}	R_ξ	T_e (s)	W_e (kN)	K_e (kN/m)	V_d (kN)
1	146.30	240.00	1.64	1.64	0.105	0.105	0.105	0.748	1.72	23238	31610.6	7586.5
0.8	93.63	192.00	2.05	1.31	0.122	0.084	0.105	0.748	1.72	22878	31120.9	5975.2
0.65	61.81	156.00	2.52	1.07	0.135	0.059	0.105	0.748	1.72	22608	30753.6	4797.6

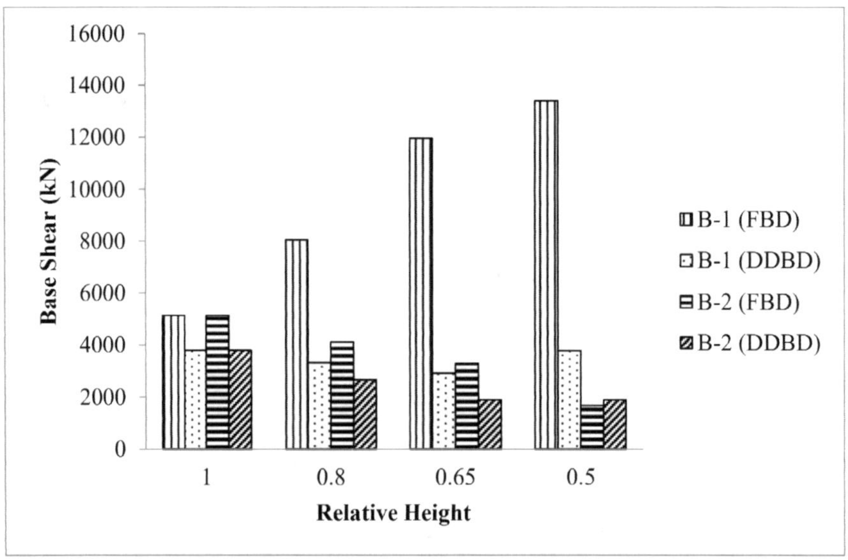

Figure 7. Variation of base shear at bents with design methodology.

In the case of moment variation at bents, a similar trend as that of the base shear value is obtained in case of FBD methodology. This means a higher amount of steel reinforcement at shorter columns and a comparatively lower quantity at longer columns. It is seen that, the moment value at B-1 for an RH of 0.5 is lesser than that of RH of 0.65 in case of bridges designed as per FBD. This is due to the fact that, the spectral acceleration value corresponding to the elastic time period obtained from the elastic response spectrum for the bridges with both RH values remain same i.e. Sa/g = 2.5 (maximum value). But, an increased column height at B-1 for bridge with RH = 0.65 increases its seismic weight, and hence the total base shear for bridge with RH = 0.65 remains higher than that of bridge with RH = 0.5. This results in a higher moment at B-1 for bridge with RH = 0.65, as design moment is estimated as the product of distributed inelastic base shear and half the column height.

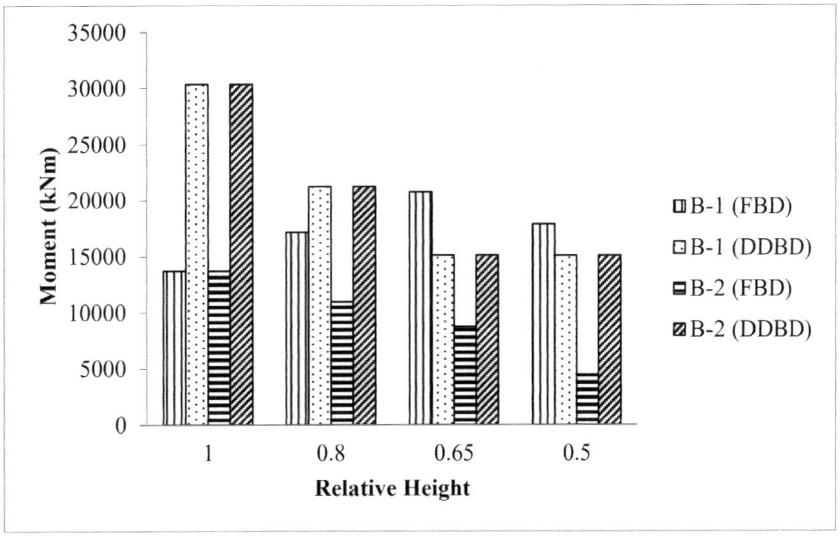

Figure 8. Variation of moment at bents with design methodology.

As per IRC 112, the minimum longitudinal reinforcement required is 0.2% A_c or 0.1 N_{ED}/f_{yd}, where A_c is the gross cross-sectional area of concrete, N_{ED} is the design axial compression force and f_{yd} is the design yield strength of the reinforcement; and the minimum transverse reinforcement volumetric ratio for M40 concrete and Fe 415 steel is 0.593%, with a maximum spacing of 200 mm. As per CalTrans (2013), the minimum longitudinal reinforcement ratio is 1% of gross sectional area and the maximum allowable spacing of confinement reinforcement is 200 mm or six times the nominal diameter of longitudinal rebar, whichever is less.

Table 3, 4 and 5 shows the reinforcement required as per the three cases discussed in section 3.3. The minimum longitudinal reinforcement percentage specified for FBD as per IRC 112 is very small compared to CalTrans. For a column section size of 1.5 m × 3 m, the minimum reinforcement required is only 9000 mm^2, which means nearly 30 numbers of 20 mm diameter bars in total. This is proven to be insufficient for achieving 'life safety' level of performance at MCE level of seismic hazard (refer table 7). Also, the performance level of columns at both bents varies considerably even with slight reduction in RH ratio. Hence, based on the present study, the minimum percentage of longitudinal reinforcement in IRC 112 following FBD methodology is recommended to be modified to 1% as in AASHTO (2012), ACI 318 (2008) and CalTrans (2013). The studies conducted by Halvorsen (1987) further proved that, column with longitudinal rebar percentage less than 1% has not exhibited sufficient ductility. The minimum confinement reinforcement specified by IRC 112 is found to be higher than CalTrans, which is better from ductility point of view.

Table 3. Reinforcement at columns of bridges designed with actual moment and shear values.

RH	Design Method	Bent-1		Bent-2	
		Longitudinal Reinforcement	Transverse Reinforcement	Longitudinal Reinforcement	Transverse Reinforcement
1	FBD	52-25 mm φ	12 mm φ @ 172 mm c/c	52-25 mm φ	12 mm φ @ 172 mm c/c
	DDBD	108-25 mm φ	12 mm φ @ 200 mm c/c	108-25 mm φ	12 mm φ @ 200 mm c/c
0.8	FBD	64-25 mm φ	12 mm φ @ 110 mm c/c	42-25 mm φ	12 mm φ @ 196 mm c/c
	DDBD	78-25 mm φ	12 mm φ @ 200 mm c/c	78-25 mm φ	12 mm φ @ 200 mm c/c
0.65	FBD	76-25 mm φ	12 mm φ @ 78 mm c/c	34-25 mm φ	12 mm φ @ 231 mm c/c
	DDBD	58-25 mm φ	12 mm φ @ 200 mm c/c	58-25 mm φ	12 mm φ @ 200 mm c/c
0.5	FBD	67-25 mm φ	12 mm φ @ 59 mm c/c	28-20 mm φ	12 mm φ @ 673 mm c/c
	DDBD	58-25 mm φ	12 mm φ @ 200 mm c/c	58-25 mm φ	12 mm φ @ 200 mm c/c

Note: φ denotes diameter of rebar.

Table 4. Reinforcement at columns of bridges designed as per DDBD and FBD.

RH	Design Method	Bent-1		Bent-2	
		Longitudinal Reinforcement	Transverse Reinforcement	Longitudinal Reinforcement	Transverse Reinforcement
1	FBD	52-25 mm φ	12 mm φ @ 118 mm c/c#	52-25 mm φ	12 mm φ @ 118 mm c/c#
	DDBD	108-25 mm φ	12 mm φ @ 200 mm c/c#	108-25 mm φ	12 mm φ @ 200 mm c/c#
0.8	FBD	64-25 mm φ	12 mm φ @ 110 mm c/c	42-25 mm φ	12 mm φ @ 95 mm c/c#
	DDBD	92-25 mm φ#	12 mm φ @ 200 mm c/c#	92-25 mm φ#	12 mm φ @ 200 mm c/c#
0.65	FBD	76-25 mm φ	12 mm φ @ 78 mm c/c	34-25 mm φ	12 mm φ @ 77 mm c/c#
	DDBD	92-25 mm φ#	12 mm φ @ 200 mm c/c#	92-25 mm φ#	12 mm φ @ 200 mm c/c#
0.5	FBD	67-25 mm φ	12 mm φ @ 59 mm c/c	30-20 mm φ#	12 mm φ @ 68 mm c/c#
	DDBD	92-25 mm φ#	12 mm φ @ 200 mm c/c#	92-25 mm φ#	12 mm φ @ 200 mm c/c#

Note: # denotes minimum value required as per IRC 112 for FBD and CalTrans (2013) for DDBD.

Table 5. Reinforcement at columns of bridges designed with revised minimum P_t for FBD.

RH	Design Method	Bent-1		Bent-2	
		Longitudinal Reinforcement	Transverse Reinforcement	Longitudinal Reinforcement	Transverse Reinforcement
1	FBD	92-25 mm φ#	12 mm φ @ 100 mm c/c#	92-25 mm φ#	12 mm φ @ 100 mm c/c#
	DDBD	108-25 mm φ	12 mm φ @ 200 mm c/c#	108-25 mm φ	12 mm φ @ 200 mm c/c#
0.8	FBD	92-25 mm φ#	12 mm φ @ 100 mm c/c#	92-25 mm φ#	12 mm φ @ 100 mm c/c#
	DDBD	92-25 mm φ#	12 mm φ @ 200 mm c/c#	92-25 mm φ#	12 mm φ @ 200 mm c/c#
0.65	FBD	92-25 mm φ#	12 mm φ @ 100 mm c/c	92-25 mm φ#	12 mm φ @ 100 mm c/c#
	DDBD	92-25 mm φ#	12 mm φ @ 200 mm c/c#	92-25 mm φ#	12 mm φ @ 200 mm c/c#
0.5	FBD	92-25 mm φ#	12 mm φ @ 88 mm c/c	92-25 mm φ#	12 mm φ @ 100 mm c/c#
	DDBD	92-25 mm φ#	12 mm φ @ 200 mm c/c#	92-25 mm φ#	12 mm φ @ 200 mm c/c#

Table 6, 7 and 8 describes the performance level attained by the bridges at DBE and MCE level of seismic hazard for the three cases studied. It is clear that, a minimum P_t of 1% makes the FBD bridges having a similar performance as that of the DDBD bridges in the present study, and hence, achieving a 'damage control' level of performance. Figure 9 depicts the hinge formation at the bridge columns designed as per DDBD with RH values of 1 and 0.5.

Table 6. Performance levels of bridges designed with actual moment and shear values.

RH	Design Method	Performance Levels			
		Bent-1		Bent-2	
		DBE	MCE	DBE	MCE
1	FBD	O	C	O	C
	DDBD	O	IO	O	IO
0.8	FBD	O	CP	O	LS
	DDBD	O	IO	O	IO
0.65	FBD	O	CP	O	O
	DDBD	O	LS	O	O
0.5	FBD	O	CP	O	O
	DDBD	O	LS	O	O

Table 7. Performance levels of bridges designed as per DDBD and FBD.

RH	Design Method	Performance Levels			
		Bent-1		Bent-2	
		DBE	MCE	DBE	MCE
1	FBD	O	C	O	C
	DDBD	O	IO	O	IO
0.8	FBD	O	CP	O	LS
	DDBD	O	IO	O	IO
0.65	FBD	O	CP	O	O
	DDBD	O	LS	O	O
0.5	FBD	O	CP	O	O
	DDBD	O	LS	O	O

Table 8. Performance levels of bridges with revised minimum P_t for FBD.

RH	Design Method	Performance Levels			
		Bent-1		Bent-2	
		DBE	MCE	DBE	MCE
1	FBD	O	IO	O	IO
	DDBD	O	IO	O	IO
0.8	FBD	O	IO	O	IO
	DDBD	O	IO	O	IO
0.65	FBD	O	IO	O	O
	DDBD	O	LS	O	O
0.5	FBD	O	LS	O	O
	DDBD	O	LS	O	O

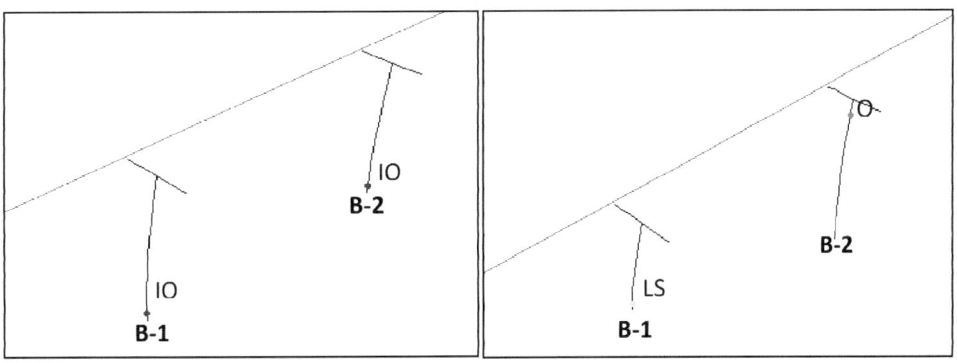

(a) (b)

Figure 9. Hinge formation in DDBD bridge columns-POA: (a) RH = 1, (b) RH = 0.5.

Note: O, IO, LS, CP and C represents respectively, the Operational, Immediate Occupancy, Life Safety, Collapse Prevention and Collapse level of performance.

4. Conclusions

The inherent discrepancies associated with the fundamental assumptions of FBD methodology makes it least preferred among the research community especially while dealing with seismic loads. The modern design codes are moving towards displacement-based design or rather DDBD methodology for achieving the performance objectives targeted. The present study gives a detailed review of the two most commonly adopted design methodologies (FBD and DDBD), along with their comparison based on the performance evaluated on bridge models with and without seismic regularity. Pushover analysis using SAP2000 NL is adopted for the seismic performance evaluation of the bridge models studied. Seismic irregularity is achieved by providing columns of unequal height along the length of the bridge with relative height ratio varying from 1, 0.8, 0.65 and 0.5.

In the present study it is found that, the column irregularity in bridges will impart uneven seismic response in bridges. The shorter columns dictate the seismic performance of such bridges, particularly while adopting FBD methodology. The comparative study on FBD and DDBD methodology reveals the inherent discrepancies associated with FBD compared to DDBD, particularly for bridges with irregular column height.

5. References

[1] IRC 112: 2011 *Code of Practice for Concrete Road Bridges, Indian Roads Congress* (New Delhi)
[2] IRC 21: 2000 *Standard specifications and code of practice for road bridges, Indian Roads Congress* (New Delhi)
[3] American Association of State and Highway Transportation Officials LRFD Bridge Design Specifications 2012 *AASHTO LRFD* (Washington D.C.)
[4] California Department of Transportation 2013 Seismic Design Criteria Version 1.7. (Engineering Service Center, Earthquake Engineering Branch, Sacramento, CA)
[5] IS 13920 (1993) Ed. 1.2 2002 *Ductile detailing of reinforced Concrete structures subjected to Seismic forces - code of practice* (New Delhi)
[6] Priestley M J N 2000 Performance based seismic design *Proc. of the 12th World Conference on Earthquake Engineering* vol 2831 (New Zealand) pp 1-22
[7] Akbari R 2010 Cyclic response of RC continuous span bridges with irregular configuration in longitudinal direction *Structure and Infrastructure Engineering* DOI: 10.1080/15732479.2010.510528 111.
[8] Priestley M J N 1993 Assessment and Design of Joints for Single-Level Bridges with Circular Columns *Department of Applied Mechanics and Engineering Sciences, University of California* San Diego, CA.
[9] ATC 40. 1996 Seismic evaluation and retrofit of existing concrete buildings *Applied Technology Council* California, U.S.A.
[10] Kowalsky M J et al. 1995 Displacement based design of RC bridge columns in seismic regions *Earthquake Engineering and Structural Dynamics* **24(12)** 1623–43
[11] Calvi G M and Kingsley G R 1995 Displacement based seismic design of multi degree of freedom bridge structures *Earthquake Engineering and Structural Dynamics* **24(9)** 1247–66
[12] Kowalsky M J 2002 A Displacement based approach for the seismic design of continuous concrete bridges *Earthquake Engineering and Structural Dynamics* **31(3)** 719–47
[13] Ortiz J 2006 Displacement-based design of continuous concrete bridges under transverse seismic excitation *Advanced Studies in Reduction of Seismic Risk* (ROSE School)
[14] Suarez V and Kowalsky M 2006 Implementation of displacement based design for highway bridges *Fifth National Seismic Conference on Bridges & Highways* (San Francisco, Canada)
[15] Priestley M J N et al. 2007 Displacement-based seismic design of structures *IUSS Press* (Pavia, Italy) ISBN:978-88-6198-000-6
[16] IRC 6: 2010 *Standard Specifications and Code of Practice for Road Bridges Section: II, Loads and Stresses, Indian Roads Congress* (New Delhi)
[17] Montes E H and Aschleim M 2003 Estimates of yield curvature for the design of reinforced concrete columns *Magazine of Concrete Research* **55(4)** 373–83
[18] Priestley M J N et al. 1996 Seismic Design and Retrofit of Bridges *John Wiley & Sons, Inc.* (New York)

IOP Conf. Series: Materials Science and Engineering **936** (2020) 012013 doi:10.1088/1757-899X/936/1/012013

Analysis of Slab Culvert Bridges using Conventional Method and Grillage Method

Amit Kumar Pandey[1*] and Praveen Nagarajan[2]

[1] PG Student, Department of Civil Engineering, NITC, Kerala, India

[2] Associate Professor, Department of Civil Engineering, NITC, Kerala, India

*amitpandey1715@gmail.com

Abstract. IRC 112:2011, the Indian standard pertaining to concrete bridge design is based on the limit state design principle. It suggests the use of effective width method for the analysis of the structure. For many years, the analysis part of concrete bridge structures design has been based on two-dimensional (2D) analysis and the results obtained from longitudinal analysis were supposed to be valid over the whole bridge width. In this paper, the grillage method has been used to model and analyse the super-structure to get the three-dimensional effect, and a comparison of results between effective width method and grillage method is presented.

1. Introduction

The superstructure in a bridge is the portion of the structure that directly receives the live load. While the part of the bridge below superstructure as pier, abutment, and other support structures are known as the substructure. If the length of the superstructure is less than 6m, it is known as culvert. The design of a civil engineering structure includes two major steps: analysis part and detailing part. Analysis part of the structure includes the evaluation of forces at different locations for possible number of load cases and in detailing part we have to find the required area of steel and cross-section of the structure considering different design philosophy e.g. ultimate load condition and serviceability load condition. Slab culvert is designed as one-way slab for unit width. In IS 456 two-way slab design, analysis can be done by moment coefficient method. In the case of bridge two-way slab analysis can be done using Pigeaud's curve and shell element method for finite element method.

The grillage analogy includes the description of a three-dimensional (3-D) concrete structure by a two-dimensional collection of discrete one-dimensional interconnected beams in bending and as well as torsion. Chithra et al (2019) [1] did a comparison among three conventional method i.e. sandwich model (IRC 112-2011), Wood Armer method (EN1992-1-1:2004) and the moment coefficient method as per IS 456-2000. This paper gives a comparison of reinforcement using effective width method and grillage analogy.

2. Design Method

2.1. *Effective width method*

This method is used to find the effective width of slab for resisting the bending moment due to concentrated and patch load when the slab is spanning in the longitudinal direction. The calculation of shears and moments for these loads is a statically indeterminate problem and to find an accurate solution in real practice is very time consuming and difficult because the deflection of slab is bi-directional. So, under a vertical load, a slab will have curvature in both directions i.e. along the span and perpendicular to the span. A semi-empirical method can be used when a slab is supported on two opposite edges only. Thus, bending moments developed in the slab will be in span direction and normal to it.

Therefore, the effective width method considers the distribution of load on a width, i.e. width of load and both side of this width. So, it can be assumed that the load is carried by a certain width of the slab, known as the effective width. If this effective width is known, the forces developed along the span can be analysed using the line beam analysis i.e. effective unit width method.

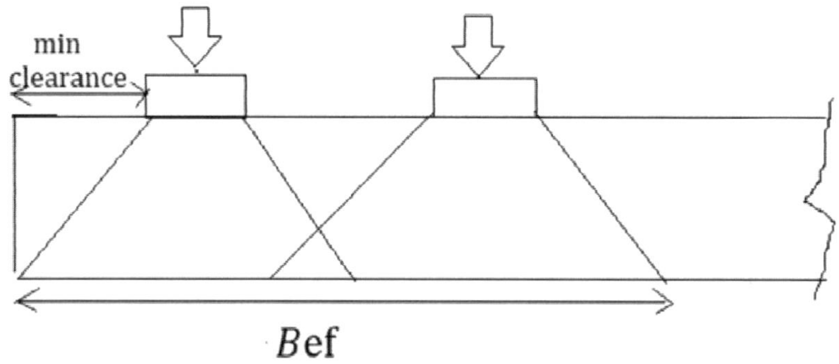

Figure 1. Effective width of load.

a) For a single concentrated load, the effective width, B_{ef} = K x (1- x/L) + b_w
b) Solid cantilever slab, b_e = 1.2 x + b_w
c) Dispersion of loads along the span, l_e= x +2(D + H)

Note: Notations according to IRC112: 2011.

2.2. M. Pigeaud method

When bridge behaves like a two-way slab i.e. in case of Tee beam and slab bridges, the slab forces have to be analysed by moments coefficients that will depend upon the aspect ratio. In the case of bridge, most common method to find the moment coefficients is Pigeaud's curve. This method is relevant to rectangular slabs supported freely on all the four sides and slab should be symmetrically loaded.

2.3. Grillage analysis

A finite number of grillage of interconnected beams can be modelled to give a satisfactory dispersion of loads and deformations inside the concrete structure. For any structure to be modelled by an equivalent grillage, the elastic rigidities of the given grillage members must be equal to the elastic rigidities of the structure so that the grillage member has to deform or rotate in the same amount as that of the actual structure under similar loading conditions. Another main design principle that must be taken into consideration while modelling is that, every grillage member position should coincide with the centre line of the actual member. Once all the members are modelled into a grillage structure, all the boundary conditions i.e. support condition should be provided to best suited in reality. These boundary conditions are to be applied at the centre line. Then, the loads should be idealized as line or point loads that can be applied to the grillage members and then the analysis is performed to get the final output of force.

2.3.1. Grillage of geometry. Grid of longitudinal and transverse beam is known as Grillage. Grillage analogy is most widely used to analyse slab and beam-slab type bridge decks. The spacing of transverse beam should be about 1.5 times the spacing of the main longitudinal members but may extend up to 2.1. Transverse beams are provided at the diaphragm and intermediate cross girder positions. It is necessary to be an odd number of transverse beams to get an intermediate member at mid-span. When the deck is overhanging from the edge of the outer main girder, the grillage will be extended to the parapet beam. In the case of skew-slabs, the transverse members should be modelled perpendicular to the main members to get the correct value of forces and deflections.

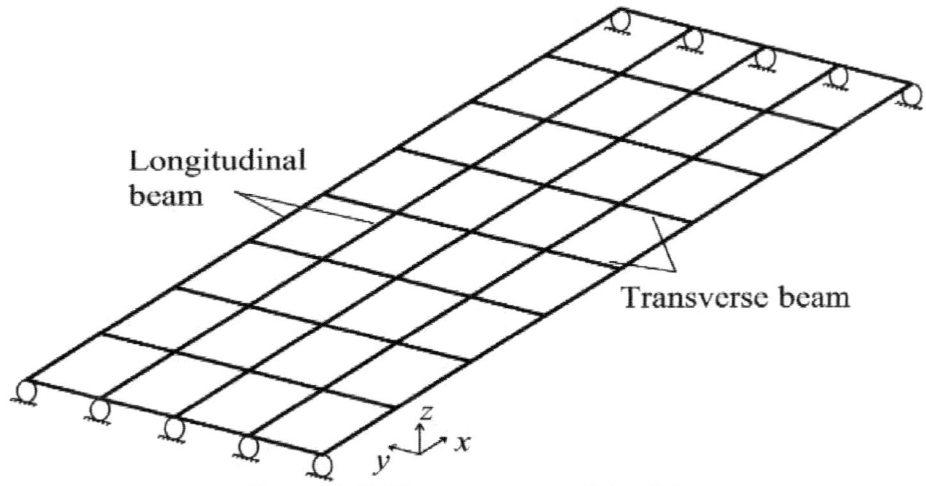

Figure 2. Grillage geometry of the slab.

2.3.2. Grillage for loading. Self-weight of the bridge can be achieved by giving a density to the cross-sectional area of the main longitudinal members. While modelling for self-weight, designers should be careful to avoid double accounting for the self-weight of the deck slab at the junction of beams. Superimposed dead load (wearing coat, footpath load and parapet loads) are idealised as uniformly distributed loads along the length of the longitudinal grillage members. For live loads, the hypothetical vehicle specified in IRC 6: 2017 can be used.

3. Analysis of solid slab bridge:

The following data were used for the analysis of slab bridge:

Clear carriage width = 7.5m
Width of footpath = 0.6m on both side
Width of bearing = 0.4m
Live load = IRC CLASS-AA Track and CLASS-A
Grade of steel Fe415
Grade of concrete M25
Exposure condition: Moderate (Kozhikode)

3.1. Calculation of forces

A bridge is designed for the different partial factor of safety for different load conditions and for different design purposes e.g. ultimate design and design for serviceability. Table 1 shows the maximum bending moment without any factor of safety using effective width method and using grillage method (STAAD software). Self-weight of the member, super imposed dead load and live load is applied in this case.

Table 1. Maximum bending moment due live load calculated without any factor of safety.

Length of bridge (m)	Bending moment (KN-m)			
	Effective width method		Grillage STAAD result	
	CLASS-AA track	CLASS-A wheel	CLASS-AA track	CLASS-A wheel
5.00	100.00	78.54	90.67	63.21
6.00	122.98	96.06	114.24	77.16
7.00	142.36	120.55	134.66	90.00
8.00	159.97	141.44	153.49	103.60
9.00	175.27	162.66	170.23	117.64

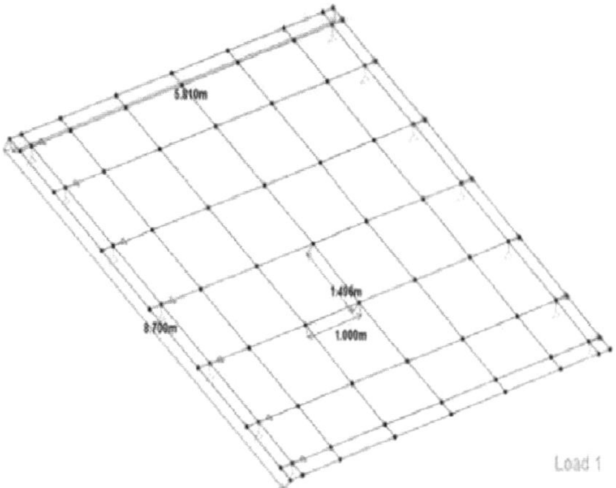

Figure 3. Grillage of 5m span bridge.

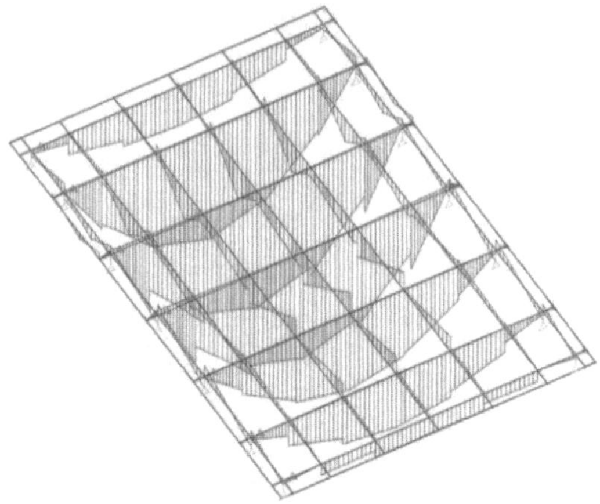

Figure 4. BMD for Class -AA track.

Table 2. Different parameter to calculate bending moment by effective width method.

Span	Load	l_{eff}(m)	b_1(m)	a	B/L	α	b_{eff}(m)	load intensity (kN/m^2)
5m	Class-AA track	5.37	0.96	2.69	1.62	2.88	6.69	28.42
	Class-A wheel	5.37	0.61	1.96	1.62	2.88	8.44	31.30
6m	Class-AA track	6.45	0.96	3.23	1.35	2.74	6.97	25.50
	Class-A wheel	6.45	0.61	2.63	1.35	2.74	8.74	26.08
7m	Class-AA track	7.55	0.96	3.78	1.15	2.62	7.23	22.76
	Class-A wheel	7.55	0.61	3.18	1.15	2.62	9.02	21.57
8m	Class-AA track	8.62	0.96	4.31	1.01	2.48	7.43	20.78
	Class-A wheel	8.62	0.61	3.71	1.01	2.48	9.23	18.99
9m	Class-AA track	9.70	0.96	4.85	0.90	2.35	7.61	18.96
	Class-A wheel	9.70	0.61	4.25	0.90	2.35	9.41	16.74

3.2. Design for ultimate condition

To design the member for ultimate condition the partial factor of safety for different load can be taken from the Table B3 from IRC:6 loads.

Table 3. Ultimate design moment (DL+SIDL+LL).

Length of bridge (m)	Area of steel (mm^2)			
	Effective width method		Grillage STAAD result	
	CLASS-AA track	CLASS-A wheel	CLASS-AA track	CLASS-A wheel
5.00	208.9	176.7	194.9	153.7
6.00	283.4	243.1	270.3	214.7
7.00	373.2	340.4	361.6	294.6
8.00	470.0	442.2	460.3	385.4
9.00	586.0	567.0	578.4	499.5

Table 4. Required tension reinforcement(mm^2).

Length of bridge (m)	Area of steel (mm^2)			
	Effective width method		Grillage STAAD result	
	CLASS-AA track	CLASS-A wheel	CLASS-AA track	CLASS-A wheel
5.00	1691	1412	1570	1218
6.00	2375	2000	2252	1744
7.00	3275	2936	3152	2482
8.00	4379	4045	4260	3406
9.00	6004	5705	5883	4753

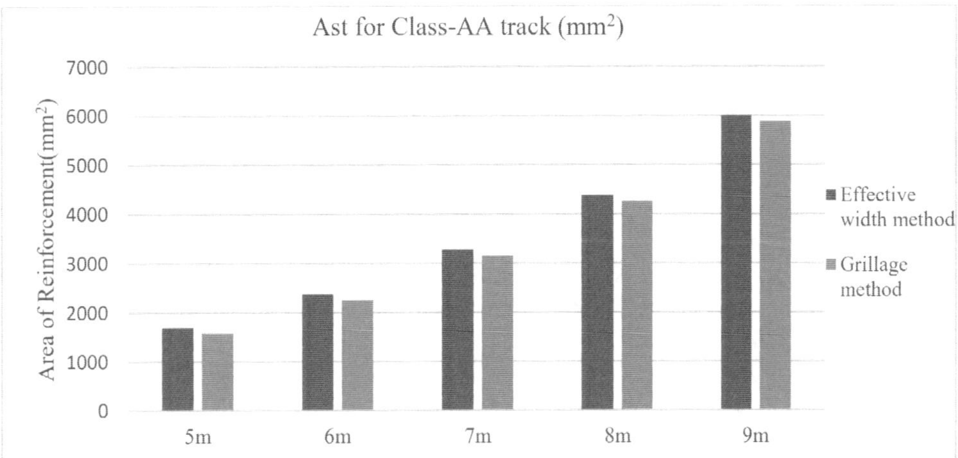

Figure 5. Comparison of required area of reinforcement for Class- AA track loading.

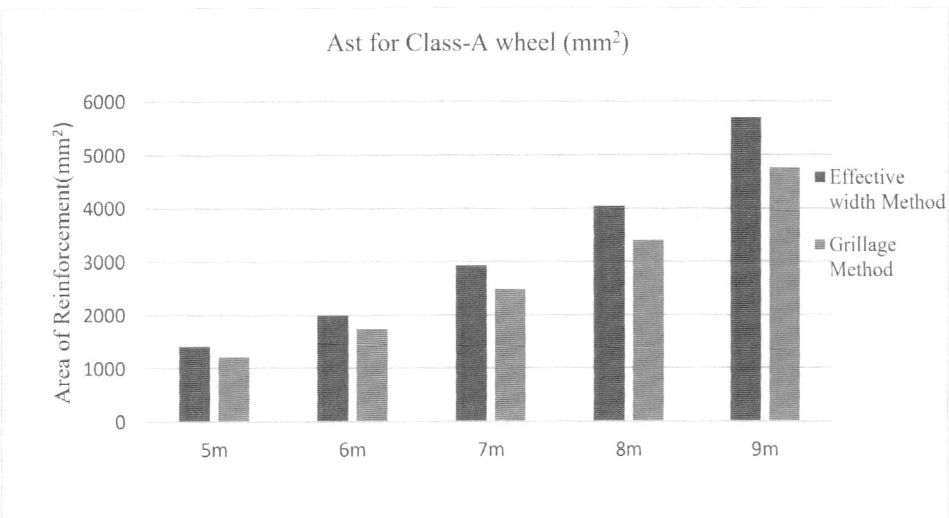

Figure 6. Comparison of required area of reinforcement for Class- A wheel loading.

The comparison of required main reinforcement for Class-AA track and Class -A wheel using effective width method and grillage method (STAAD) can be find from Figure 5 and Figure 6 and it can be observed that effective width method gives higher value of area of steel for the same type of load.

Table 5. Total designed torsional moment and its comparison with total design BM.

Length of bridge (m)	Design Torsion (KN-m)		% Torsional of total design BM	
	CLASS-AA track	CLASS-A wheel	CLASS-AA track	CLASS-A wheel
5	51	22.5	26.2%	14.6%
6	64.5	26.55	23.9%	12.4%
7	75	29.28	20.7%	9.9%
8	82.5	31.8	17.9%	8.3%
9	87	35.1	15.0%	7.0%

It can be observed from Table 5 that as the span of bridge increases the torsional moment increases but the % of torsional moment with respect to design bending moment decreases.

4. Conclusion

Effective width method, which is recommended in IRC:112 to analyse the solid slab type bridges, leads to safer design as it gives more bending moments than STAAD using grillage method which is most widely adopted by bridge designers. The variation of bending moment for IRC class -AA track loads decreases from 9.33% for 5 m of bridge span to 2.8% for 9 m of bridge span but for IRC class -A wheel loads the variation increases from 19.5 % for 5 m to 27.6 % for 9 m of span. IRC:112-2011 recommends the used of 20% of main steel reinforcement as distribution reinforcement to counter torsional moment. However, from grillage analysis it is seen that 20% reinforcement will not be sufficient in some cases. So, it is needed to check the torsional moment while using the effective width method.

It can also be observed that the effective width method leads the structure to be safe but sometimes it is uneconomical. So some factor is necessary to limit this variation.

5. References

[1] Chithra J, Nagarajan P, Sajith A S and Roshan R A 2019 Critical review of design procedures for reinforced concrete slabs as per IRC 112-2011 *Materials Science Forum* **969** 349–54
[2] Jaeger L G and Bakht B 1982 The grillage analogy in bridge analysis *Canadian Journal of Civil Engineering* **9(2)** 224–35
[3] IRC 112:2011 *Code of Practice for Concrete Road Bridges*
[4] IRC 6:2017 *Standard Specifications and Code of Practice for Road Bridges, Section II –Loads and Stresses (Fourth Revision)*
[5] IRC 5: 2015 *Standard Specifications and Code of Practice for Road Bridges, Section I (General Features of Design) (Seventh Revision*).
[6] Hambly E C 1991 *Bridge Deck Behaviour* (CRC Press)
[7] Surana C S and Agrawal R 1998 *Grillage Analogy in Bridge Deck Analysis* (Alpha Science Int'l Ltd.)
[8] Victor D J 2017 *Essentials of Bridge Engineering* (Oxford and IBH Publishing)

Seismic Fragility Analysis of Bridge Pier

Ankit Jain[1*], Robin Davis[2] and C G Nanda Kumar[3]

[1] M Tech, National Institute of Technology, Calicut, India

[2] Assistant Professor, National Institute of Technology, Calicut, India

[3] Professor, National Institute of Technology, Calicut, India

*ankitjain19111993@gmail.com

Abstract: Bridges are classified as lifeline structures as they need to be functional in an earthquake event. The performance-based analysis of the existing bridges is important for the stakeholders. Information on seismic performance in terms of fragility of existing bridges in the country can provide valuable information to the decision-makers. This study focuses on the development of seismic fragility curve, which is the probability of exceedance of a defined damage parameter of the bridge pier under a given ground motion intensity and development of damage index function of bridge piers. Existing bridge piers are considered and peak ground acceleration is taken as ground motion intensity measure and drift at the pier top level is considered as the damage parameter.

1. Introduction

Due to an increase in the frequency of seismic events, there is a need for the evaluation of the existing lifeline structures for their seismic response. Bridges are one of the lifeline structures that are most likely to be affected during any seismic event. The bridge consists of two parts i.e. superstructure and substructure. Bridge pier is a part of the substructure of any bridge which is responsible for the transfer of load from the superstructure to the foundation. Past earthquakes (1985 Mexico; 2001 Bhuj; 2005 Islamabad; 2006 Sikkim; 2011 South Island; 2015 Nepal) have revealed that bridge pier is the most susceptible part of the bridge, failure of which can lead to failure of the complete bridge. Thus, the seismic response of bridges greatly depends on the seismic behaviour of its piers.

The seismic fragility curve represents the probability of exceedance of the defined damage state parameter with the variation of intensity of Peak Ground Acceleration (PGA). Such fragility curves can be used to extract information about the seismic performance of a bridge, loss estimation due to expected damage in an earthquake, emergency preparedness, and response after an earthquake, pre and post-earthquake rehabilitation tasks, etc. Number of studies [1-9] have been done for the development of the seismic fragility curve of bridge. Many studies [4] [12] are conducted on the definition of damage states required for the development for fragility curves. Damage Index Function (DIF) is a function relating the variation of damage states with PGA for a given probability of exceedance. These curves are also capable of estimating the structural damage state for a particular pier.

In this paper the seismic fragility curves are developed based on an accepted methodology [11] for selected existing bridge piers. This study also focuses on the development of DIF of each of the selected piers.

2. Methodology

A Probabilistic Seismic Demand Model (PSDM), which is a relationship between the drift and PGA, is required to obtain from a Probabilistic Seismic Demand Analysis (PSDA). PSDA involves performing a series of dynamic time history analysis on the given structural model to estimate the probability of exceedance of structural demand. In the present study, the demand parameter (*D*) is selected as drift percentage of bridge piers subjected to a series of accelerograms. As per [11], PSDM for the selected bridge pier can be represented through a power-law expression given as:

$$D = a(PGA)^b \quad (1)$$

where '*a*', '*b*' = constant coefficients, PGA = Peak Ground Acceleration. The drift percentage is known to be log normally distributed about the median with a standard deviation of $\beta_{(D/PGA)}$ [11]. The values of *a*, *b*, $\beta_{(D/PGA)}$ are calculated after performing a series of nonlinear time-history analyses and then performing the regression analysis on the data obtained.

Seismic fragility curve represents the probability of exceeding the seismic demand parameter (*D*) for the given performance level (*C*) for the given PGA. As given by [12] fragility curve can be drawn for each of the predefined damage states in closed form as:

$$P(C - D \leq 0/PGA) = \phi\left(\frac{\ln\left(\frac{D}{C}\right)}{\sqrt{\beta^2_{\left(\frac{D}{PGA}\right)} + \beta_c^2 + \beta_m^2}}\right) \quad (2)$$

where β_c, β_m= dispersion in capacity and modelling respectively. '*D*', '*C*' is the median of the drift ratio demand and capacity at chosen performance level respectively, 'ø' represents the standard normal distribution function. Combining equations (1 and 2), we can rearrange the equation 1 to be:

$$P(C - D \leq 0/PGA) = 1 - \phi\left(\frac{\ln(c) - \ln(a(PGA)^b}{\sqrt{\beta^2_{\left(\frac{D}{PGA}\right)} + \beta_c^2 + \beta_m^2}}\right) \quad (3)$$

The value of β_c is assumed to be 0.25 based on the moderate construction quality. Damage index function is a function capable of predicting the damage index for the pier when subjected to a series of accelerograms. It can be represented by a Weibull distribution function with a cumulative probability distribution. The equation is as follows:

$$F(s) = 1 - e^{-\left(\frac{s}{\alpha}\right)^\beta} \quad (4)$$

where *F*(s) represents damage index, '*s*' represents PGA, α, β represents the Weibull distribution factors.

3. Details of the Bridge

Three bridge piers are selected from an existing four-span simply supported slab-girder highway bridge with carriageway width of 10.5 m (overall width 12.9 m). The deck consists of a concrete slab having 230 mm thickness resting on 1.015 m deep steel-I beams. The bridge is located in seismic zone V (seismic zone factor i.e. Z = 0.36). The selected piers are circular in shape with a diameter of 1.8 m. The total height of the piers (pier cap to the top of foundation) is 11.977 m and the piers rests on an isolated sloped foundation of size 10.20 m x 10.20 m. The characteristic strength of concrete and steel used for the pier is M35 and Fe500 respectively.

3.1. Seismic weight

The seismic weight at the top of piers are about 437t (for pier 1), 425t (for pier 2), 413t (for pier 3) and they are applied as lumped mass at top of pier. This mass is taken as combination of self-weight plus 20% of live-load of superstructure on either side and the self-weight (50%) of pier.

3.2. Structural Modelling

The bridge pier selected is modelled and analysed in FEM software [15]. The pier is modelled as an equivalent single degree of freedom system (SDOF). The element type used to model pier is inelastic force-based frame element and the section is divided into 400 fibre sections i.e. (fibre approach). The point of maximum curvature is found to be at a depth of 3.6 m below the ground level. Hence the piers are fixed at the depth of 3.6m below ground level. The base is assumed to be fixed at an unsupported length of 7.8m from top for pier 1, 6.0 m for pier 2 and 4.0m for pier 3.

IOP Conf. Series: Materials Science and Engineering **936** (2020) 012014 doi:10.1088/1757-899X/936/1/012014

The random parameters considered for all three bridge piers are characteristic strength of concrete, yield strength of rebars, global damping ratio (Table 1). Latin hypercube sampling (LHS) approach which is a technique to generate samples of random variables from a multi-dimensional distribution. This method is utilised to generate the 44 sets of samples of random parameters. Using the generated sets of random parameters 44 models are developed. The accelerograms used in this paper are collected from [14], each having unit PGA.

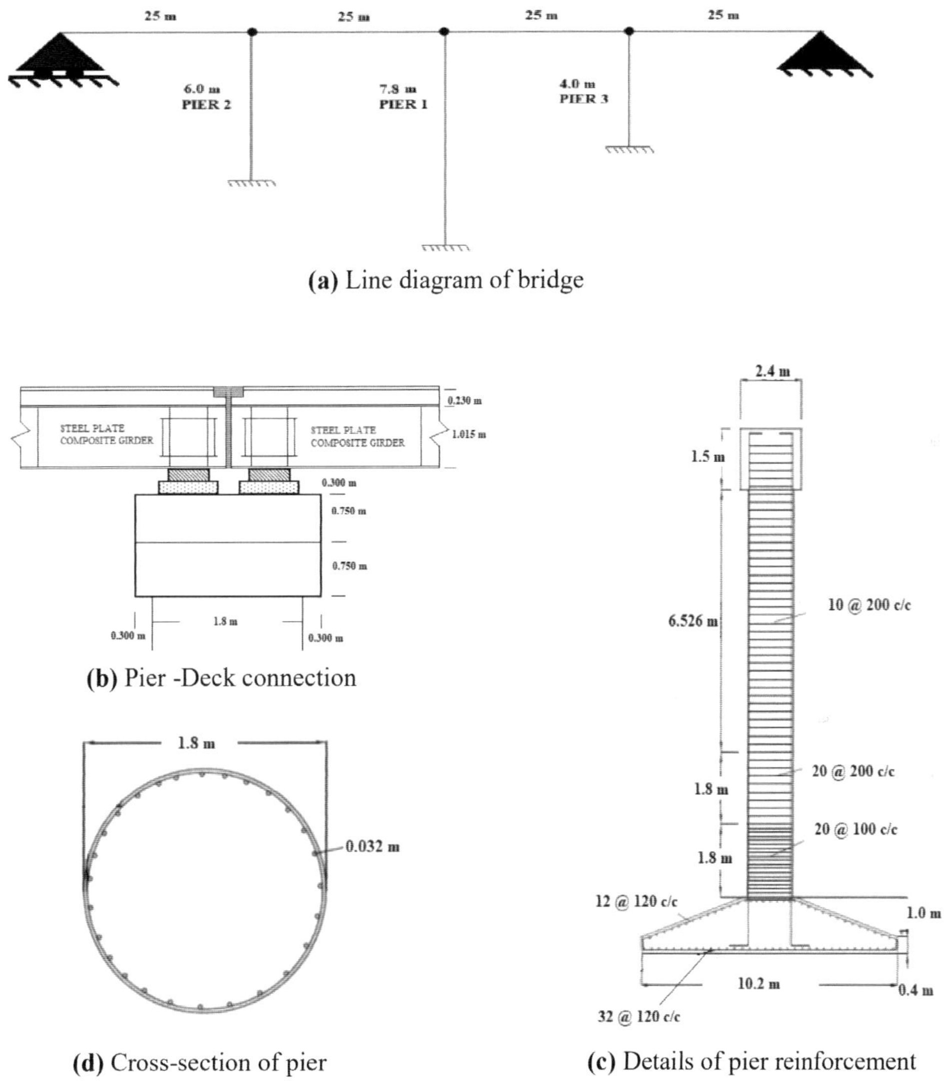

(a) Line diagram of bridge

(b) Pier -Deck connection

(d) Cross-section of pier

(c) Details of pier reinforcement

Figure 1. Geometric and cross -sectional details of bridge

Table 1. Random variable Details

Random parameters	Mean	COV (%)	Source
Concrete compressive Strength	45.0 MPa	21	Ranganathan (1999)
Yield strength	581.5 MPa	10	Ranganathan (1999)
Global damping ratio	5%	40	Davenport and Carroll (1986)

IOP Conf. Series: Materials Science and Engineering **936** (2020) 012014 doi:10.1088/1757-899X/936/1/012014

4. Probabilistic Seismic Demand Model

Non-linear dynamic analysis is performed on the developed models in which each model is subjected to a PGA ranging from 0.1g to 1.0g. Maximum drift percentage for each model is recorded and a graph is plotted between drift percentage and the applied ground motions for the selected model of bridge piers (Figure 2a, 2b and 2c). Then after performing the regression analysis on the obtained data, power-law expression (i.e. Eq. 1) is developed for the selected piers (Table 2). The PSDM model obtained will give the most probable maximum drift ratio for a given PGA.

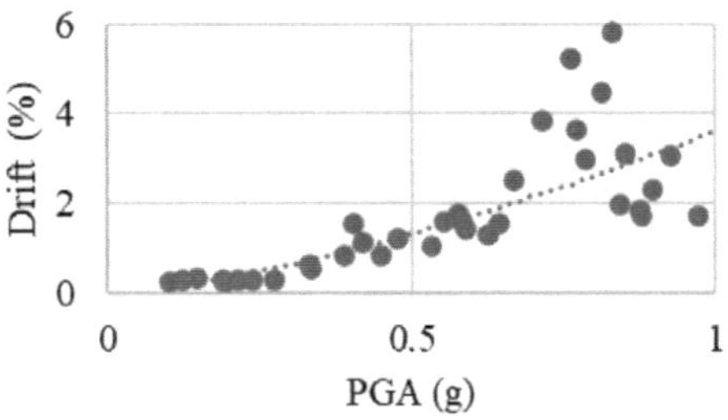

a) PSDM for Pier 1

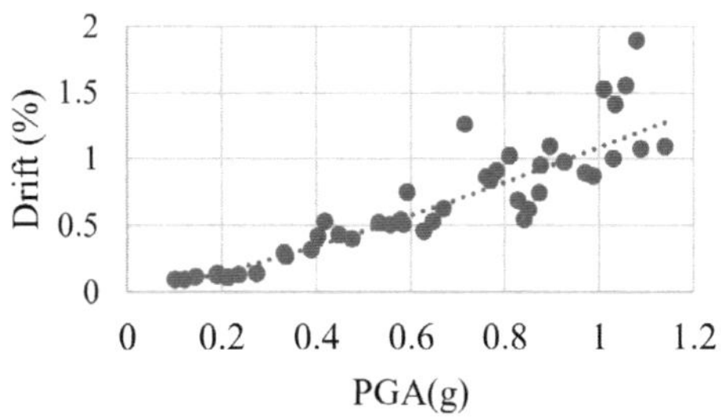

b) PSDM for Pier 2

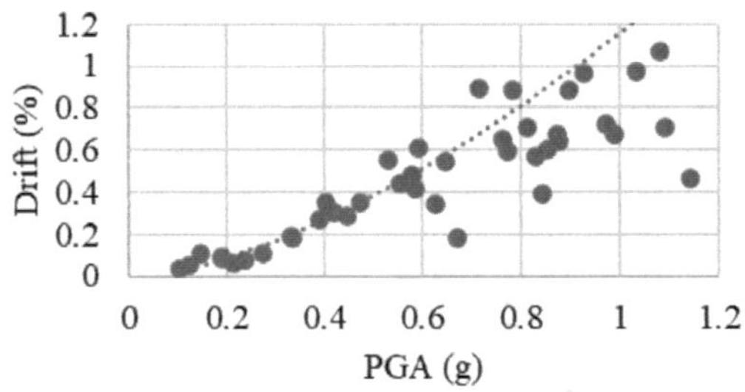

c) PSDM for Pier 3

Figure 2. PSDM model for Pier 1, 2 & 3.

Table 2. PSDM models

PIER	PSDM	R^2	$\beta_{(\frac{D}{PGA})}$
Pier 1 (7.825 m)	$3.5937(\mathrm{PGA})^{1.4533}$	0.8041	0.4713
Pier 2 (6.0 m)	$1.0863(PGA)^{1.2433}$	0.9169	0.2442
Pier 3 (4.0 m)	$1.1563(PGA)^{1.5954}$	0.6595	0.3669

5. Seismic Fragility Curve

The obtained PSDMs are utilised to develop seismic fragility curve as per Eq. 3. Drift ratios at five damage states, namely almost no, slight, moderate, extreme and complete, are taken from Dutta and Mander (1999) as shown in Table 3. Figures 3-7 represents the fragility curves obtained for the five different damage states. Pier 1 is found to more vulnerable for all damage states compared to that of other piers. This can be attributed to the high slenderness ratio of the pier 1 compared to that of other piers. Damage index values for the piers are computed based on drift (%) limits and shown in Table 3. Complete collapse of the structure is considered to be at a damage index of 0.9, and based on this, damage index for other damage states are computed.

Table 3. Damage state parameter limits

Damage State	Description	Drift (%) limits (Sc)	Damage Index
Almost no	First yield	0.5	0.09
Slight	Spalling, cracking of cover	0.7	0.13
Moderate	Anchorage loss	1.5	0.27
Extreme	Beginning of pier collapse	2.5	0.45
Complete	Collapse of pier	5.0	0.90

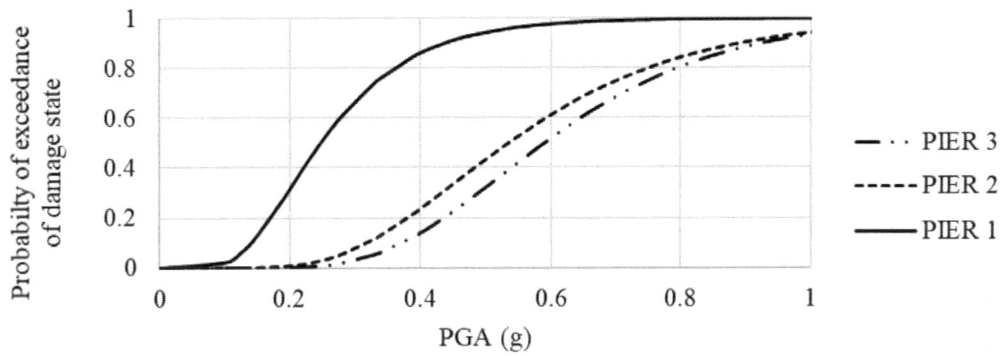

Figure 3. Fragility curves at first yield (S_c = 0.5%)

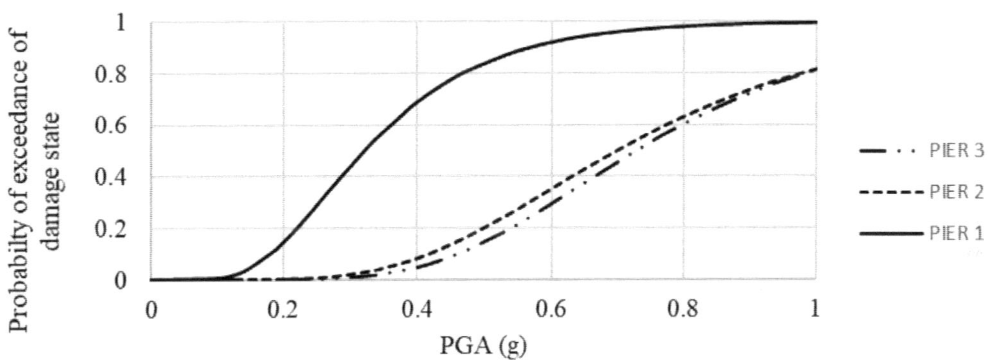

Figure 4. Fragility curves at spalling of cover (S_c = 0.7%)

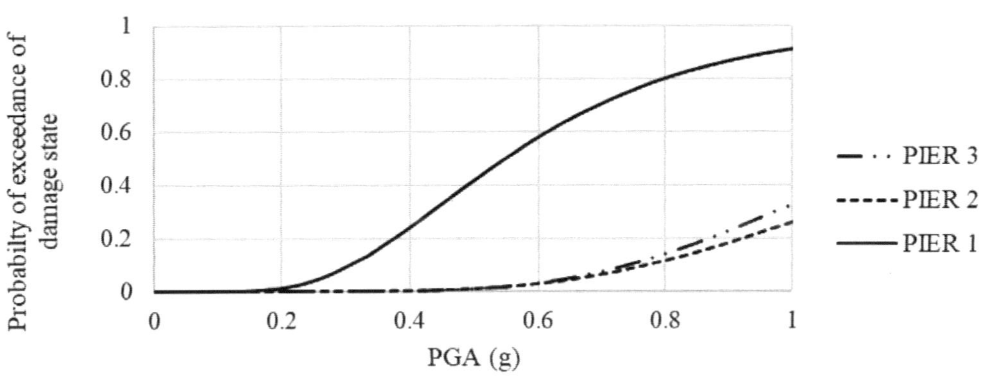

Figure 5. Fragility curves at anchorage loss (S_c = 1.5%)

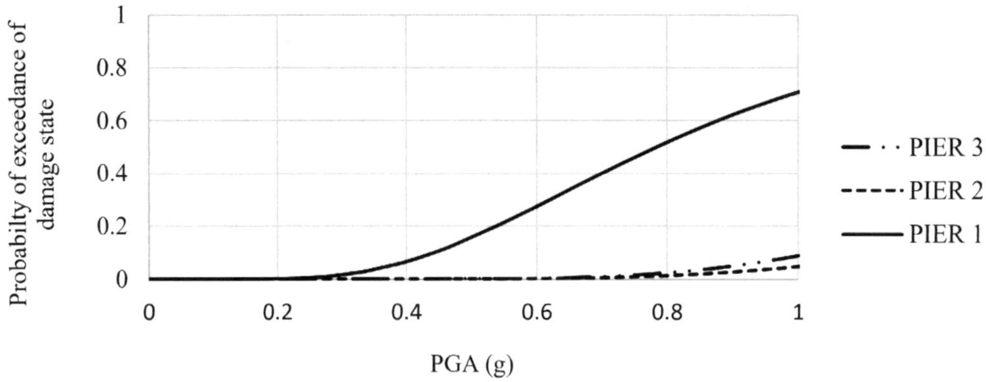

Figure 6. Fragility curves at Incipient collapse (S_c = 2.5%)

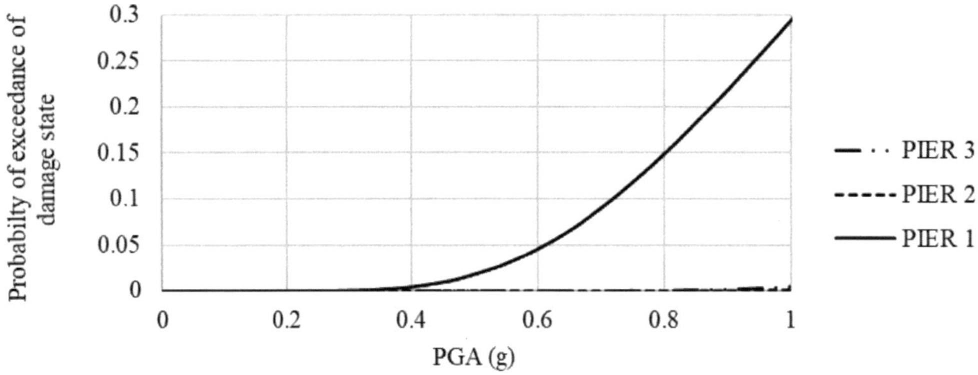

Figure 7. Fragility curves at complete collapse (S_c = 5.0%)

6. Damage Index Function

The fragility curves shown in figures 3-7 are transformed to damage index function using Weibull distribution. The damage index function is a function of damage index versus PGA which is contour corresponds to a probability of exceedance of 50% for every damage state. Pier 1 is found to have more damage index compared to other piers for a given PGA value. The procedure demonstrated in this study can be utilised to develop the fragility curves and damage functions of any type of existing bridges.

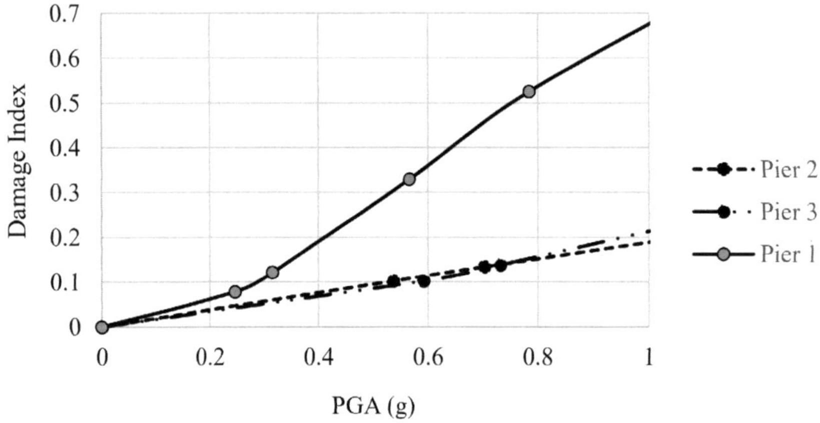

Figure 8. Damage index vs. PGA (for probability of exceedance of 50%)

7. Conclusion

In this present study, existing bridge piers are considered and their seismic fragility curves have been developed for five performance levels. It is concluded that as the height of the pier increases the probability of failure of pier also increases. The same trend can be observed from the results of damage index function. It also indicates that as the height of the pier increases it is more likely to be damaged in an earthquake event.

The information obtained from the damage index function can be used for damage cost evaluation, disaster preparedness etc. Though the present study considered only the damage of the most important component of the bridge, pier, it can be extended by considering failure of other components such as bearing, connection between superstructure and substructure, diaphragm abutment, expansion joints etc.

8. References

[1] Mackie K and Stojadinovic B 2001 Probabilistic seismic demand model for the California highway bridges *J. Bridge Eng.* **6(6)** 468–81

[2] Neilson B G and DesRoches R 2007 Seismic fragility methodology for highway bridges using a component level approach *Earthq. Eng. Struct. Dyn.* **36(6)** 823–39

[3] Pan Y, Agarwal A K and Ghosn M 2007 Seismic Fragility of continuous steel highway bridges in New York State *J. Bridge Eng.* **12(6)** 689–99

[4] Banerjee S and Shinozuka M 2008 Mechanistic quantification of RC bridge damage states under earthquake through Fragility analysis *Probab. Eng. Mech.* **23(1)**12–22

[5] Seo J and Linzell D G 2012 Horizontally curved steel bridge seismic vulnerability assessment *Eng. Struct.* **34** 21–32

[6] Ranganathan R 1999 *Structural reliability analysis and design* (Mumbai, India: Jaico)

[7] Dimitrakopoulous E G and Paraskeva T S 2015 Dimensionless fragility curves for rocking response to near-fault excitation *Earthq. Eng. Struct. Dyn.* **44(12)** 2015–33

[8] Monteiro R, Delgado R and Pinho R 2016 Probabilistic seismic assessment of RC bridges: part 1-uncertainty model *Structures* **5** 258–73

[9] Rogers LP and Seo J 2017 Vulnerability sensitivity of curved precast-concrete-I girder bridges with various configurations subjected to multiple ground motions *J. Bridge Eng.* **22(2)** 04016118

[10] C A Cornell, F Jalayer, R O Hamburger and D A Foutch 2002 The Probabilistic basis for the 2000 SAC/FEMA steel moment frame guidelines *J. Struct. Eng.* **128(4)** 526–33

[11] Shome N and C A Cornell 1999 *Probabilistic Seismic Demand Analysis of Nonlinear Structures (Stafford* University)

[12] Celik O, and B R Ellingwood 2010 Seismic fragilities for the non-ductile reinforced concrete frames – Role of aleatoric and epistemic uncertainties *Struct. Saf.* **32(1)** 1–12

[13] A Dutta and J B Mander 1999 Seismic fragility analysis of highway bridges *Proceeding of the Center-to-Center project workshop on Earthquake Engineering in Transportation Systems* (Tokyo)

[14] Haselton C B, A S Whittaker, A Hortacsu, J W Baker, J Bray and D N Grant 2012 Selecting and scaling earthquake ground motions for performing response-history analyses *Proc. 15th World conf. on Earthquake Engineering*

[15] Seismosoft 2018 *SEISMOSTRUCT – A computer programme for static and dynamic nonlinear analysis of framed structure*

[16] IRC 6:2015 *Standard specifications and code of practice for road bridges-section II- Loads and Stresses* (Fourth Revision)

Highway Steel Composite Bridge Design using Limit State Design Method

Chinnu Sabu[1*], Praveen Nagarajan[2] and Robin Davis[3]

[1] Research Scholar, National Institute of Technology, Calicut, India

[2] Associate Professor, National Institute of Technology, Calicut, India

[3] Assistant Professor, National Institute of Technology, Calicut, India

[*]chinnusabu10june92@gmail.com

Abstract. Bridges are the pathways constructed over the obstacles without changing the alignment of the way beneath. While considering the design of steel composite bridges, Indian code is in the revision process; from Working Stress Method, it is updated to Limit State Design under IRC 24:2010. This paper illustrates a design example of a steel composite highway bridge with both longitudinal girders and cross girders under Limit State Design using IRC 24:2010 and IRC 22:2015. The concrete deck slab and steel girders together act as a composite structure which is connected using shear connectors. The effective span of the bridge is taken as 35m.There are four longitudinal girders having a spacing of 2750 mm c/c and cross girders are considered with spacing 5500mm c/c.

1. Introduction

Steel composite bridges are widely used in different structural forms with different spans. They are commonly found in the form of highway bridges, railway bridges and footbridges. The strength, ductility, easy fabrication and rapid construction makes the structural steel to be versatile compared to other construction materials. It has high strength in both tension and compression. In addition, it has good strength to cost ratio and stiffness to weight ratio. Steel is an effective material that provides better solutions for long-span bridge construction. Compared to concrete construction, steel composite superstructure is lightweight and facilitates faster construction. Prefabrication of the parts of the superstructure reduces the construction operations at the bridge site.

A steel composite highway bridge consists of the deck slab with reinforced concrete and longitudinal steel girders resting on transverse girders. These are generally adopted for bridges in the medium span. These types of bridges are economical and also reduce construction time considerably. In a composite bridge deck comprising steel and reinforced concrete, the individual materials are utilised efficiently since concrete is strong in compression and steel is strong in tension. The fast erection of the prefabricated steel girders serves as supports for the cast-in-situ concrete deck with minimum formwork. The flexural stiffness of a composite beam will be about 2 to 4 times that for a corresponding steel beam resulting in reduced deflections and vibrations under moving loads. Shear connectors are provided at the junction of the concrete slab and longitudinal steel girders which transfer the horizontal shear force along the contact surface without slip.

2. Design example of a composite steel bridge under limit state design

Generally, the steel bridge structures are designed under Working Stress Method with the consideration of load factor and factor of safety. It is now required to consider the design of bridge structures under Limit State Design with reference to the revised code specifications under IRC 24:2010 and IRC

22:2015. The load factor and partial safety factor are required to be evaluated. The design procedure of a typical steel composite bridge superstructure is illustrated as follows.

2.1. Salient features

Clear span	35m
Carriage way width	9m
Centre to centre distance between expansion joints and bearings	0.75m
Effective span	33.5m
No. of main girders	4
Spacing of main girders	2.75m
Spacing between cross girders	5.5m
Width of crash barrier	0.5m
Thickness of deck slab	0.28m
Thickness of wearing coat	0.08m
Grade of steel for girder (From Table 2 of IS 2062-2011)	E350
Grade of concrete for deck slab	M 40
Modulus of elasticity for concrete, E_c	33000 MPa
Poisson's ratio, μ	0.3
Density of steel	78.5 kN/m^3
Density of concrete	25 kN/m^3
Density of wearing coat	22 kN/m^3
Yield strength of steel used for girders (For Plate Sizes <20mm)	350 MPa
Yield strength of shear studs (From Table 8 of IRC 22:2015)	385 MPa

2.2. Sectional properties of plate girder

Yield stress ratio, $\epsilon = \sqrt{\left(\frac{250}{fy}\right)}$

ϵ for top flange = 0.87

ϵ for web = 0.845

ϵ for bottom flange = 0.87

Top flange, b/t_f = 217/20 = 10.85 (Semi compact section)

Web = d/t_w = 2040/16 = 127.5 (Slender section)

Bottom flange = b/t_f = 267/25 = 10.68 (Semi compact section)

$y_b = \frac{\sum a_i y_i}{A} = 956$ mm

$y_t = 2085 - 956 = 1129$ mm

$Z_{top} = \frac{I_{zz}}{y_t} = 3.1E+07\text{mm}^3$

$Z_{bottom} = \frac{I_{zz}}{y_b} = 3.67E+07\text{mm}^3$

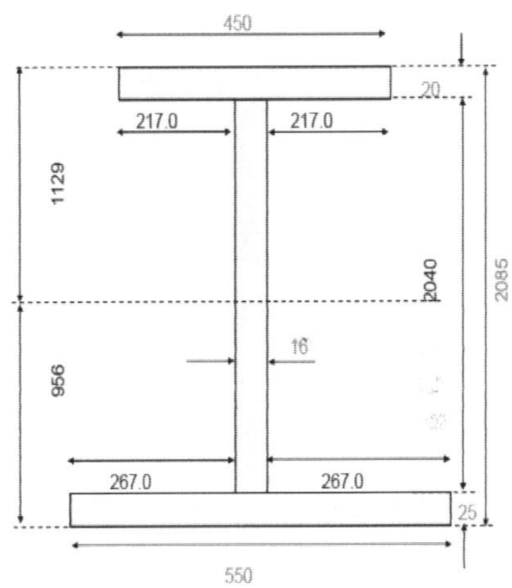

Figure 1. Cross section of plate girder.

Table 1. Sectional properties of plate girder.

Description	B mm	D mm	y mm	Area mm^2	Ay	$A(y_b - y)^2$	I_{self} mm^4	$I_{zz} = I_{self} + A(y_b - y)^2$ mm^4	I_{yy} mm^4
I section									
Top flange	450	20	2075	9000	1.87E+07	1.13E+10	3.00E+05	1.13E+10	1.52E+08
Web	16	2040	1045	32640	3.41E+07	2.58E+08	1.13E+10	1.16E+10	6.96E+05
Bottom flange	550	25	12.5	13750	1.72E+05	1.22E+10	7.16E+05	1.22E+10	3.47E+08
				55390	5.30E+07	2.38E+10		3.51E+10	4.99E+08

2.3. Sectional properties of composite section

2.3.1. Composite section for permanent loads.

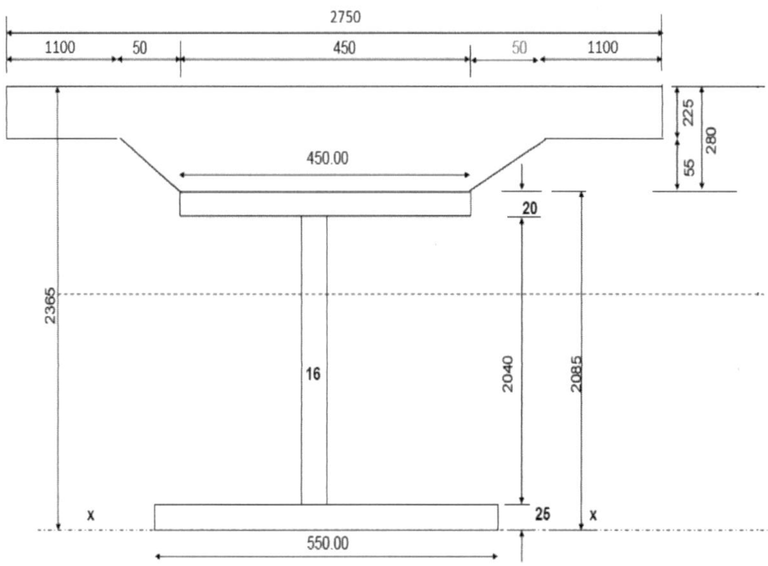

Figure 2. Sectional properties of composite section.

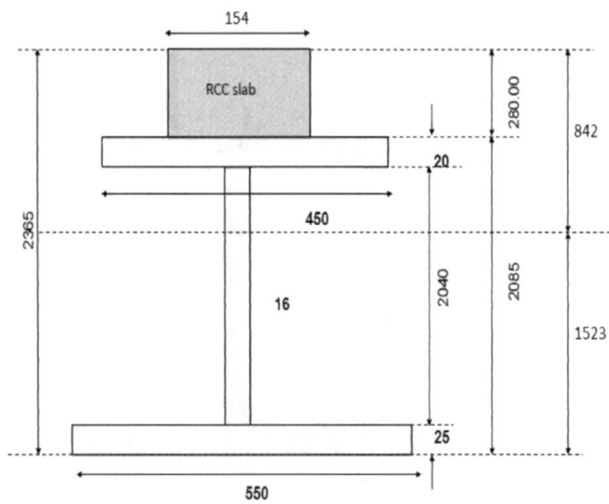

Figure 3. Composite section for permanent loads.

Modular ratio for permanent loads, $m = \frac{E_s}{K_c E_c} = 12.12 < 15$

Effective equivalent width of steel for permanent loads like SIDL = 154mm

Modular ratio for transient loads, $m = \frac{E_s}{E_c} = 7.5$

Effective equivalent width of steel for loads like LL = 308mm

$y_b = \frac{\sum a_i y_i}{A} = 1522.68$ mm

$y_t = 2365 - 1522.68 = 842.32$ mm

$Z_{top} = \frac{I_{zz}}{y_t} = 8.84E + 07\text{mm}^3$

$Z_{bottom} = \frac{I_{zz}}{y_b} = 4.9E + 07\text{mm}^3$

Table 2. Sectional properties of composite section for permanent loads.

Description	B mm	*D* mm	*y* mm	Area mm²	A	$A(y_b - y)^2$	I_{self} mm⁴	$I_{zz} = I_{self} + A(y_b - y)^2$ mm⁴	I_{yy} mm⁴
I section									
Top flange	450	20	2075	9000	1.87E+07	2.75E+09	3.00E+05	2.75E+09	1.52E+08
Web	16	2040	1045	32640	3.41E+07	7.45E+09	1.13E+10	1.88E+10	6.96E+05
Bottom flange	550	25	12.5	13750	1.72E+05	3.14E+10	7.16E+05	3.14E+10	3.47E+08
Concrete section									
Slab	154	280	2225	43120	9.60E+07	2.13E+10	2.82E+08	2.16E+10	8.52E+07
				98510	1.50E+08	6.30E+10	1.16E+10	7.45E+10	5.85E+08

2.3.2. Composite Section for transient loads

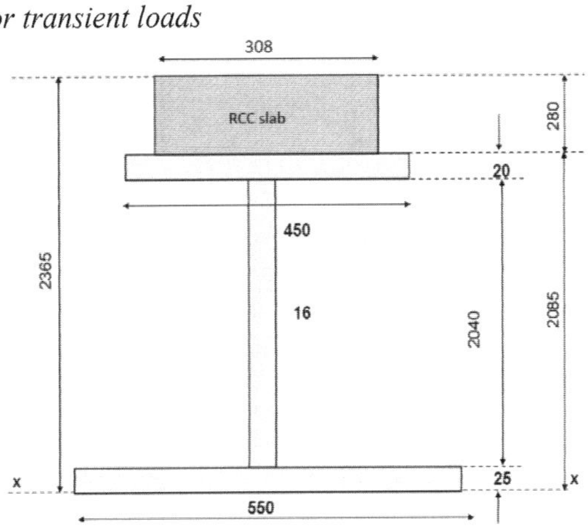

Figure 4. Composite section for transient loads.

$y_b = \frac{\sum a_i y_i}{A} = 1729.86$ mm

$y_t = 2365 - 1729.86 = 635.14$ mm

$Z_{top} = \frac{I_{zz}}{y_t} = 3.65E + 09$ mm³

$Z_{bottom} = \frac{I_{zz}}{y_b} = 1.34E + 09$ mm³

Table 3. Sectional Properties of Composite Section for transient loads.

Description	*B* mm	*D* mm	*y* mm	Area mm²	Ay	$A(y_b - y)^2$	I_{self} mm⁴	$I_{zz} = I_{self} + A(y_b - y)^2$ mm⁴	I_{yy} mm⁴
I section									
Top flange	450	20	2075	9000	1.87E+07	2.23E+12	3.00E+05	2.23E+12	1.52E+08
Web	16	2040	1045	32640	3.41E+07	1.53E+10	1.13E+10	2.66E+10	6.96E+05
Bottom flange	550	25	12.5	13750	1.72E+05	4.05E+10	7.16E+05	4.05E+10	3.47E+08
Concrete section									
Slab	308	280	2225	86240	1.92E+08	2.11E+10	5.63E+08	2.17E+10	6.82E+08
				141630	2.45E+08	2.31E+12	1.18E+10	2.32E+12	1.18E+09

2.4. Sectional properties of diaphragm section

Yield stress ratio, $\epsilon = \sqrt{\left(\frac{250}{fy}\right)}$

ϵ for top flange = 0.85

ϵ for web = 0.85

ϵ for bottom flange = 0.85

Top flange, b/t_f = 142/16 = 8.87

Web = d/t_w = 1468/16 = 91.75

Bottom flange = b/t_f = 142/16 = 8.87

$y_b = \frac{\sum a_i y_i}{A} = 750$ mm

y_t = 1500 – 750 = 750 mm

$Z_{top} = \frac{I_{zz}}{y_t} = 1.28E + 07\text{mm}^3$

$Z_{bottom} = \frac{I_{zz}}{y_b} = 1.28E + 07\text{mm}^3$

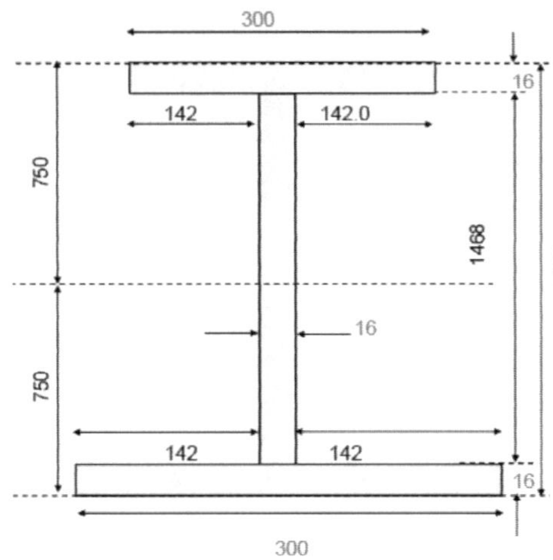

Figure 5. Sectional properties of diaphragm.

Table 4. Sectional properties of diaphragm section

Description	B mm	D mm	y mm	Area mm^2	Ay	$A(y_b - y)^2$	I_{self} mm^4	$I_{zz} = I_{self} + A(y_b - y)^2$ mm^4	I_{yy} mm^4
I section									
Top flange	300	16	1492	4800	7.16E+06	2.64E+09	1.02E+05	2.64E+09	3.60E+07
Web	16	1468	750	23488	1.76E+07	0.00E+00	4.22E+09	4.22E+09	5.01E+05
Bottom flange	300	16	8	4800	3.84E+04	2.64E+09	1.02E+05	2.64E+09	3.60E+07
				33088	2.48E+07	5.29E+09		9.58E+09	7.25E+07

2.5. Load calculation

2.5.1. Dead load calculation of girder.
DL of girder considering 38% extra for fabrication and diaphragm = 0.05539 × 78.5 × 1.38 = 6kN/m

2.5.2. Dead load calculation of diaphragm.
DL of diaphragm considering 38% extra for stitch plates & bolts = 0.033088 × 78.5 × 1.38 = 3.58 kN/m

2.5.3. Dead load calculation of deck slab.
Load per meter run = 646250 × 25 /1000000 = 16.16 kN/m

2.5.4. Super imposed dead load calculation.
Load per meter run due to crash barrier = 0.56 × 25 = 13.88 kN/m
Load per meter run due to wearing coat = 2.75 × 0.08 x 22 = 4.84 kN/m

2.5.5. Live load calculation
The vehicle loads are applied according to the load specifications mentioned in IRC 6:2017.
Impact factor for single lane 70R tracked vehicle for 33.5 m = 1.1
Congestion factor = 1.217

Table 5. Factored shear force and bending moments of main girders.

Description	Load Factor	End Girder	Intermediate Girder
DL BM of steel girder at c/l of span in kNm DL Shear force steel girder at support in kN	1.35	1136.35 135.67	1136.35 135.67
DL BM of slab - in kNm DL Shear force of slab at support in kN	1.35	3060.45 365.42	3060.45 365.42
SIDL BM due to wearing coat - in kNm SIDL SF due to wearing coat at support -in kN	1.75	1297 148.22	1173 127.47
SIDL BM due to crash barrier - in kNm SIDL SF due to crash barrier at support -in kN	1.35	2628.6 313.86	1314.3 209.3
LL BM - 1L-70R loading - with Impact & Congestion factor LL SF - 1L-70R loading - with Impact & Congestion factor	1.5	2970.92 314.75	4168.3 883.25
Total bending moment in kNm Total shear force in kN		11093.32 1277.92	10852.4 1721.11

From the above table it is observed that end girder is subjected to maximum bending moment and intermediate girder is subjected to maximum shear force.

Table 6. Factored shear force and bending moments of end diaphragm.

Description	Load Factor	End Diaphragm
DL BM of steel girder at c/l of span in kNm DL Shear force steel girder at support in kN	1.35	172.35 135.67
DL BM of slab - in kNm DL Shear force of slab at support in kN	1.35	446.75 365.42
SIDL BM due to wearing coat - in kNm SIDL SF due to wearing coat at support -in kN	1.75	191.6 148.22
SIDL BM due to crash barrier - in kNm SIDL SF due to crash barrier at support -in kN	1.35	559.32 313.86
BM due Self weight of End Diaphragm - kNm Shear force for End diaphragm - kN	1.35	3.67 5.96
Total bending moment in kNm Total shear force in kN		1373.7 969.13

3. Check for sectional adequacy

3.1. Design bending strength of composite section

Depth of neutral axis, $x_u = 318$ mm

According to IRC-22-2015, Plastic moment capacity,

$$M_p = \{[A_s\,(d_c + 0.08d_s) - b_f(x_u - d_s).(x_u + 0.16d_s)]\}\Upsilon_m$$

$M_p = 16230$ kNm >Max design bending moment on the section = 11093 kNm

Hence ok.

3.2. Design shear strength of composite section

As vertical shear force is assumed to be resisted by the web of the steel section, the shear check is same for outer and inner girder.

$$V_d = V_n/\Upsilon_{mo}$$

$V_d = 3046$ kN > Maximum shear force in the critical girder = 1721 kN

Hence ok.

IOP Conf. Series: Materials Science and Engineering **936** (2020) 012015 doi:10.1088/1757-899X/936/1/012015

3.3. Check for deflection
The deflection in the member at the critical section, i.e. mid span is checked for frequent combination as per IRC 6:2017. The final deflection in the member is calculated separately.
Total deflection in end girder = 43.1 mm
Total deflection in intermediate girder = 51.7 mm
According to IRC 22-2015, Permissible total deflection = Span / 600 = 55.83 mm
Hence ok.

4. Design summary for stud connectors
Diameter of stud = 20 mm
Number of studs provided in the cross-sections = 3 No's
Spacing of stud connectors = min (SL_1,SL_2)
Limiting criteria for Spacing of Shear connector (IRC 22:2015)
The slab is in contact over the full length
Therefore, maximum spacing of the shear connector
$S_L = 21\ t_f (250/f_y)^{1/2}$
Thickness of top flange t_f = 20 mm
Yield strength of flange f_y = 330 MPa
Therefore, S_L = 366 mm
Maximum spacing of stud connectors in longitudinal direction shall not exceed:
-Three times the thickness of the deck slab
-4 times the height of the stud connector
-600mm.

Provide stud shear connectors 20 mm diameter with 150 mm height @300 mm spacing along the length of the girder.

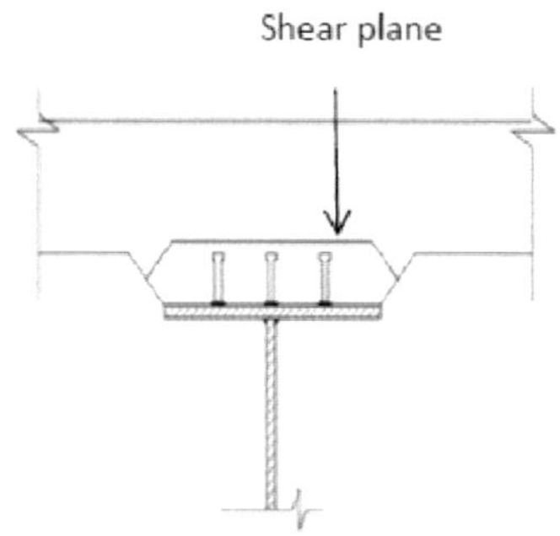

Figure 6. Stud shear connectors.

5. Design of connections
A welded connection is provided between the web and top flange as well as between web and bottom flange of steel girder with 8 mm fillet weld. However, bolted connections are provided for splice cover plates on flanges and webs of girders. In the case of top flange, a cover plate of size 550 × 1000×20 mm is provided at splice locations with 24 numbers of bolts with 22 mm diameter. Whereas, the bottom flange is provided with a cover plate of size 450 × 680 × 20 mm at splice location with 16 numbers of bolts with 22 mm diameter. The cover plate of size 680 × 1900 × 20 mm is provided on both sides of the web with 32 numbers of bolts of 22 mm diameter on each side of the web.

6. Conclusions
The design methodology of a typical steel composite highway bridge, with longitudinal girders and cross girders, is illustrated under Limit State Design using IRC 24:2010 and IRC 22:2015. The given steel composite highway bridge is proven to be structurally safe under the considered loading conditions as per IRC 6:2017. The concrete deck slab and steel girders, connected together using shear connectors, act as a composite structure.

IOP Conf. Series: Materials Science and Engineering **936** (2020) 012015 doi:10.1088/1757-899X/936/1/012015

7. References

[1] IRC 5 2015 *Standard Specifications and Code of Practice for Road Bridges Section I*
[2] IRC 6 2017 *Standard Specifications and Code of Practice for Road Bridges Section II*
[3] IRC 22 2015 *Standard Specifications and Code of Practice for Road Bridges Section VI - Composite Construction (Limit States Design)*
[4] IRC 24 2010 *Standard Specifications and Code of Practice for Road Bridges - Steel Road Bridges (Limit State Method)*
[5] Krishna Raju N 2018 *Design of Bridges-Fifth Edition* (CBS Publishers and Distributors)
[6] Johnson Victor D 2001 *Essentials of Bridge Engineering-Fifth Edition* (Oxford and IBH Publishing)
[7] Edgar Tapia-Herna´ndez, Tiziano Perea and Marco A. Islas-Mendoza 2016 Design assessment of short-span steel bridges *Int. J. Civ. Eng.* **15** 319–32

IOP Conf. Series: Materials Science and Engineering **936** (2020) 012016 doi:10.1088/1757-899X/936/1/012016

Optimal Design of Tuned Mass Damper for Base-Excited Structures

Sarranya Banerjee* **and Aparna (Dey) Ghosh**

Department of Civil Engineering, Indian Institute of Engineering Science and Technology, Shibpur, West Bengal-711103, India

*banerjeesarranya@gmail.com

Abstract. The tuned mass damper (TMD) is a well-established vibration control device that has been implemented in several structures worldwide. The performance of passive TMD is greatly dependent on its design parameters. A key issue for the design of the optimal TMD is the identification of the parameters of the structure to which the TMD is to be attached. Here, it is important also to consider the perturbations that may arise in the structural frequency due to different practical issues. The present work focusses on this and aims to obtain the optimum TMD parameters in a Genetic Algorithm (GA) framework. The results of the proposed optimal strategy, in terms of optimal tuning ratio and optimal damping ratio, for a given value of the mass ratio, are compared with some of the existing optimal solutions for linear TMD under base excitation. The effectiveness of all these optimally designed TMDs in reducing the structural response is further demonstrated when the structure-TMD system is subjected to recorded earthquake ground motions. It is seen that in the tuned condition, the control performance obtained by the different designs of the TMD are similar to each other, although having some minor differences in the values of the optimum parameters. Also, the effectiveness of the different TMD designs in the detuned condition is investigated, in which the present design technique is found to be the most effective.

1. Introduction

Tuned mass damper (TMD) is a passive vibration control device, extensively used in many important structures around the world for the mitigation of structural responses due to various kinds of environmental loads, such as wind, wave and earthquake excitations[1]. The performance of the passive TMD is greatly dependent on its design parameters, namely mass ratio, tuning ratio and damping ratio, out of which the first is generally fixed from practical considerations. In view of this, there has been considerable research on the determination of the optimal tuning ratio and damping ratio of the linear TMD, for different objectives of control. Den Hartog first proposed the fixed-point theory to find the optimum parameters of a TMD attached to an undamped structure [2]. For the structure with damping characteristics, the closed-form expression of the optimum TMD parameters, in terms of structural parameters is not derivable directly; however, the values of the parameters could be estimated numerically. Warburton (1982) was the first to provide optimum parameters of a TMD attached to a damped single-degree-of-freedom (SDOF) system subjected to white-noise excitation as base acceleration [3]. He focussed on the reduction of the variance in displacement response of the main structure in order to get the optimum parameters. Tsai and Lin (1993) performed a numerical search to find the optimum parameters of the TMD when the primary structure is subjected to a support motion assumed to be harmonic with fixed-acceleration amplitude [4]. The objective of the optimization procedure adopted by them was to minimise the peak displacement response of the primary structure. Sadek *et al.*(1997) investigated a different method of estimating the optimum parameters of the TMD, when a SDOF structure is subjected to seismic loading, the criterion used by them was to obtain the parameters that would result in equal and larger modal damping in the first two modes of vibration [5]. Apart from this, Hoang *et al.* provided site-specific optimum TMD parameters when a SDOF structure is subjected to seismic loading [6]. Researchers have also studied the effectiveness of the linear TMD and found that it is not so efficient to reduce the maximum structural displacement response under seismic applications compared to that when the structure is subjected to wind excitations [7,8].

In the present paper, a procedure is proposed to optimally design the TMD, when a damped SDOF system is subjected to base excitation. A key issue for the design of the optimal TMD is the identification of the parameters of the structure to which the TMD is to be attached. Here, it is important also to consider the perturbations that may arise in the natural frequency of the primary structure, due to different practical issues such as difficulty in the accurate computation of structural frequency due to variation in structural stiffness, mass or both, from that considered in design, or problem in ensuring exact linear behaviour of the structure throughout its design life time. Keeping this in mind, the objective of the optimization procedure is set to maximise the reduction of the displacement response of a SDOF structure with a linear TMD, having a specified variation in the natural frequency of the primary structure, when the structure is subjected to a harmonic base acceleration over a range of input frequencies. The optimum parameters of the TMD are hence found using the well-known optimization procedure of Genetic Algorithm (GA), and are compared with some of the existing optimal solutions for linear TMD for seismic applications. Further, the optimally designed TMDs found using this method, as well as, those suggested by previous researchers, are checked for effectiveness in reducing the structural response when the structure-TMD system is subjected to recorded earthquake ground motions. Finally, the performance of the TMD designs is investigated when the TMDs connected to the base-excited structures are in the detuned condition.

2. Modeling of the structure-TMD system

The structure is modelled as a linear SDOF system with mass, m_1, stiffness, k_1, and damping coefficient, c_1, and is subjected to ground acceleration, $\ddot{x}_g$. A tuned mass damper of mass, m_2, is connected to the structure with a spring of stiffness, k_2, and a linear viscous damper of coefficient, c_2. The displacement of the structure with respect to the ground is x_1, and the displacement of the TMD with respect to the structure is x_2. A schematic diagram of the model of the structure-TMD system is shown in figure 1.

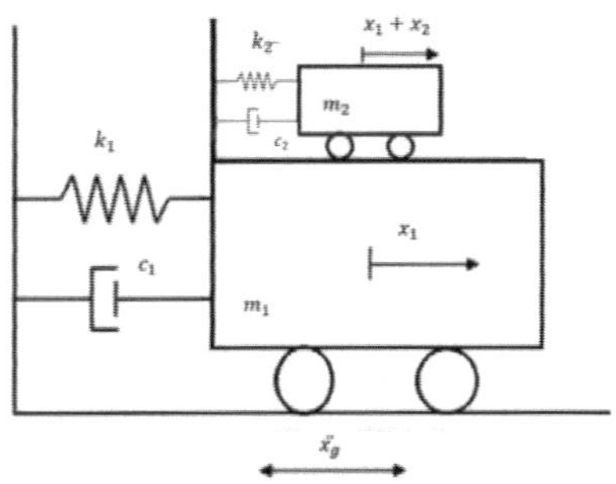

Figure 1. Schematic model of the structure-TMD system.

3. Optimal design of TMD using Genetic Algorithm

For the system described in figure1, the equations of motion for m_1 and m_2 are given by the following two equations respectively:

$$m_1\ddot{x}_1 + c_1\dot{x}_1 + k_1x_1 - c_2\dot{x}_2 - k_2x_2 = -m_1\ddot{x}_g \tag{1}$$

$$m_2\ddot{x}_1 + m_2\ddot{x}_2 + c_2\dot{x}_2 + k_2x_2 = -m_2\ddot{x}_g \tag{2}$$

Normalization of Eqns. (1) and (2) w.r.t m_1 and m_2 respectively leads to

$$\ddot{x}_1 + 2\xi_1\omega_1\dot{x}_1 + \omega_1^2x_1 - 2\mu\xi_2\omega_2\dot{x}_2 - \mu\omega_2^2x_2 = -\ddot{x}_g \tag{3}$$

$$\ddot{x}_1 + \ddot{x}_2 + 2\xi_2\omega_2\dot{x}_2 + \omega_2^2x_2 = -\ddot{x}_g \tag{4}$$

where, ξ_1 and ω_1 are the damping ratio and natural frequency of the primary structure, respectively, and, ξ_2 and ω_2 are the damping ratio and the natural frequency of the TMD, respectively. μ and γ are the mass ratio and the tuning ratio of the TMD, respectively, expressed as

$$\mu = \frac{m_2}{m_1} \quad \text{and} \quad \gamma = \frac{\omega_2}{\omega_1} \tag{5 a, b}$$

For the optimal design of the TMD, the different parameters of the TMD are to be optimised, namely, the mass ratio, μ, the tuning ratio, γ, and the damping ratio, ξ_2. Generally, the mass ratio is fixed according to the restrictions due to practical considerations. Therefore, the optimization problem here, assumes the mass ratio to be a preset parameter, while the other parameters are optimised for the best design of the TMD, under the stated objective.

Here, an optimization problem is framed for the design of the TMD, which will be effective over a wide range of input frequencies, because, seismic excitations are usually broad-banded in nature. With this, the ground acceleration is assumed to be harmonic with constant amplitude, a, and input frequency, $\bar{\omega}$, varying over a specified range. The ground excitation is thus expressed as

$$\ddot{x_g} = a\sin(\bar{\omega}t) \tag{6}$$

Moreover, to make the design more robust, a variation in the structural frequency is incorporated in the design procedure. Hence, the objective here would be to maximise the volume under the surface plot of the reduction in peak structural displacement over a specified variation of the natural frequency of the primary structure and over a range of excitation frequencies.

The optimization is carried out by using the well-known optimization procedure, Genetic Algorithm (GA). This search algorithm is based on the principles of natural selection and genetics. GA has been proved to be useful in the optimization of complex problems where the objective function is not known explicitly. The problem is solved using the GA solver provided in the MATLAB (R2018a) optimization toolbox. The following parameters are used to characterise the GA in the solver.

- Population size of 50 individuals
- Fitness scaling according to rank
- Crossover function scattered with a crossover fraction of 0.8
- Stopping criteria of 200 generations or function tolerance of 1e-6

4. Numerical study and comparison with existing optimal solutions

The design of the optimal TMD using GA is numerically illustrated in two example structures with a natural frequency of 0.5 seconds each, and damping ratios of 2% and 5%. Two different TMD mass ratios are considered as 1% and 3%. The structure-TMD system is subjected to a base acceleration assumed to be harmonic, given by Eqn. 6, with peak ground acceleration of 0.02g, and external frequency, $\bar{\omega}$ varied over a range of $0.85\omega_1$ to $1.15\omega_1$. A variation of $\pm 5\%$ in the time period of the main structure is considered for the design procedure.

The optimum parameters of the TMD obtained from this method are listed in table 1, and they are compared with the optimum parameters obtained from the previous studies on the optimal design of TMD for seismic applications. Here, three previously found optimal solutions of the TMD parameters are referred, as follows, firstly, the optimum parameters of a TMD attached to a SDOF system subjected to white-noise excitation as base acceleration provided by Warburton (1982)[3]. Secondly, the optimum parameters of TMD for support-excited structures given by Tsai and Lin (1993)[4], and thirdly, when a SDOF structure is subjected to seismic loading, the optimum parameters of the TMD presented by Sadek *et al.*(1997) [5] are considered for the performance study. All the optimum parameter sets obtained from the previous studies and from this method are listed in table 1.

From table 1, it is clear that the tuning ratios of the optimum TMDs obtained by optimization using GA are considerably lesser than the optimum tuning ratios obtained from the other methods. Further, the damping ratios of the TMDs obtained from this optimization procedure is close to the optimum damping ratios obtained by Tsai and Lin (1993). However, unlike Tsai and Lin (1993), the values of the optimum damping ratio of the TMDs obtained from this method have very slight differences due to the change in the structural damping ratio. Further, the optimum parameters obtained by Sadek *et al.* (1997) is seen to have slightly greater values than the other three optimal design parameter sets for all the example structure-TMD systems. The whole design procedure using GA was repeated for some different amplitudes of the harmonic base acceleration, and all the designs yielded the same optimum parameters, which validates that the optimum parameters of the linear TMD do not depend on the intensity of the load applied.

Table 1. Optimum TMD parameters obtained from different studies.

Structural Damping Ratio	Mass ratio of TMD	Optimum TMD parameters	Methods used in TMD design			
			Warburton (1982)	Tsai and Lin (1993)	Sadek *et al.* (1997)	Current study
2%	1%	γ	0.9819	0.9826	0.9881	0.9724
		ξ_2	0.0498	0.0635	0.1193	0.0660
	3%	γ	0.9545	0.9555	0.9676	0.9401
		ξ_2	0.0857	0.1080	0.1900	0.1130
5%	1%	γ	0.9704	0.9732	0.9852	0.9630
		ξ_2	0.0498	0.0680	0.1488	0.0710
	3%	γ	0.9380	0.9416	0.9626	0.9275
		ξ_2	0.0857	0.1134	0.2190	0.1120

The effect of these differences in the optimum parameters obtained from the four different designs of the TMD is studied through the time-domain displacement response analysis of these structure-TMD systems when subjected to three recorded seismic accelerograms, as described in table 2.

Table 2. Classification of earthquakes.

Event name	Year	Station name	PGA
Loma Prieta	1989	Golden Gate Bridge	0.23g
Imperial Valley-01	1940	El Centro	0.35g
Kobe, Japan	1995	Abeno	0.22g

Table 3 shows the reductions obtained in the variance of the displacement responses of the primary structures by the four different designs of the TMD. It is observed that the reductions obtained from the four TMD designs are close to each other; however, the TMD designed by optimization using GA has performed better for the Imperial Valley-01 earthquake. TMD design given by Warburton (1982) has worked the best for Kobe earthquake, and design given by Tsai and Lin (1993) has proved to be the most effective for Loma Prieta earthquake. Reductions obtained by the TMD designed according to Sadek *et al.*(1997) are always seen to be slightly lesser than the reductions obtained from the other designs.

Table 3. Reduction in variance of structural displacement response using the different optimal designs, when subjected to the three earthquakes.

Structural Damping Ratio	Mass ratio of TMD	Earthquake name	Reduction in variance of displacement response achieved by			
			Warburton (1982)	Tsai and Lin (1993)	Sadek *et al.*(1997)	Current study
2%	1%	Loma Prieta	38.42	39.83	37.90	37.49
		Imperial Valley-01	32.50	33.90	29.10	33.93
		Kobe, Japan	54.08	51.69	39.92	53.25
	3%	Loma Prieta	49.55	51.54	51.40	50.00
		Imperial Valley-01	50.82	51.32	45.20	51.34
		Kobe, Japan	70.84	68.65	56.87	68.33
5%	1%	Loma Prieta	17.70	22.62	26.12	21.69
		Imperial Valley-01	16.11	17.67	13.85	18.13
		Kobe, Japan	32.98	28.63	16.80	29.70
	3%	Loma Prieta	21.23	23.90	23.60	22.92
		Imperial Valley-01	32.95	33.14	25.66	33.43
		Kobe, Japan	47.64	43.80	31.20	44.51

Although all the TMD designs are seen to provide similar reductions in structural displacement responses due to the recorded seismic excitations in the tuned condition, it is worth reviewing their performance in the detuned condition. A sample structure-TMD system from the above examples is chosen with the primary structure's time period of 0.5 seconds and 2% structural damping, having a TMD of 3% mass ratio. Next, this structure-TMD system is used to illustrate the performance of the TMD designs in the detuned condition. Figure 2 provides contours showing the reduction in peak structural displacement response against a variation of the natural frequency of the primary structure and with the change in the excitation frequency, when the structure is subjected to a harmonic base excitation. From this figure, it is clear that the TMD performance in the detuned condition deteriorates by a considerable amount. In the worst detuned condition, Warburton's (1982) TMD design slightly amplifies the displacement response; Tsai and Lin's (1993) design gives no reduction at the same configuration of the system; however, the designs proposed by the current study and by Sadek *et al.* (1997) still give a small reduction to the tune of 5% in this condition of the TMD. The design of the TMD given by Sadek *et al.* (1997), even though works fine in detuned condition, in the tuned condition the reductions achieved by this design is almost 10% lesser than those achieved by the other three designs. Therefore, from figure. 2, it may be inferred that the TMD designed by optimization using GA works more efficiently than the other three designs, when there is a perturbation in the natural frequency of the main structure.

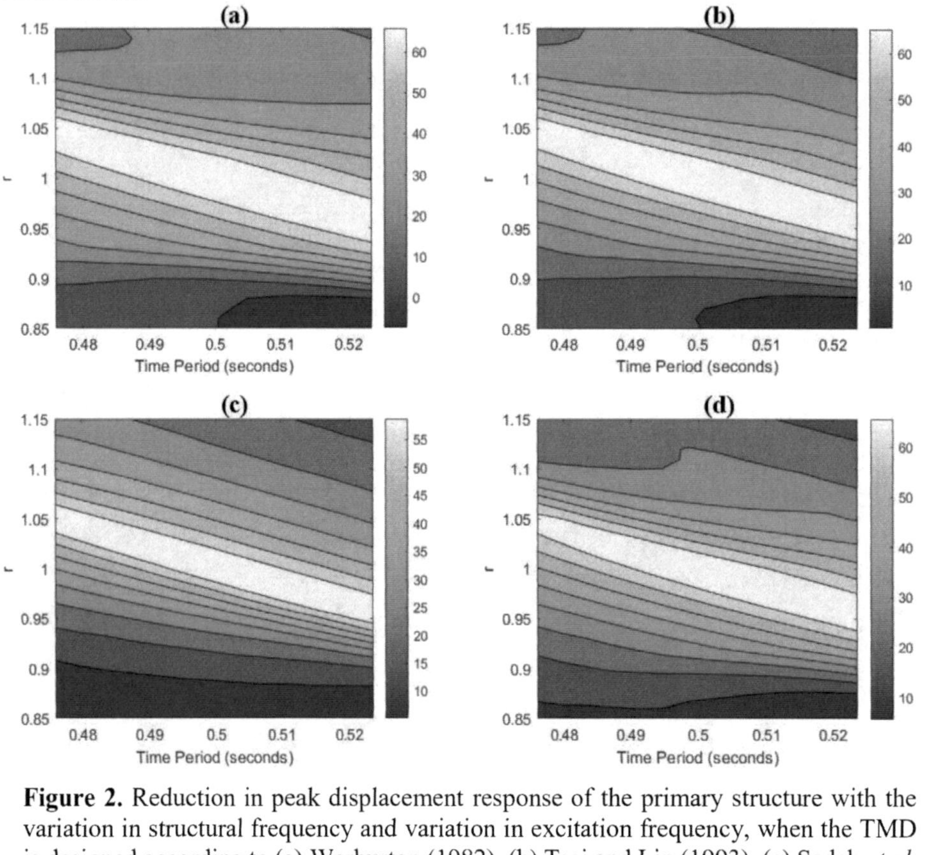

Figure 2. Reduction in peak displacement response of the primary structure with the variation in structural frequency and variation in excitation frequency, when the TMD is designed according to (a) Warburton (1982), (b) Tsai and Lin (1993), (c) Sadek *et al.* (1997), and (d) Current study.

5. Conclusion

Many authors have provided optimal design strategies to find the parameters of a linear TMD which is to be attached to a structure subjected to base excitation, where different performance criteria are optimised for the best performance of the TMD. However, it must be noted that the optimum parameters of the TMD are designed for a certain set of structural parameters, and a variation in these structural parameters adversely affects the performance of the linear TMD. This paper proposes a new design strategy, where a specified variation in the natural frequency of the primary structure is considered in the design optimization problem, and the parameters of the optimal TMD are numerically searched. The optimum parameters found using this method is then compared with the optimum parameters of a TMD designed for similar applications given by some previous researchers. Further, to check the effectiveness of the TMD, time-history analysis of some example structure-TMD systems are studied under three actual recorded seismic accelerograms. It is found that all the TMD results yield similar reductions in the variance of structural displacement responses under these three earthquakes. Thereafter, to study the effectiveness of the TMD designs in the detuned condition, the contours showing the reduction in peak structural displacement response are plotted against the variation of structural frequency over a range of input frequencies. From this study, it is confirmed that the TMD designed by this method, exceeds the level of performance achieved by the other three TMD designs when the structure is subjected to a harmonic load as base excitation.

6. References

[1] Gutirrez Soto Mariantonieta and Adeli Hojjat 2013 Tuned Mass Dampers *Archives of Computational Methods in Engineering* **20** 419–31
[2] Den Hartog J P 1947 *Mechanical Vibrations* (New York: McGraw-Hill)
[3] Warburton G B 1982 Optimum absorber parameters for various combinations of response and excitation parameters *Earthquake Engineering Structural Dynamics* **10** 381–401
[4] Tsai Hsiang-Chuan and Lin Guan-Cheng 1993 Optimum tuned-mass dampers for minimizing steady-state response of support-excited and damped systems *Earthquake Engineering and Structural Dynamics* **22** 957–73
[5] Sadek Fahim, Mohraz Bijan, Taylor Andrew W and Chang Riley M 1997 A method of estimating the parameters of TMD for seismic applications *Earthquake Engineering and Structural Dynamics* **26** 617–35
[6] Hoang N, Fujino Y and Warnitehai P 2008 Optimal TMD for seismic applications and practical design formulas *Engineering Structures* **30** 707–15
[7] Sladek J R and Klinger R E 1983 Effect of tuned mass dampers on seismic response *Journal of Structural Engineering* **109(8)**
[8] Murudi M M and Mane S M 2004 Seismic effectiveness of tuned mass damper for different ground motion parameters *Proceedings of the 13th World Conference on Earthquake Engineering (Vancouver, B.C. Canada)*

IOP Conf. Series: Materials Science and Engineering **936** (2020) 012017 doi:10.1088/1757-899X/936/1/012017

Bayesian Approach of Model Updating of Structural Parameters with Simulated Modal Data

Partha Sengupta* **and Subrata Chakraborty**

Department of Civil Engineering, Indian Institute of Engineering Science and Technology Shibpur, West Bengal 711103, India

*parthasengupta18.rs2018@civil.iiests.ac.in

Abstract. The present study investigates the effectiveness of Bayesian approach of model updating of structure, using simulated modal data sets as evidence. The importance of the prediction of error implemented in the proposed Bayesian framework to assess its capability in sampling posterior values with improved accuracy has been highlighted. The natural frequencies and mode shapes of the structure are taken only as the evidence. The study highlighted the importance of the likelihood functions corresponding to the prediction of error between the experimental and predicted data. The study also implemented the likelihood functions corresponding to the prediction of error variances of natural frequencies only in sampling posterior values with higher accuracy. The efficacy of the proposed approach in dealing with the issue of incomplete data sets, that makes the problem typically ill-conditioned, is duly addressed. The present work also implemented model reduction technique to synthesise the mode shape ordinates at unmeasured degrees of freedom (DOF) from the reduced-order model. In addition, the efficiency of the proposed model reduction algorithm is explored by adding noises of varying percentages to the measured mode shape values. The proposed methodology is illustrated numerically to update the stiffness parameters of an eight-storey shear building model, considering simulated datasets contaminated by Gaussian error as evidence. Results with both incomplete and complete data sets using the proposed Markov Chain Monte Carlo approach are studied. The accuracy on the synthesis of mode shapes from unmeasured DOF using model reduction algorithm is also studied.

1. Introduction

Bayesian inference is widely used to reduce both computational burdens and increased experimental investigations for updating model parameters in a probabilistic framework [1-4]. The Bayesian inference of model parameter updating using data sensitivity [5] and employing the best achievable eigenvectors [6] is found to be appealing, provided the damage locations of the structure are known. The applications of techniques such as transitional Markov Chain Monte Carlo (MCMC) method and Gibbs Sampling, derived from the basic MCMC technique, are also notable in this regard [10, 13-14]. The Gibbs sampling approach becomes computationally expensive and inefficient, especially when the parameter values depend on each other. The inverse problems in system identification require the solution of a family of plausible values of model parameters rather than the search for a single optimal parameter vector [10-14]. The MCMC simulation approach using Metropolis Hastings (MH) algorithm enables the propagation through the parameter space and converges to the globally optimum region [7-9]. In this regard, the re-analyses of full models can be suppressed by introducing model-reduction techniques such as Guyan's reduction, substructure coupling and component mode synthesis [11-12]. The model reduction between the measured and the synthesised modal quantities such as mode shapes is fruitful in comparison to the direct mode matching for uncertainty quantification of model parameters [15].

The present study investigates the efficacy of Bayesian approach of updating model parameters in determining their posterior distribution using MCMC technique, based on the estimates of likelihood functions of the evidences conditioned on the concerned parameters. The prediction of the likelihood function of the evidences is implemented through Gaussian pdfs, to assess its effectiveness in determining posterior values by the proposed Bayesian framework. The model reduction technique using Guyan's reduction is used to determine the unknown responses in terms of the known responses, i.e. at the measured DOFs. This study is performed considering the number of modes equivalent to the number of measured DOFs. The algorithm for model reduction is tested for incomplete measurements by considering two different sensor placement combinations. The sensor placement is considered by reducing the number of measured DOFs. The accuracy of determination of unknown mode values using model reduction algorithm is also studied. The proposed techniques are illustrated numerically by considering an eight-storey shear building model to study the accuracy of estimating model parameters and associated uncertainties considering both complete and incomplete modal data sets.

2. Bayesian approach of Model Updating

In the Bayesian approach, subjective judgements based on instinct, experience or incidental information is systematically incorporated with the measured data to obtain a balanced estimate of model parameters. The primary rule that governs Bayesian model updating is Bayes theorem, which states that given the measured data, the posterior pdf of the model depends on the likelihood function, the prior distribution and the evidence [7]. This can be mathematically expressed as,

$$P(\boldsymbol{\theta} \mid \mathbf{D}) = \frac{P(\mathbf{D} \mid \boldsymbol{\theta}) P(\boldsymbol{\theta})}{P(\mathbf{D})} \tag{1}$$

where, $\boldsymbol{\theta}$ is the vector of updating parameters, $\mathbf{D}$ is the available data, $P(\mathbf{D} \mid \boldsymbol{\theta})$ is the likelihood that represents the possibility of the data $\mathbf{D}$ when a trust of $\boldsymbol{\theta}$ is taken as true, $P(\boldsymbol{\theta})$ is the prior pdf that contains the knowledge about $\boldsymbol{\theta}$ before observing the data, $P(\boldsymbol{\theta} \mid \mathbf{D})$ is the posterior pdf and $P(\mathbf{D})$ is the normalization factor (the evidence given by the data). The related formulation of likelihood function to obtain the posterior statistics of the model parameter are presented in the following.

2.1. Likelihood function

The difference between the measured responses and the predicted outputs of evidences from a model is the prediction error, which can be defined as,

$$e(\boldsymbol{\theta}) = Y - X(\boldsymbol{\theta}) \tag{2}$$

where, $e(\boldsymbol{\theta})$ is the prediction error, Y is the measured response and $X(\boldsymbol{\theta})$ is the predicted output of the evidences from the model. The likelihood function of evidences for the outcome (Y) can be expressed as:

$$p(Y \mid \boldsymbol{\theta}, \sigma) = \frac{1}{\sigma\sqrt{2\pi}} \exp\left\{-\frac{[e(\boldsymbol{\theta})]^2}{2\sigma^2}\right\} = \frac{1}{\sigma\sqrt{2\pi}} \exp\left\{-\frac{[Y - X(\boldsymbol{\theta})]^2}{2\sigma^2}\right\} \tag{3}$$

The evidences ($\mathbf{D}$) in the present study are the modal data of a structure comprising modal frequencies. In case of incomplete modal data, i.e. the number of measured degrees of freedom of mode shape, n_o is less than the actual number of degrees of freedom (n). Assuming that the frequency of an energy mode is statistically independent to each other, the likelihood of $\mathbf{D}$ is expressed as:

$$p(\mathbf{D} \mid \boldsymbol{\theta}) = \prod_{i=1}^{m} p(\omega_i \mid \boldsymbol{\theta}, \sigma^2_{\omega_i}) \tag{4}$$

The unknown parameters in the model updating problem considering natural frequencies as evidence are increased to $[d + m]$ in which d is the extent of the vector $\boldsymbol{\theta}$ and m is the number of modes considered.

2.2. Prior Distribution
In Bayesian deduction, the prior distributions of parameters describing a model are necessary. The prior pdf of the unknown parameters $\boldsymbol{\theta}$ i.e. $P(\boldsymbol{\theta})$ is assumed to be exponential, and the different components of $\boldsymbol{\theta}$ is assumed to be independent. The likelihood functions of the error variances are assumed as inverse gamma (IG) for natural frequencies.

2.3. Posterior Distribution
The posterior distribution of the updating parameters can be now obtained by Bayes theorem using equation (1). It may be noted that apart from model parameters ($\boldsymbol{\theta}$), the updating parameters vector involve the variances of the frequency. Thus, the posterior values of the parameters in equation (1) is redefined as, $\bar{\boldsymbol{\theta}} = \left\{\boldsymbol{\theta}^T, \left\{\sigma^2_{\omega_i}\right\}_{i=1}^{m}\right\}^T$. In the present study, the MCMC technique is implemented to obtain samples from the posterior distribution.

3. MCMC based Model Parameter Updating
The MCMC allows efficient generation of random samples as a series of a "Markov Chain" according to an arbitrarily given probability distribution. The MH based MCMC approach is used. The proposal distribution, f governs the distribution of the candidate sample and affects the passage of the Markov chain from one state to another. In the present study, Gaussian distribution is used as the proposal distribution to draw samples from the high dimensional posterior distribution. The acceptance criteria ratio of MH algorithm in the following manner,

$$r = \left\{ \frac{p(\omega_i \mid \bar{\theta}, \sigma^2_{\omega_i}) p(\sigma^2_{\omega_i} \mid \bar{\theta}) p(\bar{\theta})}{p(\omega_i \mid \theta, \sigma^2_{\omega_i}) p(\sigma^2_{\omega_i} \mid \theta) p(\theta)} \right\} \tag{5}$$

The posterior parameter value $\bar{\boldsymbol{\theta}}$ is then accepted either with probability α_a or rejected with a probability of $1-\alpha_a$. The initial acceptance ratio is kept low to avoid any bias of updating the parameter value right from the start of the algorithm.

4. Model Reduction
The model reduction technique evades the full model run for each iteration using dynamic response data. In the present study, the model reduction is implemented by partitioning the mass $\mathbf{M}$ and stiffness $\mathbf{K}(\boldsymbol{\theta})$ matrices into matrices based on the master and slave DOFs. The mode shapes corresponding to slave nodes s is represented in terms of master nodes for first m modes with m being equal to the number of master nodes as:

$$\Phi_{\mathbf{sm}} = \mathbf{t}\Phi_{\mathbf{mm}} \tag{6}$$

where, $\mathbf{t} \in \mathbf{R}^{\mathbf{s \times m}}$ is a transformation parameter. To determine the unknown parameter $\mathbf{t}$ in the above, the iterative procedure suggested by Friswell et al. (1995) is applied. The choice of master DOFs depends on the underlying assumption in the Guyan reduction that at the slave coordinates, the inertia forces are insignificant compared with the elastic forces [16]. The present study explores the choice of master DOFs using a different combination of measured DOFs after solving a dynamic eigenvalue problem.

5. Numerical Study
An eight-storied shear building model, as shown in figure 1, is shown to elucidate the proposed model updating technique using synthetic modal response data. The building has uniform nominal stiffness, k_i and mass m_i associated with each storey. An updating parameter θ_i (with i=1, 2….8) is imposed on each of these stiffness parameters. The multiplication of these two factors represents the actual stiffness of the i th storey i.e. ($K_i = k_i \times \theta_i$). The unknown parameter θ_i are obtained under Bayesian inference. The frequencies of the initial three modes are considered as the mean data (ω_m, ϕ_m). The modal responses are artificially simulated from a Gaussian distribution with the estimated modal data as mean and 5 %

coefficient of variation i.e. $\omega_j \sim N(\omega_m, 0.05\omega_m)$, where, j represents jth sample of measurement. The total number of modal data sets generated in this manner are 15, 30 and 45. The spread of the random walk is adopted as 5% of the current state of the parameter.

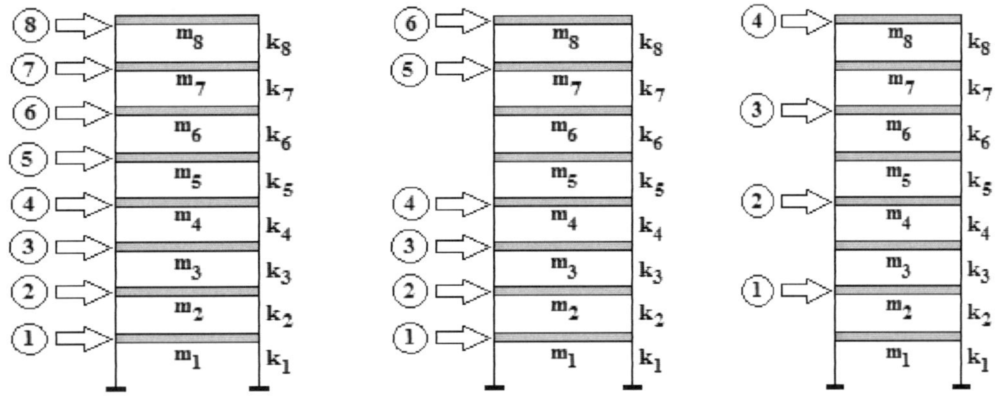

Figure 1. Schematic diagrams showing the location of measured degrees of freedom i.e. (a) complete measurements, (b) measurements at 6 locations and (c) measurements at 4 locations.

For the first case, the modal data were considered to be available for all the DOFs, as shown in figure 1(a). The posterior samples were derived by MH algorithm, and 10000 samples are simulated as the seeds of the Markov Chain. The first 500 samples are discarded as burn-in-samples as the initial fluctuations of the parametric values were high. The acceptance ratio at the start is kept low to avoid any bias of updating the parameter value. The posterior distributions of parameters θ_1 and θ_2 are shown in figure 2 in the form of a histogram. The same is observed for the other parameters also. It has been noticed from the histogram plots that the maximum occurrence of the parameters is almost around the actual value of the same and the maximum spread of the parameter value is 8.24%.

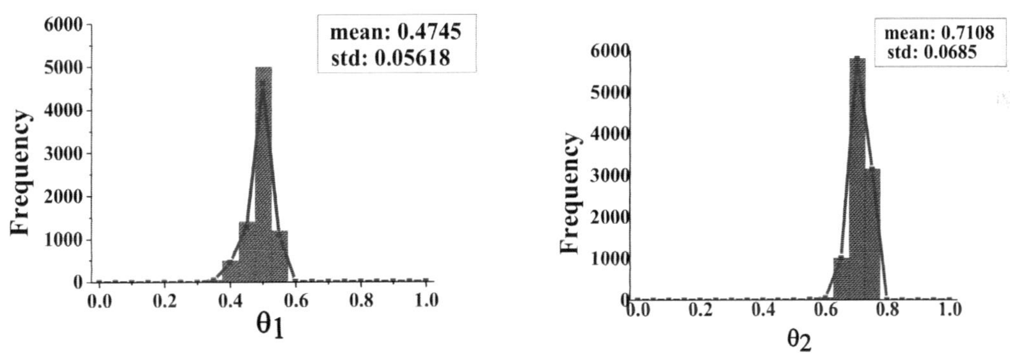

Figure 2. The posterior distribution of the identified model parameters.

Now the performance of the parameter updating algorithm is studied considering incomplete measurements. Two such cases are considered, for the first case, six measurements of modal data are considered to be available, as shown in figure 1(b). In the next case, only four measurements of modal data are considered to be known, as shown in figure 1(c). The mean and standard deviation of the posterior distribution of parameters are shown in table 1 with the percentage of error in bracket considering 45 sets of simulated incomplete data. It can be noted that the mean values of the posterior distribution of all the parameters are estimated with reasonable accuracy. The maximum percentage of error for posterior values of parameters is 7.82% and 12.11% for six measurements and four measurements respectively.

Table 1. Comparison of actual and predicted parameters.

Parameters	Actual values	Posterior value of mean and standard deviation of parameters for 45 data sets with percentage of error [shown in bracket]		
		Complete measurement	6 measurement	4 measurement
θ_1	0.5	0.474, 0.05 [5.10]	0.535, 0.09 [6.94]	0.542, 0.11 [8.42]
θ_2	0.7	0.683, 0.08 [2.37]	0.661, 0.11 [5.54]	0.644, 0.12 [7.96]
θ_3	0.2	0.194, 0.04 [2.85]	0.193, 0.07 [3.35]	0.221, 0.05 [10.50]
θ_4	0.9	0.843, 0.08 [6.29]	0.843, 0.09 [6.30]	0.794, 0.11 [11.72]
θ_5	0.6	0.588, 0.06 [2.02]	0.561, 0.08 [6.47]	0.626, 0.11 [4.30]
θ_6	1.0	0.945, 0.08 [5.46]	0.922, 0.11 [7.82]	0.879, 0.11 [12.11]
θ_7	0.8	0.764, 0.09 [4.46]	0.784, 0.09 [1.98]	0.831, 0.13 [3.85]
θ_8	0.4	0.389, 0.06 [2.58]	0.386, 0.07 [3.58]	0.379, 0.12 [5.28]

Now the mode shapes are measured for two cases of incomplete measurements at the measured DOFs as manifested in figure 1(b) and figure 1(c). The measured DOFs are taken as master nodes. The iterative model reduction algorithm is implemented to synthesise the mode shape ordinates at the unmeasured or slave DOFs. The efficiency of the model reduction approach is tested by adding noises of varying percentages to the measured mode shape values. For brevity, the mode shape ordinates at the unmeasured DOFs having a maximum percentage of errors are shown in table 2. It can be observed that the modal ordinates at the slave DOFs are estimated quite well using model reduction technique even after applying varied percentages of noise to the modes at the measured DOFs.

Table 2. Comparison of synthesised modes based on model reduction.

Mode no.	Maximum Percentage of errors of synthesised unknown modes based on model reduction technique									
	6 measurement					4 measurement				
	No Noisy data	2% Noisy data	4% Noisy data	6% Noisy data	8% Noisy data	No Noisy data	2% Noisy data	4% Noisy data	6% Noisy data	8% Noisy data
1	0.0001	1.5591	2.3706	3.2613	4.5699	0.0003	2.0746	3.9691	4.7572	5.1347
2	0.0002	1.7479	1.9867	3.9161	5.2298	0.0001	2.1478	2.3285	4.7768	5.0859
3	0.0001	1.6918	2.0692	2.6586	3.7282	0.0008	2.7277	3.6517	5.2321	5.8218
4	0.0008	1.0291	1.6476	2.6935	3.5953	0.0004	2.0541	2.9913	5.7953	6.0451
5	0.0004	1.1435	2.0115	3.7576	4.3811	--	--	--	--	--
6	0.0007	1.6619	2.2014	3.5871	3.5951	--	--	--	--	--

6. Summary and Conclusions

An effective MCMC technique using MH algorithm based on likelihood estimates of evidences and proper choice of prediction error variance from Bayesian inference is presented to update structural model parameters considering both complete and incomplete data sets. The proposal distribution of the sample for performing a random walk is guided based on physical possibilities of damage to a structure. The algorithm is elucidated to update the stiffness parameters of an eight-storey shear building model with simulated modal data as evidence. As expected, the accuracy for completely known modal data and a higher number of data sets are very high. However, the results of incomplete data sets also estimate the parameters quite well, and the maximum percentage of error of posterior value of parameters is within 12%. The model reduction technique determines mode shape ordinates at the unmeasured DOFs with high accuracy in comparison to the percentage of noise applied to the measured data. Thus, the model reduction approach appears to be a robust choice. It has been generally noted that the MCMC method is quite effective without the additional requirement of any transitional algorithms and sensitivity analysis of data points. However, the degree of accuracy achievable and the iterations required to obtain a converged posterior distribution of the model parameters depend on the type of problems and the interdependency of parameters, which needs to be explored further.

7. References

[1] Beck J L and Katafygiotis L S 1998 Updating models and their uncertainties I: Bayesian statistical framework *J. Eng. Mech.-ASCE* **124(4)** 455–61

[2] Muto M and Beck J L 2008 Bayesian updating and model class selection for hysteretic structural models using stochastic simulation *J. Vib. Control* **14(1-2)** 7–34

[3] Ramos L F, Miranda T, Mishra M, Fernandes F M and Manning E 2015 A Bayesian approach for NDT data fusion: The Saint Torcato church case study *Eng. Struct.* **84** 120–9

[4] Simoen E, Roeck G D and Lombaert G 2015 Dealing with uncertainty in model updating for damage assessment: A review *Mech. Syst. Signal Proc.* **56-57** 123–49

[5] Prajapat K and Ray-Chaudhuri S 2016 Prediction error variances in Bayesian model updating employing data sensitivity *J. Eng. Mech.-ASCE* **142(12)** 04016096

[6] Prajapat K and Ray-Chaudhuri S 2018 Detection of multiple damages employing best achievable eigenvectors under Bayesian inference *J. Sound Vib.* **422** 237–63

[7] Mares C, Dratz B, Mottershead J E and Friswell M I 2006 Model updating using Bayesian estimation *In: International Proceedings on Noise and Vibration Engineering Leuven, Belgium* 2607–16

[8] Abhinav S and Manohar C S 2015 Bayesian parameter identification in dynamic state space models using modified measurement equations *Int. J. Non-linear Mech.* **71** 89–103

[9] Straub D and Papaioannou I 2015 Bayesian updating with structural reliability methods *J. Eng. Mech.-ASCE* **141(3)** 04014134

[10] Ching J and Chen Y 2007 Transitional Markov Chain Monte Carlo method for Bayesian model updating, model class selection, and model averaging *J. Eng. Mech.-ASCE* **133(7)** 816–32

[11] Friswell M I, Garvey S D and Penny J E T 1995 Model reduction using dynamic iterated IRS techniques *J. Sound Vib.* **186(2)** 311–23

[12] Jensen H A, Muñoz A, Papadimitriou C and Vergara C 2016 An enhanced substructure coupling technique for dynamic re-analyses: Application to simulation-based problems *Comput. Methods Appl. Mech. Engrg.* **307** 215–34

[13] Cheung S H and Bansal S 2017 A new Gibbs sampling based algorithm for Bayesian model updating with incomplete complex modal data *Mech. Syst. Signal Proc.* **92** 156–72

[14] Huang Y, Beck J L and Li H 2017 Bayesian system identification based on hierarchical sparse Bayesian learning and Gibbs sampling with application to structural damage assessment *Comput. Methods Appl. Mech. Engrg.* **318** 382–411

[15] Sun H and Büyüköztürk O 2016 Probabilistic updating of building models using incomplete modal data *Mech. Syst. Signal Proc.* **75** 27–40

[16] Friswell M I, Garvey S D and Penny J E T 1998 The convergence of the iterated IRS method *J. Sound Vib.* **211(1)** 123–132

IOP Conf. Series: Materials Science and Engineering **936** (2020) 012018 doi:10.1088/1757-899X/936/1/012018

Computational and Experimental Modal Analysis of a Three Storied Building Model

M Murshida Thasneem[1*], Prince Thankachan[2] and T M Madhavan Pillai[3]

[1] PG Scholar, Department of Civil Engineering, National Institute of Technology Calicut, India

[2] Research Scholar, Department of Civil Engineering, National Institute of Technology Calicut, India

[3] Professor, Department of Civil Engineering, National Institute of Technology Calicut, India

*murshidathasneem96@gmail.com

Abstract. Vibration-based health monitoring assists in determining the modal parameters and damage identification of a structure. This paper presents the computational and experimental modal analysis of a single bay three-storied building with steel slabs and aluminium columns. Three-dimensional finite element (FE) model of the structure is analysed using ABAQUS, and natural frequencies in the first three translational modes are determined. These results are compared with those obtained from field impact testing. A small variation in analytical and experimental results may be due to the simplified assumptions in FEA, physical and numerical uncertainties.

1. Introduction

The pace and complexity in construction of civil infrastructure have shown a rising trend towards the beginning of the 21st century. It demands the need for appropriate monitoring and proactive maintenance system to ensure safety and longer lifespan of the structure.

Structural health monitoring (SHM) is the continuous evaluation of the state of a structure with the help of data collected from different types of transducers attached to the system on a real-time basis. On comparing with other non-destructive monitoring methods, SHM has the advantage that it can be deployed even in non-accessible locations, and it needs no downtime [1][2]. Vibration-based techniques in SHM use modal analysis as the basic tool for the identification of damage and health evaluation of the structure [4]. Modal analysis is the method of determining the dynamic characteristics of a structure under vibration excitations. The basic idea of vibration-based health monitoring is that changes in structural properties such as mass, stiffness, and damping affect the dynamic properties of the structure. Thus by examining the changes in dynamic characteristics, identification of damage, and appropriate maintenance can be implemented [3][5].

In this study, Numerical as well as experimental modal analyses are carried out to extract the modal parameters, mainly natural frequencies from dynamic responses. A three-storied building model with steel slabs and aluminium columns is selected as the test specimen. Numerical analysis is done by developing and analysing the 3D FE model of the structure using FEM software, ABAQUS[6]. Further, the structure is excited experimentally, and dynamic responses are extracted by means of a system of transducers attached to the model. Data collected from the transducers are processed to get the required modal parameters. The accuracy of the results from the experimental modal analysis is ensured by comparing it with the numerical results.

2. Description of the test structure

An acceptable member/structure compatible with the available laboratory conditions has to be identified before starting with the modal analysis. This study concentrates on the analysis of a single bay-three storied building model with mild steel slabs and aluminium columns. This material combination is selected as it accounts for the real-time condition of flexible columns compared to the stiffer slab system. Material properties of steel and aluminium such as Young's modulus, density, and Poisson's ratio used in modelling of the structure are shown in table 1.

Table 1. Material properties of Steel and Aluminium.

Property	Steel	Aluminium
Young's Modulus (N/mm^2)	210000	70000
Density ($tonne/mm^3$)	7.85×10^{-9}	2.7×10^{-9}
Poisson's Ratio	0.3	0.3

Rectangular sections of cross sectional dimensions 25mm × 6mm are used as columns while 12 mm thick slabs of plan area 300mm × 255mm are used as floors. Floor to floor height is kept as 300mm and fixed boundary condition is given at the bottom.

3. Computational modal analysis

With the development of computational structural dynamics, modal analysis using FEM is widely used to predict the vibration parameters of a structure under study. The FE model of the structure is developed using ABAQUS, a software suite for FE Analysis and computer-aided engineering. The columns and slabs are meshed with ten noded quadratic tetrahedron elements of sizes 5mm and 20mm, respectively, as shown in figure 1. Mesh sizes are finalised by conducting the mesh convergence study from a coarser mesh to a finer one in such a way that the analysis produces acceptable results with least simulation time. Tie connection corresponding to the required bolt area is provided between each slab-column connection, and base slab is fixed to the ground.

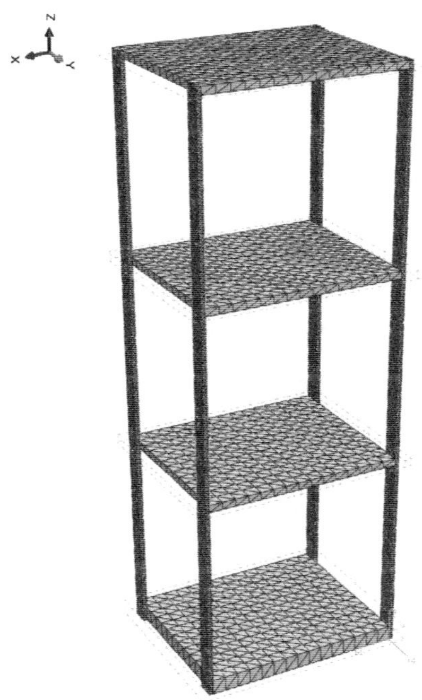

Figure 1. Finite element model showing meshing.

Natural frequencies and mode shapes of first three translational modes in x coordinate direction obtained from numerical modal analysis are shown in table 2 and figure 2, respectively.

Table 2. Natural frequencies from computational modal analysis.

Mode	Natural frequency (Hz)
M1	5.67
M2	16.15
M3	23.84

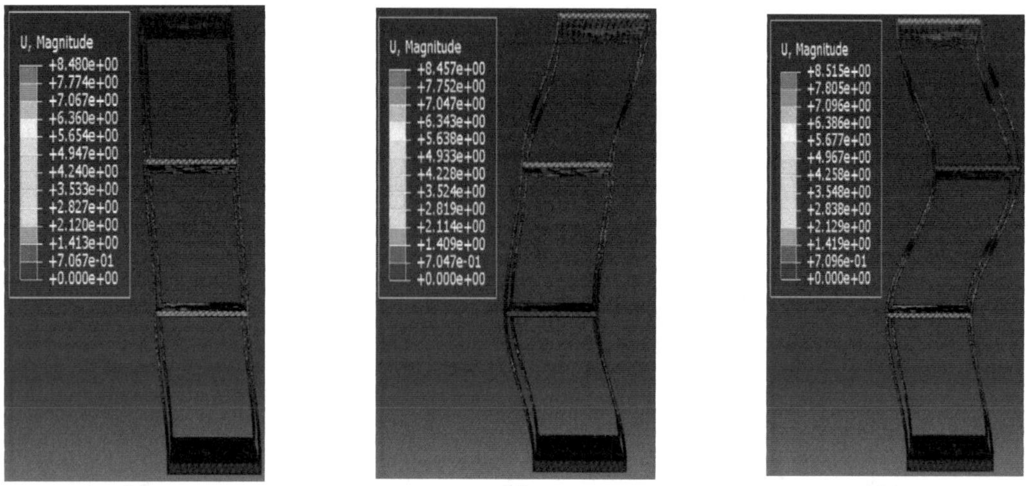

(a) (b) (c)

Figure 2. Mode shapes from computational analysis a) First mode b) Second mode 3) Third mode.

4. Experimental modal analysis

Experimental modal analysis (EMA) deals with measurement and analysis of the structural dynamic response when excited by known stimulus. It is useful in verifying numerical results as well as in determining the vibration parameters of the structure [5].

The specimen is placed at a level surface and tightly fixed at the bottom with the ground to get a fixed boundary condition. A uni-axial accelerometer is fixed at the top floor and is connected to the Data Acquisition System with controlling software through coaxial cables. Known excitations are imparted at different floor levels using impact hammer with sensitivity 2.25mV/N and is also connected to the DAQ system. The time-domain dynamic signals extracted from the data analyser are exported to the processing software, ModalVIEW for modal parameter identification. Figure 3 and figure 4 shows the experimental set-up and Frequency Response Function (FRF) obtained from the processed data, respectively.

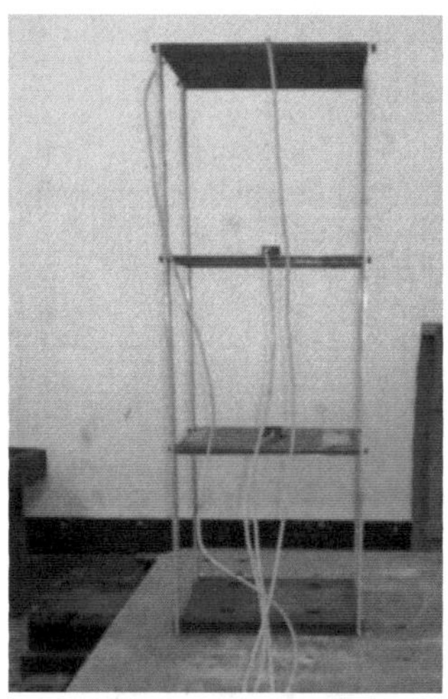

Figure 3. Experimental setup.

IOP Conf. Series: Materials Science and Engineering **936** (2020) 012018 doi:10.1088/1757-899X/936/1/012018

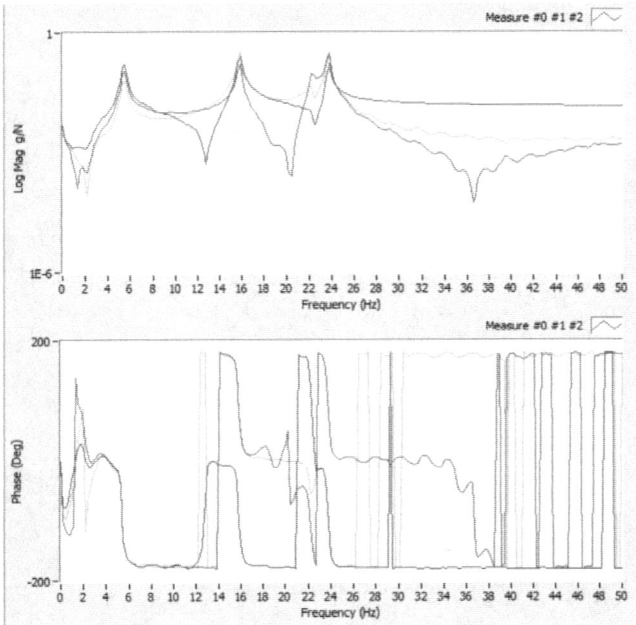

Figure 4. Frequency Response Function from the software.

Peak picking modal parameter extraction algorithm is used to identify the natural frequencies from the FRF. The procedure is shown in figure 5. The first three Natural frequencies of translational modes in the x- coordinate direction obtained from EMA is shown in table 3 along with percentage damping. Figure 6 shows the first three mode shapes obtained from the experimental analysis.

Table 3. Natural frequencies and percentage damping from EMA.

Mode	Natural frequency (Hz)	Damping (%)
M1	5.39	1.92
M2	15.84	0.83
M3	23.67	0.37

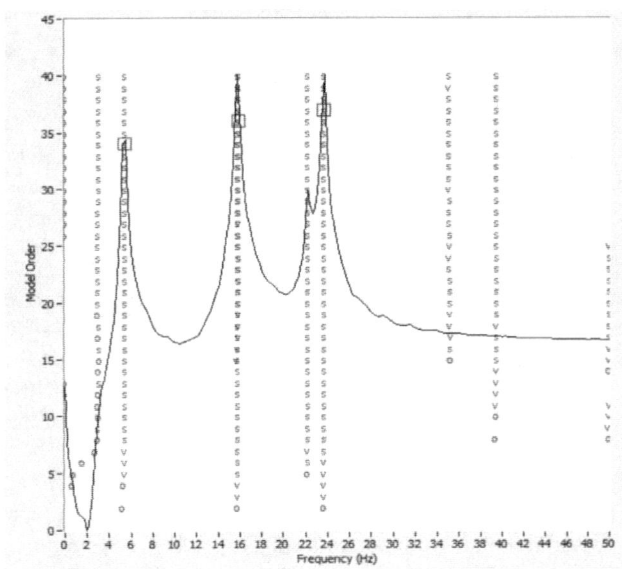

Figure 5. Natural frequency extraction by peak picking.

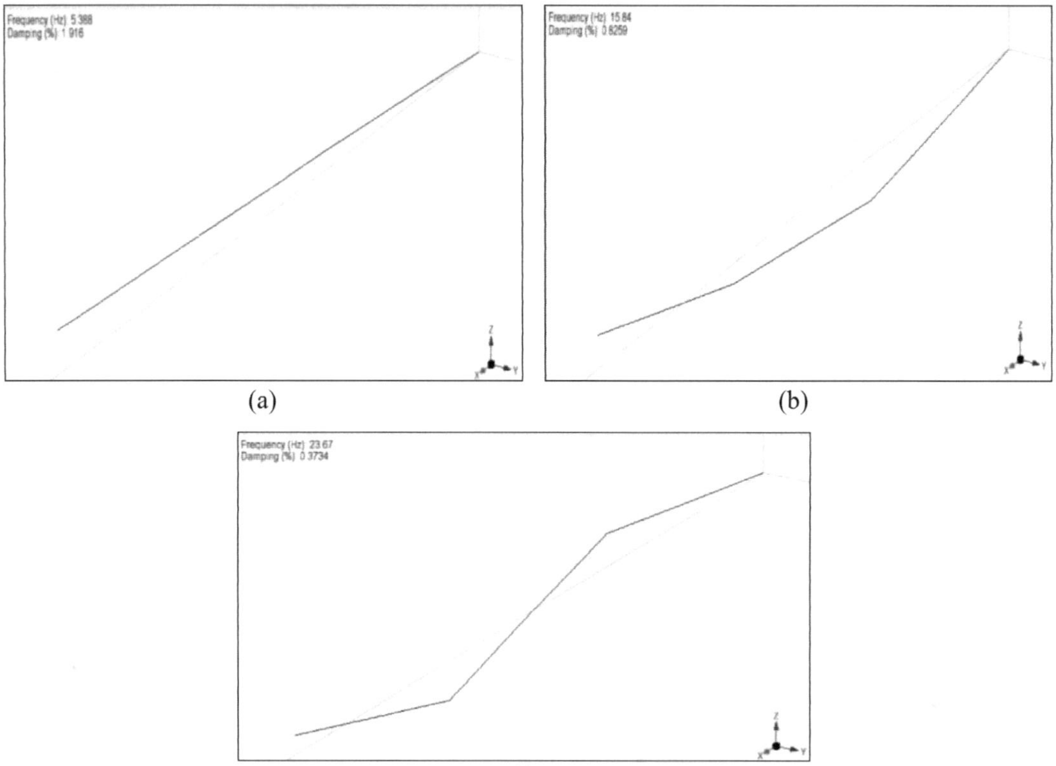

(c)

Figure 6. Mode shapes from experimental analysis a) First mode b) Second mode c) Third mode.

5. Results and discussions

Natural frequencies obtained from the computational and experimental modal analysis are compared. Percentage error between the results of both methods is summarised in table 4.

Table 4. Comparison of Natural frequencies.

Mode	Natural frequencies (Hz)		% change
	Computational modal analysis	Experimental modal analysis	
M1	5.67	5.39	4.94
M2	16.15	15.84	1.92
M3	23.84	23.67	0.71

6. Conclusion

Modal frequencies from the experimental modal analysis are having smaller values when compared with those obtained from numerical analysis. This variation in modal parameters may be due to the physical and numerical uncertainties like boundary conditions, joint stiffness, and material properties. The numerical models give the responses based on ideal assumptions of material homogeneity like mass distribution, isotropy, and Poisson's ratio. However, in real cases, the materials may not behave like the ideal one. Comparing numerical and experimental analysis, a minimum of 0.71% and a maximum of 4.94% percentage error in modal frequencies are observed.

To improve the consistency between the results, it is recommended to calibrate the FE model realistically with respect to experimentally obtained results so that the errors resulting from simplified assumptions and material uncertainties are minimised. Also, advanced comparative study of mode shapes will further enhance the study.

7. References

[1] Farrar C R and Worden K 2007 An introduction to structural health monitoring *Philosophical Transactions of The Royal Society A: Mathematical, Physical and Engineering Sciences* **365(1851)** 303–15

[2] Brownjohn James M W, De Stefano Alessandro, Xu You-Lin, Wenzel Helmut and Aktan A E 2011 Vibration-based monitoring of civil infrastructure: challenges and successes *Journal of Civil Structural Health Monitoring* **1(3-4)** 79–95

[3] Raoa P S and Ratnam C 2012 Experimental and analytical modal analysis of welded structure used for vibration based damage identification *Global Journal of Researches in Engineering* **12(1-A)**

[4] Thankachan P, Garg R K and Raj A 2016 Structural health monitoring using experimental modal analysis *Proceedings of Recent Advances in Civil Engineering and Technology* **2** pp 120–4

[5] Dammika A J, Kawarai K, Yamaguchi H, Matsumoto Y and Yoshioka T 2015 Analytical damping evaluation complementary to experimental structural health monitoring of bridges *Journal of Bridge Engineering* **20(7)** 04014095

[6] ABAQUS US. Version 6.13 Dassault Systémes Simulia Corporation, Providence, RI, USA 2013

IOP Conf. Series: Materials Science and Engineering **936** (2020) 012019 doi:10.1088/1757-899X/936/1/012019

Effect of Water-Cement Ratio on Mix Design and Mechanical Strength of Copper Slag Aggregate Concrete

Swetapadma Panda[1*], Pradip Sarkar[2] and Robin Davis[3]

[1] Department of Civil Engineering, Institute of Technical Education and Research, Siksha O Anusandhan University, Bhubaneswar 751030, Odisha, India

[2] Department of Civil Engineering, National Institute of Technology, Rourkela 769008, Odisha, India

[3] Department of Civil Engineering, National Institute of Technology, Calicut 673601, Kerala, India

*swetapadma.nit2017@gmail.com

Abstract. In concrete, the distinct influence on most or the entirety of the properties is the measure of water utilised in the matrix. In concrete mix design, the water-cement ratio is responsible for binding all constituents of concrete together. If the water-cement proportion is higher, it brings about wider spacing between the cement aggregates and thus, influences the compaction. Correspondingly, concrete's durability and compressive strength are decreased due to increased dampness levels. Especially with added fine aggregates the water demand increases because of the increased surface area of concrete which prompts a higher water-cement ratio. Wider spaces are formed between the aggregate materials with the addition of extra water and when the moisture vaporises, voids are filled up with air. The subsequent insufficient compaction lessens the concrete's strength. Sets of basic experimental investigations were conducted to examine the impact of various water-cement proportion on mix design and mechanical strength of concrete incorporated with fine aggregate in terms of copper slag. Natural sand replaced with copper slag with four different percentage and correlation coefficients between mechanical properties of copper slag concrete were also evaluated.

1. Introduction

Concrete is now the biggest used composite structural material in today's era of the construction world. The concrete matrix comprises of hard and chemically inert substances usually fine and coarse aggregates and binder typically cement when added with water works as adhesive, which hold the concrete matrix together. These ingredients of concrete making, influence the properties of concrete largely. The wish for the structure to be strong and durable exclusively depends on the mix design of the structural concrete at the rudimentary stage. The design mix of concrete is the method of choosing and arranging all constituents of concrete and deciding their relative quantities to produce concrete of desired strength, durability and workability under the circumstances as economic as possible [1,2]. There are some common factors that affects not only the mix design but also the strength and durability of structural concrete to a great degree such as (i) water/cement ratio, (ii) compaction (iii) ingredients of concrete (v) shape and size of aggregate (vi) weather condition and casting temperature [3-6]. The potential for immense strength overly indicates the quality and grade of concrete, which is related to the long-term performance of the concrete structure. The evaluation of ability of the concrete structure to withstand various external loads can be done through different laboratory tests for compression, tension and flexure, void content measurement test etc.

Apart from this, curing of concrete and the age of concrete are also influential in achieving strength and durability [7,8]. Out of all stated factors, utilization of water is significant on the grounds that the proportion of water to cement is the most basic factor for producing "perfect" concrete. The working mechanism of water in the concrete mix is a two-stage process, one is bonding and the second one is workability. Water establishes and increases the bond between all constituents of a concrete mix by behaving as a lubricant. At a perfect water-cement ratio, the perfect fluidity of the mix can be achieved [9-12]. If the percentage of water is increased in the concrete mix, the mass concrete becomes permeable and penetrable and in the event that it is less, concrete gets dry and inefficient to achieve full strength due to poor binding quality. At the same time, the type and quality of aggregates used in concrete production are also very important [13,14]. An aggregate provides body i.e. volume or mass to the concrete but ultimately does contribute significantly to the final properties of concrete. Research and development of new materials are continuously going on to replace the conventional concrete not only in order to achieve the requisite properties for application on the different situations but also to set back ecological and environmental inconsistency. Many types of industrial by-products and waste materials are now being used in concrete production [15-18]. These waste materials are being incorporated in concrete by replacing cement, coarse aggregates or fine aggregates. This incorporation of industrial waste in concrete making, not only boost better strength but also imparts better imperviousness and durability by making dense and homogeneous concrete.

Copper slag is one such trash material generated at the time of refining and smelting process of copper ore. Naked disposal of copper slag is very hazardous for the environment as it contains highly toxic elements like arsenic, barium, cadmium, copper, lead, and zinc [19]. Some of the researchers have examined copper slag as binder material where many have used as filler material by replacing conventional sand. Recent studies revealed usages of copper slag in concrete as partial or full replacement of fine aggregate, because it has definite positive impacts on enhancements in the strength of structural concrete leading to sustainability in the construction industry [20-23]. But, because of deficiency in appropriate concrete mix design utilizing copper slag as fine aggregate, in the field for structural concrete is very little. Though the outward form of copper slag is very similar to natural sand its physico-chemical parameters are much different from conventional sand [24]. So, its impact on mix design of concrete incorporating copper slag as fine aggregates cannot be ignored. For the present investigation, Ordinary Portland Cement (OPC) is taken as a binder, both copper slag and natural sand used as fine aggregate and coarse aggregates are collected from the nearby crusher. Copper slag is collected from Hindustan Copper Pvt. Ltd., Jharkhand, India and natural sand is collected from basin of Brahmani river, Rourkela, Odisha, India.

2. Characterization of copper slag

Characterization of material is the practice of quantifying and defining the mechanical, chemical, physical and micro structural properties of materials. The material characterization provides the essential constituents, its chemical and structural composition with material defects that are substantial for research, utilization or a particular preparation, and study of properties. Table 1 describes the important materialistic aspects of fine aggregate and their significant usability in the concrete mix design.

Table 1. Summary of aggregate properties and their significance.

SL#	Characteristics	Significance
1.	Grain structure of particles	Affects workability of fresh concrete
2.	Relative density	Optimise a mixer, unit weight of concrete, yield of concrete
3.	Bulk density	Useful for checking uniformity of aggregate
4.	Absorption and surface moisture	Affects the mix design, fixing water/cement ratio
5.	Deleterious substance	Affects hydration of cement
6.	Grading	Economises cement content and improves workability

2.1. Physical characterization

The physical characteristics such as shape/texture and particle size distribution of copper slag, natural sand, and coarse aggregates were examined and evaluated by following the code, IS 2386-Part-III 1963 [25]. In the process of smelting of copper slag, slag is formed from the impurities that float on the molten state of metal and after dampen down with water, waste is then disposed in the form of angular granules. Copper slag is black in colour, glassy in texture and has angular sandy grains in nature and has almost similar/higher particle size range like conventional sand. Figure 1 shows the physical appearance of copper slag both at the raw and microscopic level.

(a) (b)

Figure 1. (a) Copper slag, (b) Copper slag under microscope.

Relative density and water absorption of copper slag and natural sand were obtained by following the procedure as per IS Code: 2386-Part-III 1963 [25]. Similarly, the particle size distribution curve and corresponding fineness index of both copper slag and natural sand were evaluated by IS 2386-Part-I 1963 [26]. All materialistic properties of both copper slag and natural sand are given in table 2. Gradation curve and bulking of both copper slag and natural aggregate are shown in figure 2 and figure 3.

Table 2. Physical properties of Copper slag and Natural sand.

Properties	Copper slag	Natural sand
Relative density	3.89	2.60
Water absorption	0.37%	0.80%
Fineness modulus	3.68	2.89

Figure 2 shows the grading limits for both the fine aggregates which are similar to zone III sand and the grading of copper slag can be termed as continuous grading. Figure 3 indicate that copper slag is a coarser aggregate, its increment in volume due to the presence of free water is insignificant as the thickness of the dampness layer is very small when compared with its particle size, whereas natural sand is medium grade sand. The mix design of concrete with copper slag will slightly differ from conventional concrete design because its relative density is high and water absorption capacity is low.

2.2. Chemical characterization

Chemical characterization and X-ray Fluorescence (XRF) tests for both copper slag and natural sand and X-ray Diffraction (XRD) test was conducted only for copper slag. The XRF was conducted through Zetium PAN Analytical XRF spectrometer with WROXI as a quantitative archive and XRD was conducted through Rigaku XRD Ultima IV with cross beam optics (CBO) technology and automatic alignment capability. XRD test of copper slag is conducted to identify the presence of minerals with phase angel, and XRD plot of copper slag is shown in figure 4. Copper slag mainly contains the oxides of iron and silica. The results of the XRF test data of both copper slag and natural sand are presented in table 3.

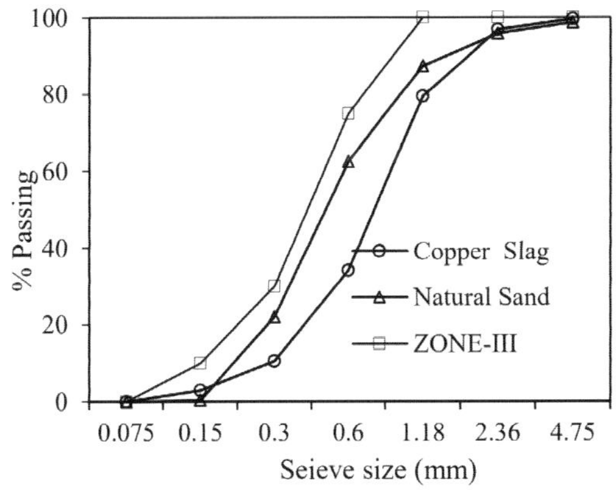

Figure 2. Gradation of Copper slag and Natural sand.

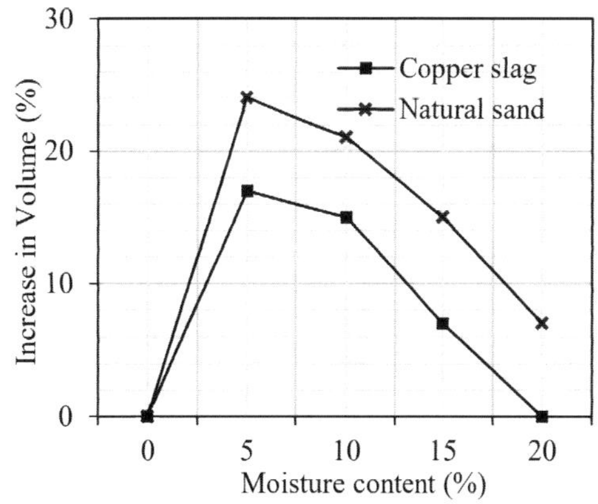

Figure 3. Effect of moisture content on the bulking of Copper slag and Natural sand.

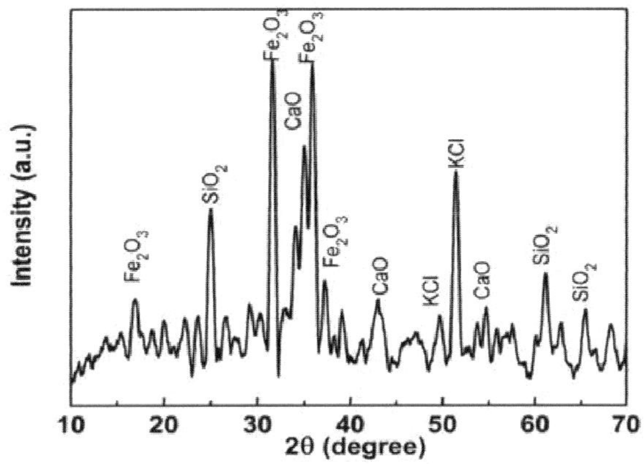

Figure 4. XRD plot of Copper Slag.

As per the given data in table 3 copper slag contains natural pozzolanas of class 'N' in accordance with ASTM C 618-2019 [27] as the summation of oxides of silicon, aluminium and iron exceed 70%.

Table 3. Chemical composition of Copper slag and Natural sand (%).

	SiO_2	Al_2O_3	Fe_2O_3	TiO_2	MgO	CaO	K_2O	SO_3	Na_2O	CuO
Copper slag	24.87	3.09	64.27	0.21	0.89	1.05	0.56	1.45	1.4	0.04
Natural sand	98.3	-	-	-	-	-	-	-	-	-

3. Mix design with copper slag and specimen preparation

The mix design of copper slag concrete is slightly complex than conventional concrete as more design parameters such as low water absorption capacity of the material, high relative density, higher fineness modulus and wide range of particle size, etc. are needed to be considered. At present as a result of a deficiency in suitable mix design procedures, the developed concrete blended with copper slag shows relatively low performance. For the current study, M_{40} grade of concrete, OPC of 53 grade, 20mm and 10mm graded angular coarse aggregate of nominal size and natural sand confirming to Zone III were used. Three different water-cement ratios were taken for the design mix with copper slag as fine aggregate. As per IS 456-2000 [28], the maximum water-cement ratio of M_{40} grade of the concrete is obtained as 0.45. So, for the present investigation, three water-cement ratios were taken such as 0.55, 0.45, and 0.35 in a reducing manner for respective exposure conditions as per IS 10262-2000 [29]. Four different percentage of copper slag were considered along with control specimens from 0% to 80% (MCS0, MCS40, MCSS60, and MCS80). The schematic diagram for design mix concrete with copper slag is shown in figure 5.

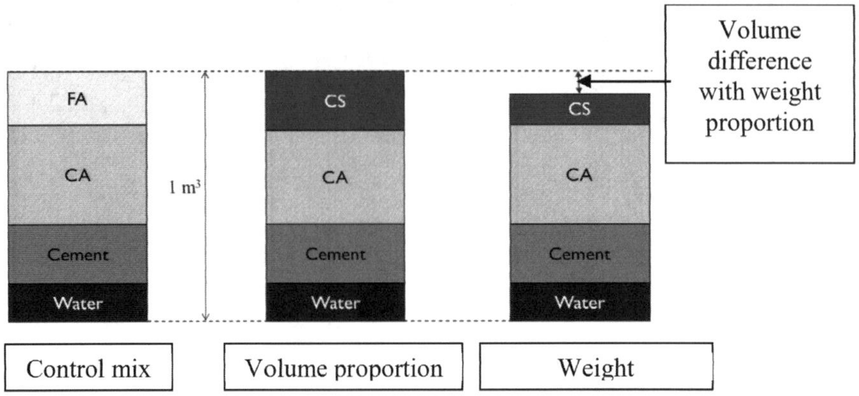

Figure 5. Schematic diagram for mix design with copper slag.

To study the effects of reduced water-cement ratios on the fresh concrete properties and hardened concrete properties of copper slag blended concrete, all the basic tests have been conducted by following the Indian Standard code. The workability of copper slag blended concrete was studied through a slump test for each mix with four selected water-cement ratios by following IS 1199-1959 [30]. Similarly, compressive, split tensile and flexure strengths were evaluated experimentally following the guidelines of IS 516-1959 [31], ASTM C496-2017 [32] and ASTM C293-2016 [33] respectively. Concrete cubes of 150 mm, cylinders of 150 mm diameter × 300mm height and prisms of 100mm × 100mm × 700mm were cast respectively for compressive, split tensile and flexure strength test. After completing one day of casting all specimens were removed from the mould and kept in water tank for conventional curing of 28 days. For M_{40} grade of concrete a mix proportion 1: 1.35: 2.7 was determined as per IS 10262-2000 [29]. Cement of 450 kg/m^3, fine aggregate of 608 kg/m^3, coarse aggregate of 1215 kg/m^{3}with three selected water-cement ratios 0.55, 0.45, and 0.35 were taken for preparing the mix.

4. Experimental results

The initial ability of the concrete mix to adopt virtually any shape can be evaluated from its slump test. Figure 6 shows the graphical presentation of all respective slump values for each copper slag blended concrete with three selected water-cement ratios. With 0.55 and 0.45 water-cement ratio, the copper slag blended concrete was found to have high to medium workabilities. For a water-cement ratio of 0.35, the degree of workability is medium and perfectly plastic for copper slag concrete mixes (MCS40, MCS60 and MCS80). For each selected water-cement ratio the workability of copper slag blended concrete is increasing with the increase in the percentage of copper slag on concrete. For water-cement ratios of 0.55 and 0.45, some segregation and bleeding were observed for the mix MCS60 and MCS80. For water-cement ratio of 0.35, segregation was not observed for MCS60 and MCS80 because of properly proportioned cohesive concrete mix with copper slag.

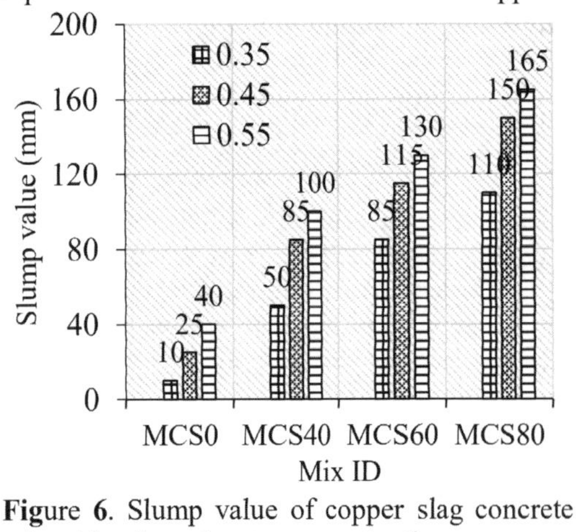

Figure 6. Slump value of copper slag concrete mix with selected water-cement ratio.

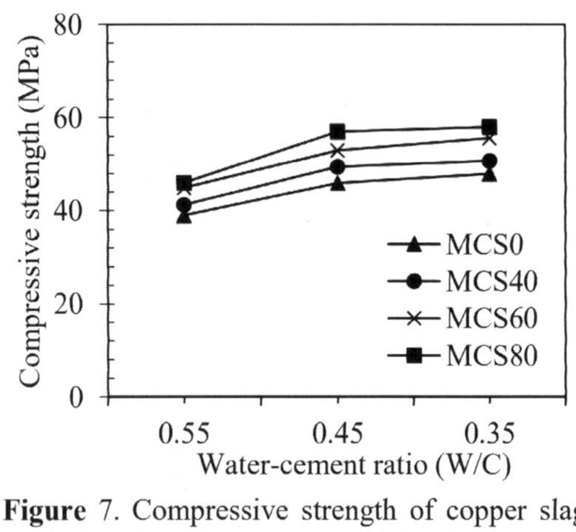

Figure 7. Compressive strength of copper slag concrete mix versus water-cement ratio.

Figure 7 shows compressive strength of all copper slag blended concrete mix for all three selected water-cement ratios. Compressive increases as the copper slag percentage increases and it is inversely proportional to water-cement ratios. As the water-cement ratio increases, the compressive strength of copper slag blended concrete also decreases.

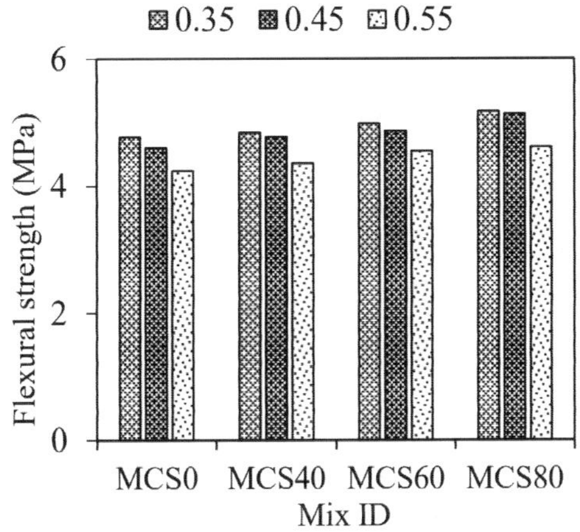

Figure 8. Flexural strength of copper concrete mix with selected water-cement ratio.

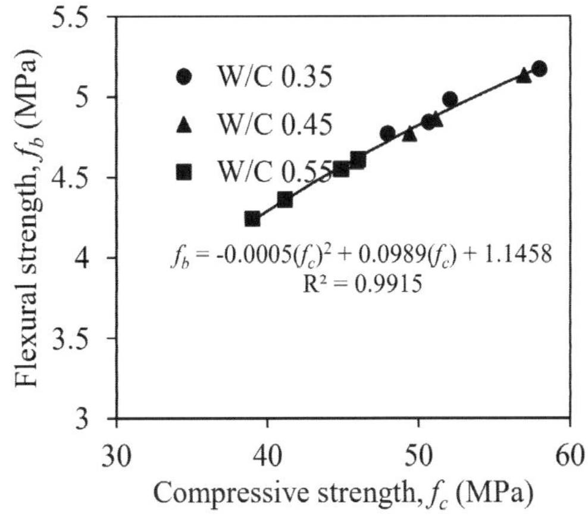

Figure 9. Correlation between compressive and flexural strength of copper slag concrete.

Figures 8 and 10 show strength of copper slag blended concrete for flexure test and split tensile test cured at 28 days for all three selected water-cement percentages. In figure 9, a second order polynomial correlation is established between the compressive strength and flexural strength for all selected water-cement ratios. Similarly, another second order polynomial correlation is established between compressive strength and split tensile strength shown in figure 11. Figure 9 and 11 indicates a good correlation between compressive, split tensile and flexural strength of copper slag blended concrete mixes for all selected water-cement ratios.

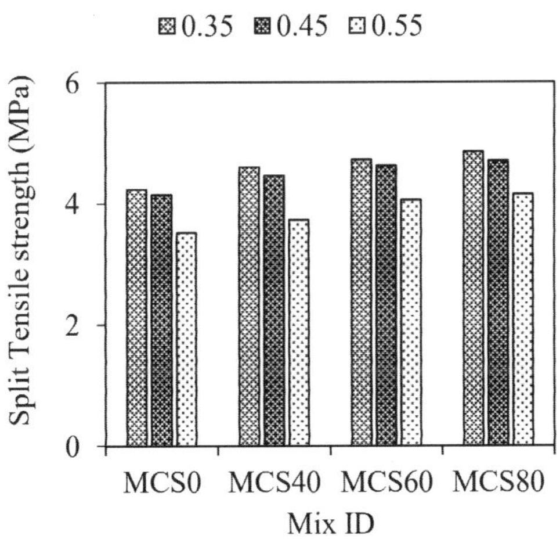

Figure 10. Split tensile strength of copper slag concrete mix with selected water-cement ratio.

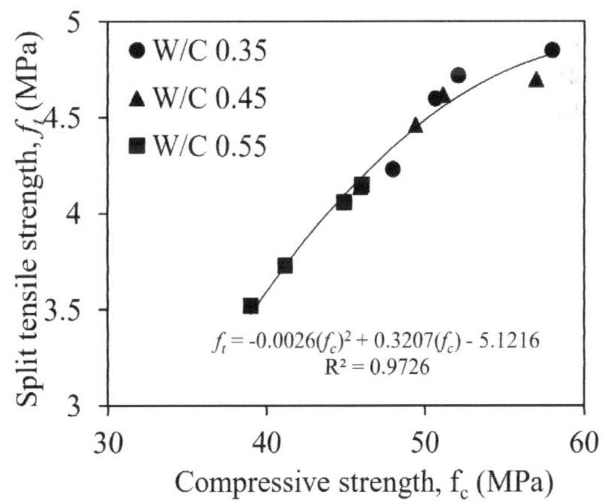

Figure 11. Correlation between compressive and split tensile strength of copper slag concrete.

5. Conclusions

The following conclusions are drawn concerning mix design with copper slag as fine aggregate in concrete making.

1. The addition of copper slag as fine aggregate improves mechanical strength of resulting concrete in comparison with the control concrete specimens.

2. Due to low water absorption capacity, copper slag incorporated concrete performs better with low water-cement ratio.
3. The bending strength and split tensile strength of copper slag incorporated concrete can be calculated from the compressive strength using the proposed relationships.

6. References

[1] Malkapur Santhosh M, Ashish Anand, Amit Prakash Pandey, Alok Ojha, Nimesh Mani and Narasimhan Chandrashekara Mattur 2014 Effect of Mix Parameters on the Strength Performance of Waste Plastics Incorporated Concrete Mixes *Journal of Structures*
[2] Nadesan Manu S and Dinakar P 2017 Mix design and properties of fly ash waste lightweight aggregates in structural lightweight concrete *Case Studies in Constr. Mat.* **7** 336–47
[3] Esmaeilkhanian B, Khayat K H, Yahia A and Feys D 2014 Effects of mix design parameters and rheological properties on dynamic stability of self-consolidating concrete *Cem. and Conc. Compos.* **54** 21–8
[4] Ilinoiu George 2004 Concrete mix design optimised approach *Civil Engineering Dimension* **6(1)** 49–56
[5] Wardeh George, Elhem Ghorbel and Hector Gomart 2015 Mix design and properties of recycled aggregate concretes: applicability of Eurocode 2 *Int. J. of Conc. Struc. and Mat.* **9(1)** 1–20
[6] Tang Chao-Wei, Chiu-Kuei Cheng and Ching-Yuan Tsai 2019 Mix Design and Mechanical Properties of High-Performance Pervious Concrete *Materials* **12(16)** 2577
[7] Al-Gahtani A S 2010 Effect of curing methods on the properties of plain and blended cement concretes *Constr. and Build. Mat.* **4(3)** 308–14
[8] Raheem Akeem Ayinde, Aliu Adebayo Soyingbe and Amaka John Emenike 2013 Effect of curing methods on density and compressive strength of concrete *Int. J. of Applied Sci. and Tech.* **3(4)** 55–4
[9] Vu Xuan Hong, Yann Malecot, Laurent Daudeville and Eric Buzaud 2009 Effect of the water/cement ratio on concrete behaviour under extreme loading *Int. J. for Num. and Ana. Methods in Geomech.* **33(17)** 1867–88
[10] Sen Sidharth and K Balakrishna Rao 2017 Effect of water cement ratio on the workability and strength of low strength quarry dust concrete *Int. J. of Civil Eng. and Tech.* **8(10)** 1448–55
[11] Rahmani Kamal, Abolfazl Shamsai, Bahram Saghafian and Saber Peroti 2012 Effect of water and cement ratio on compressive strength and abrasion of microsilica concrete *Middle-East J. of Scientific Research* **12(8)** 1056–61
[12] Salema Mohammed, Abas Abdela, and R. K. Pandeyb 2015 Effect of Cement-Water Ratio on Compressive Strength and Density of Concrete *Int. J. of Eng. Research and Technology* **4(2)**
[13] Abdullahi M 2012 Effect of aggregate type on compressive strength of concrete *Int. J of civil and structural engineering* **2(3)** 791–800
[14] Kang Ma and Li Weibin 2018 Effect of the aggregate size on strength properties of recycled aggregate concrete *Adv. in Mat. Sci. and Eng.*
[15] Tavakoli Davoud, Masoumeh Hashempour and Ali Heidari 2018 Use of waste materials in concrete: a review *Pertanika J. of Sci. & Tech.* **26(2)** 499–522
[16] Batayneh Malek, Iqbal Marie and Ibrahim Asi 2007 Use of selected waste materials in concrete mixes *Waste management* **27**(12) 1870–6
[17] Patel Ankit J, Harsh N Patel, Harsh R Patel and Hardik R Patel 2014 Review on use of Waste Material in Concrete *Int. J. for Tech. Res. in Eng.* **2** 307–10
[18] Tiwari Anshuman, Sarbjeet Singh and Ravindra Nagar 2016 Feasibility assessment for partial replacement of fine aggregate to attain cleaner production perspective in concrete: A review *J. of Cleaner Prod.* **135** 490–507
[19] Gorai B, and Jana R K 2003 Characteristics and utilization of copper slag—a review. *Resources, Conservation and Recycling* **39(4)** 299–313
[20] Al-Jabri K S, Al-Saidy A H and Taha R 2011 Effect of copper slag as a fine aggregate on the properties of cement mortars and concrete *Const. and Build. Mat.* **25(2)** 933–8
[21] Al-Jabri K S, Hisada M, Al-Saidy A H and Al-Oraimi S K 2009 Performance of high strength concrete made with copper slag as a fine aggregate *Constr. and Build. Mat.* **23(6)** 2132–40

[22] Wu W, Zhang W and Ma G 2010 Optimum content of copper slag as a fine aggregate in high strength concrete *Materials & Design* **31(6)** 2878–83
[23] Shi C, Meyer C and Behnood A 2008 Utilization of copper slag in cement and concrete *Resources, Conservation and Recycling* **52(10)** 1115–20
[24] Nataraja M C, Chandan G N and Rajeeth T J 2014 Concrete mix design using copper slag as fine aggregate *Int. J. of Civil Eng. and Tech.* 5-9
[25] BIS 1963 *Methods of Test for Aggregates for Concrete* (Bureau of Indian Standards IS 2386-1963 (Part III): New Delhi)
[26] BIS 1963 *Methods of Test for Aggregates for Concrete* (Bureau of Indian Standards IS 2386-1963 (Part I): New Delhi)
[27] ASTM 2019 *Standard Specification for Coal Fly Ash and Raw or Calcined Natural Pozzolanas for Use in Concrete* (ASTM C618-19: West Conshohocken, PA)
[28] BIS 2000 *Plain and Reinforced concrete: Code of Practice* (Bureau of Indian Standards IS 456-2000: New Delhi)
[29] BIS 2009 *Concrete Mix Proportioning: Guidelines* (Bureau of Indian Standards. IS 10262-2009: New Delhi)
[30] BIS 1959 *Methods of Sampling and Analysis of Concrete* (Bureau of Indian Standards IS 1199-1959: New Delhi)
[31] BIS 1959 *Method of Tests for Strength of Concrete* (Bureau of Indian Standards IS 516-1959: New Delhi)
[32] ASTM 2017 *Standard Test Method for Splitting Tensile Strength of Cylindrical Concrete Specimens* (ASTM C 496-17: West Conshohocken, PA)
[33] ASTM 2016 *Standard Test Method for Flexural Strength of Concrete Specimens* (ASTM C -293-16: West Conshohocken, PA)

IOP Conf. Series: Materials Science and Engineering **936** (2020) 012020 doi:10.1088/1757-899X/936/1/012020

Finite Element Modelling of Confined Masonry Wall under In-plane Cyclic Load

Bonisha Borah[1*], Hemant B Kaushik[1] and Vaibhav Singhal[2]

[1] Department of Civil Engineering, Indian Institute of Technology Guwahati, Assam, India

[2] Department of Civil and Environmental Engineering, Indian Institute of Technology Patna, Bihar, India

[*]bonisha@iitg.ac.in

Abstract. Confined masonry (CM) is an effective building construction technology for low-tech construction in seismic-prone regions. This paper focuses on developing a numerical model to study the in-plane seismic behaviour of CM wall. A half-scaled CM wall specimen is designed based on Mexican practice and tested under a combination of vertical load and in-plane cyclic load up to failure. Based on that a three dimensional (3D) finite element macro-modelling technique is developed using the commercial software package ABAQUS, where the masonry panel, RC tie-frame, and reinforcements are discretely modelled. The strength and stiffness degradation of masonry and concrete is defined using Concrete Damaged Plasticity (CDP), which is a damaged plasticity-based continuum constitutive model. All the input data in the numerical model are based on experimental test results and data obtained from past literature or inverse fitting, with justification. From the analysis, it is observed that the numerical results are in good agreement with the experimental. Therefore, the proposed model can be successfully practiced to model confined masonry walls.

1. Introduction

Confined masonry (CM) construction has emerged as a building technology that offers an alternative to both unreinforced masonry (URM) and infilled reinforced concrete (RC) frame construction in several countries or regions of high seismic risk including North and South American, European, African and Asian countries. The uniqueness of the CM building primarily lies in the construction technique, where masonry walls of the building are constructed prior to the casting of small RC tie-elements (columns and beams) around the wall and some other specific locations. The improved seismic performance is mainly due to this confinement and bond formation at wall-to-tie interface [1]. Although basic design guidelines of CM building have been addressed in various country codes, modelling and analysis of CM structure appears to be a challenging task. The finite element (FE) modelling method is a universally accepted analysis method, which provides opportunity to study the structure more thoroughly. FE analysis of masonry structures can be performed in two different level - micro or macro. Micro-modelling approaches are mostly suited for small structural elements in order to closely represent the heterogeneous states of masonry using the properties of each element and interface [2, 3]. However, macro-modelling approaches are used to represent the global structural behaviour, where the distinction between the individual units and the joints is not made and the material parameters are obtained from the masonry tests under homogeneous states of stress [4, 5, 6]. In this study, an experimental test was conducted on a typical CM wall and based on that a FE model was proposed and validated.

IOP Conf. Series: Materials Science and Engineering **936** (2020) 012020 doi:10.1088/1757-899X/936/1/012020

2. Experimental study to understand the In-plane behaviour of CM wall

A half-scaled model of a typical CM wall was considered for the study. The wall was designed and detailed (figure 1) as per Mexican guidelines [7]. Material properties, as given in table 1, were evaluated using relevant standards by conducting tests on masonry, concrete, and reinforcing bars. Masonry panel was constructed atop the foundation beam using solid clay bricks in English bond pattern with 1:1:6 (lime: cement: sand) mortar. Confining RC (frame) elements were constructed after the masonry panel. An RC slab of 75 mm thickness was constructed over the specimen to study its influence on the behaviour. As shown in figure 2a, the imposed (dead and live) load was applied on the test specimen by placing steel plates over the RC slab (around 17 kN) to simulate the loading of a typical two-story residential building on an exterior ground floor wall. The wall was tested under pseudo-static cyclic lateral loading applied at the slab level using servo-controlled hydraulic actuator of ±125 mm stroke length and 250 kN load capacity. The in-plane loading consisted of displacement-controlled slow cycle of gradually increasing story drifts from 0.10% till failure as per ACI 374.1-05 [8]. Three cycles of each displacement level were applied and the response was recorded using a data acquisition system. The obtained response of the specimen in the form of hysteretic for first cycle of every displacement level along with the envelope curve is shown in figure 2b.

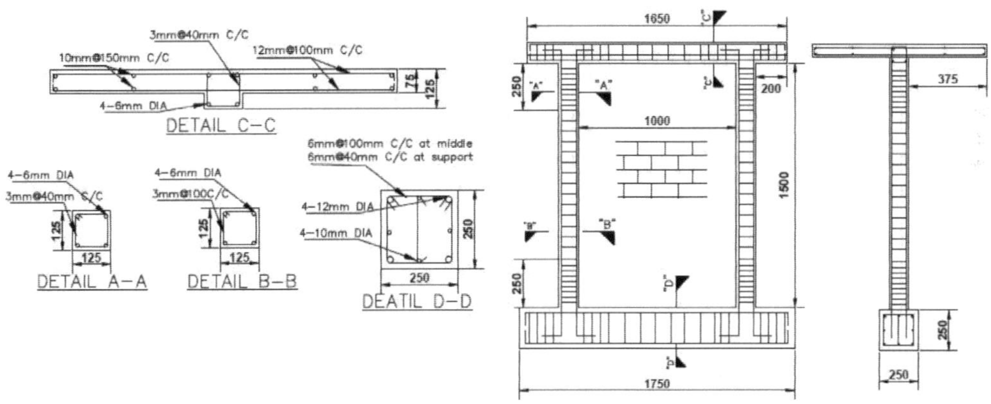

Figure 1. Geometry and details of reinforcement of the considered CM wall.

Table 1. Properties of the CM wall specimen.

Compressive strength of concrete (f_{ck})	25MPa
Compressive strength of masonry prism (f_m')	5.2 MPa
Elastic modulus of masonry (E_m)	2000 MPa
Yield strength of reinforcing bars (f_y)	446MPa(3φ); 549MPa(6φ)

Note: φ represents diameter of bar

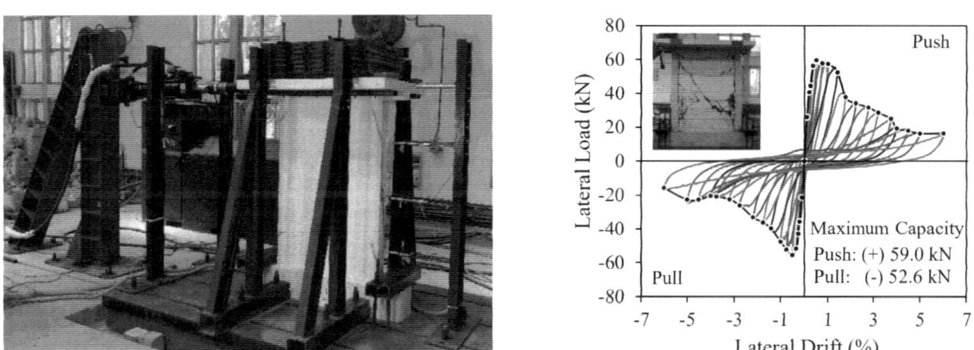

(a) (b)

Figure 2. Experimental study: (a) Test set-up, and (b) Hysteretic response with envelope.

3. Finite element (FE) modelling scheme

To capture the behaviour of the tested specimen with FE method, the selected discretization scheme and constitutive model are explained below:

3.1. *Discretization scheme for masonry and RC tie-elements*

In the present study, macro-modelling technique was considered for masonry to avoid material property definition of individual unit and joint. The FE analysis was carried out using the commercial software package ABAQUS [9]. 8 noded three-dimensional solid continuum elements with reduced integration (C3D8R) from the explicit element library were used for the concrete frame and masonry. Due to the selected reduced integration, hourglass control was activated to avoid zero-energy deformation.

3.2. *Discretization scheme for reinforcement*

Reinforcement bars were modelled using wire feature associated with 2 noded three-dimensional truss elements (T3D2). A proper bonding between reinforcement and concrete was ensured by applying embedded body constraints. The constraints allow the reinforcement nodes to displace exactly equal to the nodes of the surrounding concrete.

3.3. *Constitutive model - concrete damage plasticity model (CDP)*

In this study, the nonlinear behaviour of concrete and masonry were modelled using the CDP model (figure 3) selected from the ABAQUS material library [10]. The CDP model is designed to simulate the behaviour of concrete and other quasi-brittle materials, such as masonry, in any type of structure subjected to monotonic as well as cyclic loading. This model uses the concepts of isotropic damaged elasticity along with isotropic compressive and tensile plasticity to account for the multiple damage states. The evolution of yield surface is controlled by two hardening variables, i.e., equivalent plastic strain in compression and tension, which are linked to the compressive crushing and tensile cracking failure mechanisms, respectively. The transition from tension to compression and vice versa is another important aspect that is considered in modelling the response under cyclic loading. The cracks occurred during stress reversal from compression to tension, results in reduction of stiffness and these cracks are expected to close in the next stress reversal cycle. The change in load from tension to compression leads to partial recovery of the compressive stiffness. Also, total or partial recovery of tensile stiffness is expected when uncrushed or partially crushed concrete undergo tensile loading. Therefore, the stiffness degradation variable d depends on the stress state; recovery parameters w_c and w_t; and the degradation (or damage) parameters d_c and d_t as shown in figure 3. w_c= 1 and w_t= 0 will be used for this FE analysis. d_c and d_t were determined using Park Pivot rule model [11]. Among the input parameters for CDP, the plasticity characteristics of the materials need different types of experimental tests and in absence of such data, the parameters were determined by calibrating with the experimental results, and by using common values recommended in the literature. The dilatancy in CDP model is controlled by the parameter 'dilation angle' in the plastic potential function. The values of dilation angle as 30° and 12° were found to be appropriate for concrete and masonry, respectively, in order to generate the wall behavior close to experimental result. The other parameters, e, σ_{b0}/σ_{c0}, K_c and viscosity were taken with their default values from ABAQUS as: 0.1, 1.16, 0.667, and 0.00001, respectively. The compressive and tensile stress-strain relationships of materials are explained in the next section.

4. Calibration of material models

4.1. *Masonry*

The stress-strain curve specified for masonry in compression was developed using the parabolic model proposed by Kaushik et al. [12] as shown in figure 4a. The tensile behaviour of masonry was assumed to be linearly elastic till it reaches its tensile strength, followed by a linear softening curve as shown in figure 4b. The tensile strength is estimated as a fraction of its compressive strength and for the present study, the tensile strength was assumed as 8% of the compressive strength. The post-peak behaviour was modelled in such a way that the masonry completely loses its tensile strength at ultimate strain (equal to 10 times the strain corresponding to the peak tensile strength).

4.2. *Concrete and reinforcement*

The Kent and Park model of concrete under compression without confinement [13] was used for the compression behaviour of concrete as shown in figure 4c. It has to be noted that the unconfined concrete model was used because the steel stirrups were modelled separately in the FE program. The tensile response of concrete was modelled, as shown in figure 4d, based on the tension stiffening model proposed by Wahalathantri et al. [14]. Elastic-perfectly plastic model was used to simulate the uniaxial response of embedded steel bars.

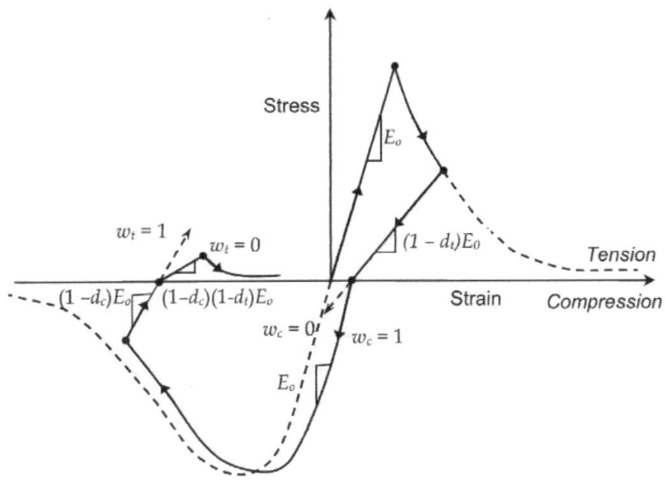

Figure 3. Concrete Damage Plasticity (CDP) Model of ABAQUS [12].

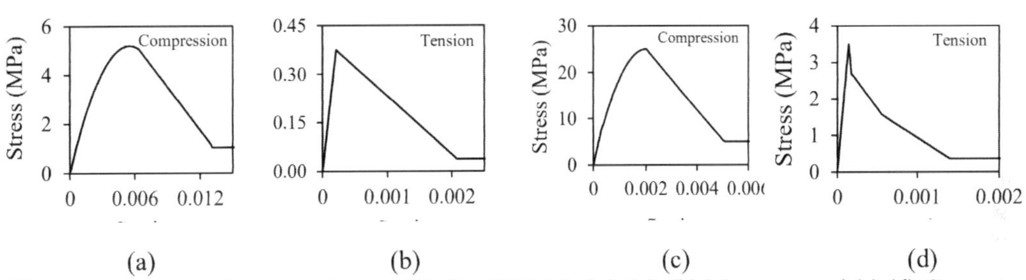

Figure 4. Stress-strain properties specified in CDP Model: (a), (b) Masonry, and (c) (d) Concrete.

5. Abaqus analysis procedure

To simulate the experimentally obtained behaviour of the considered wall in the numerical model, a three-dimensional (3D) FE macro-modelling strategy was adopted as shown in figure 5. Masonry panel, RC confining frame, and reinforcements were discretely modelled as different units. The masonry modelled at macro level was defined using macro-masonry properties and the confining RC elements were modelled considering the tension stiffening effect using the CDP model as explained earlier. The steel reinforcement was modelled as an embedded truss element located within the concrete using perfectly bonded constraint. The contact between the masonry panel and RC confining frame was simulated by assigning suitable interface properties. The interface was modelled with frictional properties to account for tangential slip and hard contact was assigned in the normal direction to avoid penetration of finite elements. The mesh was continuously refined to achieve convergence and a finer mesh size of 25 mm was chosen for this study. The boundary conditions were simulated in an idealised manner; the bottom face of the bottom beam was assumed to be fixed using the 'Encastre' boundary condition. To model the pseudo-static lateral load phenomenon, in-plane cyclic displacement-controlled loading was applied at the center of the top beam and cyclic analysis was carried out in ABAQUS explicit solver. To maintain the pseudo-static condition, the lateral displacement was applied at a reduced rate, ensuring the kinetic energy of the model does not become more than 5% of the total internal energy.

The analysis was carried out in three steps in ABAQUS. First, the initial step is mandatory and default, in which the boundary condition was initialised. In the next step, gravity load was applied along with pre-compression load, which was followed by the in-plane loading stage.

5.1. *Sensitivity study*

Numerical simulations are parameter sensitive. Therefore, a sensitivity study was carried out to understand influence of some parameters. As shown in figure 6a, it is observed that the results are quite stable for the mesh size below 30mm. Hence, a mesh size of 25mm, which discretise confining elements into 5 parts, was selected considering computational power required into account. Similarly, dilation angle of masonry was selected as 12 (figure 6b). At masonry panel and RC confining element interface, different permutations of possible interface scenarios were tested. As shown in figure 6c, the interfaces were modeled with penalty friction with frictional coefficient of 0.8.

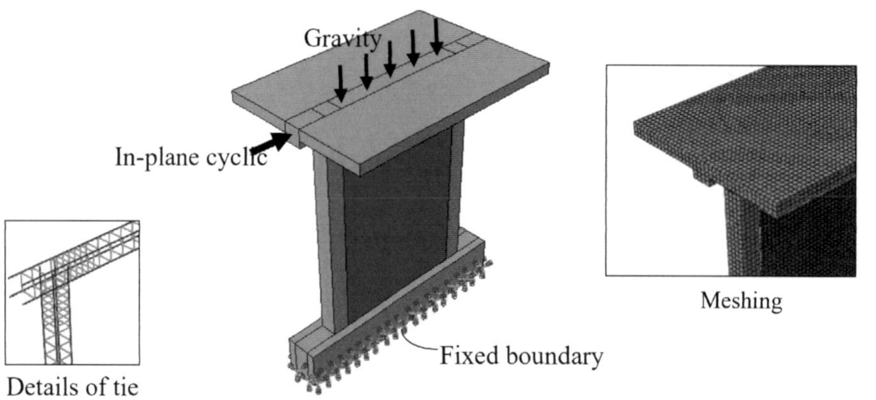

Figure 5. Numerical model.

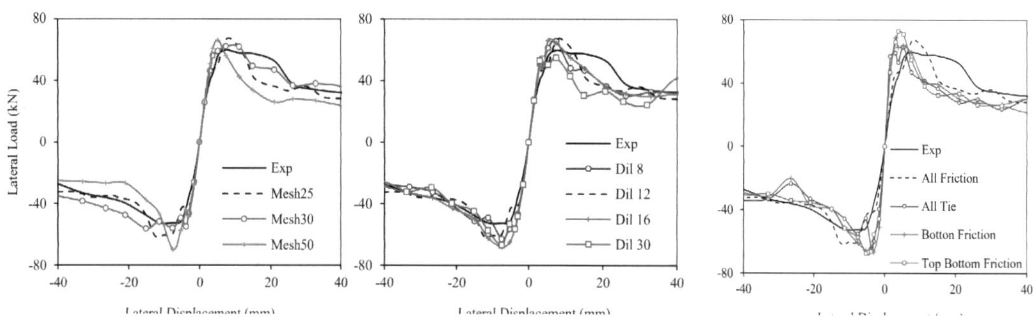

(a) (b) (c)

Figure 6. Sensitivity study for: (a) Mesh, (b) Dilation angle, and (c) Boundary condition.

6. Validation of the proposed FE model

The load-displacement relation obtained from the FE analysis was compared with the experimental as shown in figure 7. The pre- and post-peak behaviour of the numerical model matched the experimental results quite well. The estimated maximum in-plane strength from the numerical model (average value 63.6kN) almost matched with the experimental (average value 56.1 kN) with an error of 13.5%.

Moreover, figure 8 shows the equivalent plastic strain contour (PEEQ) in the wall obtained from the FE analysis and it was compared with the crack pattern in wall obtained from the experimental study. Dotted lines were drawn for visualization of major crack as per simulation. In the FE model, wall was modelled as a continuum (single unit) using CDP model from ABAQUS library. The discontinuity in the form of mortar joints was not explicitly considered while modelling. The obtained PEEQ signifies the diagonal cracking of wall, which provides a reasonable concordance with its experimental.

Therefore, the validation study confirmed that the developed FE macro model could effectively predict the lateral capacity and crack pattern of CM wall.

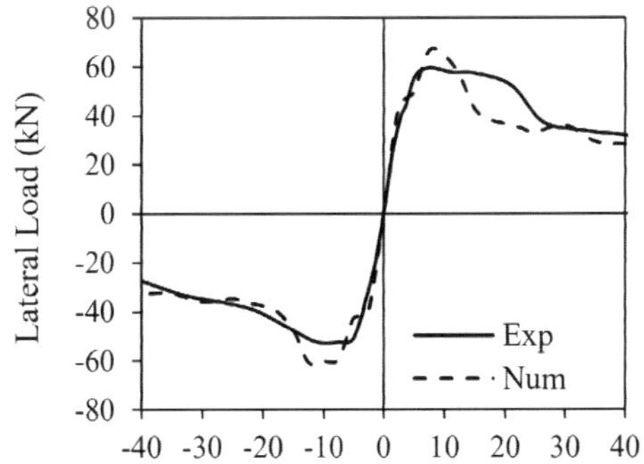

Figure 7. Comparison of numerical lateral load-displacement behaviour with experimental.

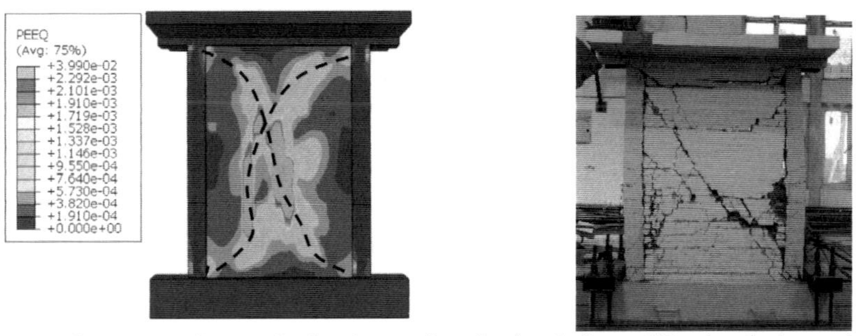

Figure 8. Comparison of plastic strain obtained from FE simulation with experimental crack pattern

7. Summary and conclusion

Confined masonry constructions have been documented as an alternative to URM and infilled RC structural system. For having a better understanding of the behaviour of CM structure, a solid CM wall with aspect ratio 1.5 was designed using Mexican guidelines and tested in the laboratory under pseudo-static cyclic lateral loading. Based on that, a 3D finite element model was developed using macro-modelling approach. The developed numerical model was validated with the experimental study. Therefore, the proposed numerical model is able to capture the nonlinear response and cracking pattern of CM wall subjected to cyclic lateral loading. It will be helpful in understanding the effect of parametric variations, such as, geometric variation, material variation, and other possible uncertainties, which will reduce the number of required experimental tests by many folds.

8. References

[1] Borah B, Singhal V and Kaushik H B 2019 Sustainable housing using confined masonry buildings *SN Appl. Sci.* **1** 983 https://doi.org/10.1007/s42452-019-1020-4
[2] Smoljanović H, Živaljić N, Nikolić Ž and Munjiza A 2017 Numerical model for confined masonry structures based on finite discrete element method *Int. J. for Engg. Modelling* **30(1-4)** 19–35 https://hrcak.srce.hr/201719
[3] Tabrizi M A and Soltani M 2017 In-plane response of unreinforced masonry walls confined by reinforced concrete tie-columns and tie-beams *Adv. in Struc. Engg.* **20(11)** 1632–43 https://doi.org/10.1177%2F1369433216689569
[4] Eshghi S and Pourazin K 2009 In-plane behaviour of confined masonry walls–with and without opening *Int. J. of Civil Engg.* **7(1)** 49–60

[5] Janaraj T and Dhanasekar M 2014 Finite element analysis of the in-plane shear behaviour of masonry panels confined with reinforced grouted cores *Const. and Build. Mat.* **65** 495–506 https://doi.org/10.1016/j.conbuildmat.2014.04.133
[6] Yacila J, Camata G, Salsavilca J and Tarque N 2019 Pushover analysis of confined masonry walls using a3d macro-modelling approach *Engg. Struct.* **201** 109731
[7] NTC-M 2004 *Technical Norms for Design and Construction of Masonry Structures- Mexico City Building Code* (Mexico D.F)
[8] ACI 374.1 2005 *Acceptance Criteria for Moment Frames Based on Structural Testing and Commentary* (Farmington Hills MI)
[9] Simulia D S 2016 Abaqus v. 6.16 [Software] Providence RI: Dassault Systèmes Simulia Corp
[10] Dassault Systèmes D 2010 *Abaqus Analysis Users' Manual ABAQUS Documentation* **6**
[11] Park Y J, Reinhorn A M and Kunnath S K 1987 *IDARC: Inelastic Damage Analysis of Reinforced Concrete Frame-Shear-Wall Structures* (State University of New York at Buffalo, Buffalo, NY) NCEER Rep. No. 87-008
[12] Kaushik H B, Rai D C and Jain S K 2007 Stress-strain characteristics of clay brick masonry under uniaxial compression *J. of Mat.in Civil Eng.* **19(9)** 728–39
[13] Kent D C and Park R 1971 Flexural members with confined concrete *J. of Struc. Division* **97(7)** 1969–90
[14] Wahalathantri B L, Thambiratna D P, Chan T H T and Fawzia S 2011 A material model for flexural crack simulation in reinforced concrete elements using ABAQUS *Proc. First International Postgraduate Conference on Engineering, Designing and Developing the Built Environment for Sustainable Wellbeing* Queensland University of Technology pp 260–4

Acknowledgments
Authors gratefully acknowledge the Grant received from the Science and Engineering Research Board, Department of Science and Technology, Government of India (Grant No. ECR/2018/000489).

Non-linear Behaviour of Infilled RC Frame

Nikhil P Zade[1*], Shivprasad A Koparde[2], Pradip Sarkar[3] and Robin Davis[4]

[1] Ph.D. Scholar, National Institute of Technology Rourkela, India-769008

[3] M Tech Scholar, National Institute of Technology Rourkela, India-769008

[3] Associate Professor, National Institute of Technology Rourkela, India-769008

[4] Assistant Professor, National Institute of Technology Calicut, India-673601

[*]nikhilzade648@gmail.com

Abstract. The reinforced concrete (RC) infilled frame is very common in exercise. In general, only infill weight is considered in the design of the RC infilled frame, and the strength of masonry is ignored. However, the prominence of masonry infill on the seismic behaviour is highlighted during past earthquakes. Nowadays, applications of different types of masonry such as fly ash, and lightweight masonry in infill, are increasing due to their superior properties over the traditional clay brick masonry. The exact behaviour of the infilled RC frame during an earthquake cannot be predicted if the strength of infilled is ignored. A nonlinear static method i.e. pushover analysis for the seismic evaluation, is commonly used due to its simplicity and high accuracy. A typical four storey three bay plane infilled RC frame is considered in the present study to investigate efficacy of several formulas available in the literature for the modelling of infill employing pushover analysis method. In addition to that sensitivity analysis also conducted which revealed that the parameters, masonry wall width and masonry shear strength are significant that control the base shear at yield level. Significant decrease in the lateral stiffness is observed for a higher percentage of openings in the infilled. However, up to 5% of openings in infilled can be ignored due to its marginal change in the lateral strength of infill.

1. Introduction

Infilled masonry wall reinforced concrete framed structure, is widely spread and generally used in construction throughout the world. Normally, the strength and stiffness of the infilled masonry wall are ignored, considering it as a non-structural component. However, in the seismic prone region, avoiding the contribution of an infilled masonry is not a noble choice. Under the action of seismic load, the infilled wall stiffness enhances by diagonal compression action; it results in a possible change of the lateral demand. Masonry infilled wall can be modelled as a diagonal strut element (Figure 1).

In general, the infill used in the RC framed masonry structures is constructed of the material available in the vicinity, which results in a change in engineering properties from region to region. The variation in engineering properties might cause a difference in strength and stiffness behaviour. A non-linear static method of seismic evaluation is conducted in this study to evaluate the effect of properties of an infilled masonry wall on the seismic response of infilled RC frame.

The nonlinear static pushover (NSP) can be primarily categorised as NSPs method and N2 method. NSPs method consists of capacity spectrum method, implemented in [1], introduced by Freeman [2], and N2 method suggested by Fajfar [3] which is later included in Eurocode 8 [4]. Nonlinear analyses are essential to catch the exact behaviour of structure as it behaves inelastically when subjected to seismic loading under seismic excitation. Due to extensive proof studies that have verified the exactness and simplicity of NSP, pushover analyses are commonly adopted for the seismic assessment of existing structures and design of new structures. Hinges, where nonlinearity will occur, can be defined manually or as default hinge, as per ATC-40 [1] and FEMA-356 [5] guidelines which are available in analysis software such as SAP2000.

The motivation of the present work is to assess the design guidelines suggested by IS 1893 [6] for infilled modelling. Different formulas are available in the literature to evaluate the diagonal strut width. In the present study efficacy of several formulas available in the literature for the modelling of infill is checked along with the various opening percentage in infilled RC frame. A sensitivity analysis to understand the significant parameters that affects the base shear response is also carried out as part of this study.

2. Modelling of infilled masonry wall

With due importance for the infilled masonry, many experimental and analytical efforts have been made to study the behaviour of the infilled RC frame. A brief literature review has revealed that the earlier version of the design code has no sufficient guideline to reflect the influence of infilled masonry on the response or design of infilled RC framed buildings.

Previous studies [7-9] has performed several experimental tests with various aspect of the infilled masonry wall. The wall is represented by the diagonal struts carrying compressive forces. Smith and Carter [10] have suggested analytical methods to ascertain the viable width of the diagonal strut. The study also concluded that cracking and crushing strength of diagonal strut depends upon the contact length, angle α between infill and frame in its deformed pattern. Contact length was expressed as a function of non-dimensional factor (λ_h), as presented in Eq. (1) and Eq. (2).The width of equivalent strut reduces as the loading increases, to consider the change in the stiffness of the infill due to cracking [10].

$$\frac{\alpha}{h} = \frac{\pi}{2\lambda_h} \tag{1}$$

$$\lambda_h = h\left(\frac{E_w t \sin 2\theta}{4E_c I_c h_w}\right)^{0.25} \tag{2}$$

Here, E_w, h_w, and t are modulus of elasticity, the height of the infill and thickness respectively. Also, column properties are E_c (Young's modulus of column) and I_c (moment of inertia of column). θ, is the gradient between the horizontal and the diagonal of the infill.

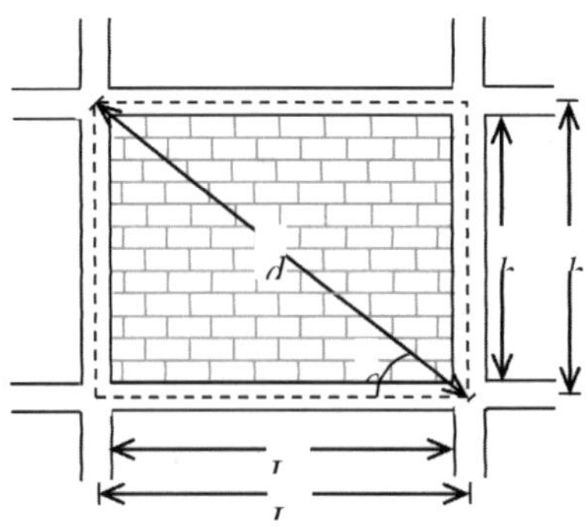

Figure 1. Schematic presentation of an equivalent strut.

Modified equation is given by [11] to calculate the maximum compressive strength of infill, as shown below:

$$F_{max} = 0.818\frac{l_w t_w \tau_{cr}}{C_1} \tag{3}$$

where,

$$C_1 = 1.925^{\frac{l_w}{h_w}} \tag{4}$$

The model is formally the same as that of the one by Panagiotakos and Fardis [12] with the major difference of no residual strength ($F_u = 0$), shown in figure 2. Post-capping branch and elastic hardening are the three tri-linear response of single equivalent diagonal strut. Studies [13, 14] reported the change in lateral behaviour of an infilled frame in line for the openings in the masonry and [14] also suggested a reduction factor (ρ_w), as presented below:

$$\rho_w = 1 - 2.6\alpha_{co} \tag{5}$$

where α_{co} is the Opening-Area-Ratio.

Different formulas are available in the literature to estimate the effective equivalent strut width (w) as given in table 1.

Table 1. Various models for the width of the strut.

Equivalent strut width (w)	Reference
$d/3$	[8]
$d/4$	[15]
$d(0.175\lambda_h^{-0.4})$	[16]
$d(0.475\lambda_h^{-0.5}\sin 2\theta)$	[17]

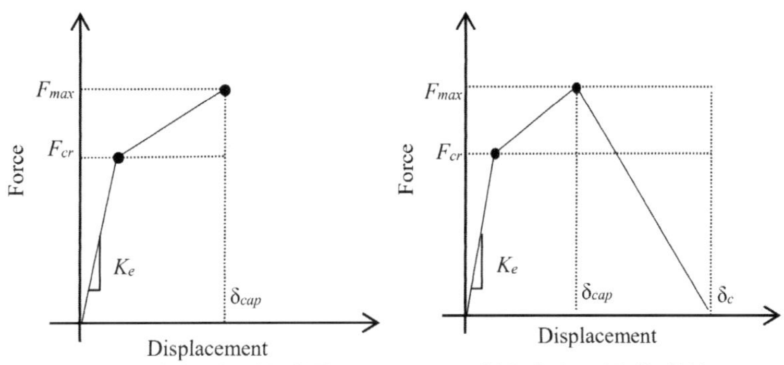

(a) Saneinejad and Hobbs [18] (b) Dolsek and Fajfar [11]

Figure 2. Representation of the backbone curve of an infill strut model.

2.1. Building Geometry

A typical four storey three bay plane infilled RC frame with (230 mm wall thickness), with and without openings is considered in the present investigation for the analysis. The selected building is a residential building, having slab thickness 150 mm, located in the seismic zone II and designed as per IS 456 (2000). Live Load on the building is 3kN/m^2, unit weight of Clay masonry and Fly ash brick masonry is considered 18.84 kN/m^3 [19]. The lateral load profile is taken as per IS per 1893 [6].

3. Seismic Assessment

NSP and sensitivity analyses are carried out to evaluate the nonlinear behaviour of the infilled frame under lateral load. Parabolic lateral load pattern is used in the present study [6]. The lateral load is increased up to the failure of the structure. Schematic presentation of the pushover curve is shown in figure 3. To check the parameters of frame and masonry infill, which are sensitive to the lateral loading, sensitivity analysis is conducted by replacing one property with its percentile value and keeping all other properties as mean. The change in the response of the structure is assessed for the various input mechanical and physical properties presented in table 2. Similar properties for clay brick masonry are available in the literature [19]. Sensitivity is defined by Celarec [20] as given below.

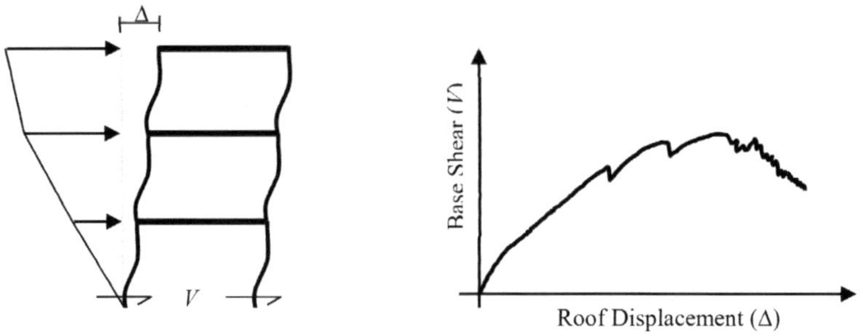

a) Sructural model

b) Capacity curve

Figure 3. Schematic presentation of NSP.

$$\Delta y = \frac{y(P_n) - y(P_{mean})}{y(P_{mean})} \times 100(\%) \tag{6}$$

where $y(P_n)$ and $y(P_{mean})$ are the response at percentile and mean value of the random variables.

Table 2. Details of input variables for fly ash masonry [21].

Property	Variable	Mean	COV	Distribution[a]
Masonry compressive strength (MPa)	f_m	5.45	0.27	N
Masonry shear strength, (MPa)	τ	0.125	0.12	LN
Elastic modulus of masonry, (MPa)	E_m	3275	0.3	N
Yield strength of steel (MPa)	F_y	393	12	LN
Depth of column, (mm)	d_{col}	400	1.5	LN
Width of infill, (mm)	w	-	1.5	LN

[a]N: Normal and LN: Lognormal

4. Results and Discussion

Different formulae to calculate the width of the equivalent strut as per the previous literature listed in table 1, are used to model the infill, and NSP curves are obtained as shown in figure 4. Sensitivity analysisand pushover analysis are carried out on the selected plane RC frame. 5%, mean and 95% percentile values of random variables for the clay and fly ash infill masonry are considered to observe the change in the seismic response of infilled RC frame. Tornado diagram (*TD*) is used to present the result of the sensitivity analysis as shown in figure 5. Ultimate base shear is considered as a sensitive parameter in the present study.

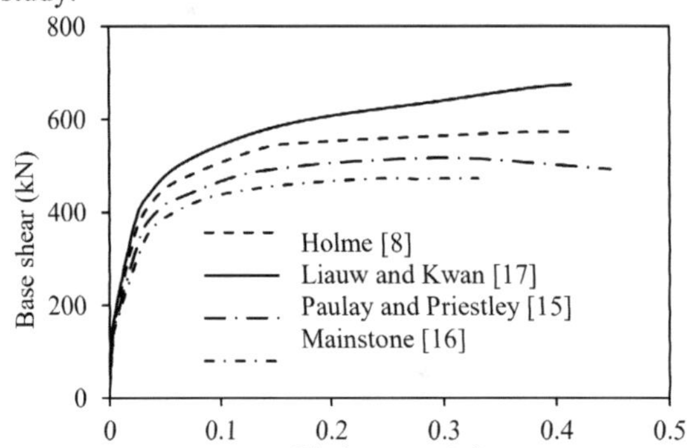

Figure 4. Effect of strut width.

It can be observed from the TDs that τ (shear strength of infill) and w (width of masonry) greatly influence the response of structure for both the type of infills among all other properties considered.

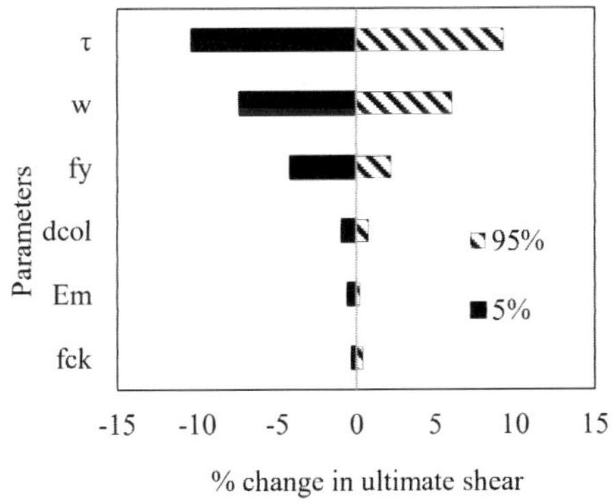

(a) Clay masonry

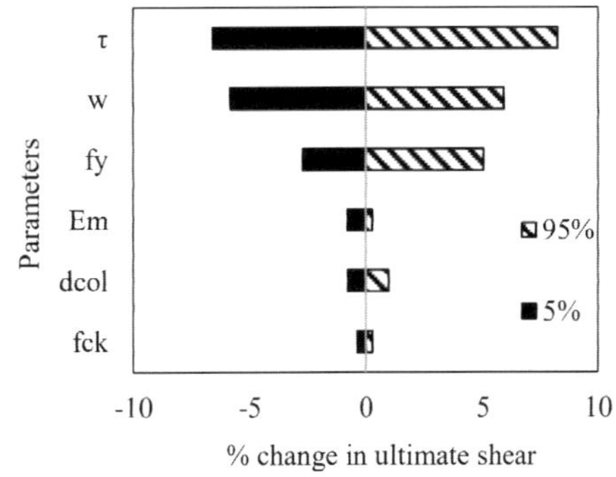

(b) Fly ash masonry

Figure 5. TD masonry wall.

NSP curves are obtained for the considered plane RC frame. Figure 6 presents the NSP curves for clay and fly ash masonry infilled RC frame along with the different percentage of openings respectively. Pushover curves of a bare frame are also shown for the reference.

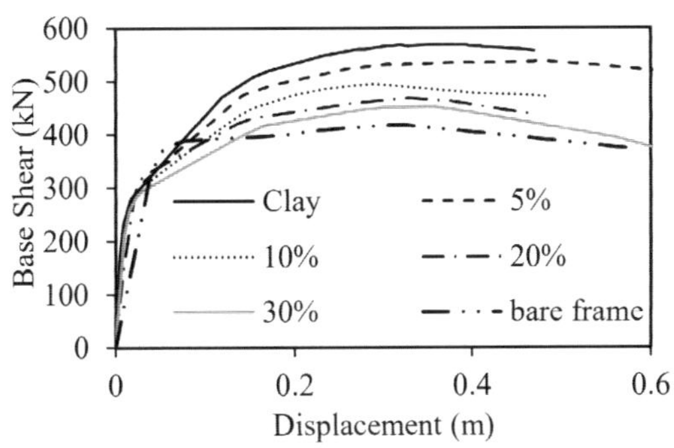

(a)

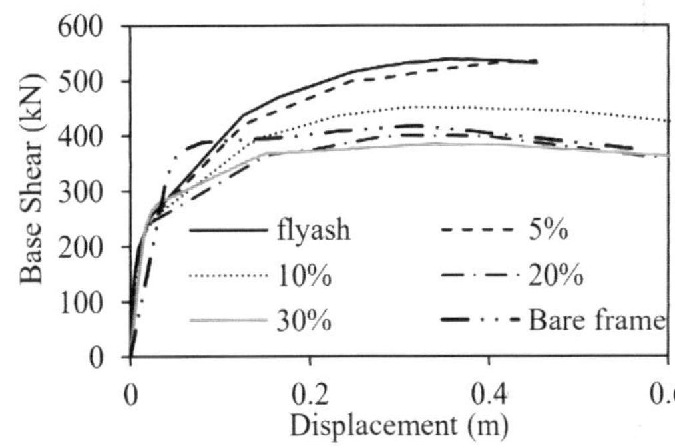

(b)

Figure 6. Capacity curve for a different opening percentage (a) Clay brick infilled and (b) Fly ash brick infilled.

5. Conclusions

The main findings of the present work are summarised as follows:

- Results of pushover analysis show that reduction in the value of base shear (V_b) increases as the percentage of opening in the infill wall increases. However, the effect of 5-10% openings in the infill can be ignored due to insignificant change in the ultimate base shear.
- Sensitivity analysis shows that the width and shear cracking strength of the infill are the most sensitive parameters in the analysis of masonry infilled RC frames.
- Among all the formulations to calculate the width of equivalent diagonal strut, Smith and Carter model [16] gives the lower bound results which is conservatively included in Indian standard IS 1893 [6].

6. References

[1] ATC-40 1996 Seismic evaluation and retrfit of concrete building *Applied Technology Council* (Redwood City, California)
[2] Freeman S A 1998 Development and use of capacity spectrum method *6th U.S. National Conf. Earthquake Engineering* (Seattle Oakland USA)

[3] Fajfar P 2000 A nonlinear analysis method for performance-based seismic design *Earthquake spectra* **16(3)** 573–92
[4] EC 8 1998 Design provisions for earthquake resistance of structures. *European Committee for standardisation* Eurocode 8
[5] FEMA 356 2000 Pre-standard and Commentary for the Seismic Rehabilitation of Buildings (Washington DC) Federal Emergency Management Agency
[6] IS 1893 (Part 1) 2016 Indian Standard Criteria for Earthquake Resistant Design of Structures (New Delhi) Bureau of Indian Standards
[7] Polyakov S V 1956 On the interactions between masonry filler walls and enclosing frame when loaded in the plane of the wall *Translations in Earthquake Engineering Research Institute* (Moscow)
[8] Holmes M 1961 Steel frames with brickwork and concrete infilling *Institution of civil Engineers* **19(4)** 473–8
[9] Smith B S 1962 Lateral stiffness of infilled frames *Journal of the Structural Division* **88(6)** 183–226
[10] Smith B S and Carter C 1969 A method of analysis for infilled frames *Proceedings of the institution of civil engineers* **44(1)** 31–48
[11] Dolsek M and Fajfar P 2008 The effect of masonry infills on the seismic response of a four-storey reinforced concrete frame-A deterministic assessment *Engineering Structures* **30(7)** 1991–2001.
[12] Panagiotakos T B and Fardis M N 1996 Seismic response of infilled RC frames structures *11th world conference on earthquake engineering* (Acapulc, Mexico)
[13] Govindan P and Vivekanandan D 1990 Effect of Opening in Brick Infilled Square Panel *Indian Concrete Journal* **64** 298–303
[14] Mondal G and Jain S K 2008 Lateral stiffness of masonry infilled reinforced concrete (RC) frames with central opening *Earthquake spectra* **24(3)** 701–23.
[15] Paulay T and Priestley M N 1992 *Seismic design of reinforced concrete and masonry buildings*
[16] Mainstone R J 1971 Summary of paper 7360- On the stiffness and strengths of infilled frames *The Institution of Civil Engineers* **49(2)** 230
[17] Liauw T C and Kwan K H 1984 Nonlinear behaviour of nonintegral infilled frames *Comput. Struct* **18** 551–60
[18] Saneinejad A and Hobbs B 1995 Inelastic design of infilled frames *Journal of Structural Engineering* **121(4)** 634–50
[19] Ravi Teja P 2015 Studies on Mechanical Properties of Brick Masonry *M. Tech Research ethesis* (NIT, Rourkela)
[20] Celarec D, Ricci Pand Dolsek M 2012 The sensitivity of seismic response parameters to the uncertain modelling variables of masonry-infilled reinforced concrete frames *Engineering Structures* **35** 165–77
[21] Bhosale A, Davis R and Sarkar P 2016 Sensitivity and Reliability Analysis of Masonry Infilled Frames. *International Journal of Civil, Environmental, Structural, Construction and Architectural Engineering* **10(12)** 1531–5

A Study on Challenges in the Rehabilitation of Large Masonry and Concrete Dams

K Jiji Panicker[1,2*], Praveen Nagarajan[3] and Santosh G Thampi[4]

[1] Research Scholar, National Institute of Technology, Calicut, Kerala, India

[2] Asst. Executive Engineer, Kerala State Electricity Board Limited, Kottayam, India

[3] Associate Professor, Dept. of Civil Engineering, National Institute of Technology, Calicut, Kerala, India

[4] Professor, Dept. of Civil Engineering, National Institute of Technology, Calicut, Kerala, India

*kjpanicker12@gmail.com

Abstract. Large dams retain enormous volume of water intended for various purposes. The design, construction, operation and maintenance of these special structures are of paramount importance considering its uses and safety. Safety of the dam is to be ensured throughout its' life from structural, hydrological and operational perspectives as its failure would be catastrophic. Periodical safety inspections, review on safety, dam health monitoring as well as risk assessments are measures to ensure safety of these large structures. Ageing of dams is a concern in dam safety as there can be inadequacies or deficiencies with respect to structural, hydrological and operational requirements. Revision of relevant codes of practice/standards also may necessitate review and strengthening. As such, these structures are to be rehabilitated to the prevalent standards and maintained healthy to continuously derive the benefits envisaged during construction. Rehabilitation of dams aims to bring it back to a good condition and ensure that it does not pose a risk to the stakeholders downstream. The process involves various systematic procedures such as review of structural design, review of design flood and diagnosis using non-destructive tests to detect possible deterioration in its health, etc. As per recent statistics, more than half of the large dams in India are more than 25 years old. In this paper, a review of aspects such as challenges, considerations as well as various factors involved in the effective implementation of rehabilitation/strengthening is presented. Certain case studies of rehabilitation work undertaken under the on-going Dam Rehabilitation and Improvement Project in India are cited to bring forth the importance of such measures. As dams play crucial role in the prosperity of the nation, they are to be maintained for an endured life without jeopardizing the safety of the structure. In that perspective, rehabilitation of dams or its strengthening process is extremely important.

1. Introduction

Development of Water resources is a portfolio of high priority for all governments as this resource having vast potential in many areas is limited. Sustainable economic growth, through improvement in irrigation, hydro-electric power generation, flood control, and potable and industrial water supplies, etc., besides many other uses can be derived from the water available to a country by careful and scientific manner. Dams play a vital role in the precipitous and persistent growth in the above sectors. Having identified the potential of dams, the respective Nations are keen to maintain these vital structures contributing positively to the economy, safely. Globally, India is in the third position for having most number of large dams. China and the United States rank first and second respectively. Though the initial constructions for impounding the water were earthen embankments and masonry, the twentieth century found a growth in the number of concrete and masonry dams by virtue of advancements in construction materials, methodology and technology. As per the statistics (Table 1) by ICOLD, 65% of the reported large dams in the world are earthen dams [1]. The operation and maintenance protocols of concrete and masonry dams are different from earthen dams and the present discussion is confined to concrete and masonry dams.

Table 1. Synthesis of large dams by type [1].

Sl. No.	Type of dam	Number
1.	Earthen dams	37537
2	Rockfill dams	7729
3	Gravity dams	8115
4	Buttress dams	415
5	Barrages	299
6	Arch dams	2352
7	Multiple arch dams	129
8	Others	1409
	Total	57985

2. Dam engineering

Dam construction using stones dates as early as 2900 B.C.~2600 B.C., in Egypt. Later, the Romans and the Mongols started constructing dams. The 55m high arch dam constructed in rubble masonry in 1300 AD at Qum is believed to the oldest surviving arch dam. Civilizations and Nations realised the need of dam construction and accordingly the world saw more dams coming in. Till, 1850, there were no rational criteria for design of dams but the theories by M.de Sazilly, in 1853 that the pressures within the dam have to be within limits and dimensions are important for prevention of sliding, and the conceptualization of the middle third criteria after another 25 years, by W. J. M. Rankine paved way for more gravity dams. The construction of large dams capable of storing large volumes of water is a result of technological advancement imbibing of relevant theories in the conception, analyses, design and execution stages. The dam engineering involves selection of suitable site, investigation of geology for foundation preparation, and structural as well as hydrological analyses. The calculation of risks involved, its' mitigation measures, inspection and reviews for confirming whether the structure would safely route the flood from the rainfall intercepted by the catchment, maintaining seepages, stresses, uplift and movements within limits, etc. are relevant in dam engineering. Evaluation of behaviour of dams requires knowledge from a spectrum of technologies and practices exercised by Engineers over past many years. The dam engineering takes into consideration new perspectives on earthquake engineering and its' dynamics, improved methodology for foundation treatment, advancement in soil mechanics/ geotechnical Engineering, seepage analyses using state of the art methods, stability analyses, advances in concrete dam technology, refined techniques for hydrologic and hydraulic analyses, state of the art instrumentation, employment of newer materials for construction as well as for rehabilitation such as synthetic membranes, real time data acquisition, dam operations and management etc.[2].

Dam engineering forms a vital part in the history of civilization as reservoirs were constructed by various civilization for water supply and irrigation using different material, viz., earth, wood, stones, fills of various length, width and height. The failures of the past helped devise new theories and efficient methods. Dams of various type and shapes were constructed over the past. Rectification and rehabilitation works were attempted on the dams of the past. De-siltation, control of seepage, drainage systems for reducing uplift etc. were identified as a rehabilitation measure of the existing dam. The Couzon dam constructed in 1811, was to be rehabilitated in 1896 to control seepage and sliding and later it remained in service without trouble. The Aswan dam in Egypt, 20m high, constructed in 1902 was of quarried granite mass, which was raised to 27m in 1912 and in 1933 to 53m height above foundation [3]. As various types of deficiencies are identifiable, remedies can also be systematically executed as the rehabilitation and it has become an indispensable part in dam engineering.

2.1. Dam safety

Dam safety comprises of structural safety, surveillance, maintenance, and emergency planning which will help minimise the risks if it cannot be eliminated. ICOLD recognises certain aspects as the pillars of dam safety, they are, structural integrity of dams, the routine surveillance and maintenance programme for early detection of risks, instrumentation and monitoring plan throughout the life-span of the dam, addressing of design-intrinsic risks, regular review and updating of natural hazard risks as they change with time, emergency planning, sharing lessons learned to benefit the entire industry, comprehensive dam safety approach for minimizing the risk, the owners' ultimate responsibility of the dam, prominent role of regulatory authorities etc., besides emphasizing the international perspective to dam safety as it would be able to impart the experiences as well as best practices worldwide to the designers, dam owners and authorities concerned [4].

3. Rehabilitation of large dams

Safety of these structures contributing vitally to the economy and well-being of the respective nation is most important on two aspects, viz., for deriving the targeted benefits uninterruptedly and to get rid of the risks to the life and belongings of stakeholders at the downstream. Constructing a new dam involves many challenges and impediments, and this emphasises that the existing are to be managed healthy and safe for prolonged use. Many existing large dams are having various deficiencies such as structural and hydrological besides deteriorations due to ageing and operational shortcomings. Rehabilitation of a dam is often referred to as, bringing it back from a distressed state to the original state and certain improvement for meeting requirements by virtue of changes in the safety criteria over the time. Methodology of rehabilitation will vary for each dam. Implementation and planning of rehabilitation of a concrete or masonry dam depends on field investigations, materials for rehabilitation etc. The safety evaluation of an existing dam should recognise problems and suggest remedial measures, restrictions in operations, modifications (if needed), besides doing analyses and studies to determine solutions to the problems [5]. Each year, the number of aged dam is increasing; India, the 3rd Country having most number of dams, presently have 5334 large dams[6], of which 67 are constructed prior to the year 1900 whereas there are 304 dams constructed during the period 1901 to 1950 (Figure 1) and every year this count is increasing considerably. Over 75% of its' large dams are more than 20 years of age. These ageing structures require a systematic rehabilitation, consisting of deciding their needs, design and implementation of rehabilitation activities with adequate control on the quality.

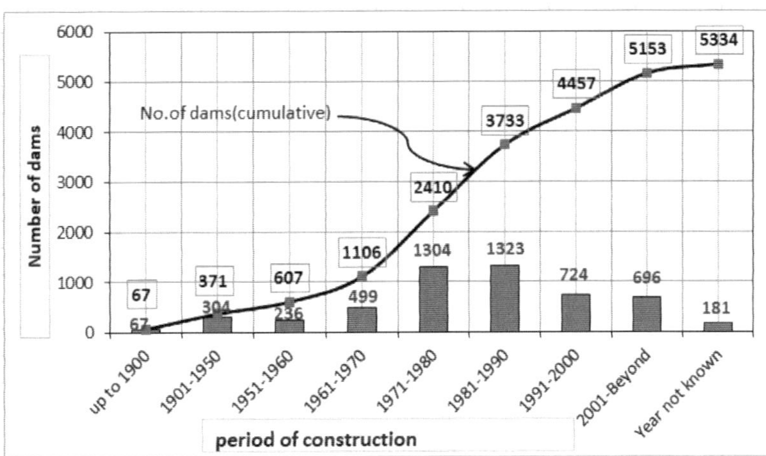

Figure 1. Growth in the count of large dams in India [6].

Suitable rehabilitation measures after adequate studies can negate the deficiencies due to ageing and the inadequacies due to revisions in standards/criteria/guidelines on dam design, operation and their safety operations. Deterioration due to environmental actions, deterioration of equipment due to continuous use, becoming unserviceable due to prolonged operation, destruction caused by natural actions such as earthquakes, floods, or land-slides, destruction caused by sabotage and war etc. too are deficiencies or inadequacies.

3.1. Need of rehabilitation of dams

Deterioration of concrete and masonry due to ageing is a major concern to dam owners and stakeholders downstream and it warrants for rehabilitation. The thermal effects due to the temperature gradient between various parts of the body, alkali-aggregate reaction and other chemical reaction due to presence of various chemicals in aggregates and cement, leaching actions, climatic actions such as freezing-thawing, etc. are various factors that cause deterioration of dam body health. Deterioration of foundation may also be caused due to the loads acting, properties of the foundation strata/geological conditions etc. Deterioration of dam body and foundation may lead to cracking, stability problems, increased seepage and uplift pressures. These causes and effect are to be investigated thoroughly before rehabilitations are planned and implemented. As per a study of ICOLD [7] on the deterioration of concrete and masonry dams, summarised the following scenarios (Figure 2-Figure 4):

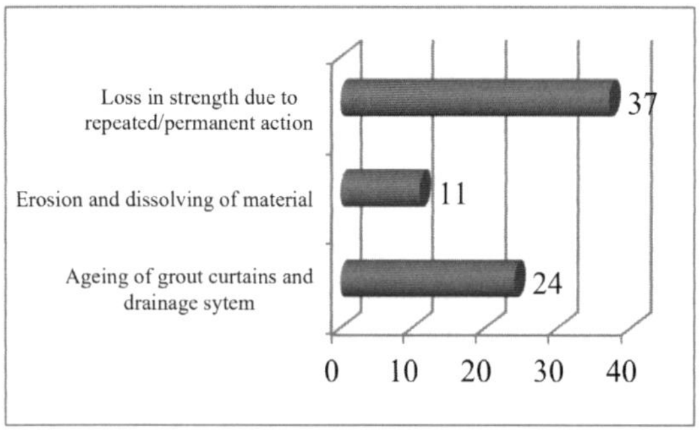

Figure 2. Foundation rock mass cases.

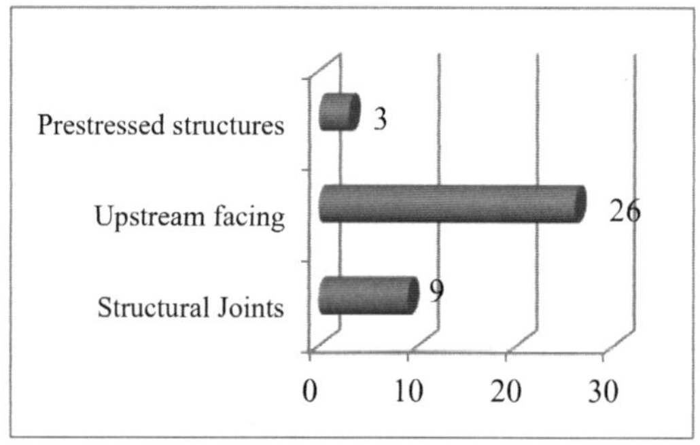

Figure 3. Ageing scenario pertaining to:

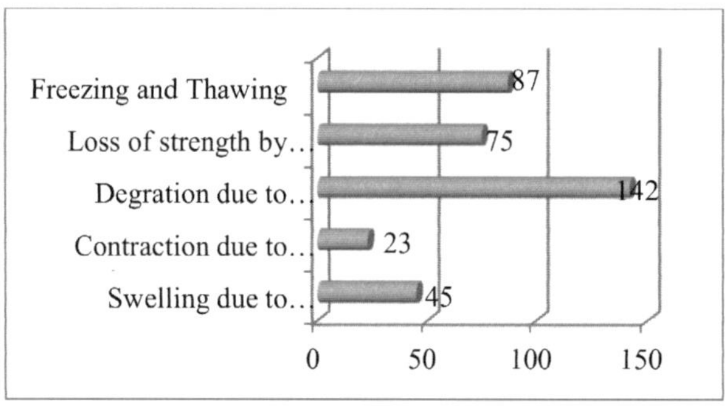

Figure 4. Dam body (Concrete/Mortar/Stone) cases.

Dams require rehabilitation mainly for two reasons, viz., for dealing with the effect of the ageing process and for conforming with the requirements of new standards introduced. Large numbers of dams worldwide were subjected to rehabilitation to meet the stability criteria against flood and earthquake as well as uplift criteria. Mainly areas subjected to rehabilitation, are foundation, dam body and the increased stability. Loss of strength for the rock mass foundation is a major concern related to stress reversals in the foundations due to fluctuations in hydraulic gradient due to the water level in the reservoir. Stress reversal may cause the joint infillings to wash away, deformation to foundation, moving of joints in rock, and crack initiation and propagation. Monitoring is necessary as this will happen over a large period of time and as such, seepage, uplift pressure, and inelastic foundation displacements are to be monitored using instrumentation data. Review of stability owing to revision in design flood, review for increase in spillway length to route the increase in flood, if any, change of hazard zones and revision of standards for earthquake resistant design, etc. are important aspects in deciding rehabilitation. Seepage in concrete and masonry dams is a problem often met by dam owners and this is caused by factors such as cracks, poor lift joints, deterioration of concrete material, deterioration in the masonry pointing at the upstream face causing the joints to open up, poor block joints/transverse joints, etc. Though evidence to seepage can often corroborate with wetness, sprouting, calcination, increased flow in the drainage gallery, in many cases, the exact source could not be easy to detect. Seepage if unchecked, undetected and not remedied can lead to undesirable results.

3.2. International practices
Various measures for the rehabilitation of large dams use technology, materials, processes, methodology, and literature available from acclaimed sources. International organizations such as, ICOLD, United States Society on Dams, Bureau of Reclamation, US Dept. of the Interior (USBR), U.S. Army Corps of Engineers, Federal Emergency Management Agency, Federal Engineering Regulatory Commission, British Dam Society, ASCE, Institution of Civil Engineers, Bureau of Indian Standards, the Central Board of Irrigation & Power, India(CBIP) etc. give useful information on various loads acting, criteria, guidelines for analyses and investigations/inspections to be done as well as rehabilitation aspects of dams through their websites and publications.

4. Materials for rehabilitation
Repair works, for the rehabilitation and methodology to be adopted shall be decided with suitable material and for the same, experience, guidance based on relevant codes such as, ASTM, ACI, EN, BIS as well as relevant literature from ICOLD and other internationally acclaimed authorities cited in paragraph 3.2 can be made use of. The selection of a repair material and the methodology depends on the causes of distress, type and extent of deterioration in concrete or masonry and their symptoms, etc. For a concrete dam, materials such as high performance concrete, cementitious mortar, free flowing micro-concrete, epoxy based compounds, fiber-reinforced as well as polymer concrete, coatings, geo-synthetics and sealing system with geo membrane etc. can be used. Various bonding agents matching to the materials, grouts, materials for patching and resurfacing, steel liner, etc. will be used as necessitated by the rehabilitation. Likewise, the materials for rehabilitation of masonry dams depending on the problem, the material can be decided.

High performance concrete having high erosion resistance finds use in repair of spillways (crest, and glacis, etc.) structures for energy dissipation, silt excluder tunnels and tunnel spillways, etc. As such, the coarse aggregates should have the required abrasion value, crushing value, and impact value. The cementitious mortar used for the structural and non-structural use should possess the required physical and mechanical properties as well as bonding characteristics matching for the parent concrete, in case of repair application. The European standards, such as EN 1504-1 to EN 1504-10, EN:12190, EN:1015-17, EN:1542, EN:13295 and EN:13412 etc. stipulates various performance characteristics, test methods and requirements in those cases. Such cementitious materials possessing UV resistance, compatibility in thermal properties, high strength in compression and impermeability are now being used in dams of masonry also for pointing of the damaged upstream face to control seepage[8]. Quality control for its' storage, proportioning, application (methodology), testing etc. are areas of concern and are matters requiring utmost care and diligent supervisory control. The cementitious repair mortar, satisfying with these properties is of crystalline technology or polymer based additives with the cement. These materials are often used in rehabilitation works at upstream sides of the dam, rectifying damages in spillway piers, etc.

Various types of micro-fine concrete can be used for the repairs of deteriorated reinforced concrete zones where accessibility and compaction by vibration are hindered. The cement, graded aggregates, additives, fillers etc. used must be capable of imparting the strength and non-shrink characteristics. Various types of epoxy compounds find its' use as sealants, bonding agents and patching materials, grouts, binders, etc. due to their favorable performance characteristics such as adhesion, high strengths in compression and tension, rapid hardening, bonding characteristics in damped concrete as well as under water, chemical resistance & moisture resistance etc. ASTM C881 gives the technical specifications of various types of epoxies for load bearing areas /non-load bearing situations whereas ASTM D 1084, ASTM D695, ASTM D638 and ASTM C882 gives the mechanical properties such as viscosity, compressive, tensile and bond strength etc. [9-13]. Difference in thermal expansion of epoxy compounds and concrete, peeling off the repaired layer from surface due to temperature stresses at the boundary between concrete & epoxy mortar, decomposition due to ultraviolet rays when exposed to sunlight are challenges in using these kind of repair material and hence require strict quality control.

Steel liners and fiber reinforced concrete can be used for the repair or rehabilitation works where cavitation, erosion and abrasion etc. are likely. Steel liners even though offers resistance to cavitation and abrasion, fixing of plates against old concrete surfaces are challenging task as fixing using anchor rods and back filling/grouting the voids behind the plates, quality of the steel plate against pitting, etc.

are factors affecting prolonged life. Various types fibers can be added to the concrete to increase the strength and life of the repaired portion. Polymers can also be used in repair concrete, to improve their properties such as bonding and accordingly there are polymer, polymer modified and polymer impregnated concrete. The improvement in properties depends on the additives and admixtures used. The challenge in using these types of concrete is that they are hazardous and expertise is required in preparation and application.

Geo-synthetics or Geo-membrane sealing systems are used in rehabilitation of water retaining structures for checking seepage through the body. This technology can be used in various types of dams, viz., embankments, masonry as well as concrete. Primary requirements for these materials to work efficiently are impermeability, flexibility to use, mechanical strength, resistance to frost and heat, easiness for industrial production and easy workability [14].

5. Rehabilitation planning

Rehabilitation on dams should be preceded by preliminary as well as detailed field investigations (for remedial actions) which include inspection of the expert panel for evolving recommendations and the minimum design studies. The minimum design studies include the design flood review (DFR)(In India, it is for the PMF, Standard Project Flood or 100 Year return Period flood). The ICOLD list the above criteria for various other countries [15]. Maximum water level, free board available and the demand for additional spillways, if any shall be obtained through routing of revised flood. Selection of site for additional spillway, if necessitated, investigations for assessing topographical and geological conditions, stability analyses of the dam including seismic review of the dam-spillway structure etc., have to be carried out using actual material properties.

Worldwide, it is a good practice in monitoring for dam safety that the dam inspections of the components, viz., dam, reservoir, downstream areas etc., are being carried out on a regular basis and in case of any unusual behaviour, expert panel will have independent inspections. Dam owners shall have monitoring equipment for re-assessment of the structural behaviour of the body and foundations, regular evaluation of data measurement and the structural behaviour reports. Periodic inspections & maintenance will help in reducing major unexpected rehabilitation works. In case of large scale rehabilitation, financing arrangements for the rehabilitation work is equally important.

5.1. Design aspects

Rehabilitation in dams is generally based on the recommendations brought out as per the observations in various inspection reports. Reports of earlier inspections, viz.,comprehensive and others, physical condition and performance of the dam body, appurtenant structures as well as hydro-mechanical components, measurements taken from instrumentation in the dam, photographs taken periodically, designs & construction drawings, properties of materials used, particulars of any earlier dam incident and corresponding rehabilitation done so far, risk analysis and risk assessment if any, etc. are factors affecting the choice of a rehabilitation. Continuous structural monitoring utilizing additional instrumentation, inspections etc. will also be necessitated. Further studies & investigations to determine the actual area affected & rehabilitation plan, measures to be taken for lowering of reservoir level & for handling of emergent situations, repairs of spillway gates, if required, etc. are also pertinent. Review of design flood studies which decide the spillway capacity, revised water availability studies which decide the hydro-electric potential, irrigation, water supply, impact of projects built recently on the upstream(if any), experience of the hydro-mechanical parts operating personnel, changes, if any, in river morphology etc. are relevant. Rehabilitation works shall take into account construction techniques, the effect of the methods like blasting, pre-stressing and grouting, etc. on the existing structure, effect of using materials of different properties vis-à-vis the original construction materials on the dam, measures for environmental impact mitigation etc.

5.2. Construction aspects

Rehabilitation work requires a good contract in which the work is unequivocally defined so that surprises are avoided to the extent possible. Rehabilitation work can be implemented through item rate contract, turnkey contract or EPC contract. Pre-qualification documents are to be structured carefully for arranging the rehabilitation work without conflicts. Items of works of special nature should preferably be implemented with specialised agencies having expertise. Performance warranty for the specified duration, construction supervision, Quality Control and Quality Assurance are aspects for careful consideration during the implementation stage.

5.3. Risk management

Through risk assessment in dams which are to be rehabilitated helps the dam owner to prioritise them for effectively reducing the risk, integrating all the dam safety aspects in the assessment (i.e. aspects such as dam safety monitoring, health of the dam, operation and maintenance, mitigation measures such as emergency action plan, rules of operation etc.) which normally are analysed as separately. Various methodologies of risk analysis [16] are to be adopted for supplementing with the direct approach of analysis, and reporting after visual inspection.

5.4. Field investigation and minimum design studies

Inspections evaluating the overall condition of the dam along with detailed studies and additional exploration or analyses substantiating the general assessment already conducted are necessary to evaluate the safety of the dam before proceeding with the rehabilitation process. Minimum design studies, as mentioned in paragraph 5.1, would be necessitated.

5.4.1. Investigation and testing. Tests or investigations to be carried out for salient parameters depend on the inspection observations and considerations according to the problem. It would sometimes require extraction of representative cores from the concrete/ masonry dam for carrying out certain tests. Indications on certain parameters could be obtained through non-destructive testing. Test for deciding parameters such as density, compressive strength of concrete/ mortar in masonry, static as well as dynamic modulus of elasticity, Splitting tensile strength of concrete, permeability of concrete / masonry, chloride content & pH of concrete, corrosion activity in concrete, analysis of the reservoir water / seepage water, investigations for seepage path or permeable locations in the dam for planning directional grouting or conducting a seismic study for the site-specific seismic parameters, are a few important activities to name. The tests shall conform to relevant national/federal/international standards as applicable. The Indian standards such as IS 13311-Part1, IS 516, IS 5816, IS 11216, IS 456, IS11216, IS 5529 etc. can be referred to for the tests mentioned.

5.4.2. Geophysical explorations. In masonry or concrete dams, determination of seepage zones or seeping paths is an important investigation that decides the objective and methodology of rehabilitation. Various types of geological exploration methods (seismic refraction method etc.) and other methods utilizing the principles of geophysics are capable to determine seismic parameters as well as relevant details pertaining to the existing conditions of the concrete/masonry dam. Electrical resistivity, Sonic tomography (Plate 1 &2), ground penetrating radar, streaming potential are a few methods used in analyzing the condition of existing dam. In sonic tomography, the velocity contour maps in longitudinal and cross section would be able to interpret for desirable remedial action for rehabilitation works.

6. Role of Instrumentation in Large dams

Lack of proper and timely instrumentation of dams is a real challenge in the dam safety monitoring and implementation of timely rehabilitation measures. Instrumentation in dams helps the dam owner or the administration, in safety evaluation of the existing dams, for obtaining early warnings of any emergent distress on the dam health, in verification of the design and the actual behaviour of the dam with respect to the designed and for more realistic analyses, etc., using the actual observations from the instrumentation. The type of measurements from the instrumentation in dams include strain and stress measurements, differential displacements with respect to vertical and horizontal, relative and rotational movements, temperature measurements, seepage measurements, uplift pressure measurements, piezometer measurements for pore-water pressure, geodetic measurements for movement of dam, abutments, foundation and seismic measurements that give valuable inputs to study the ground motion due to any seismic event and effects of the same on the dam. ICOLD bulletin [17, 18] summarises the different types of dam instrumentation.

Plate 1. Geophysical investigation crew member for sonic tomography imparting source of sound waves at the upstream face of the masonry dam.

Plate 2A. Acquisition station.

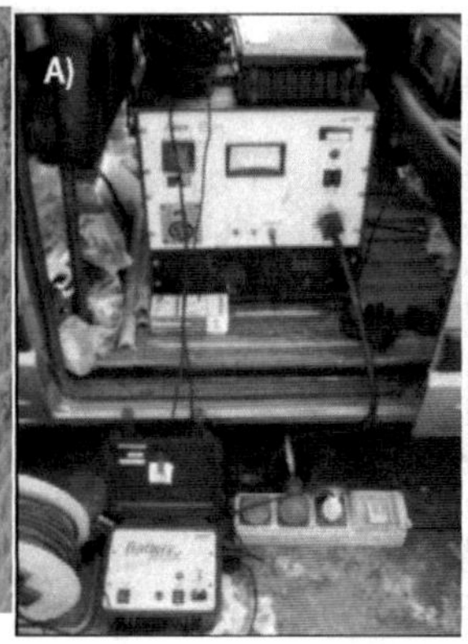

Plate 2B. Accelerometer sensors.

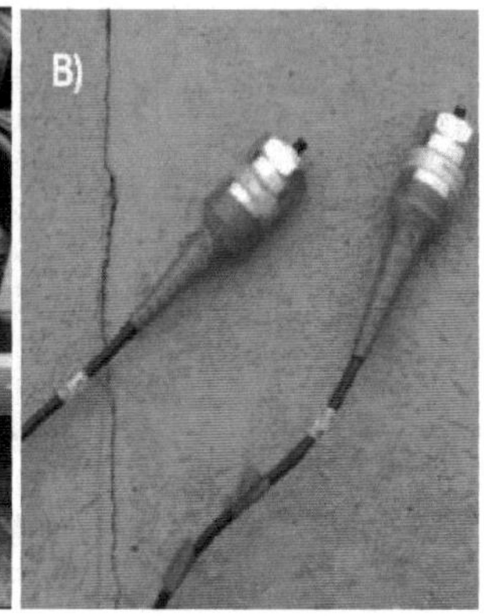

In order to monitor the loss of strength of the foundation, seepage, uplift pressure, and inelastic foundation displacements are to be measured. The quantity of seepage, origin, quality, presence of solids in the water, the origin of the same, etc. are to be gathered and analysed. Reliable and stable piezometers shall be installed at distinct points to measure the uplift pressure on a long term basis. Geodetic surveys, observation by pendulum, inverted pendulum moored deep bottom of dam, etc. are useful in detecting the displacement of the rock foundation due to loss of strength. Foundation problem, if any, have to be identified timely through expert intervention. Grouting is a remedial measure for strengthening foundation rock whereas problems due to uplift can be improved by grouting and appropriate drainage measures. Seepage and uplift pressure measurement during the operation of the dams will give light to erosion and solution of foundation rock also [8].

The selection and installation of various types of instruments, location, their measurement and frequency of measurement and analyses [19] are important aspects to the dam safety management (Figure 5-6). The faulty or non-working instruments, absence of periodic measurements as per the desired frequency, lack of proper maintenance as well as analyses by qualified Engineers are impediments in getting timely data for evaluating the safety assessments and deciding the rehabilitation measures. In order to have accurate and reliable analysis for a sensitive structure like double curvature arch dam, comprehensive analyses for chemical and physical properties of concrete need to be done. Field investigations include NDT such as tests on UPV, carbonation, in-situ moisture content, resistivity, half-cell potential, air permeability etc., whereas laboratory tests include tests for strength in compression and tension, poisson's ratio, modulus of elasticity, drying shrinkage, moisture movement and thermal expansion of the aged concrete [20].

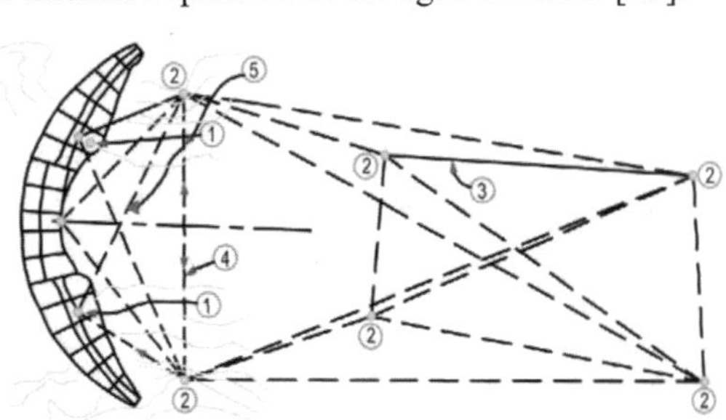

Figure 5. Layout for triangulation measurement: (1) targets fixed on dam surface; (2) theodolite; (3) base line, measured; (4) base line, computed; (5) sight lines [19].

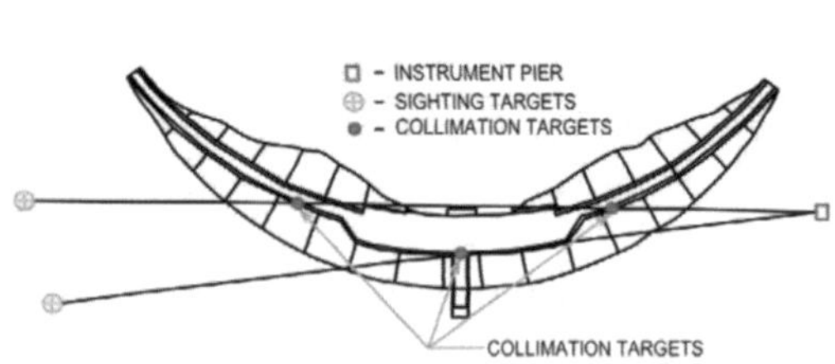

Figure 6. Layout for collimation measurements used for an arch dam [19].

IOP Conf. Series: Materials Science and Engineering **936** (2020) 012022 doi:10.1088/1757-899X/936/1/012022

7. Some noteworthy past instances of dam rehabilitation

Revision of data, guidelines, standards, methodology, etc., may require an increase in storage capacity; there can be decrease in useful storage due to silting and there can be situations where the ageing had caused deterioration or loss in strength. In all these situations, strengthening of the dam or raising the height of the dam would be necessitated to conform to the safety requirement insisted by the standards, as realised through the safety evaluation. If these vital structures are not strengthened or rehabilitated, for exploiting their uses, they will stay vulnerable to failures and will be a potential source of risk to the stakeholders downstream of the dam. Various methods and techniques can be used for the rehabilitation. Some of the methods used in the past for raising the concrete dams are, adding weight to the dam on its' top so as to bring the resultant within the middle third, give earth backing, by providing an embankment downstream for the parallel heightening or cable anchoring from the foundation by pre-stressing through the Coyne method (Andre Coyne, 1932) in which the heel region is subjected to compression by applying a vertical force at top of the dam. As per the literature, earth backing or embankments at downstream side were provided for 61m high Talakalele dam in Sharavathi Hydel Project, 30m high Stettynskoolf Dam, South Africa whereas there are many instances of adding concreting at downstream side(parallel heightening), viz., Aswan dam, 42m high Loskop Dam(Figure 7)which is raised by 9m, the Grand Dixence Dam(Figure 8), which had increased its' height up to 167m first and then to the full height of 280m, Guri dam in Venezuela which was raised from 100m to 162m height from the lowest foundation level[21]. The 103m high Koyna dam in India, a gravity dam with rubble-concrete construction, which suffered an earthquake measuring 6.5-7.0 in magnitude on December 11, 1967 were to be subjected to short term and long term rehabilitation measures. The long term rehabilitation was by providing buttresses at the downstream and the contact between the buttress and old surface was grouted with cement for monolithic action [21]. The 59.3m high(max.) constant radius concrete arch, Gibraltar dam in California, was strengthened against the possible tensile stresses and cracking against MCE and to compensate for the loss in storage due to sedimentation, by providing a roller compacted concrete (RCC) gravity section at the downstream.[21].

In all those cases, the issues addressed were the load distribution between the old and new concrete, bonding and the shear stress at the interfacing plane. The difference in Elastic modulus of old and new concrete, differential stresses due to temperature effect, increased uplift pressure due to increased height etc., also are matters of concern. The solution to uplift issue requires proper and sufficiently deep drainage system. The surface preparation for sufficient bonding, contraction joints in old and new structure, differential settlement due to addition of weight etc. are to be studied using appropriate modelling techniques before being implemented. FEM techniques are capable of simulating the scenario and predicting the structural behaviour of the combined structure before the rehabilitation is implemented. Though pre-stressing by cable anchoring are used in various dams, the maximum height of dam recommended is 60m, beyond which the application of pre-stressing force would increase considerably besides the practical difficulty in constructing anchorages[21].

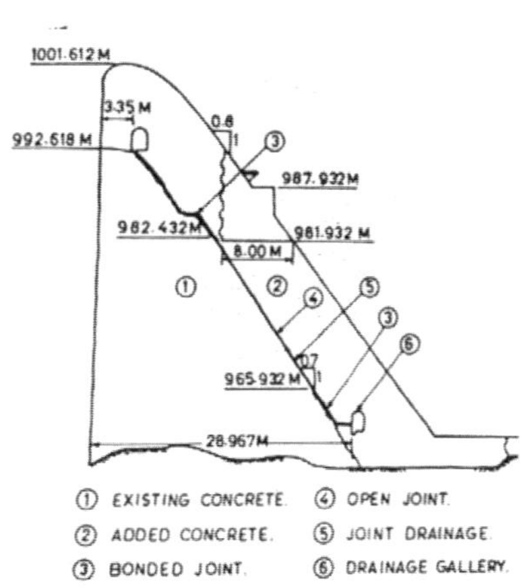

Figure 7. Loskop dam[21].

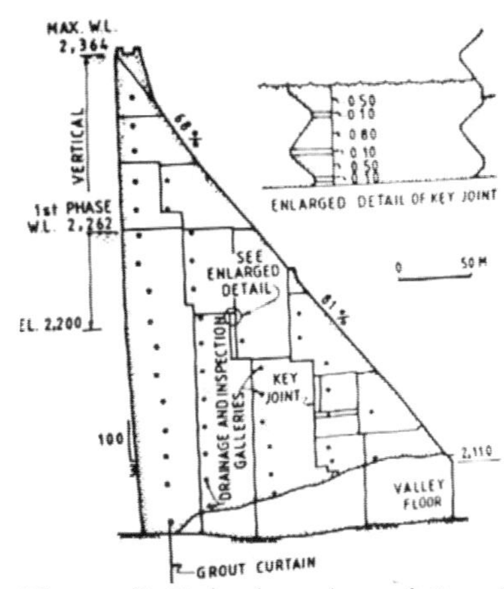

Figure 8. Raised section of Grand Dixence dam[21].

8. Some recent cases of rehabilitation from india

Certain rehabilitation works require extensive field investigation, studies, comprehensive analyses and approvals & clearances as it involves administrative, engineering, land, forest, environmental aspects before entering into real implementation for which more time, money and other resources are required. Examples of such important and challenging rehabilitation attempts in India in recent past include, de-siltation of Kundahpalam, Pillur, Papanasam dams in Tamil Nadu, spillway requirement for the acclaimed Hirakud dam, Odisha, rehabilitation of gates and hoists –almost 100 year old- of the famous Krishnarajasagar dam, Karnataka and analyses for unusual behaviour in Idukki dam, etc.[22]. Environmental and social management framework requiring satisfactory mitigation measures is an important and challenging task in major rehabilitation work such as construction of additional spillways to cater to the increased design flood; the rehabilitation and resettlement of the projected affected people, as well as the environmental issues involved in the de-siltation of reservoirs to remove the sediments deposited that reduces the useful storage are challenging task requiring careful implementation.

8.1. Geo-membrane in masonry dams

PVC composite geo-membrane and geo-textile was used for the waterproofing system in many concrete and masonry dams against the leakage of upstream reservoir water through the dam body. There are various instances where the upstream faces of the dam were to be installed with a PVC composite geo-membrane or geotextiles as a rehabilitation measure against the seepage related issues in such dams all over the world[14]. The Kadamparai dam in TamilNadu, installed with PVC geo-membrane system were reported to have reduced the seepage by a large measure. In Kadamparai dam waterproofing for an area of 17,300m2 was reported to have completed in 3 months and six weeks [23]. The important challenging aspects to the project authorities and installation team would be the need of meticulous planning for activities such as depletion of the reservoir as well as thorough upstream face preparation before installation of the system, ensuring of strict quality control in the material and its' installation besides proper design and installation of the drainage, perimeter sealing and anchoring system, After Kadamparai dam India, the Servallar dam (Figure 9) and Upper Bhavani dam in Tamilnadu are also being rehabilitated using geo-membrane sealing system(GSS) [24].

Figure 9. Kadampara dam upstream face, before, during and after installation [23].

8.2. Servalar dam

Heavy seepage of 742.65 litres/min.(lpm) with water level at +263.18 m were reported through the left flank portion besides leaching in the servalar dam, in Tamil Nadu, India. Raking and pointing using special mortar, guniting, etc. done were not able to control dam body seepage and to address the seepage issue through the left bank masonry joints, rehabilitation using PVC geo-membrane sealing system was adopted (Figure 10). Many challenges were experienced in the implementation which included, difficulties in depletion of the dam, excavation and de-silting of the sediments, unprecedented rain, monsoon, cyclonic rain, wind, adverse working conditions, agitation of local public etc. The work was delayed and implemented different compartments and different phases for obtaining dry working sessions, before it was completed. After completion, the project authorities reported that the leakage could be brought down to 30 lpm at water level +257.35m [24].

(a) Earthwork excavation (b) Work in progress at scour & power vent area (c) After completion of left bank area

Figure 10. Servalar dam during different phases of implementation [24].

8.3. Cementitious material

Apart from, PVC GSS, there are various other treatment system of cementitious materials for upstream face treatment for checking seepage of water from upstream side to downstream through the dam body. They include crystalline technology and certain polymer based materials. The important aspects in the effective implementation of these methods too are depleting the reservoir for treatment to extend up to the foundation level, strict quality control in material physical and chemical properties, storage, proportioning, curing as well as application. The Alamatti dam & Krishnaraja Sagar Dam in Karnataka and Anathode dam of Kerala are a few examples where such upstream treatments are taken up as rehabilitation measures. The upstream face treatments along with appropriately designed and implemented grouting are often suggested in such situations for getting better results.

8.4. Structural Analyses for rehabilitation

The rehabilitation process to the large dam structure many a times would require comprehensive structural analyses. The following cases are cited.

8.4.1. Konar dam. The 3682m long earth-cum-concrete Konar dam, Jharkhand, India completed in October 1955, consist of a 58 m high concrete gravity dam, 277m in length, flanked by earthen dam on both sides, three galleries, i.e. inspection gallery (+408.432 m), Access gallery (+ 391.668 m, and a gallery at lower elevations in foundation. About seven years after the completion, in 1962-63, the dam developed cracks on both d/s and u/s faces of the galleries in all the blocks. Cracks in the top gallery were wider ones than those in the two galleries below, and it expanded as time passed by, until it reported stabilised in 2010. Downstream slope of non-overflow portions of the dam developed surface cracks. Several studies to deduce the cause, nature, and behaviour of the cracks were conducted. Accordingly several measures adopted to deal with the situation. In order to device suitable rehabilitation measures, under the Dam Rehabilitation & Improvement Project (DRIP) in India and an exhaustive 3D and 2D finite element coupled thermo-mechanical analyses in ABAQUS v6.14-4 software environment, with the dam-foundation modelled as linear-elastic for materials, but geometry and contacts having modelled as non-linear, for investigation of its cracks, has been carried out[25,26].

The time consuming analyses simulating the existing condition confirmed the temperature effects as the predominant reason of cracking and suggested rehabilitation measures such as repairing of the cracks with low pressure normal grouting during the summer during which period the cracks are expected to be in fully open position with cementitious materials/mortar Class R-4 having specific properties conforming to British & European Standards BS EN-1504, restricting admixtures to shrinkage reduction only. The study recommended special type of painting on the exposed faces as well as downstream face to reduce the effects of solar radiation besides recommending for instrumentation for reservoir temperature measurement and for crack monitoring [27,8].

8.4.2. Idukki dam. Idukki dam, India's only double curvature arch dam, is installed with 18 groups of instruments in electronic, mechanical, geodetic and seismic devices for measuring temperature, deformation, and seismic disturbances influencing the dam. Monument triangulation, Geodetic targets installed at predefined points, suspended pendulums, crest leveling, gallery leveling and crest collimation helps understand the dam behaviour[28]. Still, there were field investigations and lab investigations for identifying the various properties of the concrete and these parameters were used for the FEM study. The behaviour in stress and displacement at various salient locations were studied using 3D FEM analyses using ABAQUS software, and the thermo-static analyses ascertained the dam behaviour conforming to the instrumentation data besides the observation that results are more accurate when the effects of vertical joints are taken into consideration[29]. The study also recommended certain rehabilitation measures including painting with reflective properties on the intrados surface to control the thermal effects on the dam. A modernised instrumentation for real time structural health monitoring was also evolved after the study which will help the project authorities, a comprehensive dam monitoring.

8.4.3. Anathode- kakki dam. Seepage is to be thoroughly investigated at areas such as abutment and foundation areas, upstream and downstream faces of the dam and in galleries. Where the upstream faces of the dam and other suspected areas are not accessible even by divers, a remotely operated vehicle (ROV) can be deployed to investigate. Since the location of seepage was from deep and inaccessible areas of control gates and sluice gates under water, exploration using RoV will be able to locate the faults and designing suitable rehabilitation measures. Exploration of details of deterioration at the sluice area was done in Kakki, through the RoV(Figure 11& Figure 12) using expertise. In case of severe seepage in masonry dams, geophysical investigations often will be able to locate seepage and seepage path for remedial measures, in which they use the principle of geophysics and mapping the structure using the state of the art technologies. There are instances in high concrete dams where installation of passenger lifts are impeded by leakages, in which cases, leakages into lift well are suspected to be from upstream block joints. Remedial measures can be done by systematically designed grouting (and reaming of drainage holes)in adjacent blocks from the dam top to the depth below seepage is detected. This kind of rehabilitation is attempted in 110m Kakki dam in Kerala, India [30].

8.4.4. Hirakud dam. Hirakud Dam, India, a multipurpose project comprising of a composite dam constructed of earth, masonry and concrete, built across Mahanadi river, was commissioned in the year 1957. The project authorities studied various options such as increasing the height of dam, construction of additional stretch of spillway on the dyke at left /right bank, diminishing the crest level of the spill way, etc. to route the revised flood without endangering the safety of the dam. After analysing various options, and consultation with various authorities, the proposed intervention is to construct an additional length of spillway on the left dyke, spillway with five radial gates having a capacity to release about 3.20 lakh cusecs and other related structures. This construction will pass through and include habitations causing to considerable resettlement impacts necessitating a Resettlement Action Plan based upon an Environmental and Social Management Framework approved by various competent authorities. Such kind of rehabilitation requires extensive planning and execution and challenges are multifaceted [31].

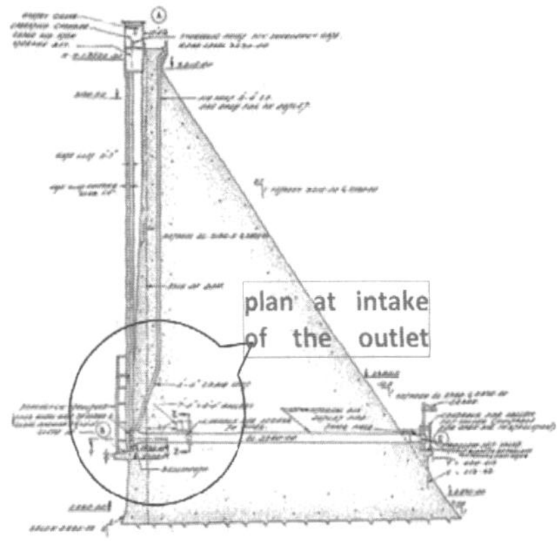

Figure 11. Kakki dam-section through C/L of sluice outlet [30].

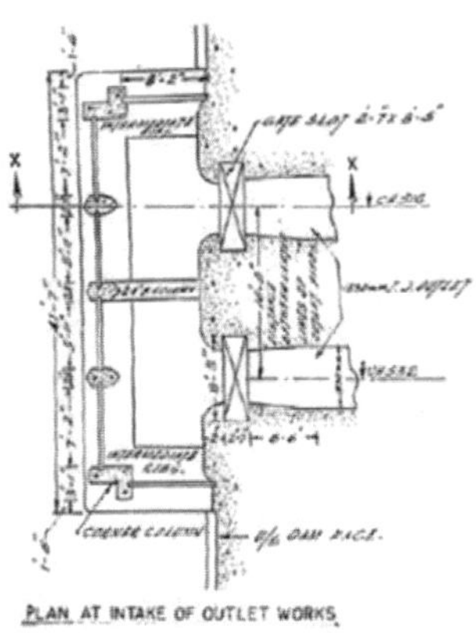

Figure 12. Plan at intake of the outlet works(detailed) where the RoV located deterioration in concrete [30].

9. Conclusion

Concrete and masonry dams will be subjected to deficiencies and deterioration primarily due to ageing and due to various structural, hydrological and operational aspects besides factors such as revision of guidelines. The reasons for rehabilitations are not typical but vary for each dam. Selection of the most suitable remedial measure is decided by the problem. The necessity for rehabilitation of large dams of these categories is important considering their demand for future benefits without any risk on safety of the stakeholders downstream. The structural, hydrological, operational and management aspects are to be reviewed periodically through comprehensive safety evaluations and based on meticulous analyses and planning the rehabilitation have to be implemented to alleviate the risk that may result in the absence of such intervention.

10. References

[1] International Commission on Large Dams (https://www.icold-cigb.org)
[2] Jansen Robert B 1988 *Advanced Dam Engineering for Design, Construction and Rehabilitation* (Van Nostrand Reinhold: New York)
[3] Jansen Robert B 1983 *Dams & Public Safety, A Water Resources Technical Publication* (U.S. Dept. of the Interior, Bureau of Reclamation: Denver)
[4] International Commission on Large Dams 2019 Better Dams for a Better World *World declaration on dam safety* Porto
[5] A Water resources technical publication 1995 *Safety evaluation of existing dams* (U.S. Dept. of the Interior, Bureau of Reclamation: Denver)
[6] National Register of Large Dams, Central Water Commission, Govt.of India, June 2019 web access: www.damsafety.in/damsafety/NRLD_2019.php
[7] International Commission on Large Dams Bulletin 93 *Ageing of dams and appurtenant works-review and recommendations* 1994
[8] Central Water Commission, Ministry of Water Resources, RD & GR, Govt. of India 2018 *Manual for Rehabilitation of Large Dams* Doc. No. CDSO_MAN_DS_02_v1.0
[9] ASTM C881/C881M-15 *Standard Specification for Epoxy-Resin-Base Bonding Systems for Concrete* (ASTM International: West Conshohocken, PA, USA)
[10] ASTM D1084-16 *Standard Test Methods for Viscosity of Adhesives* (ASTM International: West Conshohocken, PA, USA)
[11] ASTM D695-15 *Standard Test Method for Compressive Properties of Rigid Plastics* (ASTM International: West Conshohocken, PA, USA)
[12] ASTM D638-14 *Standard Test Method for Tensile Properties of Plastics* (ASTM International: West Conshohocken, PA, USA)
[13] ASTM C882 / C882M - 13a *Standard Test Method for Bond Strength of Epoxy-Resin Systems Used With Concrete By Slant Shear* (ASTM International: West Conshohocken, PA, USA)
[14] Rajagopal K and Arnepalli D N 2015 State-of-the-art on the applications of geosynthetics for dam repair and rehabilitation *1st National Dam Safety Conference* (Chennai) pp 257–68
[15] ICOLD Bulletin 170 2018 *Flood Evaluation and Dam Safety* (Paris)
[16] U.S. Dept. of the Interior, Bureau of Reclamation, *Best practices in dam and levee safety risk analysis: Risk assessment/management* Chap.IX-1 Risk guidelines 1–35
[17] ICOLD Bulletin 68 1989 *Monitoring of the Dams and their Foundations-State of the Art*
[18] ICOLD Bulletin 23 1972 *Reports of the Committee on Observation on Dams and Models*
[19] Central Water Commission, Ministry of WR, RD & GR, Govt of India 2018 *Guidelines for instrumentation of large dams* (New Delhi)
[20] Dhawan A K et al. 2016 State-of-the art instrumentation for dam safety-case study of Idukki dam *Compendium of tech. papers, Second National Dam Safety Conference* (Bengaluru) pp 522–9
[21] Sharma, H D 1998 *Concrete dams, Publication No.266* (Central Board of Irrigation and Power :New Delhi)
[22] Kapoor V K et al. 2016 Project implementation issues in rehabilitation of existing dams- DRIP experience *Compendium of tech. papers, Second National Dam Safety Conference* (Bengaluru) pp 335–43

[23] Subramanian V 2015 Geosynthetics Water Barriers in Kadamparai Dam: Observed Behaviour of its Exposed Geomembrane after 10 Years of Installation *Compendium of tech. papers, 1st National Dam Safety Conference* (Chennai) pp 277–82

[24] Navasakthi B A and Rameshwari P 2019 Experience of TANGEDCO in completion of geomembrane lining works in servalar dam, Tamil Nadu *Compendium of tech. papers, International Dam Safety Conference 2019* (Bhubaneswar) pp 959–73

[25] Dipankar C, Dubey A K and Pandey S B 2015 An Overview of Konar Dam on Need of Rehabilitation *Compendium of Technical Papers, First National Dam Safety Conference*, (Chennai) pp 269–74

[26] Dam Rehabilitation and Improvement Project (CWC: New Delhi) www.damsafety.in

[27] Pillai B R K, Patra B K, Agarwal A and Shukla A 2017 Investigation of cracks in Konar concrete gravity dam by nonlinear thermo-mechanical analysis *Compendium of Technical Papers 3rd National Dam Safety Conference* (Roorkee) pp 240–55

[28] Dhawan A K and Patra Bikram 2016 Assessment of current material properties of aged concrete arch dam-a case study *Compendium of tech. papers, Second National Dam Safety Conference* (Bengaluru) pp 185–90

[29] Pillai B R K and Patra Bikram 2015 Understanding the unusual behavior of an arch dam using FEM approach *Second National Dam Safety Conference* (Bengaluru) pp 266–71

[30] Panicker K Jiji, Somarajan P K, Baburaj O and Joseph Bibin 2019 Potential of remotely operated vehicles in the diagnosis for defects and rehabilitation for high dams – a case study on kakki dam of Sabarigiri project, *Compendium of tech. papers, International Dam Safety Conference 2019* (Bhubaneswar) pp 1025–31

[31] Department of Water Resources, Government of Odisha 2018 *Draft resettlement action plan, Construction of additional spillway of hirakud dam, in sambalpur district, odisha* (https://www.damsafety.in/ecm-includes/PDFs/Disclosure/Hirakud_RAP.pdf)

IOP Conf. Series: Materials Science and Engineering **936** (2020) 012023 doi:10.1088/1757-899X/936/1/012023

A Software Tool to Estimate the Time-dependent Deformation in RC Walls and Columns

Varun Sahay*, **M N Shariff and Devdas Menon**

Indian Institute of Technology Madras, Chennai, Tamil Nadu, India

*varunsahay19@gmail.com

Abstract. Tall reinforced concrete (RC) buildings are sensitive to time-dependent effects which arise due to creep and shrinkage. In this study, a software tool has been developed to estimate the time-dependent deformations in RC walls and columns. The program can invoke the ACI 209, EC 2, IS 1343 and *fib* MC 10 models to determine the creep and shrinkage in plain concrete. In order to include the effects of longitudinal reinforcement, two alternative methods have been explored: (i) effective modulus method and (ii) one-dimensional boundary value problem method. The effect of variable environment is included using the principle of superposition, which is valid within the linear creep domain. The program is written on the Python platform and can be used on any Windows machine. The program is validated using tests reported in the literature. The program is capable of accepting input of a linear elastic one-step-analysis from standard commercial software like ETABS and StaadPro. This program requires minimal input from the user, such as the grade of concrete, rate of loading and time at which the deformations are to be evaluated. The time-dependent deformations are evaluated for each column or wall and the super-elevation levels for each story are presented in a chart form, which serves as a design aid for the engineers and can be readily used by the site construction team.

1. Introduction

Creep and shrinkage of the concrete depend on the dimensions of the element, the ambient relative humidity and the composition of the concrete. Additionally, creep is also dependent on the magnitude of sustained load, the maturity of the concrete when the load is first applied and the duration of the loading [1]. It is imperative to estimate the time-dependent deformation due to creep and shrinkage in reinforced concrete (RC) columns with reasonable accuracy to assure safety and serviceability of high-rise buildings, pylons of cable-stay bridges and other creep sensitive structures [2]. If not adequately accounted for in the design, these time-dependent deformations result in serviceability issues such as slab-beam deflections, cracking of partitions and failure of other non-structural elements [3]. The time-dependent deformations calculated using the existing commercial software often ignore the effect of reinforcements and hence they have a considerable deviation from the observed values in the real structures. Through this study, an effort has been made to develop a software tool that could estimate these strains with higher accuracy by incorporating the use of effective modulus method and one-dimensional boundary value problem method [4,5]. Any one of the methods can be employed as per the convenience of the design engineer. In both methods, the effect of steel is considered assuming strain compatibility (perfect bond) between concrete and steel. In the effective modulus method, steel's restraining effect on creep and shrinkage deformations is included in the equilibrium equations, and the modulus of elasticity of concrete is modified using the creep coefficient calculated at every time increment. The one-dimensional boundary value problem method is an exact method to determine the time-dependent deformations, which is recently developed, wherein a linear viscoelastic constitutive law for concrete is considered. The reinforcing steel and concrete are considered to be in a parallel arrangement. A software tool has been developed on the Python platform, which can be a useful tool while designing and construction of tall RC buildings using the above two methods.

2. Models for predicting the creep and shrinkage strains in plain concrete

The software evaluates the creep and shrinkage induced time-dependent deformation on plain concrete using the models as per ACI209, EC2, IS1343 and *fib* MC 10. The choice of using any one of the models is given to the user. The software can either use the effective modulus method or the one-dimensional boundary value problem method to include the effect of longitudinal reinforcements.

2.1. ACI 209R-92 model [6]

This model predicts the shrinkage strain $\varepsilon_{sh}(t,t_c)$, where t (in days) is the age of concrete and t_c (in days) is the age of concrete at the start of drying or the end of initial curing, using the equation below-

$$\varepsilon_{sh}(t,t_c) = \frac{(t-t_c)^{\alpha}}{f+(t-t_c)^{\alpha}}.\varepsilon_{shu} \tag{1}$$

here, f (in days) and α are constants for a given shape and size of the member and ε_{shu} is the ultimate shrinkage strain. The values of ε_{shu}, f and α depend on factors like age of curing t_c, relative humidity, V/S of the member, the slump of concrete, percentage of fine aggregates, cement content of concrete and percentage of air in concrete. The detailed expressions to evaluate can be found in Appendix A of the code.

The stress-dependent strain is evaluated as-

$$\varepsilon_{cr} + \varepsilon_i = J(t,t_0).\sigma_c \tag{2}$$

here, ε_{cr}=creep strain at the time t considered; ε_i=Initial elastic strain at the time of loading; σ_c =stress applied to the concrete at the time of loading.
$J(t,t_0)$ is the Compliance function and is defined as-

$$J(t,t_0) = \frac{1+\phi(t,t_0)}{E_{cmto}} \tag{3}$$

where, E_{cmto} is the elastic modulus of concrete (in MPa) at the time of loading t_0 (in days) and $\phi(t,t_0)$ is the creep coefficient which is the ratio of creep strain to the elastic strain at the beginning of loading t_0 (in days).
The creep coefficient $\phi(t,t_0)$ is calculated using the equation below –

$$\phi(t,t_0) = \frac{(t-t_0)^{\Psi}}{d+(t-t_0)^{\Psi}}.\phi_u \tag{4}$$

d, ψ and $\phi(t,t_0)$ depend on the time of loading t_0, relative humidity, V/S ratio, the slump of concrete, percentage of fine aggregates and air content in the concrete. The detailed expression of which can be found in the code.

2.2. EC2 model and IS 1343 model [7,8]

The total shrinkage strain (ε_{sh}), according to this model, consists of two parts- the drying shrinkage strain (ε_{cd}) and the autogenous shrinkage strain (ε_{ca}).

$$\varepsilon_{sh} = \varepsilon_{cd} + \varepsilon_{ca} \tag{5}$$

Drying shrinkage strain when the age of concrete is t (in days), is evaluated using the expression-

$$\varepsilon_{cd}(t) = \frac{(t-t_s)}{(t-t_s)+0.04\sqrt{h_0^3}}.k_h.\varepsilon_{cd,0} \tag{6}$$

Here, k_h is a coefficient which depends on the notional size $h_0=2A_c/u$ of the concrete member, A_c is the cross-sectional area of concrete, u is the perimeter of that part of the cross-section which is exposed to drying, t_s is the age of concrete at the start of drying shrinkage which is normally the end of curing. The basic drying shrinkage strain ($\varepsilon_{cd,0}$) depends on RH, Mean compressive strength of concrete and the cement type. The detailed expressions can be obtained from Annex B of the Code.

The expression for the Autogenous Shrinkage strain when the age of loading is t (in days) is given by

$$\varepsilon_{ca}(t)=(1-e^{-0.2t^{0.5}}).\varepsilon_{ca}(\infty) \tag{7}$$

$$\varepsilon_{ca}(\infty)=2.5\times10^{-6}.(f_{ck}-10) \tag{8}$$

here, f_{ck} is the characteristic compressive strength of concrete (in MPa)
The creep strain is evaluated using the following expression-

$$\varepsilon_{cr}(t,t_0)=(\varphi(t,t_0)).(\sigma_c/E_c) \tag{9}$$

where, $\sigma_c<0.45f_{ck}(t_0)$ and E_c is the tangent modulus of elasticity of concrete under compression. $\varphi(t,t_0)$ is the creep coefficient and is evaluated using the expression-

$$\varphi(t,t_0)=\varphi_0\cdot\left[\frac{(t-t_0)}{\beta_H+t-t_0}\right]^{0.3} \tag{10}$$

here φ_0 is the notional creep coefficient, which depends on RH, mean compressive strength of concrete and the age of concrete at the time of loading t_0 and β_H depends on RH, detailed expressions of which can be obtained from the respective codes. It is crucial to note that though the basic governing equations in the above two models are the same, the expressions for the calculations of certain constants such as φ_0, β_H and E_c are different in these two models.

2.3. *fib MC 10 model* [9]

The total shrinkage strain, $\varepsilon_{cs}(t,t_s)$, consists of two parts, basic shrinkage, $\varepsilon_{cbs}(t,t_s)$, which occurs even in the absence moisture loss and drying shrinkage, $\varepsilon_{cds}(t,t_s)$, which is the additional shrinkage in case of moisture loss.

$$\varepsilon_{cs}(t,t_s)=\varepsilon_{cbs}(t,t_s)+\varepsilon_{cds}(t,t_s) \tag{11}$$

$$\varepsilon_{cbs}(t)=(1-e^{-0.2\sqrt{t}}).\varepsilon_{cbs0}(f_{cm}) \tag{12}$$

$$\varepsilon_{cds}(t,t_s)=\left[\frac{(t-t_s)}{0.035h^2+(t-t_s)}\right]^{0.5}.\varepsilon_{cds0}(f_{cm}).\beta_{RH}(RH) \tag{13}$$

here, t_s is the age of concrete at the start of drying (in days), f_{cm} is the mean compressive strength (in MPa) of concrete at the age of 28 days, t is the age of concrete (in days) at which the strains are to be calculated, RH is the relative humidity of the ambient atmosphere (in %), and h is the notional size (in mm) = $2.A_c/u$, the detailed expression for $\beta_{RH}(RH)$, $\varepsilon_{cds0}(f_{cm})$ and $\varepsilon_{cds0}(f_{cm})$ can be obtained from section-5.1.9.4.4 of *fib* Model Code for Concrete Structures 2010.
The expression for evaluating the stress-dependent strain $\varepsilon_{c\sigma}(t,t_0)$ at a time t (in days) is given by-

$$\varepsilon_{c\sigma}(t,t_0)=\sigma_c(t_0).\left[\frac{1}{E_{ci}(t_0)}+\frac{\varphi(t,t_0)}{E_{ci}}\right]=\sigma_c(t_0).J(t,t_0) \tag{14}$$

here, $J(t,t_0)$ is the creep compliance or the creep function, $E_{ci}(t_0)$ is the elastic modulus of concrete at the time of loading. E_{ci} is the elastic modulus of concrete at the age of 28 days. The expression for the variation of E_{ci} with time can be obtained from the code.

The creep coefficient $\varphi(t,t_0)$, consists of two components, basic creep coefficient, $\varphi_{bc}(t,t_0)$ and drying creep coefficient $\varphi_{dc}(t,t_0)$, which are estimated by the expressions,

$$\varphi(t,t_0)=\varphi_{bc}(t,t_0)+\varphi_{dc}(t,t_0) \tag{15}$$

$$\varphi_{bc}(t,t_0)=\ln\left(\left(\frac{30}{t_{0,adj}}+0.035\right)^2.(t-t_0)+1\right).\beta_{bc}(f_{cm}) \tag{16}$$

$$\varphi_{dc}(t,t_0)=\beta_{dc}(f_{cm}).\beta(RH).\beta_{dc}(t_0).\left[\frac{(t-t_0)}{\beta_h+(t-t_0)}\right]^{\gamma(t_0)} \tag{17}$$

here, RH is the relative humidity in the ambient environment (in %), f_{cm} is the mean compressive strength of concrete at the age of 28 days (in MPa), $t_{0,adj}$ is the adjusted age at loading (in days) and h is the notional size of the concrete member (in mm). The detailed expressions for $\beta(RH)$, $\beta_{dc}(t_0)$, $\beta_{bc}(f_{cm})$ and $\beta_{dc}(f_{cm})$ can be obtained from section-5.1.9.4.3 of *fib* Model Code for Concrete Structures 2010.

3. Methods for including the effect of reinforcement in RC

The software can implement two alternative methods to include the effect of reinforcements.

3.1. Effective modulus method [4]

The effective concrete modulus is given by the expression-

$$E_{eff}=\frac{E_{ci}}{(1+\nu_t)} \tag{18}$$

where, E_{eff}=effective concrete modulus at the time t (in days) considered, E_{ci}= concrete modulus at the time of loading (taken as 28-day concrete modulus), υ_t=creep coefficient at time t, t is the time after loading (in days).

The *fib* MC 10 Model uses E_{ci} at two different time instances in its expression of the compliance function. To account for this E_{eff} has been derived considering strain compatibility between steel and concrete and the obtained expression is given by-

$$E_{eff}=\frac{E_{ci}.E_{ci}(t_0)}{(E_{ci}+\nu_t.E_{ci}(t_0))} \tag{19}$$

here, $E_{ci}(t_0)$ is the concrete modulus at the time of loading.

By this method, the basic equations for prediction of creep, shrinkage and total strains are–

$$\varepsilon_{initial}+(\varepsilon_{creep})_t=\frac{P}{[A_g(1-\rho_g)+A_g\rho_g n_{eff}].E_{eff}} \tag{20}$$

$$(\varepsilon_{shrinkage})_t=(\varepsilon_{sh})_t-\frac{(\varepsilon_{sh})_t E_s A_g\rho_g}{[A_g(1-\rho_g)+A_g\rho_g n_{eff}].E_{eff}} \tag{21}$$

$$(\varepsilon_{total})_t=\varepsilon_{initial}+(\varepsilon_{creep})_t+(\varepsilon_{shrinkage})_t \tag{22}$$

where, $(\varepsilon_{creep})_t$ = creep strain in the reinforced concrete member at time t (in days) considered, and n_{eff}=modular ratio at time t which is equal to (E_s/E_{eff}), E_s is the elastic modulus of steel, $(\varepsilon_{shrinkage})_t$=shrinkage strain of the reinforced concrete member at time t considered, $(\varepsilon_{sh})_t$ =shrinkage strain as calculated for the plain concrete specimen with other parameters being the same, $(\varepsilon_{total})_t$= total strain of the reinforced specimen at the time t considered and t is the time after loading (in days).

3.2. One-dimensional boundary value problem method [5]

The creep, shrinkage and total strain in the reinforced specimen is given by-

IOP Conf. Series: Materials Science and Engineering **936** (2020) 012023 doi:10.1088/1757-899X/936/1/012023

$$\varepsilon_{creep}(t) = \mathrm{L}^{-1}\left[\frac{s\mathrm{L}(J)[\mathrm{L}(P) - A_s E_s \mathrm{L}(\varepsilon_{sh})]}{[A_c + A_s E_s s\mathrm{L}(J)]}\right] \tag{23}$$

$$\varepsilon_{shrinkage}(t) = \mathrm{L}^{-1}\left[\frac{s\mathrm{L}(J)[-A_s E_s \mathrm{L}(\varepsilon_{sh})]}{[A_c + A_s E_s s\mathrm{L}(J)]}\right] \tag{24}$$

$$\varepsilon_{total}(t) = \varepsilon_{creep}(t) + \varepsilon_{sh} \tag{25}$$

here, L is the Laplace transform of the respective functions, s is the Laplace transform variable and L^{-1} is the inverse Laplace transform of the respective functions. ε_{sh}=shrinkage strain as calculated for the plain concrete specimen with other parameters being the same, $\varepsilon_{shrinkage}(t)$=shrinkage strain in the reinforced specimen at the time considered; $\varepsilon_{creep}(t)$=creep strain in the reinforced specimen at the time considered.

4. Results

The program (Creep And Shrinkage Estimator – CASE) has a user-friendly interface and invokes the respective predictive models as per their choice. It can accept inputs of a linear elastic one-step analysis from standard commercial software like ETABS and StaadPro. It can generate the plots showing deformations and strains in the queried section of the column as well as the deformations in the queried column at each story at a particular time. The loading is considered to be applied in a stepwise manner, which is the actual way in which columns are loaded in a real structure. These deformations can be used to provide the necessary extra elevations to the columns while construction to ensure that the actual dimensions are reasonably close to the design dimensions.

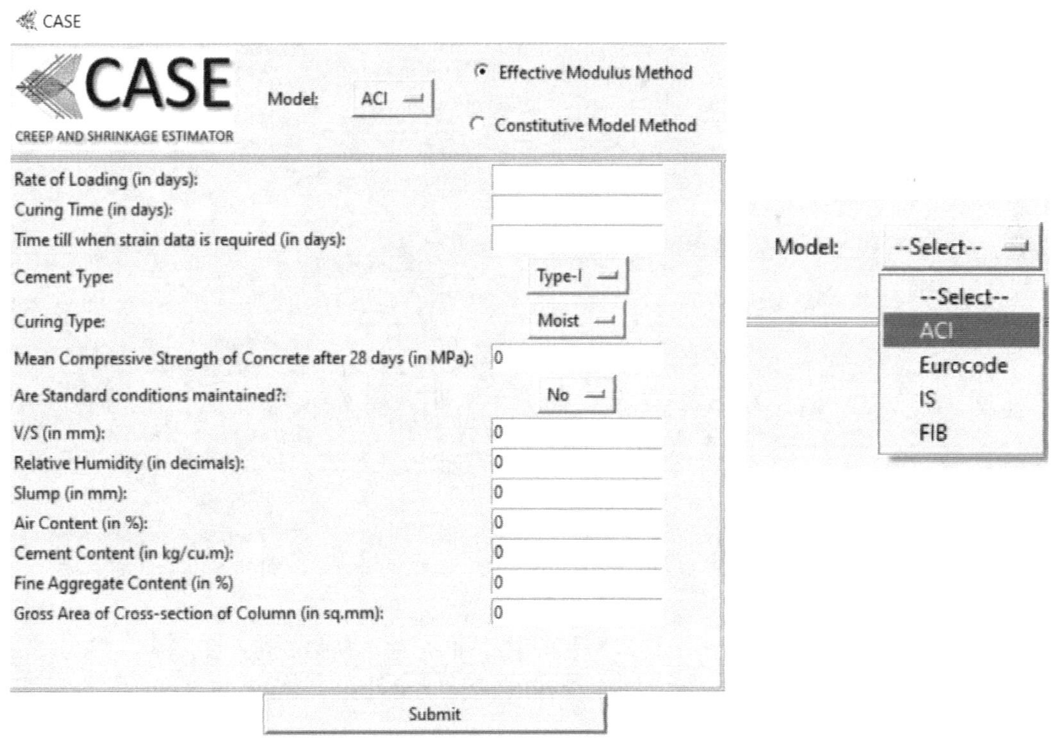

Figure 1. A snippet of input window of the program. The required inputs change according to the model selected. Here the ACI model is selected.

The CASE program is first validated using tests reported in the literature. A number of tests were conducted by Ziehl et al. (1998) [10] on reinforced concrete as well as plain concrete columns subjected to axial compression for a 450 days period. The column specimens were made using M27 concrete and compressive stress of 0.4 × f_c (10.8 MPa) was applied. The time-dependent strains are estimated using the ACI code and the effect of reinforcement is incorporated by applying the 'effective modulus method'. Figures 2a and 2b show the comparison of the experimental versus predicted time-dependent strains. It is seen that the strains predicted from the program are in good agreement with the experimental data.

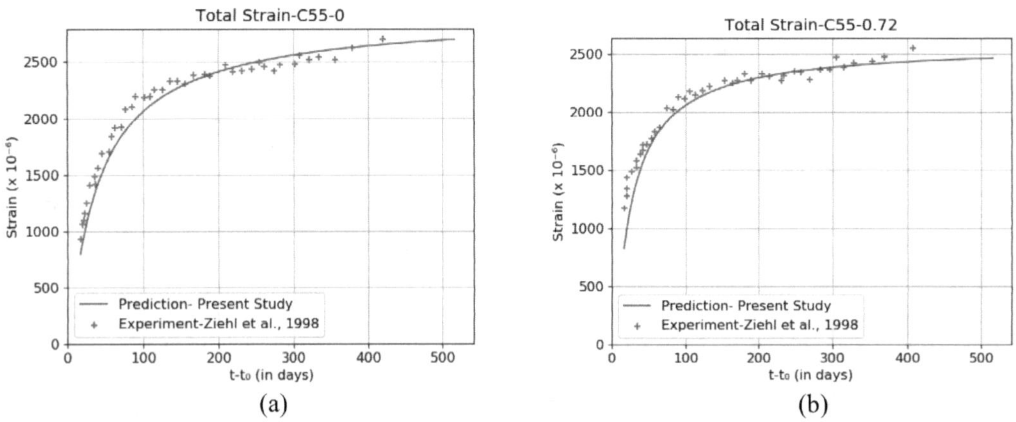

Figure 2. Comparison of experimental versus the prediction using ACI code (a) for plain concrete column and (b) for reinforced concrete column (p =0.72%).

A sample ETABS model is generated for a 10 storeyed structure. The analysis and design are carried out in ETABS and the output file (in Microsoft Excel format) is imported to the developed program. The floor construction cycle has been assumed as 10 days. Results for a typical column (having a column name 149, which is located at the outer perimeter of the building) have been extracted and the creep, shrinkage and total strains have been plotted in figures 3a and 3b.

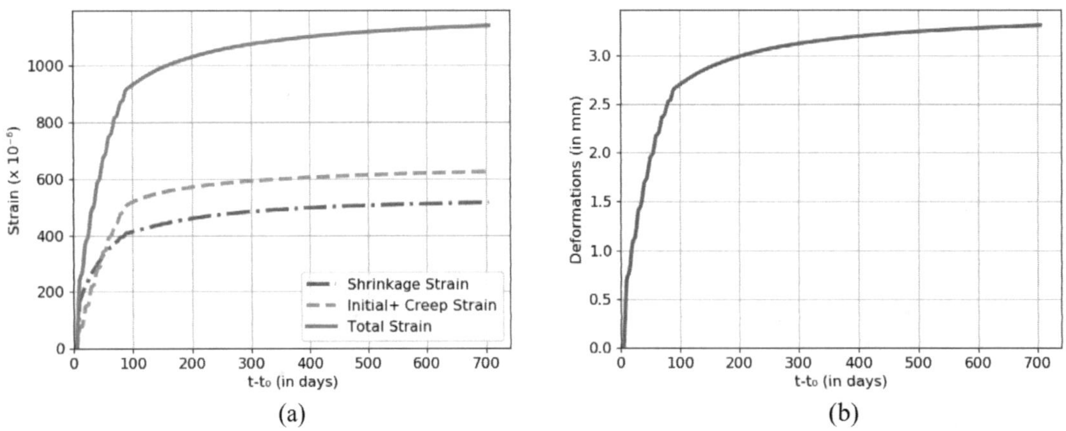

Figure 3. The strains and deformations in a particular part of the column in a story.

The designer and the construction engineer would be interested in estimating the casting level of the columns, such that the datum level is reached after the initial creep and shrinkage deformations stabilize. In this study, a stabilization period of two years has been assumed. The super-elevation levels for column '149' are plotted, both ignoring the effect of steel and including the effect of steel in figure 4. Here, the restraining effect of steel is clearly demonstrated. It can be seen that the time-dependent deformations would be over-estimated by about 15 percent if the effect of steel is ignored. This can have a significant influence on the differential axial shortening of the columns in a multi-storey structure.

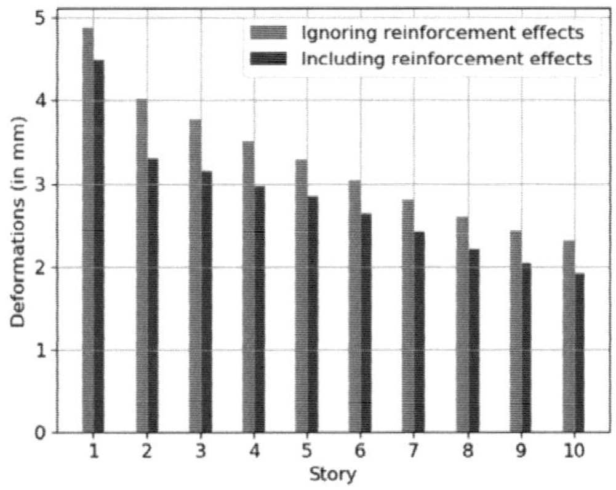

Figure 4. Deformations of a particular column in the building at various levels at the time considered.

5. Conclusions

From this study, the following conclusions can be drawn:

(a) A software tool has been successfully developed to estimate the creep and shrinkage strains (deformations) in an axially loaded reinforced concrete member, using the axial stress output generated using commercial software like (ETABS and StaadPro). The software tool requires only minimal input and can account for the construction load histories.

(b) The program has been validated using tests reported in the literature for both plain concrete and reinforced concrete. It is seen that the predictions are in good agreement with the experimental results for both the cases.

(c) The restraining effect of reinforcing steel on the time-dependent deformations and hence the error in estimating the super-elevation levels of columns is clearly demonstrated. For the example considered in the present study (10 storeyed building), it can be seen that the error in column shortening is about 15 percent if the restraining effect of reinforcement is ignored in the axial shortening calculation.

(d) This software tool can be easily used by engineers and construction managers to decide on the levels of column super-elevation to ensure the vertical elements reach the pre-planned datum level, as required in the design after they undergo creep and shrinkage deformations.

6. References

[1] Mehta P and Monteiro P J M 2006 *Concrete* (McGraw-Hill)

[2] Fintel M and Khan F R 1969 Effects of column creep and shrinkage in tall structures - prediction of inelastic column shortening *ACI J. Proc.* **66** 957–67

[3] Chiorino M A, Dougill J W, Huet C, Kanstad T and Kim J 1998 Measurement of time- dependent strains of concrete *Mater. Struct.* **31** 507–12

[4] Ziehl P H, Cloyd J E and Kreger M E 2004 Investigation of minimum longitudinal reinforcement requirements for concrete columns using present-day construction materials *Struct. J.* **101** 165–75

[5] Shariff M N, Saravanan U and Menon D 2020 Time-dependent strains in axially loaded reinforced concrete columns *J. Eng. Mech.* **146(8)** 04020076

[6] ACI Committee 1997 *Prediction of Creep, Shrinkage and Temperature Effects in Concrete Structures* (Farmington Hills, USA: American Concrete Institute).

[7] EN 1992-1-1 2007 *Eurocode 2: Design of Concrete Structures - Part 1-1: General Rules and Rules for Buildings* vol 2007

[8] BIS 1343:2012 2012 *Prestressed Concrete - Code of Practice*

[9] CEB FIB MC 2010 2013 *Fib Model Code for Concrete Structures 2010* (Switzerland)

[10] Ziehl P H, Cloyd J E, Kreger M E and River R 1998 *Evaluation of Minimum Longitudinal Reinforcement Requirements for Reinforced Concrete Columns* **7**

IOP Conf. Series: Materials Science and Engineering **936** (2020) 012024 doi:10.1088/1757-899X/936/1/012024

Experimental Study on Microbial Concentration Optimization in Cement Mortar using M-Sand as Fine Aggregate

Babitha Benjamin[1*], Anisha Sabbavarapu[2], J Sudhakumar[3] and T V Suchithra[4]

[1] Research Scholar, Department of Civil Engineering, National Institute of Technology, Calicut, India

[2] PG Scholar, Department of Civil Engineering, National Institute of Technology, Calicut, India

[3] Professor, Department of Civil Engineering, National Institute of Technology, Calicut, India

[4] Professor, School of Biotechnology, National Institute of Technology, Calicut, India

*benjaminbabitha@gmail.com

Abstract. One of the major shortcomings of conventional concrete based constructions is its gradual strength deterioration over time due to its week tensile strength. Without any effective measures to remedy this problem the alternatives we have now are periodic monitoring, retrofitting, repair & rehabilitation, etc. The above-mentioned methods are all highly unsustainable due to manpower, money and materials being spend in a cycle to upkeep. If a protective coating of microbial cement mortar is given to the structures like bridges, underground tunnels, etc. were a lot of dampness is observed, structural deterioration can be prevented. The study is aimed at optimizing the concentration of bacteria in self-healing cement mortar which uses microbiologically induced calcium carbonate precipitate to induce crack healing ability. For this purpose, bacteria *Bacillus Subtilis MTCC441* (non-ureolytic) were added in concentrations 10^4, 10^6, 10^8 cells/ml to a mix using M-sand as fine aggregate. The bacteria were cultured in a nutrient broth and were added to cement mortar by water medium during mixing. The addition to cement mortar was done when the microorganisms were still alive and active. In this study, a series of experiments like compressive strength, water absorption, and pour healing was done to optimise the concentration. The results showed that 10^6 cells/ml was the optimum quantity of microorganisms required to enhance the physical properties of concrete.

1. Introduction

Concrete usually experience cracks as its tensile strength is much lower than its compressive strength. On cracking, the stiffness of the concrete is reduced and it allows the permeation of moisture causing corrosion of steel which further leads to the deterioration of the structure before the design period. This needs extra expenditure for regular inspection and maintenance. However, sometimes it is difficult to access the damaged site. An alternative method is to induce self-healing property to concrete that is whenever crack occurs the cracks gets healed within in a short duration thus preventing further structural deterioration. There are many methods to incorporate self-healing property to concrete but this paper focuses on self-healing consummated by bacterial action through Microbiologically Induced Calcium Carbonate Precipitate (MICCP) [1]. This partial method is chosen since it is the most sustainable and convenient approach than other self-healing methods since there is no additional consumption of raw materials or other chemical substances involved as in other self-healing methods. Bacterial induced Calcium Carbonate precipitate is environment friendly, heals micro-crack, closes pore, increase the density of concrete and improves the compressive strength thus increasing the durability of structures. The present study aims at optimizing the concentration for calcite precipitating non-ureolytic bacteria *Bacillus Subtilis* MTCC441 in cement mortar.

1.1. Significance of the study
As it is known, concrete is weak in tension but strong in compression which makes the material brittle. To compensate this problem in concrete we provide steel reinforcement for concrete structure, but still the inherent brittle nature of concrete causes micro-cracks which further deteriorate the structure and reduces its design lifespan. Hence it is necessary to come up with a solution for the micro-cracks created in concrete structures. There are several solutions available for this condition, but most of which are not sustainable since it consumes additional raw materials, use of chemical polymers, etc. Biomineralization is the most sustainable solution available so far for this problem, which is use of microorganisms in concrete to self-heal the cracks whenever it occurs just like a fracture in a human bone gets healed by osteoblasts cells. This crack healing by microorganisms in concrete does not require any additional resources or manpower for crack repair thus making it sustainable by reduction in manhours, money, and material. This study is conducted in cement mortar since there is no necessity to provide self-healing for the entire structure which increases the construction cost, rather a protective plastering for the structure can be given which eliminates the additional cost factor.

1.2. Need for this study
Traditionally, the river sand used as fine aggregate for cement mortar, which is a scarce material nowadays. The over mining of river sand causes environmental issues like disturbing aquatic ecosystem, land erosion, etc., it is high time we consider this problem and start using an alternative material like M-sand which is already available in markets as a substitute for river sand. In this study M-sand is used as fine aggregate for preparing cement mortar. This makes building construction more sustainable and makes us more responsible individuals to the society.

1.3. Concept of microbial action
MICCP or biomineralization is the basic concept behind bacterial based self-healing concrete. MICCP is the process in which bacteria produces calcium carbonate as a result of their existence under suitable environmental conductions. The cell wall is negatively charged which draws the cations from its surroundings, this way Ca^{2+}reach the cell wall which acts as nucleation site where the reaction with CO_3^{2-} ions occurs to form $CaCO_3$[2].MICCP via non ureolytic bacteria is simply through the extracellular activity as shown in equation below.

$$Ca^{2+} + Cell \rightarrow Cell - Ca^{2+} \tag{1}$$

$$Cell - Ca^{2+} + Co_3^{2-} \rightarrow Cell - CaCo_3 \tag{2}$$

2. Materials
Traditionally the materials involved in the preparation of cement mortar are cement, river sand and water. Ordinary Portland Cement (OPC) Grade-53 conforming to IS 12269-1987 were used having specific gravity 3.35 with initial and final settling time as 210 minutes and 300 minutes, respectively for preparing specimens. Well-graded Manufactured Sand (M-Sand) conforming to IS 383-1970 were used having specific gravity 2.56 and fineness modulus 2.65 as fine aggregate. The additional important constituents used in this study are microorganisms to impart self-healing property to cement mortar. The bacteria *Bacillus Subtilis* MTCC441were used in this present study, bought from gene bank MTCC (Microbial Type Culture Collection), Chandigarh. This particular bacterium is chosen from available literature since it is a spore forming non-ureolytic bacterium which can survive in high pH environment and sustain mechanical stresses faced during mixing of cement mortar.

2.1. Maintenance of stock culture
The pure culture of *B.subtills* is maintained continuously on nutrient agar slants. The major components of the Nutrient Agar are peptone, beef extract, yeast extract, NaCl and Agar. The cultures were streaked on to agar slants with an inoculation loop and the slants are incubated at 37°C for 24-72 hrs. After sufficient growth on the agar slants, cultures were preserved on refrigerator at 4°C. For addition of bacterial culture into the concrete, a single colony from the nutrient agar plate is inoculated into 200ml of nutrient broth and the culture flasks are maintained at 37°C in a 125 rpm orbital shaker [3].

2.2. Concentration Control for bacteria
The concentration of bacteria in the nutrient broth is identified by comparing with McFarland Standards [4]. It standardises the concentration of bacterial suspension by comparing the turbidity of the test suspension with that of the McFarland Standard. The chemical constituents of McFarland Standard are barium chloride and sulphuric acid and the reaction between these two components results in the formation of a fine precipitate i.e. barium sulphate. McFarland standard tubes were shaken well prior to usage in order to visually compare a bacterial suspension of known concentration and the values were identified from the McFarland table.

2.3. Mix proportions and specimen preparation
The mix ratio adopted for the cement mortar study was 1:3 with a water cement ratio of 0.55. The water cement ratio was fixed by conducting the flow table test getting flow value106.5. For preparing mixes using bacteria, the bacterial culture is used instead of distilled water, by maintaining bacterial culture cement ratio as 0.55. The test specimens for compressive strength, water absorption, and pore healing were done in a laboratory type mortar mixer of capacity 10 litres. The mixing continued until a uniform mix was obtained. The cement mortar was then placed into the moulds which were properly oiled. After placing of cement mortar in moulds, proper compaction was given to the cement mortar using the mortar vibrator. Specimens were de-moulded after 24 hours of casting and were kept in a curing tank for water curing until the age of test. Since the main objective is to find the optimum concentration of bacteria *Bacillus Subtilis*, different specimens were cast using water or equivalent amount of live bacterial culture with concentrations of 10^4, 10^6, 10^8 cells/ml. Test specimens were prepared using standard cube moulds of the size 50mm [5]. The type of specimens prepared is detailed in table 1.

Table 1. Specimen details.

Sl. No	Specimen Type	Specimen Name
1	Normal Cement Mortar	NCM
2	Bacterial Cement Mortar 10^4cells/ml	BCM4
3	Bacterial Cement Mortar 10^6 cells/ml	BCM6
4	Bacterial Cement Mortar 10^8 cells/ml	BCM8

2.4. Compressive strength
To study the compressive strength of concrete, 50mm cube specimens were cast and compacted using a vibration machine. After 24 h of casting, the cubes were demoulded and then cured in water at room temperature and compressive strength testing is done at the age of 3, 7, and 28 days as per IS 516:1959. The surface water is wiped from the cubes before testing. The tests were performed in sextuplets and average value was taken as compressive strength.

2.5. Water absorption test
Water absorption tests of the control mix and bacterial live mix were performed for 3, 7 and 28 days of curing. For this test, the mortar cubes were oven dried at 110° C for 24 h and dry weight W_1 is measured. The samples were then kept in saturated state in water at room temperature for 24 h and W_2 is weighed again. The water absorption was then calculated by the formula.

2.6. Pore healing test
The pore healing was done using Non-contact Radio 3D Optical Profiler. Optical profilers are interference microscopes, with great precision using the wavelength of light as the ruler and are used to measure height variations such as surface roughness on surfaces. The cube specimens of size 50 mm were kept under the optical profiler after marking a small area on surface, for repeat revaluation on 3, 7 and 28 days.

3. Results and discussions

3.1. Compressive Strength Development
The compressive strength for the respective days as per IS:4031 part 6 obtained by applying different bacterial concentrations to the cement mortar are compared with the control specimens. It is found that

the compressive strength of microbial tested specimens of all the concentrations is higher than the control specimens. The increase in compressive strength is due the filling of pores with the calcite precipitation of bacteria. It is noted that the compressive strength of cement mortar increases with increase in live bacterial concentrations up to 10^6 cells/ml and there is a reduction at 10^8 cells/ml. The 28th day compressive strength is increased by 20%, 31% and 3% for 10^4, 10^6 and 10^8 cells/ml respectively. The figure 1 shows the bar chart for the compressive strength of different mix and table 2 shows the compressive strength values for the mixes.

Table 2. Compressive Strength.

Sample	3rd Day	7th Day	28th Day
NCM	10.36	15.93	23.88
BCM4	11.99	22.61	28.60
BCM6	12.35	23.23	31.17
BCM8	15.14	17.19	24.53

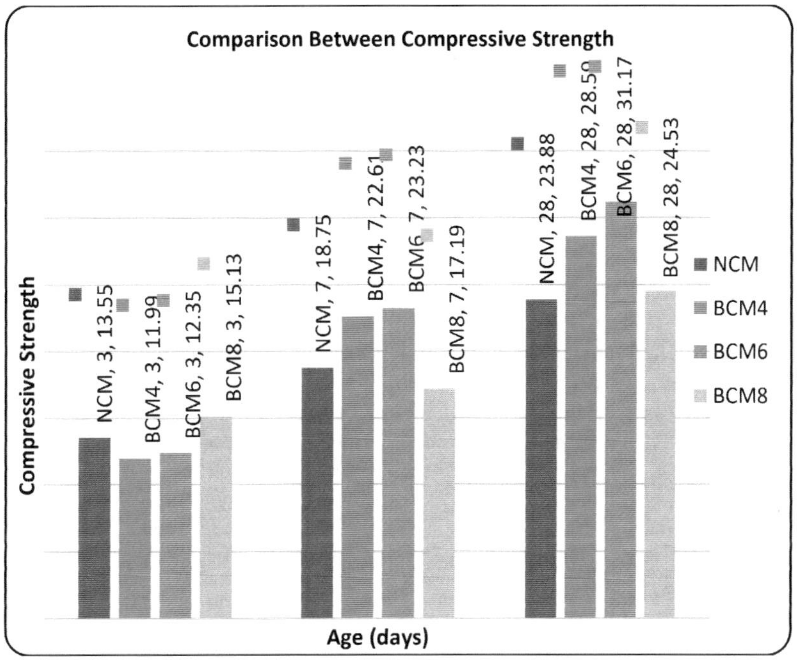

Figure 1. Comparison between compressive strength for specimens.

3.2. *Water absorption*

The water absorption for 3, 7, and 28 days are measured for control and bacterial specimens. It is found that the water absorption obtained from different bacterial concentrations is less compared with control mix. It is noted that the water absorption decreases with increase in live bacterial concentrations up to 10^6 cells/ml and there is an increase at 10^8 cells/ml. The 28th day water absorption is decreased by 13%, 90% and 12% for 10^4, 10^6 and 10^8 cells/ml respectively. Figure 2 shows the bar chart for water absorption for different specimens and table 3 shows the values for water absorption.

Table 3. Water Absorption (%)

Sample	3rd Day	7th Day	28th Day
NCM	9.41	7.31	5.93
BCM4	8.68	5.93	5.15
BCM6	7.59	1.04	0.47
BCM8	6.44	5.37	5.11

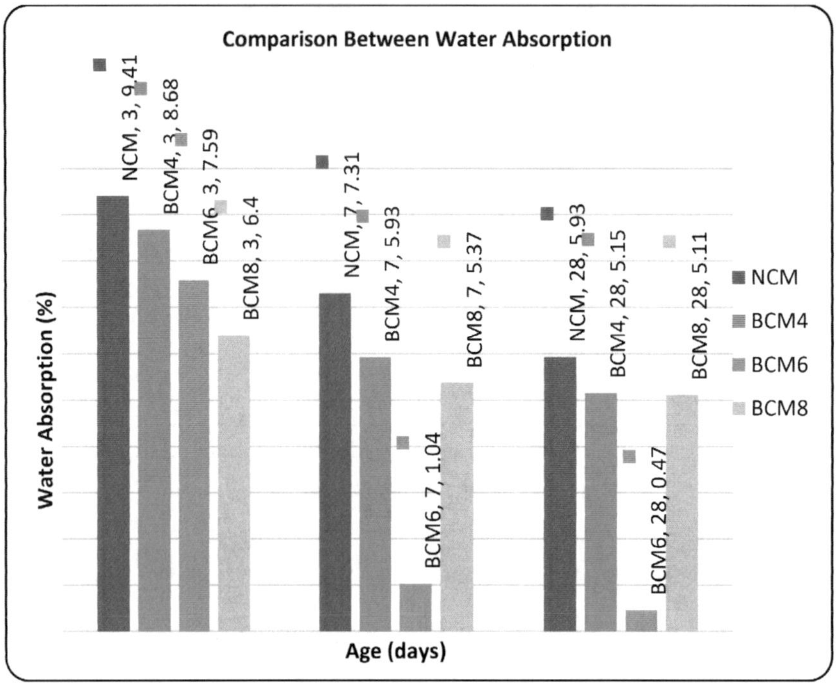

Figure 2. Graph for comparison between water absorption for specimens.

3.3. Pore healing

The table 3 below shows the surface measurement decreasing as curing days pass. The average surface measurements on 3, 7 and 28 days respectively are taken for all the specimens and concentration 10^8 showed pore healing. As the concentration increases the pore healing is found to be increasing. Figure 3 shows the bar graph for pore healing and table 4 shows values for pore healing.

Table 4. Pore depth details for the specimens in microns.

Sample	3rd Day	7th Day	28th Day
NCM	0.45	0.3	0.26
BCM4	0.22	0.14	0.09
BCM6	0.22	0.12	0.09
BCM8	0.24	0.22	0.1

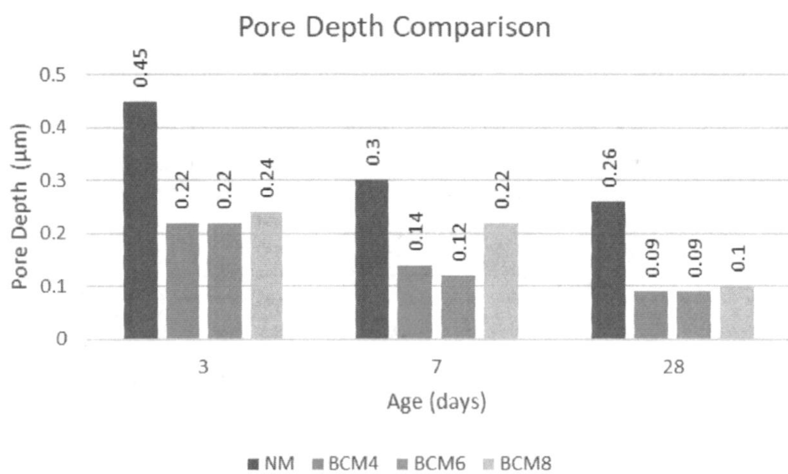

Figure 3. Graph for pore depth comparison.

4. Conclusions

Based on the experimental studies conducted, the following conclusions are obtained. The optimal concentration of bacteria *Bacillus Subtilis MTCC441* for improving the mechanical properties and to impart self-healing to cement mortar is 10^6 cells/ml. As the concentration of microorganisms increased beyond the optimal concentration of 10^6 cells/ml, the compressive strength of cement mortar specimens decreased by 21%.The compressive strength of cement mortar specimens increased by 30% as the concentration of bacteria increased up to an optimum concentration. The water absorption for cement mortar specimens at optimal concentration decreased drastically than other concentrations specimens up to 80%, i.e. to a point much less than normal cement mortar specimens without bacteria. The water absorption for cement mortar specimens with concentration more than the optimal concentration specimens showed 10% increase in water absorption than optimal concentration specimens but not greater than normal cement mortar specimens. The pours and surface roughness of cement mortar specimens decreased in diameter and depth as concentration of bacteria increased, maximum pore healing was achieved for BCM8. The decrease in surface roughness and pour size occurred beyond the optimal microbial concentration. i.e. as concentration of microorganisms increased the pour size and surface roughness decreased for the cement mortar specimens but compressive strength decreased.

5. References

[1] Thakur A, Phogat A and Singh K 2016 Bacterial concrete and effect of different bacteria on the strength and water absorption characteristics of concrete: a review *Int. J. Civ. Eng. Technol.* **7** 43–56

[2] Seshagiri Rao M, Srinivasa Reddy V and Sasikala C 2017 Performance of microbial concrete developed using bacillus subtilus JC3 *J. Inst. Eng. Ser. A* **98** 501–10

[3] Luo M and Qian C 2016 Influences of bacteria-based self-healing agents on cementitious materials hydration kinetics and compressive strength *Constr. Build. Mater.* **121** 659–63

[4] Sutton S 2011 Determination of inoculum for microbiological testing **15** 49–53

[5] Mondal S and Ghosh A 2018 Investigation into the optimal bacterial concentration for compressive strength enhancement of microbial concrete *Constr. Build. Mater.* **183** 202–14

[6] Luo M and Qian C 2016 Influences of bacteria-based self-healing agents on cementitious materials hydration kinetics and compressive strength *Constr. Build. Mate.* **121** 659–63

[7] Jose M, Carlos E, Sixto O, Rajeswari N, Gerardo F, Hector H and Nagamani B 2018 Improved strength and durability of concrete through metabolic activity of ureolytic bacteria *Environ. Sci. Pollut. Res.* **25** 21451–8

[8] Stabnikov V, Jian C and Ivanov V 2013 Halotolerant alkaliphilic urease-producing bacteria from different climate zones and their application for biocementation of sand *World J. Microbiol. Biotechnol.* **29** 1453–60

[9] Soleimani S, Ormeci B and Isgor O 2013 Growth and characterization of Escherichia coli DH5 α biofilm on concrete surfaces as a protective layer against microbiologically influenced concrete deterioration (MICD) *Appl. Microbiol. Biotechnol.* **5** 1093–102

[10] Sarayu K, Iyer N and Murthy A 2014 Exploration on the biotechnological aspect of the ureolytic bacteria for the production of the cementitious materials - a Review *Appl. Microbiol. Biotechnol.* **172** 2308–23

Acknowledgements

We acknowledge MTCC Chandigarh for supplying the bacterial culture Bacillus Subtilis MTCC441 for the project work and NIT Calicut for supporting the research work.

IOP Conf. Series: Materials Science and Engineering **936** (2020) 012025 doi:10.1088/1757-899X/936/1/012025

Influence of Fibre Geometry on the Fracture of Steel Fibre Reinforced Concrete

Komathi Murugan[1*], Stefie J Stephen[2] and Ravindra Gettu[3]

[1] Senior Project Officer (Postdoc), Department of Civil Engineering, IIT Madras, India

[2] Junior Research Fellow (Postdoc), Department of Civil Engineering, IIT Madras, India

[3] VS Raju Chair Professor, Department of Civil Engineering, IIT Madras, India

*komathimurugan@gmail.com

Abstract. Steel fibre reinforced concrete (SFRC) is used in various structural applications such as slabs, pavements, hydraulic structures, tunnel linings etc., due to their ability to limit the initiation and propagation of cracks, thereby improving the tensile load-carrying capacity and toughness. Two important parameters affecting the mechanical performance of a given SFRC are the type and dosage of fibres. The study presented in this paper examines the influence of fibre geometry on the fracture properties, considering two different types of cold-drawn hooked-ended steel fibres. Details of the pertaining experimental investigations conducted on SFRC with two different hooked-end geometries (denoted as 3D and 5D) at equivalent dosages are included; the tests conformed to EN 14651:2005+A1:2007. With the change in geometry, from the more conventional (3D) hooked-end with one kink, to that with three kinks at the ends (5D), even with a reduction of dosage from 45 to 35 kg/m^3, 5D fibres were found to exhibit a slight superior performance than 3D fibres.

1. Introduction

Steel fibre reinforced concrete (SFRC) has the toughness and crack bridging ability to encourage its use in a wide range of applications, such as shotcrete, bridge decks, tunnel linings, slabs, pavements, hydraulic structures and precast elements. The geometry of the fibre plays an important role in the SFRC response, as it affects the pull-out characteristics and crack-bridging response during crack propagation. An appropriate geometry should ensure both resistance to crack opening and dissipation of energy in controlled manner, without brittle failure. Also, the geometry should be such that entanglement and balling during fabrication are prevented.

The present study investigates the flexural properties of SFRC with cold-drawn wire fibres, with two configurations having the hooked-ends with one kink (denoted as 3D) and with two kinks (denoted as 5D); see figure 1.

(a) (b)

Figure 1. Selected types of fibres (a) 3D (b) 5D (Courtesy: www.bekaert.com).

2. Research background and motivation

The fibre parameters such as aspect ratio, tensile strength, volume fraction, dosage etc., along with concrete parameters, have been investigated previously for hooked-ended steel fibres (Johnston, 1974 [1]; Naaman, 1985 [2]; Michels *et al.*, 2013 [3]; Tiberti *et al.*, 2015 [4]; Jose *et al.*, 2018 [5]; Stephen *et al.*, 2019 [6]). Recently, the conventional hooked-end steel fibres have been modified with more kinks at the ends to increase the performance. Fibres with multiple kinks have been found to exhibit higher efficiency due to better anchorage, more stress distribution, multiple cracking and increased ductility (Abdallah *et al.*, 2016 [7]; Lee *et al.*, 2019 [8]). The study reported in this paper probes the mechanical performance of SFRC under flexure to assess the parameters used normally in design criteria for structural applications.

3. Experimental investigation

3.1. Details of specimens

SFRC specimens with nominal dimensions of 150 mm × 150 mm × 550 mm (width × depth × length) were examined; see table 1 for details of the fibres used (as available in the product specifications given by manufacturer). Note that both the types of fibres had lengths of 60 mm and diameter of 0.9 mm, and made with cold-drawn wires with modulus of elasticity equal to 210 GPa. The grade of concrete is M50, with a cement content of 340 kg/m^3.

Table 1. Details of fibres

Type of fibre	Dosage (kg/m^3)	Tensile strength (MPa)	Specimen designation
3D	45	1.16	3D - S1 to S8 (8 beam specimens)
5D	35	2.30	5D - S1 to S8 (8 beam specimens)

3.2. Tests for flexural toughness characterisation

The flexural performance of the SFRC beams was evaluated in terms of the limit of proportionality (LOP) or flexural strength, and the residual flexural strengths, using the experimentally-determined load versus crack mouth opening displacement (CMOD) curves. The tests were conducted using a closed-loop servo-hydraulic testing system of 300 kN capacity; see figure 2 for test set-up. Eight specimens were tested for each type of fibre concrete.

Figure 2. Test setup to determine flexural strength.

The test procedure, broadly conforming to EN 14651:2005+A1:2007 [9], is as follows.

- First, a 25 mm deep notch was cut at the bottom surface of a specimen at mid-span, perpendicular to the direction of casting.
- Next, the specimen was simply-supported with a centre-to-centre span of 500 mm. A clip gauge was mounted on knife-edges across the notch mouth to record the change in CMOD (horizontal displacement), with increase in load.
- The load is vertically applied at the mid-span, perpendicular to the longitudinal axis of the beam (along the notch plane). It was initially controlled based on load, with application at a constant rate of 100 N/s, up to a value corresponding to about 40% of the peak load (peak load for trial specimens with 3D and 5D fibres were found to be about 30 kN). Beyond this level, it was controlled based on displacement, by increasing the CMOD at a constant rate of 0.018 mm/min up to 0.1 mm, and then further by 0.09 mm/min. Each test was performed up to a CMOD of not less than 3.5 mm. The load and CMOD readings were recorded during the entire duration of a test.

4. Test results and discussions

The typical failure pattern of a specimen under load applied at mid-span is shown in figure 3, with a single macro-crack emanating from the notch tip, as expected. The load versus CMOD curves of all the tested specimens are given in figures 4.1, 4.2, 5.1 and 5.2. The first crack for each specimen was observed in a load range of 6 kN to 8 kN. For Specimens 3D - S7, 3D - S8, 5D - S3, 5D -S4, 5D - S6 and 5D-S8, there is a gradual reduction in load with increase in CMOD, on the widening of crack. This is attributed to the absence of fibres at immediate vicinity to effect the bridging action. With respect to the ultimate load, a wide variation is obtained between the specimens of each type (3D and 5D). This is because of the inherent variability in the specimens due to the casting and action of fibres.

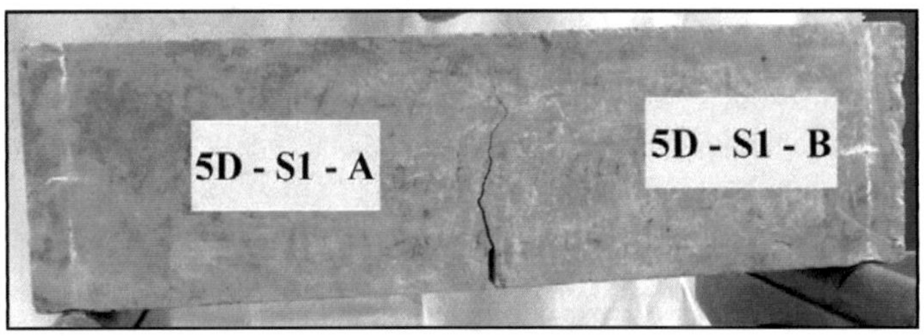

Figure 3. Typical failure pattern (shown for Specimen 5D - S1).

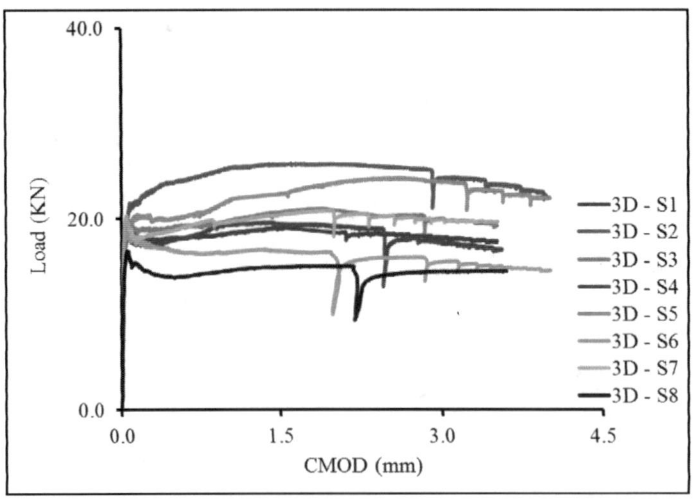

Figure 4.1. Load versus CMOD curves for specimens with 3D fibres (complete behaviour).

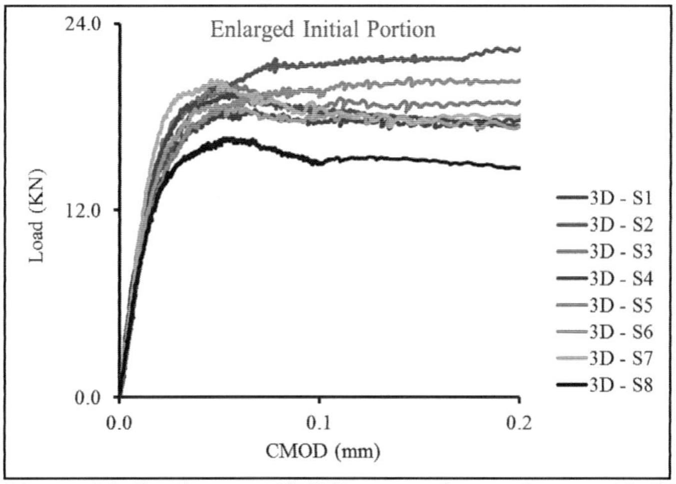

Figure 4.2. Load versus CMOD curves for specimens with 3D fibres (enlarged initial portion).

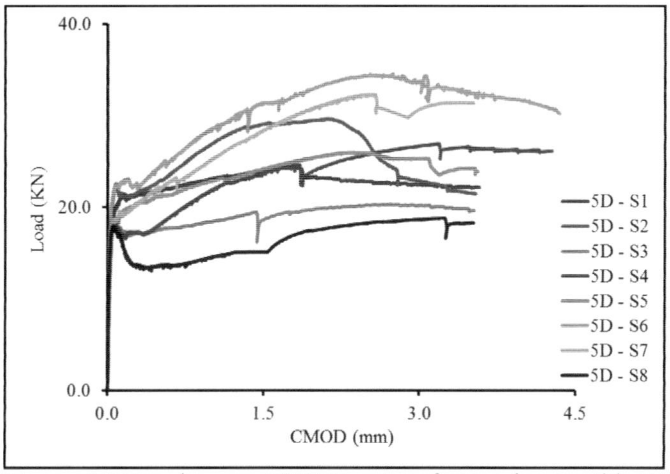

Figure 5.1. Load versus CMOD curves for specimens with 5D fibres (complete behaviour).

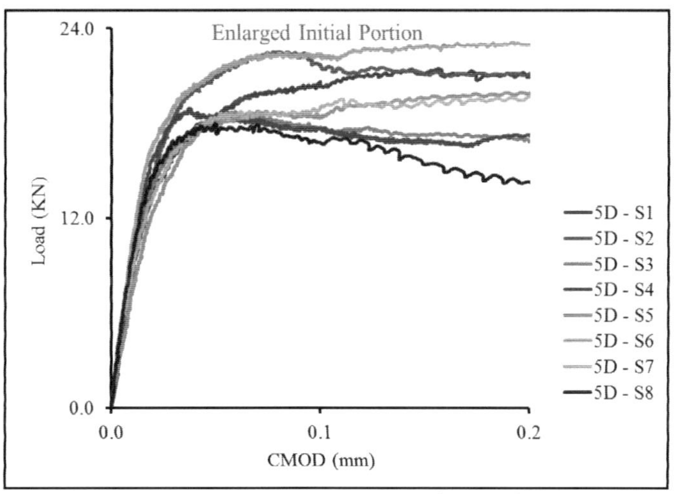

Figure 5.2. Load versus CMOD curves for specimens with 5D fibres (enlarged initial portion).

The flexural toughness parameters were computed as explained below:

- The LOP is given by the following expression (equation (1)):

$$f_{ct,L}^{f} = \frac{3F_L l}{2bh_{sp}^2} \quad (1)$$

Here,

$f_{ct,L}^{f}$ = LOP (MPa)
F_L = load corresponding to the LOP (N)
l = supported length of the specimen (= 500 mm)
b = width of the specimen (= 150 mm)
h_{sp} = distance between the tip of the notch and the extreme compression fibre at top of the specimen (= 125 mm)

The LOP for each specimen was calculated. It is expressed in terms of mean, standard deviation (s) and characteristic strength (f_{Lk} = mean – ks, where k = 2.00 for 8 samples, as per RILEM TC 162-TDF, 2003 [10]), for each type of fibre in table 2. The values of limit of proportionality indicate that there is no significant change in flexural strength or its variability due to the addition of different types of fibres.

- The residual flexural tensile strength ($f_{R,j}$) is given by the following expression (equation (2)):

$$f_{R,j} = \frac{3F_j l}{2bh_{sp}^2} \quad (2)$$

Here, F_j is the load (N) with j = 1, 2, 3, 4 corresponding to CMOD of 0.5, 1.5, 2.5 and 3.5 mm, respectively. The values of $f_{R,j}$ pertaining to each type of fibre are reported in table 2. The mean values of residual flexural tensile strength are found to be highest for the specimens with 5D fibres. However, the characteristic values are similar due to the larger scatter in the concretes with 5D fibres.

Table 2. Estimated flexural toughness parameters.

Type of fibre	LOP (mean ± standard deviation), $f_{ct,L}^{f}$ (MPa)	Characteristic flexural strength, f_{Lk} (MPa)	$f_{R,j}$ (mean ± standard deviation) (MPa)				Characteristic strength, $f_{R,jk}$ (MPa)			
			$f_{R,1}$	$f_{R,2}$	$f_{R,3}$	$f_{R,4}$	$f_{R,1k}$	$f_{R,2k}$	$f_{R,3k}$	$f_{R,4k}$
3D	6.02 ± 0.38	5.26	5.92 ± 0.92	6.36 ± 1.05	6.23 ± 1.25	5.98 ± 1.07	4.08	4.25	3.73	3.83
5D	6.03 ± 0.45	5.13	6.47 ± 1.13	7.78 ± 1.74	8.25 ± 1.76	7.82 ± 1.67	4.20	4.30	4.73	4.49

5. Summary and conclusions

The experimental investigation on SFRC with two different types of end-hook fibre geometry shows that the fibres with one kink (3D) at 45 kg/m^3 and the fibres with two kinks (5D) at 35 kg/m^3 gave similar characteristic flexural strength parameters and fracture toughness. However, in terms of the mean performance, the 5D fibres were slightly superior, even though they were used at a lower dosage.

6. References

[1] Johnston C D 1974 Steel fiber reinforced mortar and concrete: a review of mechanical properties *ACI Spec. Publ.* **44** 127–42
[2] Naaman A E 1985 Fiber reinforcement for concrete *Concr. Int.* **7** 21–5
[3] Michels J, Christen R and Waldmann D 2013 Experimental and numerical investigation on post cracking behaviour of steel fiber reinforced concrete *Eng. Fract. Mech.* **98** 326–49
[4] Tiberti G, Minelli F and Plizzari G 2015 Cracking behavior in reinforced concrete members with steel fibers: a comprehensive experimental study *Cem. Concr. Res.* **68** 24–34
[5] Jose S, Gettu R and Indhuja S 2018 Flexural toughness characterisation of steel, polymer and glass fibre reinforced concrete based on the notched beam test *The Indian Concr. J.* **92** 35–50

[6] Stephen S J, Raphael B, Gettu R and Jose S 2019 Determination of the tensile constitutive relations of fibre reinforced concrete using inverse analysis *Constr. Build. Mater.* **195** 405–14

[7] Abdallah S, Fan M, Zhou X and Geyt S L 2016 Anchorage effects of various steel fibre architectures for concrete reinforcement *Int. J. Concr. Struct. Mater.* **10** 325–35

[8] Lee S J, Yoo D Y and Moon D Y 2019 Effects of hooked-end steel fiber geometry and volume fraction on the flexural behaviour of concrete pedestrian decks *Appl. Sci.* **9** 1241–61

[9] EN 14651 2005 *British Standard Test Method for Metallic Fibre Concrete - Measuring the Flexural Tensile Strength (Limit of Proportionality (LOP), Residual)*

[10] RILEM TC 162-TDF 2003 Test and design methods for steel fibre reinforced concrete *Mater. Struct.* **36** 560–7

Acknowledgements

The authors wish to thank the T48R team of Gammon Engineers and Contractors Pvt. Ltd. for contributing the beam specimens prepared with fibres manufactured by Bekaert. The tests were conducted in the Laboratory for Mechanical Performance of Civil Engineering Materials, IIT Madras.

IOP Conf. Series: Materials Science and Engineering **936** (2020) 012026 doi:10.1088/1757-899X/936/1/012026

A Review on Fatigue Life Prediction of Plain Concrete

D R Renju[1*] and M S Keerthy[2]

[1] Student, Department of Civil Engineering, NSS College of Engineering, Palakkad, Kerala, India

[2] Assistant Professor, Department of Civil Engineering, NSS College of Engineering, Palakkad, Kerala, India

[*]renjudr93@gmail.com

Abstract. Fatigue analysis of concrete structures is important since it is the limiting factor to service life of many infrastructures. The heterogeneity in concrete and its complex behaviour upon repetitive loading makes the fatigue life prediction of concrete erratic. This paper presents a review on the fatigue life prediction models of plain concrete. Review includes the deterministic and probabilistic approach for the prediction of fatigue life of concrete. It has been observed that deterministic approach depends on certain parameters and initial conditions and is not reliable for the prediction of fatigue life of concrete. So there is a demand for a more generalised model based on probabilistic approach which includes randomness in the fatigue failure of concrete. Recent studies show utilisation of artificial neural network, a computational tool based on this approach can be utilised to predict the fatigue behaviour of concrete.

1. Introduction

The phenomenon of gradual microstructural changes that occurs permanently in a material due to cyclic loading is known as fatigue. In many infrastructures such as highway and airfield pavements, railway sleeper, bridge decks supporting structure of offshore windmills, concrete has to resist large number of cycles of varying amplitude. Basically concrete consists of three components: the cement matrix, the aggregate and matrix-aggregate interface. The heterogeneous nature of concrete result in the formation of numerous internal flaws which triggers crack initiation. These flaws are the region of high stress concentration and upon fatigue loading cracks will starts to propagate and continue until the failure. Crack growth due to fatigue is classified into three different zones namely crack initiation zone, crack propagation zone and unstable crack growth. The crack propagation will be through the weakest path in the concrete matrix. Several experimental, numerical and analytical investigation were conducted based on different approaches to understand the mechanism of fatigue failure of concrete and to develop a fatigue model.

There are basically three approaches to assess the fatigue life of a structural element viz. the S-N curve method, fracture mechanics method and continuum damage mechanics method. In the first approach, fatigue stress life data of concrete at varying stress levels from many fatigue tests are analysed and a typical empirical formula is generated using regression analysis based on S-N curve [1].But this approach only considers overall fatigue life and suffers from heterogeneity of concrete. The approach based on fracture mechanics considers the failure condition of the material by crack initiation and propagation. It depends on the stress intensity factor which is a function of stress state and crack length [1]. To investigate fatigue damage mechanics, there are two approaches including continuum damage method and the energy dissipation method. In the former, the fatigue life of concrete is predicted using damage evolution equation which is based on certain damage variables representing the damage of material. The variation in the energy dissipation during fatigue loading is analysed in energy method and the mode of energy dissipation is utilised to predict the fatigue failure of concrete [3], [4]. In this method stress level is not taken into account and is concerned only with the strength. Moreover, it is only applicable to viscoelastic concrete. This is the limitation of fatigue damage method since it depends highly on stress level for fatigue life prediction of concrete

The deterministic model does not include elements of randomness and so the models that account for it with as many variables is too cumbersome to work with it. It depends upon certain mathematical model and conditions will give a simple solution but will not be consistent. Probabilistic (stochastic) model uses distribution instead of fixed values which include variation and uncertainty. All the three approaches stated above gives a deterministic model based on parameter values and initial conditions which is not a dependable one for fatigue life of concrete since it consists of uncertainty.

Thus there is a demand for stochastic approach which can include the randomness in the fatigue of concrete. It can also incorporate the varying parameter involved in the fatigue life prediction of concrete. Artificial neural network (ANN), an effective computational tool is gaining wide acceptance for its application to engineering problems [5].All the approaches to analyse the fatigue life of concrete is based on deterministic approach and suffers from randomness in fatigue testing and variable properties of concrete. ANN which is based on probabilistic approach may be more suitable for fatigue prediction models by which randomness in fatigue failure of concrete can be included.

2. Models based on S-N curves

The approach based on fatigue stress-life relationship (S-N curve) is known as Palmgren-Miner hypothesis and it states that damage accumulates linearly with number of cycles applied at a particular stress level. The failure criterion is given by

$$\sum_{i=1}^{k} \frac{\mathrm{n_i}}{\mathrm{N_i}} = 1 \quad (1)$$

where n_i is the number of cycles at stress level i, N_i is the number of cycles to failure at stress level i, and k is the number of stress levels [1]. The applicability of Miner hypothesis of cumulative fatigue damage which assumes that fatigue damage is accumulated linearly was studied and it did not work well for concrete. Also it was concluded that by introducing rest periods into loading spectrum, there occurred a reduction in the failure probability corresponding to a specific number of load cycles and stress levels [6]. In 1958 by analysing the data from fatigue test on plain concrete, a relationship between applied stress level and cycles to failure is established [7].

To understand the hypothesis of fatigue failure of concrete, further experimental studies were conducted varying the parameters and criteria of failure. The studies included the deterioration of bond between aggregate and mortar mix, influence of material properties, difference in loading pattern, effect of stress ratio etc.The conventional S-N equation was further modified to consider the influence of frequency of loading. Two different models were developed to anticipate the fatigue strength of concrete, one for high cycle fatigue ($10^3<N<10^7$) and other for low cycle fatigue ($N<10^3$)[8].

Due to high variability in fatigue test data, probability theories were introduced to estimate fatigue life of concrete and concluded that conventional S-N curves are valid for prediction for fatigue stress level from 65 to 85 % [9]. Through the studies conducted on plain concrete under constant and variable amplitude fatigue loading, it was observed that Palmgren-Miner hypothesis is not reliable to evaluate the fatigue fracture of concrete [10].

3. Models based on fracture mechanics

The fracture mechanics method considers the condition around the crack tip and the crack growth due to fatigue as the failure criterion. In Paris formula, the crack propagation associated with variation of stress state is the basis for identifying the fatigue fracture. Several researchers extended the fatigue damage study on metals to concrete and found some of them working for concrete.Paris law turned into the first proposed model based on this approach and many researchers used these models directly to observe the crack growth behaviour of concrete. Paris law gives the relationship between crack length increment per cycle and the amplitude of stress intensity factor. The proposed law is based on fatigue crack propagation and is given by

$$\frac{da}{dN}=C(\Delta K)^m \tag{2}$$

Where N is the corresponding load cycle, a is the crack length, ΔK is the amplitude of stress intensity factor and C and m are material constants [11]. Paris law is only applicable to macroscopic cracks, that is only for very large specimens and was found to be incompatible with the heterogeneous nature of concrete and it requires some modifications.

Since Paris law is applicable only for large sized specimen, it was adjusted to apply for small sized concrete specimens and evolved size effect law. Paris law was adjusted to include the size-effect and the crack length increment depends on size adjusted stress intensity factors. A new parameter called brittleness number (ratio of structure size D to transitional size d_0) was introduced for size adjustment which distinguishes the responses governed by stress intensity factor and nominal stress according to the transitional size of specimen. The size-adjusted Paris law is given by

$$\frac{da}{dN}=C\left(\frac{\Delta K}{K_{IC}}\right)^m \tag{3}$$

$$K_{IC}=K_{If}\left(\frac{\beta}{1+\beta}\right)^{1/2}$$

where K_{If} is the fracture toughness and β is the brittleness number (D/d_0). The advantage of size-effect method is that one can obtain size independent fracture parameters by measuring the maximum load values of geometrically similar specimen of different sizes. Experiments show that d_0 for monotonic loading is much less than d_0 for cyclic loading, which means that the brittleness number for monotonic loading is much greater than the brittleness number for cyclic loading [12]. The fatigue tests conducted on single edged notched beam specimens inferred that Paris law can be applied for crack growth in plain concrete with C rely on the stress cycle ratio R and m as a constant [13].

The concrete beams tested under three point bending subjected to flexural fatigue loading was carried out. Specimens were subjected to quasi-static cyclic and constant amplitude fatigue loading and was concluded that the critical value of mode I stress intensity factor is constant along the post-peak part of quasi-static envelope and can be predicted using the following condition: K_I = critical value of $K_I = K_{IC}$. It was shown that the crack growth in the acceleration stage could be accurately modelled using the Paris law. The Paris coefficients were considered to be material constants for low cycle, high amplitude fatigue loading. The crack growth in the deceleration stage was modelled using an analytical expression which is derived from the concepts of R-curve and is given as

$$\frac{\Delta a}{\Delta N}=C_1(\mathrm{Da})^{n1} \tag{4}$$

where C_1 and n_1 are constant and exponent respectively and D_a is the total increase in crack length [14]. Based on the concepts of fracture mechanics which considers the effect of the loading pattern, size effect parameters and loading frequency a fatigue crack propagation model was proposed for concrete and was used to predict the residual strength of plain and reinforced concrete beams. An increase in brittleness is noticed as the size of plain concrete beam increases with the fatigue crack propagation. In the case of reinforced concrete (RC) beams, the stability of fracture process can be accomplished only when the beam has sufficient reinforcement [15].The prediction of crack growth rate and fatigue life of concrete was achieved through the fatigue crack propagation law. However the varying parameters make it more complex and not dependable.

4. Model based on fatigue damage mechanics

4.1. Continuum damage models

The destruction of bond between the cement matrix and aggregate, damage of material grains themselves affects the elastic properties of concrete and the material becomes anisotropic.The progression of microcracks and voids is the reason for this destruction and thus the nonlinearity in concrete. The damage due to this anisotropy which occurs in an initially isotropic material is utilised for the development of Continuum damage models (CDM) [16].

A damage model based on both monotonic and cyclic behaviour of concrete was developed. Based on continuum damage theory, certain damage variables are selected to characterise the material damage and then an equation for damage evolution was established. By applying thermodynamics theory, using dissipation function, growth rate of damage is evaluated and the damage evolution model was developed with respect to thermodynamic force conjugates of the damage variables [17]. The theory was based on certain assumptions and cannot be generalised. A nonlinear damage model which is able for predicting the fatigue life under compression for a range of concrete strength was developed. The fatigue behaviour for both high-cycle and low-cycle fatigue induced by varying cyclic stress ratios can be predicted using this model. Also a damage evolution model for concrete was developed applicable for proportionate monotonic loading conditions and uniaxial compression [18].

4.2. Energy method

In concrete, there will be an exchange of energy with the surrounding environment throughout the loading process. The total energy dissipated from the system can be divide into two, heat energy and the dissipated energy due to material damage. Plastic strain energy, the energy required during crack propagation and defect formation etc. are the usual energy dissipation due to material damage. The degree of material damage is characterised by corresponding physical quantities and is taken as governing damage variable. Then the damage evolution law of materials and its fatigue life prediction is obtained based on the energy dissipation trends [19]. An equation was proposed to determine the fatigue failure of concrete based on the stability of the single-cycle energy dissipation and uniformity of the critical energy dissipation and is validated with experimental data. However, this method is only applicable for viscoelastic concrete material and is unfit for ordinary concrete material which is quasi-brittle [20].

5. Model based on neural network

Artificial neural networks, or neural networks (NNs) are a class of computational tools developed inspiring the biological nervous system. Neurons are the basic components of ANN and is divided into three layers viz. input, hidden and output layers. The neurons in one layer is connected to the previous layers by weights and a computational model is used by these neurons for information processing. Using ANN, complex relationships between inputs and outputs can be modeled and can find the patterns of data. ANNs are able to learn from the predefined samples without possessing any prior understanding to the complex formula[21].

The compressive strength of concrete was predicted using ANN model and it shows better performance than traditional methods. Also it considers sufficient factors which influence the concrete strength development [22]. The number of reversals to fatigue failure of steel reinforcing bars was predicted using ANN and it gives better prediction results than nonlinear regression analysis with improved accuracy [23]. This enlightens the possibilities of using ANN for the fatigue life prediction of concrete which may provide a model consideringthe discreteness of concrete property and randomness in fatigue testing.

Very few researches were done using ANN for the prediction of fatigue life of concrete. A prediction model was created to predict the fatigue endurance limit of conventional asphalt concrete pavements utilising ANN. Also by using ANN extraction rules, a standalone equation was derived which is capable to predict the strain values with good prediction accuracy [24]. Fatigue cracking (FC) is a significant problem in asphalt pavements and is related with transfer function which is an empirical mathematical model based on critical mechanical responses. Utilising the similarity between FC transfer function and neural network, accuracy of FC prediction was improved considering all variables which is otherwise difficult and not reliable[25]. However, a generalised model for prediction of fatigue life of concrete is not yet established and is a scope to find it with ANN.

6. Conclusions

In this paper a review of literature on fatigue life prediction models of plain concrete is carried out. Review mainly includes the evolution of models developed using three basic approaches of fatigue life assessment of structural element and the possibilities of using artificial neural network in predicting the fatigue life of concrete. At first prediction of fatigue life of concrete was based on S-N curves. Various fatigue tests were conducted and analysed to obtain the number of cycles at different stress level, which helped in developing a failure criterion. Later further investigations showed that this approach is not sufficient for inculcating the variability in fatigue tests. Fracture mechanics based approach gained more acceptance in which Paris law was the first model proposed based on this approach. Later, certain modifications were brought by researchers and the size effect law was the major one among them. In model based on fatigue damage approach, results were predicted based on the damage of material which is associated with a non-local approach, where fracture mechanics is related to local stress intensity range at the crack tip. Damage approach produce a rational generalised behaviour but based on non-realistic behaviour of material.

All these were based on deterministic approaches and it is subjective because of the following reasons

- The fatigue life of structural elements is influenced by the uncertainties in the lifetime of structures
- Randomness in fatigue test and discreteness in concrete property upon fatigue loading
- Approaches is based on certain failure criteria and cannot be generalised

Considering all these, a conclusion can be drawn that there is necessity for a probabilistic approach to overcome the limitations of deterministic approach. ANN, a nonlinear statistical data modelling tools which can be used to model complex relationships between the inputs and outputs based on a probabilistic approach. So far it is used to predict fatigue lives of pavements and found it as a reliable approach with better prediction accuracy. This can be extended to establish a generalised model for predicting the fatigue life of concrete.

7. References

[1] Miner M A 1945 Cumulative damage in fatigue *Journal of applied mechanics* **12(3)** 159–64
[2] Slowik V, Plizzari G A, and Saouma V E 1996 Fracture of concrete under variable amplitude fatigue loading *ACI Materials Journal* **93(3)** 272–83
[3] M H Baluch 2003 CDM model for residual strength of concrete under cyclic compression *Cement Concrete Composites* **25(4–5)** 503–12
[4] D Lei 2008 On the energy dissipation in fatigue process and fatigue life prediction *Journal of Experimental Mechanics* **5(008)**
[5] Hajela P and Berke L 1991 Neurobiological computational modes in structural analysis and design *Computers and Structures* **41** 657–67
[6] Hilsdorf H K and Kesler C E 1966 Fatigue strength of concrete under varying flexural stresses *ACI Journal Proceedings* **63(10)** 1059–76
[7] McCall J T 1958 Probability of fatigue failure of plain concrete *ACI Journal Proceedings* **55(8)** 233–44
[8] Hsu T T 1981 Fatigue of plain concrete *ACI Journal Proceedings* **78** 292–305
[9] Raju N 1969 Prediction of the fatigue life of plain concrete in compression *Building Science* **4(2)** 99–102
[10] Holmen J O 1982 Fatigue of concrete by constant and variable amplitude loading *ACI Special Publication* **75** 71–110
[11] Paris P and Erdogan F 1963 A critical analysis of crack propagation laws *Journal of Fluids Engineering* **85(4)** 528–33
[12] Bazant Z P and Kazemi M T 1991 Size dependence of concrete fracture energy determined by RILEM work-of-fracture method *International Journal of Fracture* **51(2)** 121–38
[13] Baluch M, Qureshy A and Azad A 1989 Fatigue crack propagation in plain concrete *Springer New York* **16** 80–7
[14] Subramaniam K V, O 'Neil E F, Popovics J S and Shah S P 2000 Crack propagation in flexural fatigue of concrete *Journal of Engineering mechanics* **126(9)** 891–8
[15] Chandra Kishen J M and Rao P S 2007 Fracture of cold jointed concrete interfaces *Engineering Fracture Mechanics* **74(1)**122–31
[16] Ortiz M 1985 A constitutive theory for the inelastic behaviour of concrete *Mechanics of Materials* **4** 67–93

[17] Sauris W, Ouyang C and Fernado V 1990 Damage model for cyclic loading of concrete *Journal of Engineering Mechanics* **116(5)** 1020–35
[18] A H Rahman, Al-gadhib, M H Baluch, A Shaalan and A R Khan 2000 International *Journal of Damage mechanics* **9** 57–78
[19] V Dattoma S Giancane 2013 Evaluation of energy of fatigue damage into GFRC through digital image correlation and thermography *Composites Part B* **47** 283–9
[20] Dong Lei, Pei Zhang, Jintao He, Pengxiang Bai, and Feiping Zhu 2017 Fatigue life prediction method of concrete based on energy dissipation *Construction and Building Materials* **145** 419–25
[21] Abhijeet Dey, Ghanashyam Miyani, Arjun Sil 2019 Application of artificial neural network for estimating reliable service life of reinforced concrete structure book keeping factors responsible for deterioration mechanism *Soft Computing* **13(2)** 1–15
[22] Seung-Chang Lee 2003 Prediction of concrete strength using artificial neural networks *Engineering Structures* **25** 849–57
[23] Moises Jimenez-Martineza and Mariel Alfaro-Ponceb 2019 Fatigue damage effect approach by artificial neural network *Internatinal Journal of fatigue* **124** 42–7
[24] Mayzan Isied and Mena Souliman 2019 Fatigue Endurance Limit Model Utilising Artificial Neural Network for Asphalt Concrete Pavements *Airway and highway field pavements* **46(2)** 42–50
[25] Hongren Gong, Yiren Sun, Wei Hu and Baoshan Huang 2019 Neural networks for fatigue cracking prediction using outputs from pavement mechanistic empirical design *International Journal of Pavement engineering* **22** 1–10

IOP Conf. Series: Materials Science and Engineering **936** (2020) 012027 doi:10.1088/1757-899X/936/1/012027

A Study on Effect of Fluid on the Modal Characteristics of Ground Supported Water Tanks

B V Anagha[1*], P Nimisha[2] and B R Jayalekshmi[3]

[1] M.Tech Scholar, Department of Civil Engineering, National Institute of Technology, Karnataka, Surathkal, Mangalore 575025, India

[2] Research Scholar, Department of Civil Engineering, National Institute of Technology, Karnataka, Surathkal, Mangalore 575025, India

[3] Professor, Department of Civil Engineering, National Institute of Technology, Karnataka, Surathkal, Mangalore 575025, India

[*]anaghabv@gmail.com

Abstract. Liquid storage tanks are one of the essential lifeline structures which should function even after natural disasters. Water tanks in particular are essential for proper supply of water in the affected areas and for fire-fighting purposes. Sloshing is an important phenomenon occurring in partially filled tank with free surface. It can cause violent liquid motions that can cause instability in carrier-vessel and structural damage in the tank walls. To maintain the structural integrity of the carrier vessels, modal characteristics should be known as the impact of sloshing will increase when the excitation frequency matches with the natural frequency of container and result in resonant condition. This study investigates the modal characteristics such as natural frequency, mode shapes of square tank by 3-D finite element method using ANSYS 2019 R1software. The effect of impulsive and convective frequency of tanks with different fill levels and different boundary conditions are considered.

1. Introduction

Liquid sloshing is a major phenomenon occurring in stationary or moving containers that cause greater concern in various fields like civil, aerospace, nuclear engineering, design of ship and road tankers. In fluid dynamics, sloshing can be defined as the movement of two or more non-miscible fluids (liquids and gases) inside another object which is undergoing motion. Liquid storage tanks are considered as important lifeline structures. They should remain functional after natural disaster like earthquake. Seismic excitations are formed during earthquakes that cause severe damage to the tanks which in turn results in serious injury to life and loss of property. The walls of tanks are subjected to hydrodynamic pressure due to heavy sloshing during earthquake in the case of ground supported tanks. Therefore to understand the response to dynamic loads, the modal characteristics like natural frequency, mode shapes and participation factors should be investigated. This paper mainly focuses on the modal characteristics of steel square tanks. Bayraktar [1] investigated the dynamic characteristics of cylindrical steel tank filled with oil. The 3-D dynamic characteristics of partially filled elastic, rectangular tanks were investigated by Zhou and Liu [2]. An analytical method was proposed by Hashemi [6] to determine the dynamic behaviour of flexible rectangular fluid containers. Rayleigh-Ritz method is used to develop simplified solutions. But the resulting frequencies from these methods may underestimate or overestimate the exact solution due to inappropriate assumptions. Thus in order to overcome these drawbacks, finite element analysis can be effectively used. The modal characteristics can be investigated using the FEM models. 3D mode shapes can be visualised easily which are difficult to obtain via theoretical approach.

In this paper, a square tank with different fill levels are considered to investigate the modal characteristics using the finite element method. ANSYS software is used to develop finite element models. The convective and impulsive frequencies and corresponding mode shapes with various boundary condition and fill levels are determined. Three boundary conditions are used: base fixed and free at top, clamped-clamped and clamped-free ends at the bottom and top of the tank, respectively. All the translations and rotations are restricted at the base of the tank for the fixed condition whereas, the edge of the base is restricted similarly for the clamped conditions. The fixed base is given by assigning full fixity to the whole base slab, while in the clamped condition; only the edges of base slabs are clamped. The fill levels considered are empty, 25%, 50%, 75% and full tank condition. The effect of fluid level and boundary condition on modal characteristics is evaluated by carrying out modal analysis.

2. Geometric configuration

A scaled down model of square tank with height 300 mm, width 500 mm, and thickness 3 mm is selected for modal analysis.The material of tank is considered as steel. The material properties, Young's modulus of elasticity and Poisson's ratio are given as 210 GPa and 0.3 respectively. The mass density of the tank material is 7850 kg/m^3. The tank is filled with fluid to four different heights from the tank base. Water, having a mass density of 1000 kg/m^3 and a bulk modulus of elasticity of 2.2 GPa is used as the fluid. The speed of sound in water is 1483 m/sec. The tank was first analysed for empty condition and then the analysis was done for different fill condition.

3. Finite element analysis

Finite element analysis of water tanks are done using ANSYS Workbench 2019 R1. Geometric modelling of water tank is done using Design Modular (DM). Modal is used for empty tank condition and Modal Acoustics is used for all other fill levels to determine frequency and mode shapes. The Navier-Stokes equation, continuity equation and conservation of momentum equation govern the fluid motion in a tank.

- *Navier-Stokes equation:*

$$\frac{\partial(\rho u)}{\partial t}+\rho\left(u\frac{\partial u}{\partial x}+v\frac{\partial u}{\partial y}+w\frac{\partial u}{\partial z}\right)=\rho x-\frac{\partial p}{\partial x}+\frac{1}{3}\mu\frac{\partial}{\partial x}\left(\frac{\partial u}{\partial x}+\frac{\partial v}{\partial y}+\frac{\partial w}{\partial z}\right)+\mu\nabla^2 u$$

where ρ is the density, t is time, p is pressure, μ is dynamic viscosity, and u, v, w are velocity components in x, y, z direction.

Continuity equation:

$$\frac{\partial\rho}{\partial t}+\frac{\partial(\rho u)}{\partial x}+\frac{\partial(\rho v)}{\partial y}+\frac{\partial(\rho w)}{\partial z}=0$$

- *Conservation of momentum:*

$$\frac{\partial}{\partial T}(\rho\vec{V})+\nabla.(\rho\vec{V}\vec{V})=-\nabla p+\nabla(\tau)+\rho g+F$$

$\vec{\tau}$ is the stress tensor, $\rho\vec{g}$ and $\vec{F}$ are the gravitational body force and external body force respectively.

3.1. Finite element models

The square tank is generated using 3-D finite elements as shown in figure 1. SOLID187 element is used to model the square tank which is a higher order 3-D tetrahedral element, with one extra node at the middle of each side of the element. These are defined by ten nodes. Each node has three degrees of freedom, with translations in the nodal x, y and z directions. FLUID220 is a twenty nodded acoustic element which is used to divide the fluid region into identical 3-D elements. The element node has four degrees of freedom at each node, with translations in the nodal x, y and z directions and pressure. The fluid element is coupled with the tank surface by adding FSI flag at the interface, which will couple the pressure degrees of freedom of the nodes. The governing equations of acoustics, namely 3-D wave equation have been discretised taking into account the coupling of structural motions and acoustic pressure at the interface. Contact and sliding between 3-D target surface and a deformable surface is defined by the element CONTA174. The target surface is defined by 3-D TARGE170 which is used only for rigid bodies. The model consists of 15754 nodes and 6195 elements as shown in figure 2.

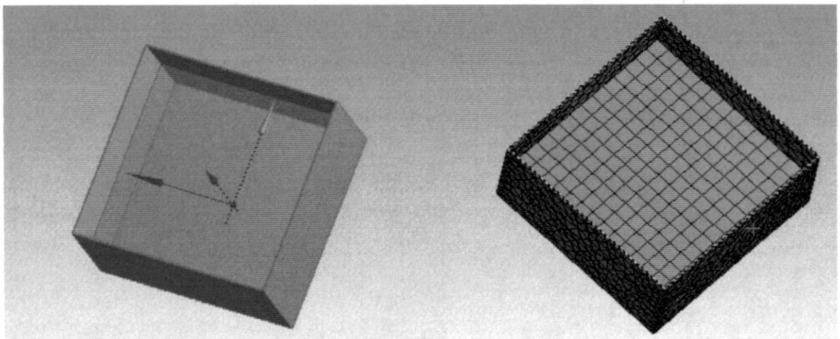

Figure 1. Square tank with 50% fill level and its finite element model.

3.2. Modal analysis

Mode-frequency (modal) analysis is done to determine natural frequencies and mode-shapes of the square tank. There are two distinct components for frequency of liquid in moving rigid containers-impulsive and convective. These two components have a direct effect on the performance and dynamic stability of moving containers. The fluid filled tank was analysed for convective and impulsive frequencies using Modal Acoustics. The frequency corresponding to the upper part of the fluid in the tank near the free surface, which vibrates in low frequencies and independent of the tank walls, is the convective frequency. The frequency of the remaining fluid, i.e., the part that moves with the tank walls, and that vibrates in higher frequencies, is the impulsive frequency.

3.2.1. Modal Acoustics. Modal Acoustics analysis usually involves modelling the fluid medium as well as the surrounding structure in order to determine frequencies and standing wave patterns within a structure. The physical region and acoustic region is assigned to the tank. For getting convective frequencies free surface boundary condition is also assigned. This boundary condition enables to specify a plane as a free surface in order to consider sloshing effects on model. Gravitational acceleration values need to be defined before assigning free surface to properly define the sloshing problem. Similarly for getting impulsive frequencies the top surface of liquid is assigned a pressure of 0 Pa. The set up for impulsive and convective frequencies are shown in figures 2 (a) and (b) respectively. In both cases fluid solid interface is assigned to the necessary contact regions.

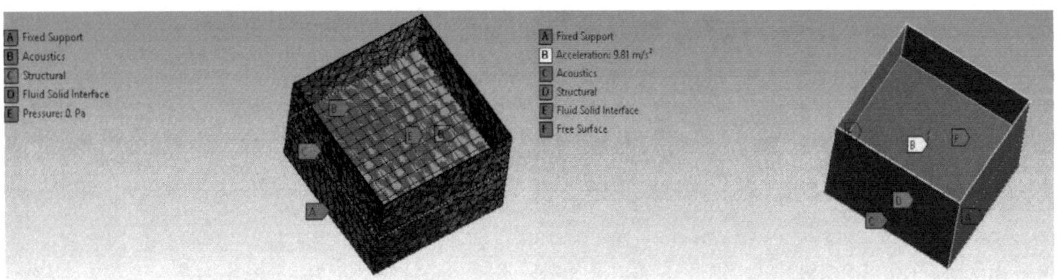

(a) (b)

Figure 2. Modal Acoustics setup. (a) Impulsive and (b) Convective frequencies.

4. Results and discussions

The first three mode shapes of half-filled tank and empty tank with clamped-free boundary condition are shown in figures 3 and figure 4, respectively. The comparison of impulsive frequencies for different boundary conditions is shown in figure 5. As the depth of fluid increases, the frequency of tank decreases for all the three boundary conditions. The fluid effect is similar for clamped-free and fixed base condition.

IOP Conf. Series: Materials Science and Engineering **936** (2020) 012027 doi:10.1088/1757-899X/936/1/012027

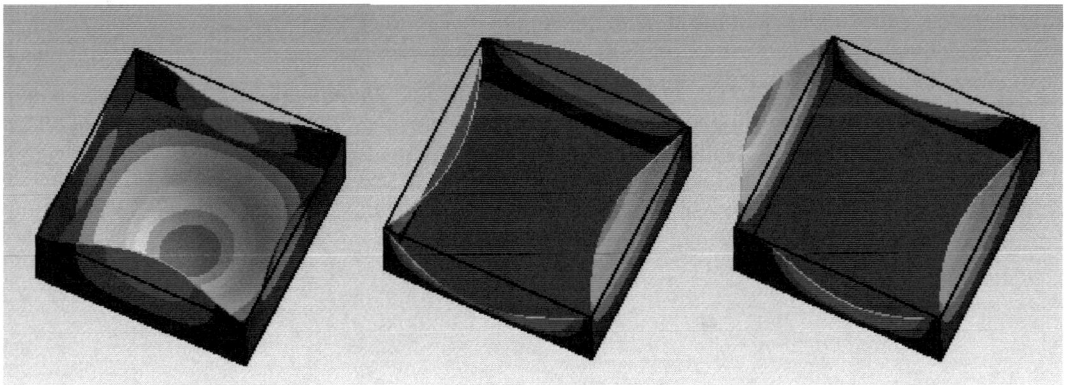

1st mode (57.64 Hz) 2nd mode (68.764 Hz) 3rd mode (90.358 Hz)

(a) Impulsive modes

1st mode (1.0722 Hz) 2nd mode (1.0722 Hz) 3rd mode (1.386 Hz)

(b) Convective modes

Figure 3. Mode shapes of half-filled tank.

1st mode (72.304 Hz) 2nd mode (96.282 Hz) 3rd mode (96.547 Hz)

Figure 4. Mode shapes of empty tank.

The frequencies of clamped-clamped condition show drastic decrease in frequencies up to 50% of fluid level which shows that fluid effect is more significant in this range of fluid level for clamped-clamped case. Above 50% fluid level, there is not much difference in frequencies for clamped-clamped case, but for clamped-free and fixed base case, there is slight variation in frequency above 50%. The fluid has more prominent effect on frequency in clamped-clamped condition as compared to other two conditions.

Thus the frequency of tank depends on both fluid depth and boundary conditions.

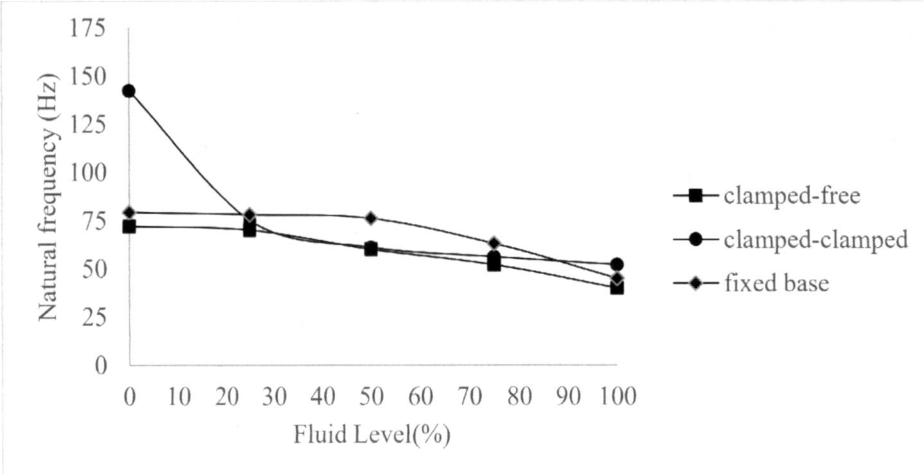

Figure 5. Comparison of impulsive frequencies for different boundary condition.

The convective frequencies for various fill conditions are shown in figure 6. The convective frequency increases with increase in the fluid depth, due to decrease in convective mass of fluid. Depending on the confinement provided by the tank, the mass of impulsive and convective fluid varies. Thus greater the confinement greater is the impulsive liquid and smaller the convective liquid. Due to this reason, the tank with lower fluid level has higher impulsive frequency and lower convective frequency.

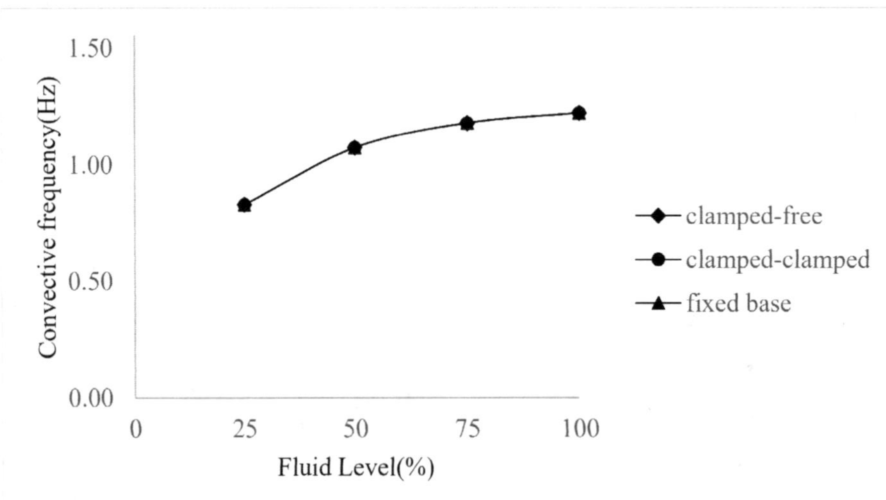

Figure 6. Comparison of convective frequencies of different fill levels.

The normalised frequency of first mode of tanks for different boundary conditions and fluid depth are depicted in figure 7. The normalised frequency can be defined as the ratio of frequency in loaded condition and that of empty condition which is expressed in percentage. The fluid inside the tank will decrease the frequency in impulsive mode and effect of fluid is significant for fluid level below 50%. The normalised frequency for full tank condition is 56% for clamped-free and fixed base conditions whereas for clamped-clamped condition the normalised frequency is 36% which is lower than that of other two boundary conditions. This confirms the fact that the effect of fluid is more prominent in the clamped-clamped condition.

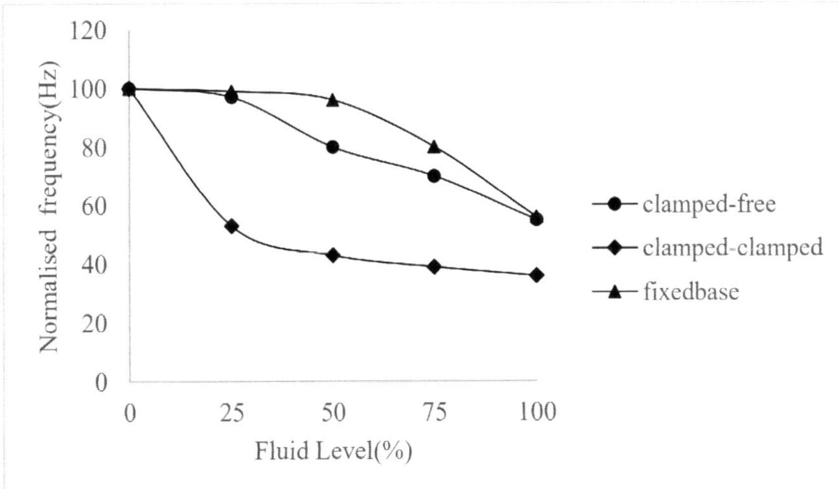

Figure 7. Normalised frequencies of square tank with respect to the fluid level.

5. Conclusions

Finite element analysis of several models with different boundary conditions and fluid levels were carried out to find the impulsive and convective frequencies and corresponding mode shapes of square tank. From the study conducted, the following conclusions can be drawn:

- The frequency of tank will decrease with inclusion of fluid inside the tank. The impulsive frequencies are more at lower fluid level which goes on decreasing with increasing the fluid level.
- The convective frequency of the tank will increase with increase in the fluid depth in the tank due to decrease in the mass of convective fluid.
- The fluid has more impact on clamped-clamped condition than clamp-free and fixed base condition for fluid depth below 50%.
- The maximum rate of decrease in frequency was in clamped-clamped condition .i.e. 36% compared to 56% in clamped-free and fixed base conditions.

6. References

[1] Bayraktar A, Sevim B, Altunışık A C and Türker T 2010 Effect of the model updating on the earthquake behaviour of steel storage tanks *Journal of Constructional Steel Research* **66(3)** 462–9

[2] Zhou D and Liu W 2007 Hydroelastic vibrations of flexible rectangular tanks partially filled with liquid *International Journal for Numerical Methods in Engineering* **71(2)** 149–74

[3] Jung J H, Yoon H S and Lee C Y 2015 Effect of natural frequency modes on sloshing phenomenon in a rectangular tank *International Journal of Naval Architecture and Ocean Engineering* **7(3)** 580–94

[4] Jhung M J and Kang S S 2019 Fluid effect on the modal characteristics of a square tank *Nuclear Engineering and Technology* **51(4)** 1117–31

[5] Jeong K H 2011 Hydro elastic vibration analysis of liquid-contained rectangular tanks *Structural Engineering and Mechanics* **40(5)** 665–88

[6] Hashemi S, Saadatpour M M and Kianoush M R 2013 Dynamic behaviour of flexible rectangular fluid containers *Thin-Walled Structures* **66** 23–38

[7] Jeong K H, Park J S and Lee W J 2010 Free vibration analysis of a simply supported rectangular tank partially filled with a liquid *ASME Pressure Vessels and Piping Division/K-PVP Conference* pp 167–74

[8] Molin B and Remy F 2013 Experimental and numerical study of the sloshing motion in a rectangular tank with a perforated screen *Journal of Fluids and Structures* **43** 463–80

IOP Conf. Series: Materials Science and Engineering **936** (2020) 012028 doi:10.1088/1757-899X/936/1/012028

Performance of Alkali-Activated Aerated Concrete using Class C Fly Ash

E Muthu Kumar*

Doctoral research scholar, IIT Madras, Chennai-600036, India

*civil.emuthu@gmail.com

Abstract. This paper focuses on the factors influencing the properties of alkali-activated aerated paste and mortar with high calcium fly ash as the source material, by comparing oven and ambient curing conditions. Design of experiments with response surface methodology is used to study the influence of each of the parameters on the properties of alkali-activated paste and mortar. To achieve a stable pore formation with aeration cooled liquid alkaline solution with a temperature below 15°C is used. The influence of each of the factor and their interaction effects on properties such as density, compressive strength and absorption of water were studied. For a constant dry density, the compressive strength achieved by ambient curing is higher compared to oven curing i.e. for a density of 900 kg/m^3 the compressive strength with ambient curing is 4.5MPa while for oven curing it is 2.75MPa. But less water absorption is observed for oven curing compared to ambient curing. While the water absorption achieved was lower for oven curing it was higher for ambient curing resulting in not less than 10%.

1. Introduction

Alkali activation in concrete helped in the production of construction material along with the use of industrial by-products which helped in the reduction of carbon emission. It also helps in generating revenue with the use of fly ash along with the conservation of landfill space [1]. The physical and chemical characteristics of fly ash are governed by the method of production and source of the coal being used [2][3][4]. Alkali-activated aerated concrete using class F fly ash has been studied in detail while much on alkali-activated aerated concrete with class C fly ash is not reported. Studies by [5] using a foaming agent in class C alkali activation showed a comparison of curing condition and it was observed that room temperature curing resulted in increased porosity and water absorption compared to oven-cured samples. Most of the studies used curing ranged from ambient to 90°C. Alkali activation with aeration process in most of the studies has shown fast aeration [6][7] due to the reaction between alkali solution and aluminium powder resulting in rapid release of hydrogen [8]. Hence by varying the mix proportions on alkaline activated aerated paste and mortar with controlled aeration, a systematic study is carried out.

2. Materials and methods

The elemental composition of class C fly ash according to ASTM C618-12a [9] is presented in table 1. Using Blaine's air permeability test, fineness value of fly ash is determined to be 390 m^2/kg. Two mixes chosen for each curing type are alkali-activated paste and alkali-activated mortar. In the case of paste, it is 100% fly ash while in case of mortar;sand passing through 300 μm is used in the ratio of 1:1 measured in weight basis.

Table 1. Elemental composition of Class C fly ash.

Elements	CaO	SiO_2	MgO	Al_2O_3	Fe_2O_3	SO_3	K_2O	Na_2O	Others
Composition mass (%)	13.76	39.89	2.28	31.46	6.17	3.19	0.12	0.59	2.54

Commercially available C50 aluminium powder (Al powder) was used as an aerating agent. Trials were conducted and it showed that there was no aeration with Al powder dosage below 0.1 % and further, when increased beyond 0.2 % the mix settled down with the inability to hold the mix. For a chosen range of Al powder and molarity of sodium hydroxide (NaOH), the liquid alkaline to ash ratio (La/ash ratio) was 0.6 to 0.65 for paste and 0.8 to 1.0 for mortar. When the liquid content is increased more than 0.65 for paste and 1.0 for mortar, the pores formed gets collapsed due to escape of the gas from the mix.The ranges of each of the varying parameters used for the alkali-activated aerated paste and mortar is shown in table 2.

Table 2. Parameters and its ranges.

Parameters	**Alkali-activated aerated paste**	**Alkali-activated aerated mortar**
Al powder dosages	0.1 – 0.2 %	0.1 – 0.2 %
La/ash ratio	0.6 – 0.65	0.8 – 1.0
NaOH	6 – 12 M	6 – 12 M
NaSi / NaOH ratio	1 – 2	1 – 2
Oven curing temperature	60 – 90°C	60 – 90°C
Ambient curing temperature	27±5°C	27±5°C

The alkaline solution is prepared using NaOH pellets and sodium silicate (NaSi) solutions. First, NaOH of required concentration was prepared and left for dissolution which dissipates the heat evolved. After 24hrs,with the prepared NaOH solution, NaSi solution was mixed in the required ratio and left undisturbed for 12hrs until the solution reaches room temperature. This prepared solution is used as an activator in the polymerisation process.

2.1. *Studies on controlled pore formation*

There observed a fast formation of hydrogen gas from the initial trial mix study of alkali-activated aerated paste. This is due to the sudden aeration process due to the reaction of Al powder with the alkaline solution. To control this sudden formation of aeration, different methods were studied namely; (a) Lowering the concentration of the NaOH solution below 6M (controlled the rate of aeration but resulted in very low compressive strength), (b) Superplasticiser as retarders (Enhances the mixing process but could not control the faster aeration), (c) Alkaline solution stored at low temperature (Could control the fast aeration process without disturbing the formation of pores). Hence, the required reaction rate was attained by lowering the temperature of the alkaline solution. The alkali-activated paste was prepared by mixing an alkaline solution having a temperature of 15°C with class C fly ash for about 2 minutes which resulted in aeration up to 10 minutes.

2.2. *Experimental procedure*

In the experimental study with an increased number of parameters, 'Design of Experiments' was used separately for alkali-activated paste and mortar using Response Surface Methodology. In case of temperature curing 50 sets and for ambient curing 40 sets of pre-set experiments were conducted separately for each combination of mixes with alkali-activated paste and mortar.Fly ash, Al powder and pulverised sand (in case of mortar) were dry mixed and then alkaline solution at a lower temperature of 15°C was used. This was mixed thoroughly for about 1-2 minutes and samples were cast. For measuring the variation in fresh density a graduated glass beaker of 250 ml was taken and is initially filled up to 2 cm graduation scale and the variation in density is measured up to 15 minutes at an interval of 5 minutes concerning the rate of aeration.Six cubes (IS 4031 – Part 6) were cast to determine the dry density, compressive strength and for water absorption, 3 cubes were cast separately, for each set of the designed experiment.The surfaces of the cubes were levelled after 12hrs and prepared for curing. In case of oven curing the samples were cured for 24 hours at pre-set temperature and for ambient curing the cubes were stored at room temperature for 28 days. The variation in properties observed in both the curing conditions for alkali-activated paste and mortar are analysed statistically using Statistical Analysis Software [11] with a quadratic response surface model and the influence of varying factors and their interaction effects on the properties were studied.

3. Results and discussions

3.1. Workability

Figure 1 shows the measured flowability of alkali-activated paste and alkali-activated mortar using the flow table test. It is observed that workability is mainly influenced by La/ash ratio in both the alkali-activated paste and mortar irrespective of the NaSi/NaOH ratio and NaOH concentration. Increase in La/ash ratio makes the mix more workable while the increase in molarity reduces the workability. This reduction in workability with the increase in molarity is due to an increase in viscosity of the alkaline solution which decreases the ability of the mortar to flow [12]. Further lowering of the La/ash ratio resulted in a stiff mix regardless of the variation in NaSi/NaOH ratio and NaOH concentration.

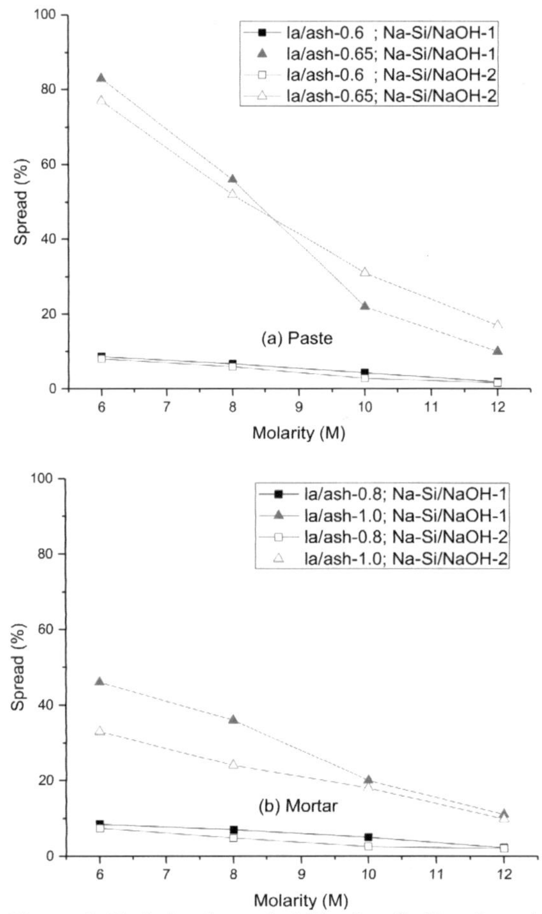

Figure 1. Variation in workability for alkali-activated aerated Class C flyash.

3.2. Fresh density

The fresh density of alkali-activated paste and mortar was measured for each combination of designed experiments. It is observed that reduction in fresh density was substantial up to 10 minutes, which

IOP Conf. Series: Materials Science and Engineering **936** (2020) 012028 doi:10.1088/1757-899X/936/1/012028

indicates that aeration is complete almost within 10 minutes from the time of samples being cast in the mould. It is also evident from figure 2 that Al powder dosage and La/ash ratio influenced the fresh density rather than other parameters.

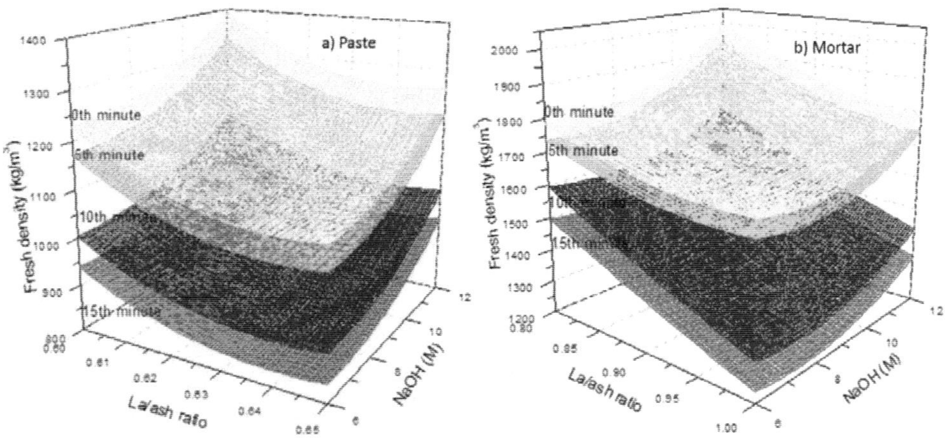

Figure 2. Fresh density variation for Al powder dosage-0.1%, NaSi/NaOH ratio–1.

3.3. Dry density

Figure 3 illustrates the difference in dry density of alkali-activated aerated paste for both the oven curing and ambient curing condition. From figure 3(a) it is witnessed that the density increases linearly with increase in the concentration of NaOH for a given range of NaSi/NaOH ratio and Al powder dosage. In the case of oven cured samples,the polymerisation is enhanced which increases the density with increase in temperature. Further, with lower NaSi/NaOH ratio resulted in gradual variation in density, whereas when the NaSi/NaOH ratio was increased, only marginal variation in density is observed. Moreover, for a given molarity when the La/ash ratio increases the workability increases which results in reduced density. In the case of ambient curing, it is observed that there is a significant increase in density with increase in molarity for a given NaSi/NaOH ratio and lower dosage of Al powder. Whereas, for the increased dosage of Al powder and NaSi/NaOH ratio, La/ash ratio greater than 0.63, resulted in very marginal density variation. It is evident that for a constant NaSi/NaOH ratio and Al powder dosage, ambient curing shows higher density compared to oven curing in the alkali-activated aerated paste.The trend observed in alkali-activated aerated mortar is similar to that of the alkali-activated aerated paste.

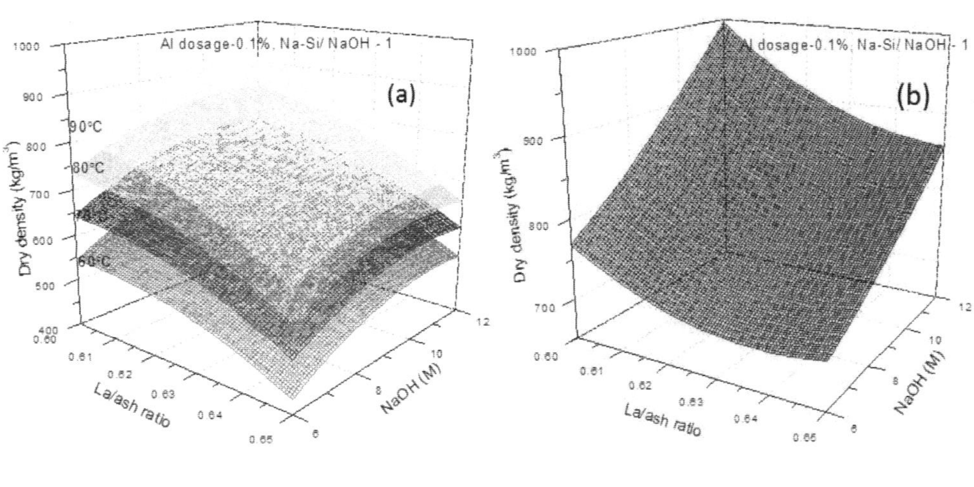

Figure 3. Dry density of alkali-activated paste for oven curing and ambient curing.

3.4. Compressive strength

The compressive strength of oven cured and air-cured samples of alkali-activated aerated paste and mortar are shown in figure 4. The lower La/ash ratio and higher NaOH concentration yielded in higher compressive strength for the chosen range of Al powder dosage. It is observed from the studies that there is a reduction in compressive strength with increase in NaSi/NaOH ratio from 1 to 2 with a reduction in density (i.e. 770 to 305 kg/m^3). This reduction is due to the variation in silica content altering the pH of the alkaline solution [12].

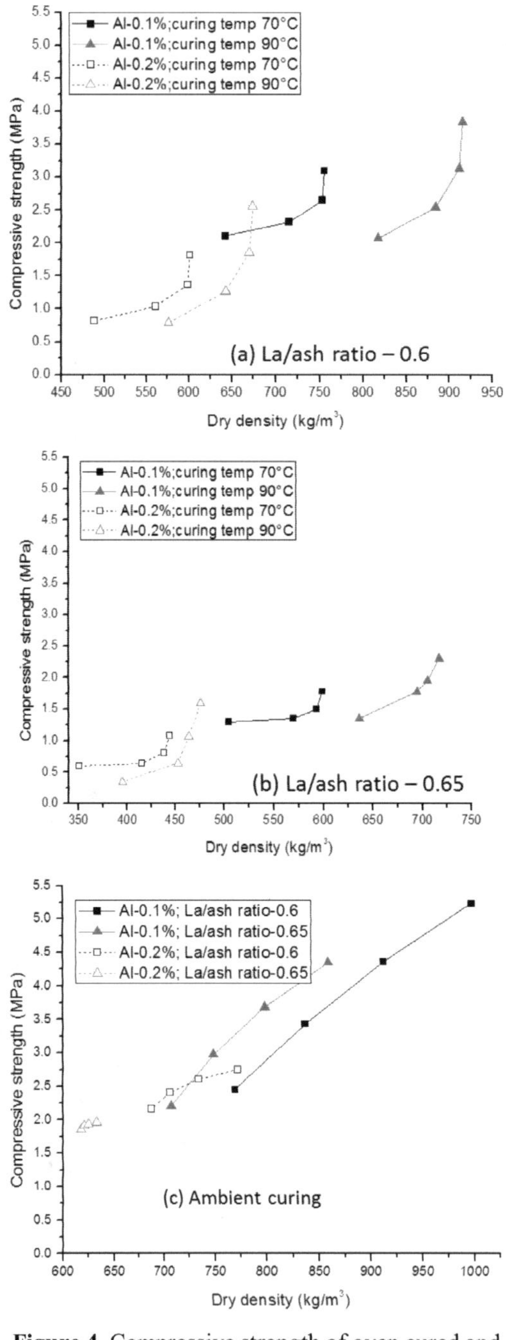

Figure 4. Compressive strength of oven cured and ambient cured samples of alkali-activated paste.

From figure 4 it is observed that as the temperature of curing increases to 90°C the compressive strength increases for a given dosage of Al powder. The air-cured alkali-activated paste resulted in higher strength of 5.25 MPa compared to oven-cured strength of 3.8 MPa. Further, when the molarity of NaOH and Al powder dosage is being constant, an increase in NaSi/NaOH ratio ensured reduced compressive strength. Considering both the curing condition it is observed that for any chosen NaSi/NaOH ratio and dosage of Al powder, reduced La/ash ratio increased compressive strength.

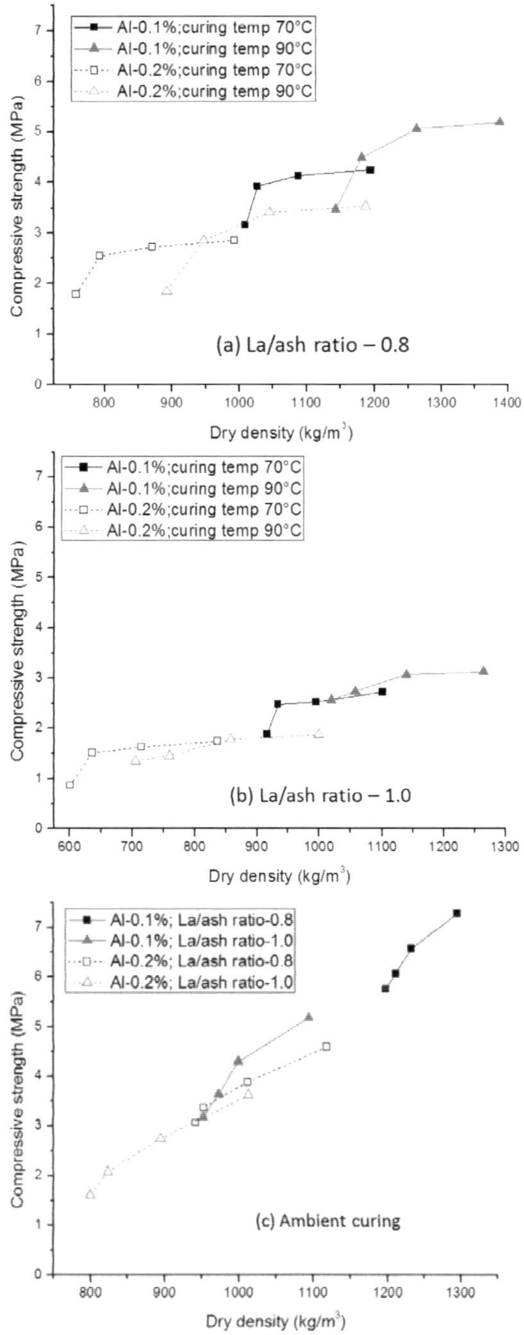

Figure 5. Compressive strength of oven cured and ambient cured samples of alkali-activated mortar.

The strength of alkali-activated aerated mortar is higher compared to that of alkali-activated aerated paste for the chosen density. Further, with the increase in NaSi /NaOH ratio from 1 to 2, compressive strength decreased with mortar which is similar to oven-cured aerated paste. From figure 5 it is witnessed that for air curing there is an increase in compressive strength for the similar range of densities to that of the oven curing for the given range of Al powder.

3.5. Water absorption

Even though water absorption plays anessential role in designing the concrete, very few research papers discusses the water absorption properties of aerated concrete.It is observed that for a given Al powder dosage and temperature, the increase in NaSi/NaOH ratio resulted in an increase in water absorption with the corresponding decrease in dry density. Also, it is observed that as the curing temperature is raised, it resulted in reduced water absorption with an increase in density. An increase in dosage of Al powder resulted in very high water absorption irrespective of the effect of curing temperature. The range of water absorption obtained is very high compared to oven curing which may be due to increased interconnected pores with incomplete polymerisation.

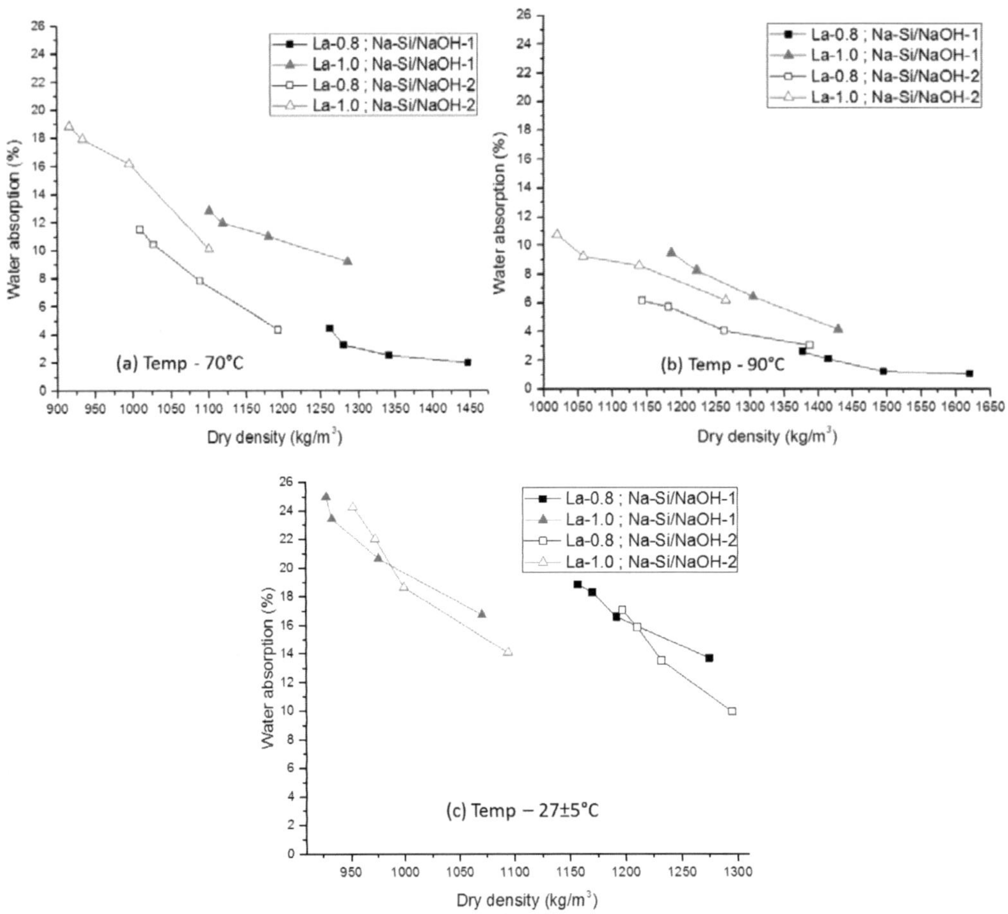

Figure 6. Water absorption for aerated alkali-activated mortar for Al dosage - 0.1%.

A similar observation as made in the alkali-activated aerated paste is observed from figure 6 for mortar, with reduced water absorption not exceeding 28%. Also, the reduction in water absorption of alkali-activated mortar with air curing is similar compared to the aerated paste. Whereas, it is seen that with an increase in NaSi/NaOH ratio, there is a marginal decrease in water absorption or is in a close-by range which is a reverse trend compared to alkali-activated aerated paste.

4. Conclusions

The factors influencing the properties of alkali-activated aerated paste and mortar with high calcium fly ash as the source material is studied at two different curing conditions, namely; ambient and oven curing.

i) Reduced alkaline solution temperature resulted in stable pore formation of alkali-activated aerated class C fly ash.
ii) For a given molarity and La/ash ratio, the density is increased in the range of 400–900 kg/m^3 for paste and 900–1650 kg/m^3 for mortar with an increase in temperature.
iii) In ambient curing, increase in NaSi/NaOH ratio there exists a wider range of density that could be achieved.
iv) As the strength range achieved in both types of curing condition yields almost the same range; hence low-temperature curing is suitable for Class C fly ash.
v) Ambient curingresults in higher water absorption compared to oven curing by about 17 to 35%more for the chosen range of Al dosage.
vi) Presence of reduced calcium along with reduced Si/Al ratio might not be influencing the properties of high calcium fly ash alkali-activated aerated concrete. Thus the suitable addition of binder may enrich the properties along with improved Si/Al ratio.

5. References

[1] Diaz E I, Allouche E N and Eklund S 2010 Factors affecting the suitability of fly ash as source material for geopolymers *Fuel* **89** 992–6
[2] Hanjitsuwan S, Hunpratub S, Thongbai P, Maensiri S, Sata V and Chindaprasirt P 2014 Effects of NaOH concentrations on physical and electrical properties of high calcium fly ash geopolymer paste *Cem. Concr. Compos.* **45** 9–14
[3] Chindaprasirt P, De Silva P, Sagoe-Crentsil K and Hanjitsuwan S 2012 Effect of SiO_2 and Al_2O_3 on the setting and hardening of high calcium fly ash-based geopolymer systems *J. Mater. Sci.* **47** 4876–83
[4] Chindaprasirt P and Chareerat T 2010 High-strength geopolymer using fine high-calcium fly ash *J. Mater.* **23** 264–71
[5] Abdullah M M A B, Hussin K, Bnhussain M, Ismail K N, Yahya Z and Abdul Razak R 2012 Fly Ash-based Geopolymer Lightweight Concrete Using Foaming Agent *Int. J. Mol. Sci.* **13** 7186–98
[6] Hlaváček P, Šmilauer V, Škvára F, Kopecký L and Šulc R 2015 Inorganic foams made from alkali-activated fly ash: Mechanical, chemical and physical properties *J. Eur. Ceram. Soc.* **35** 703–9
[7] Sanjayan J G, Nazari A, Chen L and Nguyen G H 2015 Physical and mechanical properties of lightweight aerated geopolymer *Constr. Build. Mater.* **79** 236–44
[8] Soler L, Candela A M, Macanás J, Muñoz M and Casado J 2009 In situ generation of hydrogen from water by aluminum corrosion in solutions of sodium aluminate *J. Power Sources* **192** 21–6
[9] ASTM C618-12a 2010 *Standard Specification for Coal Fly Ash and Raw or Calcined Natural Pozzolan for Use* (ASTM)
[10] IS 4031- part 6 2005 *Methods of Physical Tests for Hydraulic Cement* (BIS: New Delhi)
[11] SAS Release 8.02 1999 SAS Institute Inc, Cary, NC, USA
[12] Chindaprasirt P, Chareerat T and Sirivivatnanon V 2007 Workability and strength of coarse high calcium fly ash geopolymer *Cem. Concr. Compos.* **29** 224–9

IOP Conf. Series: Materials Science and Engineering **936** (2020) 012029 doi:10.1088/1757-899X/936/1/012029

Enhancement of the Seismic Performance of Open Ground Storeyed Building using X-bracings

Shemin T John[1], Deepak Kumar Sahu[1*], S Sujeesh[2] and Pradip Sarkar[3]

[1] Ph.D. Scholar, National Institute of Technology Rourkela, India, 769008

[2] M Tech Scholar, Mar Baselios College of Engg and Technology, Trivandrum, India, 695015

[3] Associate Professor, National Institute of Technology Rourkela, India, 769008

*deepaksahu3269@gmail.com

Abstract. High rise buildings are rapidly increasing in India, and the reason is the unavailability of land. The latest developments in the construction industry improved the utility of available space in productive ways. Perhaps the most recent pattern in the development field is Open Ground Story (OGS) encircled structures, where the ground story is kept open without infill dividers mostly because of points of interest on parking facilities. The ground storey of these buildings are considered as soft and weak, making it most vulnerable to earthquakes, and incorrect structural designing of these buildings will result in catastrophic failure. It is found that the addition of steel bracings significantly improves the seismic performance of OGS buildings in recent years. The motivation of the present study is to focus on the OGS framed buildings with and without steel bracings. In this work, a combination of conservative lateral load resisting systems such as X-bracings and multiplication factor is employed along with moment resisting frame to develop a hybrid system thus by improving the seismic performance of OGS structures.

1. Introduction

Structures nowadays are constructed with the bottom storey made vacant for increasing the parking space, i.e. there is no partition wall (RCC or masonry) between columns, and such type of buildings are called as OGS buildings. The characteristics of OGS buildings are relative flexibility in the ground floor having a more significant relative horizontal displacement of the upper storey compared to the ground storey, also known as Soft Storey condition [1]. This condition results in a relatively weaker ground storey. A large variety of OGS buildings can be found in Ahmedabad city in India. The Bhuj earthquake in 2001 caused severe damage, resulting in huge losses, have proved that these buildings can be extremely vulnerable under seismic ground motion. After the Bhuj earthquake, significant revisions have been implemented in Indian Standard IS 1893 (Part1) 2002 regarding OGS buildings [2]. It suggests to calculate the design forces of bare frame building without considering the effect of infill wall under earthquake loading and then to redesign the columns of the ground storey of an OGS building by applying a factor, 2.5 times the forces calculated from the previous analysis. The factor is termed as multiplication factor (MF) [1]. Several studies suggested for providing walls in the ground storey made up of either masonry or reinforced concrete wall (shear wall) to avoid the unfair and irregular distribution of stiffness and strength in any storey of the building [3,4,5]. The present study efforts to calculate the seismic performance of OGS frame braced with X-bracing system. It is found that the addition of steel bracings considerably improves the performance of OGS buildings [6,7,8]. In this work, a combination of conservative lateral load resisting systems such as X-bracings and multiplication factor is employed along with moment resisting frame to develop a hybrid system thus by improving the seismic performance of OGS structures. The recent evolution of performance-based design has brought the non-

IOP Conf. Series: Materials Science and Engineering **936** (2020) 012029 doi:10.1088/1757-899X/936/1/012029

linear static method to the forefront[9]. Pushover analysis is a static non-linear technique in which the magnitude of the structural loading is increased in the lateral direction of the structure according to a pre-defined pattern [10].

2. Modelling

The building used for analysis is a four-storied RC building with a floor height of 3.2 m. The building is assumed to be located in seismic zone IV. The structure is designed as a frame model with masonry infill and constraints as fixed for the ground storey columns. The load combinations specified by Indian seismic code [1] has been considered.

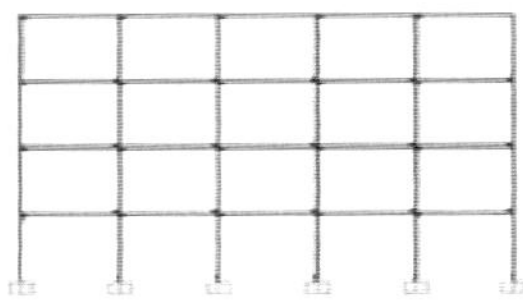

Bare Frame, BF [MF=1]

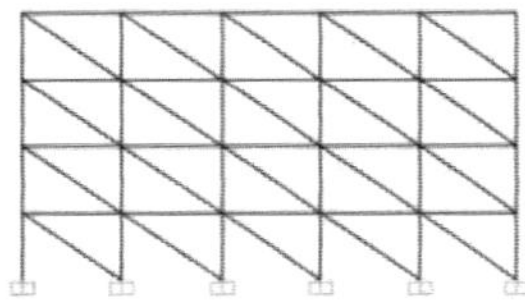

Fully Infill [FI]

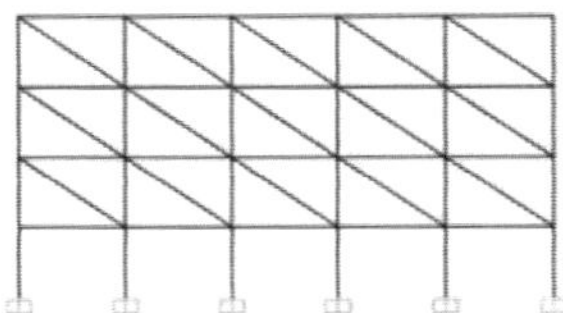

(a). OGS with MF=1 [O 1]

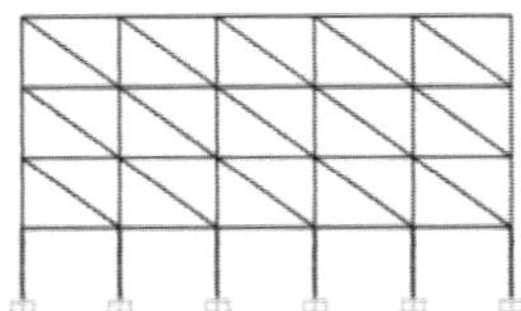

(b). OGS with MF=1.5 [O 1.5]

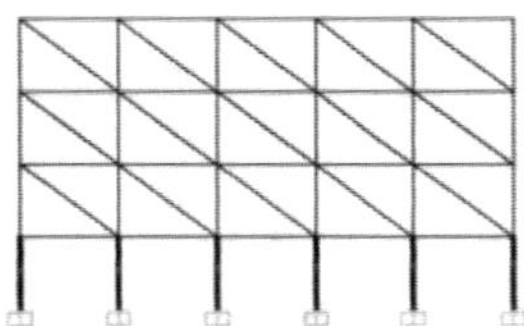

(c). OGS with MF=2.0 [O 2]

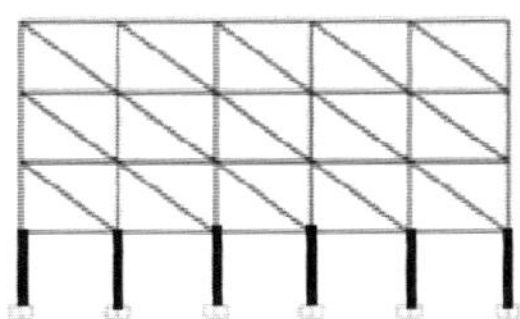

(d). OGS with MF=2.5 [O 2.5]

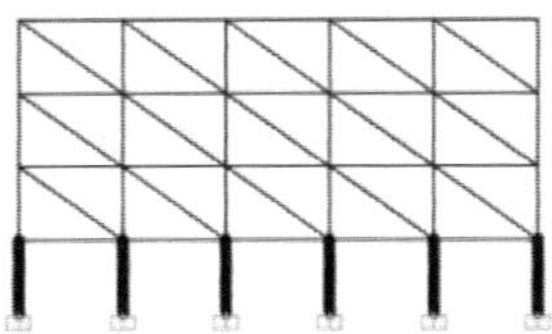

(e). OGS with MF=3.0 [O 2.5]

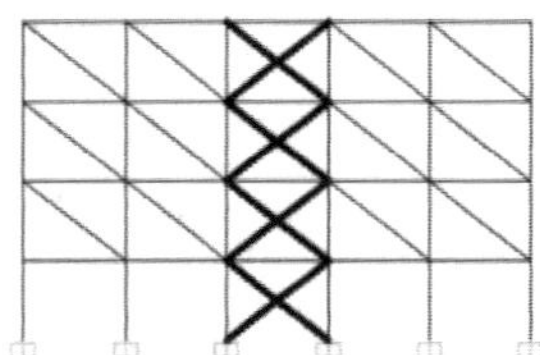

(f). OGS with MF=1 With X bracing [OX 1]

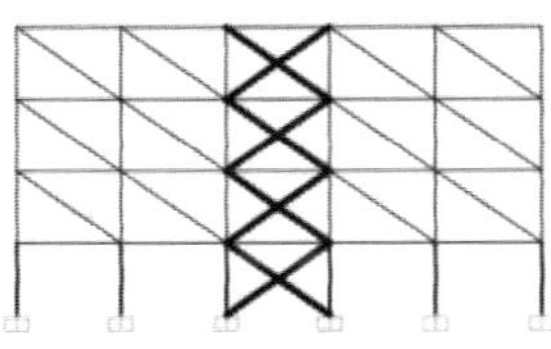

(g). OGS with MF=1.5 With X bracing [OX 1.5]

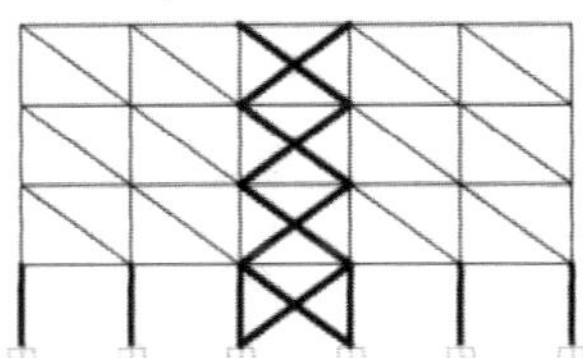

(h). OGS with MF=2 with X bracing [OX 2]

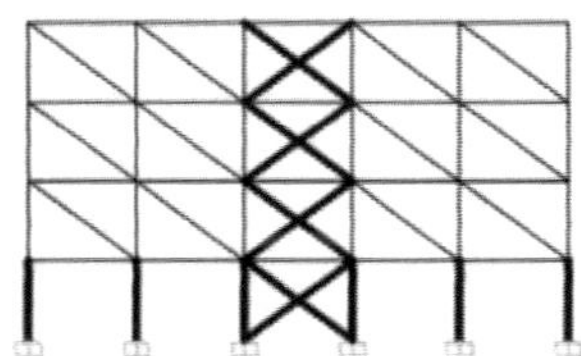

(i). OGS with MF=2.5 with X bracing [OX 2.5]

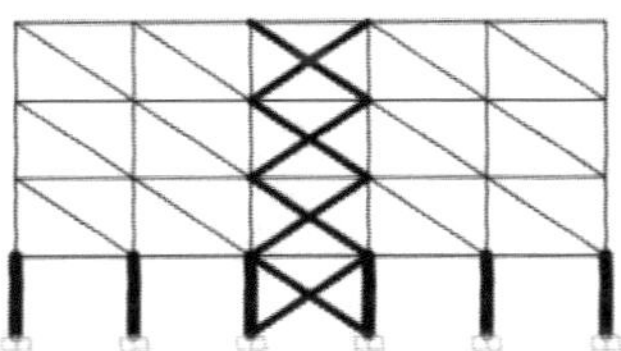

(j). OGS with MF=3 with X bracing [OX 3]

Figure 1. Frames considered for analysis.

Figure 1 shows 12 models considered for non-linear static pushover analysis. The figure shows OGS buildings with and without X-bracing with different magnification factor [11]. The column dimensions near the supports are increased with increasing magnification factor as per IS 1893 [1].

3. Results and discussions

Four storied OGS frames were analysed by non-linear static (pushover) analysis. The structure was modelled and analysed using the software SAP2000 v18. Pushover analysis is then carried out by vertical loading (gravity load) followed by a gradually increasing displacement controlled lateral load [12].

3.1. Pushover curve

The capacity curve, also known as pushover curve is a plot of base shear versus roof displacement of the structure. The pushover load case was defined as the earthquake load pattern assigned to the nodes, and the displacement was controlled with the maximum monitored value as 4% of the height of the building, which is equal to 0.512 m. The capacity curve is a good indicator of the inelastic behaviour of the building beyond the elastic stage [13,14]. The pushover curve of all 12 models is plotted in figure 2 and figure 3.

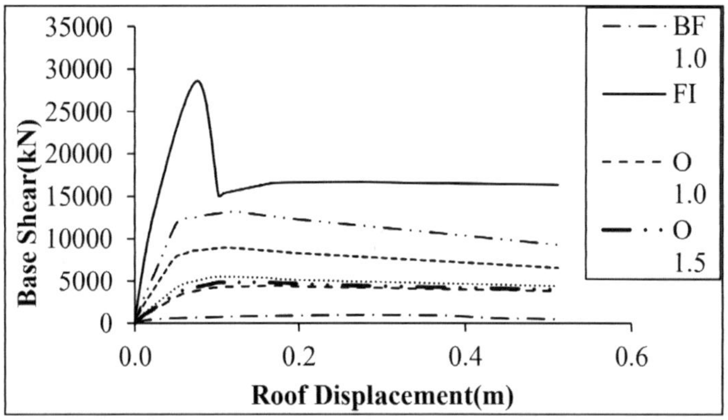

Figure 2. Pushover curve of OGS models including bare frame and fully infilled.

Table 1. Maximum base shear and corresponding displacement.

Frame model	Base Shear(kN)	Base Shear Increase (%)	Roof Displacement(m)
BF 1.0	969		0.32
FI	28607		0.07
O 1.0	4394	-	0.17
O 1.5	4826	8.94	0.13
O 2.0	5495	20	0.12
O 2.5	8966	50.99	0.11
O 3.0	13180	66.66	0.10

From the pushover curves obtained as in figure 2, the maximum base shear and corresponding roof displacement of various buildings are tabulated in Table 1. It can be concluded that by applying different multiplication factors to the ground storey of the OGS building, ranging from 1.5 to 3.0 at an interval of 0.5, the base shear capacity is increased from 8.94% to 66.66%. The base shear is comparatively more for MF 2.5 and MF 3.0; this is mainly due to the increase in the area of cross-section causing an increase in total base shear, which makes the overall weight of these buildings more than ordinary OGS. Now the OGS building with various MF is provided with X bracing throughout the centre of the building and corresponding pushover curves are being plotted in figure 3 and the base shear capacity is tabulated in table 2. It can be seen that by applying different multiplication factors to the ground storey of the OGS with X bracing ranging from 1.5 to 3.0 at an interval of 0.5 the base shear capacity is increased from 10.49% to 68.11% with a total increase of 57.62%.

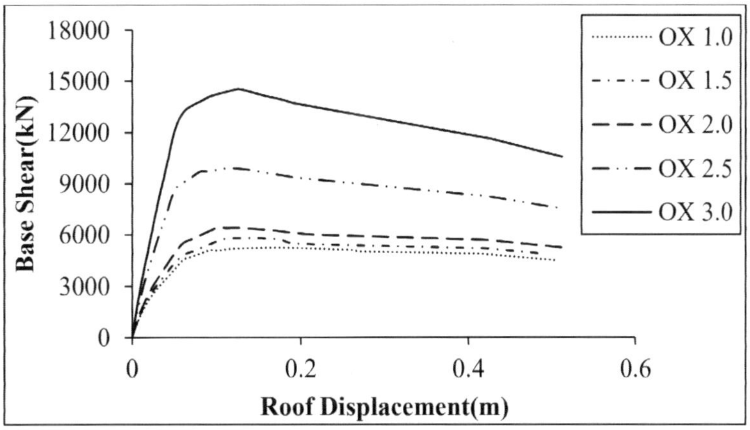

Figure 3. Pushover curve for OGS with X-bracings.

Table 2. Maximum base shear capacity of OGS with X bracing.

Frame model	Base Shear (kN)	% increase in Base Shear Capacity	Roof Displacement(m)
OX 1.0	5260.85	-	0.16
OX 1.5	5812.90	10.49	0.14
OX 2.0	6416.43	21.96	0.12
OX 2.5	9999.58	52.02	0.11
OX 3.0	14528.21	68.11	0.09

The base shear capacity of the OGS building with X bracing is relatively higher than the OGS buildings without X bracing. This indicates that the seismic ability of the building is enhanced with a redesign of OGS building with MF and with the X bracing.

3.2. *Performance level*

Three performance level namely performance level 1(PL1), performance level 2 (PL2), performance level 3(PL3) are the limits defined to assess the performance of a structure. Performance level 1(PL1) indicates only a very limited structural damage, performance level 2(PL2) shows significant damage to some structural elements and performance level 3 (PL3) shows severe damage to some structural elements. All the performance level of the OGS building with various MF (1 to 3) with and without X-bracing are plotted in figures 4 and 5. The performance points of the corresponding building are marked on the graph and are plotted below.

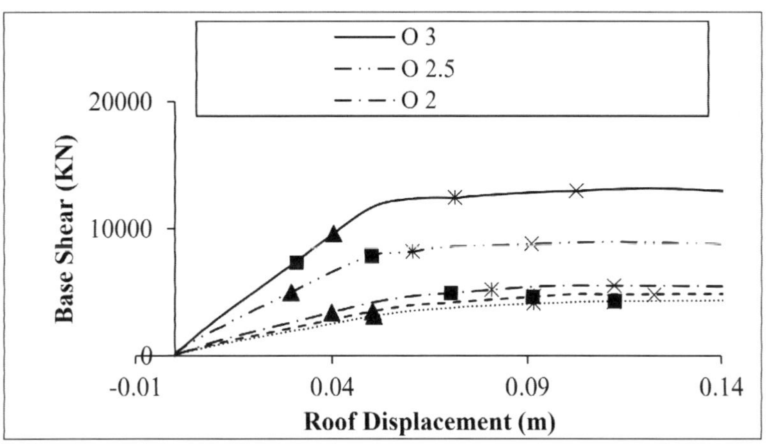

Figure 4. Performance points of OGS without X-bracings.

From figure 4, it can be seen that the OGS building with MF 1.0 is having the performance point between the PL 2 and PL 3 that means spalling of concrete is occurring at this stage. For OGS building with MF 1.5, the point lies just at the PL 2, which shows that the spalling of concrete has been started. For the OGS building with MF 2.0 and MF 2.5, the performance point lies between PL 1 and PL 2, which shows that the yielding of steel is taking place, which enables the occupants to get to a safer place before the collapse. For the OGS building with MF 3.0, the point lies below PL 1, which shows that the building is on much safer side but approaching to the yielding of steel.

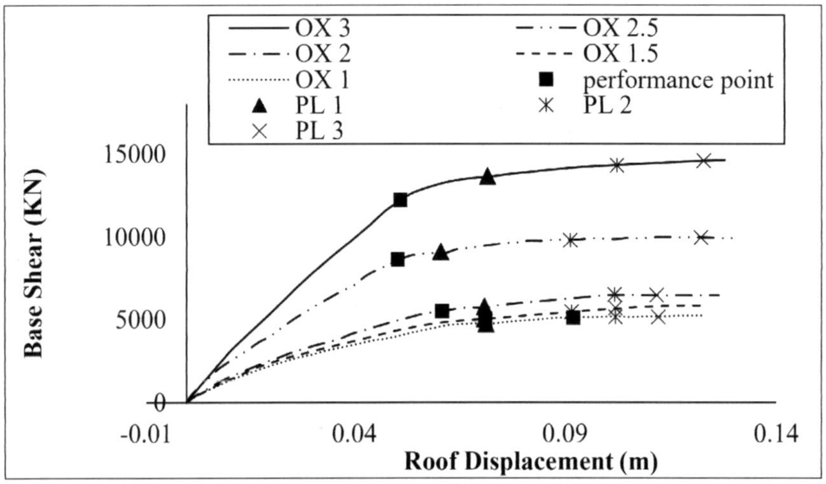

Figure 5. Performance point of OGS with X-bracings.

Figure 5, shows that the OGS building with MF 1.0 and X bracing is having the performance point between the PL 1 and PL 2 indicating yielding of reinforcement steel. For the OGS building with MF 1.5, including X bracing, the performance point lies just at the PL 1, which show that the yielding of steel has just been started. For the OGS building with MF 2.0 with X bracing, the performance point lies just below the PL 1, which indicates that the yielding of steel is only going to happen. For OGS building with MF 2.5 and MF 3.0 with X bracing, the performance point lies just below PL 1, this shows that the building is on much safer side but approaching to the yielding of steel.

By comparing the above plots, the addition of X bracing to OGS building with various MF, the performance of the building is improved to some extent, thus enhancing the seismic performance of the building. From the nonlinear analysis, it is clear that OGS building with X bracing is somewhat a suitable replacement for the RC shear wall and designing the OGS building with MF 2.5 and 3.0 enhances the seismic performance. Like-wise the OGS frame with MF 1.5 has lower base shear capacity than frame with MF 1. From the analysis, OGS building with MF 2.0 has a better performance and a comparable stiffness when compared to the MF 1.0. And the addition of X bracing to the OGS building with MF 2.0 increases the seismic performance due to larger base shear capacity.

IS 1893(part 1) 2002 suggests that the building with soft storey should be redesigned with an MF 2.5 for the ground storey columns. However, from the present study, it is clear that MF 2.5 is an overestimated value. OGS building with MF 2.0 has higher base shear capacity and lower roof displacement with or without X-bracings. Thus, MF 2.0 is observed to be an optimum value for the redesign of the ground storey columns of OGS building.

4. Conclusions

- The introduction of X-bracing increases the seismic performance of the OGS buildings considered in the present study.
- With the addition of the X-bracing to the OGS building with MF (1.5 to 3) the total base shear capacity is increased by 57.62%.
- With the addition of X- bracing, the performance level (PL 3) shifted to PL 2 for OGS buildings indicating increase in base shear capacity as well as safety.

IOP Conf. Series: Materials Science and Engineering **936** (2020) 012029 doi:10.1088/1757-899X/936/1/012029

5. References

[1] IS 1893 2002 *Indian Standard Criteria for Earthquake Resistant Design of Structures Part 1 General Provisions and Buildings (Bureau of Indian Standard: New Delhi)*
[2] Jain S K 2003 Review of Indian seismic code IS 1893 (Part 1) 2002 *Indian concrete journal* **11** 1414–22
[3] Thorat P M and Rasane A R 2015 Use of RC shear walls in strengthening of multi-storeyed building with soft story at different level *Journal of Mechanical and Civil Engineering* 98–103
[4] Maheri M R and Sahebi A 1997 Use of steel bracing in reinforced concrete frames *Engineering Structures* **19** 1018–24
[5] Rhjoo S and Mmaquani B H 2014) X-Bracing configuration and seismic response *International Journal of Civil, Environmental, Structural, Construction and Architectural Engineering* **8(6)** 642–6
[6] Bhat S A, Setia S and Sehgal V K 2015 Seismic response of moment resisting frame with open ground storey designed as per code provisions *Advances in Structural Engineering* **68** 869–83
[7] Hadad H S, Metwally I M and Betar S 2014 Cyclic behavior of braced concrete frames experimental investigation and numerical simulation *Concrete Structures Research Inst., Housing & Building Research Centre* 1–9
[8] Ghomi S S and Ebadi P 2008 Concept improvement of behavior of X-Bracing systems by using Easy-Going Steel *World Conference on Earthquake Engineering*
[9] Jain S K 2005 The Indian Earthquake problem *Current Science* **89** 1464–6
[10] FEMA 356 2000 Prestandard and Commentary for the Seismic Rehabilitation of Buildings *Federal Emergency Management Agency* (Washington D C)
[11] IS 4923 1997 *Indian Standard Hollow Steel Sections for Structural Use– Specification 2nd Ed* (Bureau of Indian Standards: New Delhi)
[12] Park R and Thomas P T 2009 *Reinforced Concrete Structures* (Wiley: India)
[13] Gobhiga S and Aravinthan K 2016 Design of earthquake resistant RCC soft storied structure with shear wall and bracing *Engineering Science and Technology* **6(2)** 51–54
[14] Agrawal P, Thakkar S K and Dubey R N 2002 Seismic performance of reinforced concrete building during Bhuj earthquake of January 26 2001 *ISET Journal of Earthquake Technology* **39** 195–217

IOP Conf. Series: Materials Science and Engineering **936** (2020) 012030 doi:10.1088/1757-899X/936/1/012030

Application of Conditional Mean Spectrum in the Seismic Assessment of a Braced Frame Steel Structure

P Jayarajan*

Department of Civil Engineering, National Institute of Technology, Calicut, India

*puttatt1@gmail.com

Abstract. Detailed engineering of critical infrastructure and tall buildings require that their response be estimated from nonlinear response history analysis (NLRHA) by subjecting the structure to suites of earthquake accelerograms. Selection of ground motion records due to future earthquake events represents an important step in the robust assessment of structural response. As per the current practice, both site-specific uniform hazard spectrum (UHS) and conditional mean spectrum (CMS) are employed as the target spectra for ground motion record selection. The CMS approach is superior considering that it represents realistic seismic events and also consider the multiple natural periods of the structure. The paper presents the application of conditional mean spectrum in the seismic assessment of an eight-story ordinary braced steel structure built on an example site. The NLRHA was conducted using the ground motions spectrally matched to the CMS at different conditioning periods. It was found that the CMS spectrally matched to larger modal period always result in larger seismic demands. A comparison was also made between seismic demands obtained from UHS & CMS indicating that the structural response parameters are almost the same in both cases except the story shears & drifts which is dominated by UHS in lower floors and by CMS in upper floors.

1. Introduction

The nonlinear response history analysis (NLRHA) is being widely used by practicing engineers for the design of tall building structures & critical infrastructure and development of suitable retrofit schemes for existing buildings. It is also extensively used in performance-based earthquake engineering (PBEE) assessment, wherein structures subjected to suites of ground motions representing multiple hazard levels are assessed for different performance objectives. The nonlinear dynamic procedures are more accurate than the static procedure as the effects of higher modes of vibration are inherently accounted and considers the material inelastic response in the performance assessment. In cases, where acceptance criteria for a linear analysis could not be achieved, nonlinear procedures can be used [1]. However, in NLRHA the structure responses are highly influenced by the characteristics of selected ground motions & the structural model used to represent inelastic behaviour.

Selection of suitable ground motion records is a challenging task in NLRHA considering that the calculated seismic demands are highly influenced by the ground motion characteristics. Further the suite of ground motions used for analysis shall be a representative of the site seismic hazard. PEER[2] provides a detailed account of various ground motion selection and modification methods including the uniform hazard spectrum (UHS) and conditional mean spectrum (CMS). The ground motion selection procedure in various codes are currently based on a target spectrum. ASCE 7-16[3] requires that either a code specified response spectrum for maximum considered earthquake (MCE) derived from a uniform hazard spectrum (UHS) or site-specific conditional mean spectrum (CMS) developed for predominant periods of the structure could be used as a target spectrum for selection of ground motion records. CMS could be used to address the conservatism inherent in UHS. Normally the periods of vibration of the structure are chosen as the conditioning periods for the development of CMS. The effect of conditioning periods on the shape of CMS and structural response is studied in detail [4]. The use of CMS for selection of ground motion records and tools available for the automation of CMS calculations are provided [5].

The present study investigates the application of conditional mean spectrum (CMS) in NLRHA to arrive at various structural response parameters for a concentric braced steel structure built in an example site. The conditioning periods for CMS were based on natural periods obtained from a modal analysis. Three suites of ground motions were used, two suites spectrally matched to the CMS at conditioning periods and the third matched to the developed site specific UHS. The structural responses obtained from two conditioning periods are compared to study the influence of higher modes. Finally, responses obtained from CMS are compared with UHS. The response parameters include the story displacement, story shear, story drift and story moment.

2. Example site

The example site has two active faults which shall be considered in the probabilistic assessment of various ground motion intensity measures. The fault F_1 at a distance $R = 10$ km from the site produce earthquakes of magnitude $M = 6.5$ with an annual occurrence rate (λ) of *0.01*. The corresponding values for fault F_2 are $M = 7.5$, $R = 20$ km & $\lambda = 0.002$. Both events have an assumed strike slip mechanism with the rupture extending the entire length of faults. The site is characterized by shallow soil followed by rock strata so that the effect of local site amplification need not be considered.

3. Description of structure

An eight-story steel braced structure considered for the study is shown in figure 1. The structure is 17x17m in plan and 28 m in height. The storey height is maintained at 3.5m at all levels. The seismic force resisting system consists of two perimeter braced frames in each direction. The interior columns of the structure are designed only for gravity loads and do not contribute to seismic resistance. The structural floor at all levels essentially consists of corrugated reinforced cement concrete (RCC) deck slab supported on secondary steel beams. The deck slab ensures a diaphragm action and is effective in transferring the seismic shear to the peripheral braced frame. Permanent gravity load (G) of 3.0 kN/m^2 and a live load (Q) of 5.0 kN/m^2 is considered on all floors. A load combination of G+0.3Q is used for the computation of lumped masses at all floor levels. The modal response spectrum analysis is performed for estimation of natural periods, mass participation ratios and the seismic base shear. The UHS developed for the site is used as a target spectrum. Considering the structural symmetry, a 2-D frame analysis is deemed to be satisfactory. The behaviour factor (q) is taken as 4.0 as per Eurocode 8(EC8)[6] corresponding to a concentric braced frame with diagonal bracings. European steel section profiles are used for the structure. The structural members are sized based on the results of elastic analysis to be verified later for capacity requirements. The modal analysis results are given in table 1.

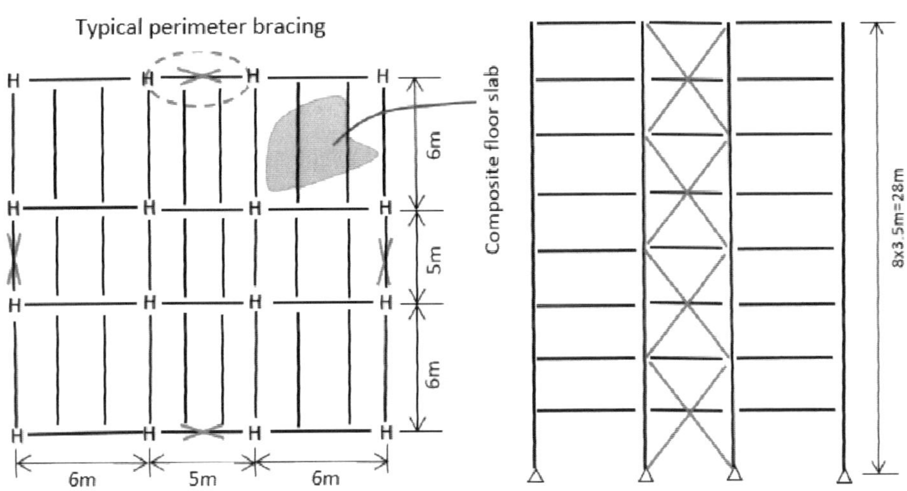

(a) Building plan (b) Elevation along braced bay

Figure 1. Structure details.

Table 1. Dynamic Characteristics of the structure.

Mode	Period (T) sec	Modal participation ratio (MPR)	Cumulative MPR	(Sa/g)-UHS	Elastic EQ shear (kN)	Design EQ shear (kN)
1	1.80	0.709	0.709	0.286	896.3	224.1
2	0.55	0.204	0.913	0.067	765.8	191.5
3	0.30	0.046	0.959	1.667	257	64.3

4. Development of site-specific uniform hazard spectrum (UHS)

A uniform hazard spectrum (UHS) represents a spectrum developed in such a way that the spectral acceleration values at different periods has an equal exceedance probability set according to the design objective. The UHS therefore has an inherent disadvantage that it does not correspond to single earthquake event and therefore would be conservative. However, a UHS would be a better choice to select the spectral acceleration corresponding to the target periods for further development of a realistic conditional mean spectrum. Different stages followed in the development of UHS are given below.

4.1. Probabilistic seismic hazard analysis (PSHA)

A detailed PSHA study is performed for the example site to arrive at the site -specific uniform hazard spectrum (UHS) for a preselected probability of exceedance. PSHA has the advantage that it considers various uncertainties in the occurrence of earthquakes and includes the effect of earthquake size onto the ground-motion characteristics to provide the seismic hazard at a site. PSHA is dealt in great detail [7] & National Research Council Report [8]. The basic PSHA equations used for the present study are

$$P(IM > x) = \int_{m_{min}}^{m_{max}} \int_{0}^{r_{max}} P(IM > x|m,r) f_M(m) f_R(r) dr dm \quad (1)$$

$$\lambda(IM > x) = \lambda(M > m_{min}) \int_{m_{min}}^{m_{max}} \int_{0}^{r_{max}} P(IM > x|m,r) f_M(m) f_R(r) dr dm \quad (2)$$

where IM indicates the intensity measure such as peak ground acceleration (PGA), spectral acceleration (SA) used in the analysis and $P(IM>x|m,r)$ denotes the probability of exceeding an IM intensity level x calculated from ground motion prediction equation (GMPE) as

$$P(IM > x|m,r) = \int_{x}^{\infty} \frac{1}{\sigma_{\ln IM}\sqrt{2\pi}} \exp\left[\frac{-1}{2}\left(\frac{\ln u - \overline{\ln IM}}{\sigma_{\ln IM}}\right)^2\right] du \quad (3)$$

$\overline{\ln IM}$ & $\sigma_{\ln IM}$ are the mean and standard deviation of the GMPE model. $f_M(m)$ and $f_R(r)$ represent probability density functions associated with magnitude and distance, $\lambda(M>m_{min})$ is the occurrence rate of earthquakes having magnitude greater that m_{min} from the source and $\lambda(IM>x)$ is the rate of $IM>x$.
For multiple sources,

$$\lambda(IM > x) = \sum_{i=1}^{n_{sources}} \lambda(M_i > m_{min}) \int_{m_{min}}^{m_{max}} \int_{0}^{r_{max}} P(IM > x|m,r) f_{M_i}(m) f_{R_i}(r) dr dm \quad (4)$$

where $n_{sources}$ represents the number of sources and M_i and R_i corresponds to the magnitude and distance for source i.

4.2. Ground motion prediction equations (GMPE)

GMPEs provide the mean or median estimate and the associated probability distributions for earthquakes characterized by their magnitude, source-to-site distance considering the soil & geologic conditions at the site. They represent a critical input parameter for the PSHA. A brief review of GMPEs developed in the period 1964-2010 is available [9]. The most desirable feature in a GMPE is its capability to furnish accurate estimate of ground motion considering a minimum set of parameters. The Abrahamson-Silva 2008 (AS08) GMPE model being more complex with the over parameterization, a much simpler and realistic AS97 model [10] is used in the present study. The values of coefficients required for median model predictions and standard deviation are taken from AS97 model. The median spectral acceleration obtained from GMPE for the site is shown in figure 2. The pronouncing effect of large magnitude earthquakes in the long period range is clearly visible. The hazard curves showing the relation between the spectral accelerations and associated rates of exceedance shall be constructed at multiple periods for the development of UHS. A typical hazard curve generated at a period of T=1 sec. is shown in figure 3.

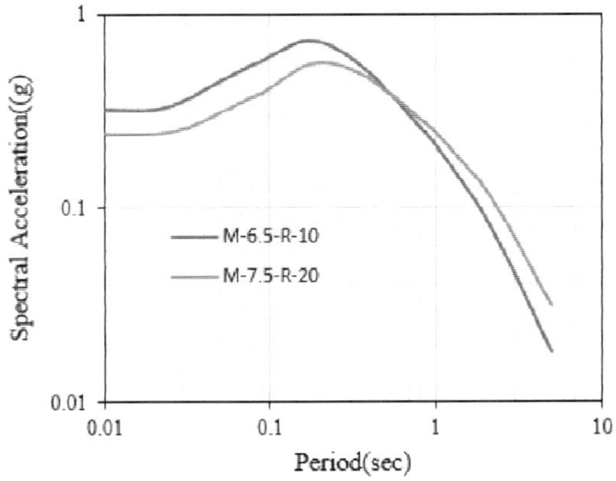

Figure 2. Median spectral acceleration-GMPE.

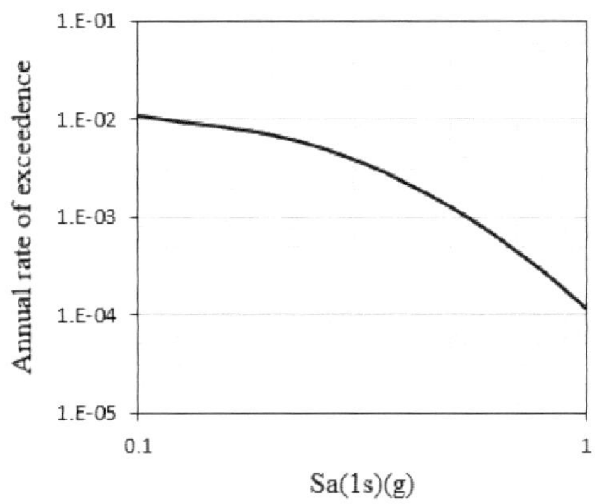

Figure 3. Hazard curve for S_a at T=1 sec.

4.3. Uniform hazard spectrum (UHS)
Using the hazard curves constructed for example site for different periods, a UHS computed for a specified rate of exceedance is developed. The target probability of exceedance equal to 10 % in 50 years (Return period=475 years) is selected to meet the "No-collapse requirement" as specified Eurocode 8[6]. The UHS is represented in figure 4. All calculations for UHS are performed using engineering math software Mathcad [11].

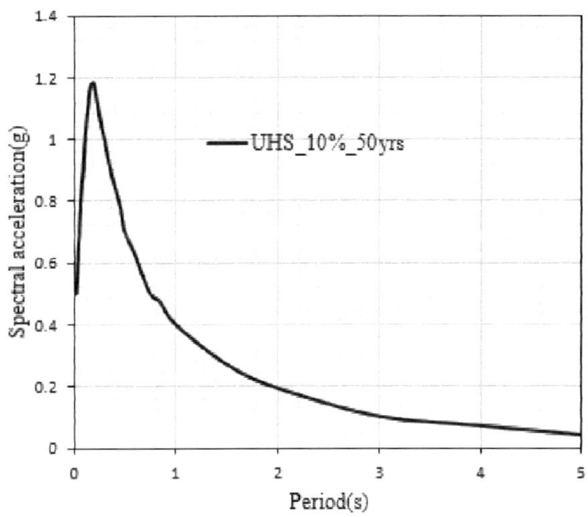

Figure 4. UHS for the example site.

5. Conditional mean spectrum (CMS)

The CMS provides the mean response spectrum associated with a ground motion corresponding to (M, R, ε) values that results in a target spectral acceleration at preselected conditioning period. The values of epsilon (ε) at a period represents the variation of the ground motion's spectral value with respect to the one obtained from GMPE. The CMS constructed for a target period therefore corresponds to a single earthquake event having the same target spectral acceleration and thereby eliminates the inherent disadvantage with UHS. The CMS incorporates both site and structure characteristics in its formulation. A detailed procedure for the construction of CMS is provided [12]. The conditioning periods (T*) for the development of CMS are taken as 1.50 and 0.50 s that are closer to the modal periods of the structure 1.80 and 0.55 s respectively. The CMS at these conditioning periods are given in figure 5.

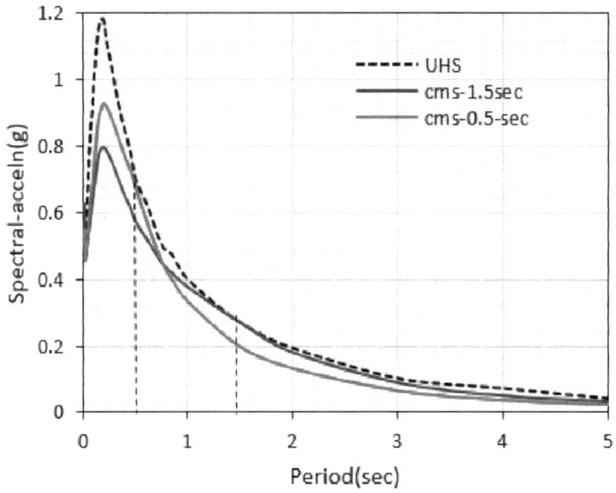

Figure 5. CMS at conditioning periods T*=1.50 and 0.50 s CMS(T_1) & CMS(T_2)

6. Nonlinear response history analysis (NLRHA)

The concentric braced frame (CBF) structure in the present study sized for elastic forces is verified for various capacity design criteria requirements laid out in chapter 6 of EC8[6] for dissipative structures. This requires identifying, designing the dissipative zones and ensuring their occurrence at predefined locations and avoidance of elastic instabilities at non-dissipative zones. The diagonal braces are identified as dissipative members and the connected column members are checked for various capacity requirements. In order to ensure sequence of plastic hinging and global yield behaviour and to arrive at a realistic estimate of behaviour factor (q) a nonlinear static analysis (pushover analysis) is first performed as per the guidelines in EC8. The generalized component force - deformation relations for modelling the inelastic behaviour in tension diagonals are taken from ASCE standard ASCE/SEI 41-17 [1] and shown in figure 6. The resulting pushover curve shown in figure 7 confirms the capacity and ductility of the designed structure. As per FEMA-356[13], the structural performance is monitored through the maximum inter-story drift with associated values of 0.5%, 1.5% & 2.0 % for braced steel frames corresponding to Immediate Occupancy (IO), Life Safety (LS) and Collapse prevention (CP) levels. The structure is capable of undergoing drifts well above 3.5 % demonstrating its deformation capacity. The development of plastic hinges as shown in figure 8 were observed only in bracing members, ensuring the energy dissipation mechanism as envisaged in the design. Further, the achieved value of behaviour factor(q) of 5.48 calculated as per ATC-19[14] is higher than the assumed value of 4.0 ensuring the superior ductility characteristics. All linear, static nonlinear and nonlinear response history analysis are performed using finite element structural analysis software SAP2000[15].

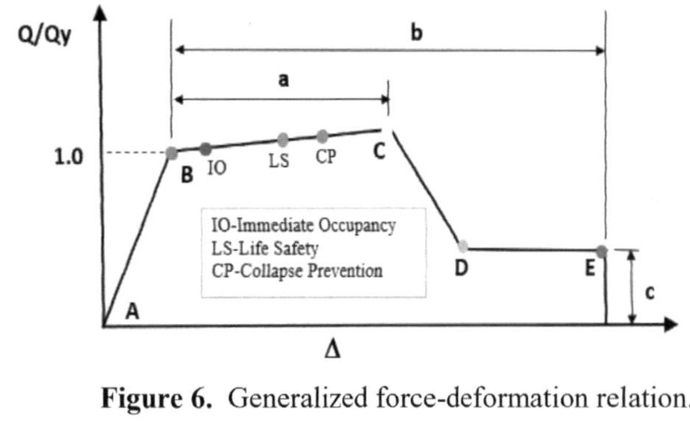

Figure 6. Generalized force-deformation relation.

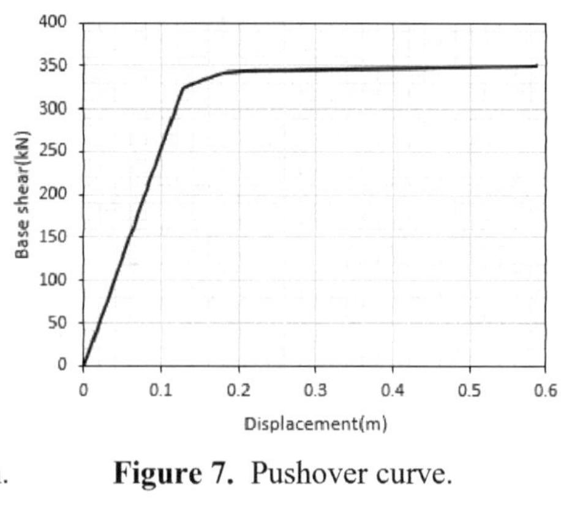

Figure 7. Pushover curve.

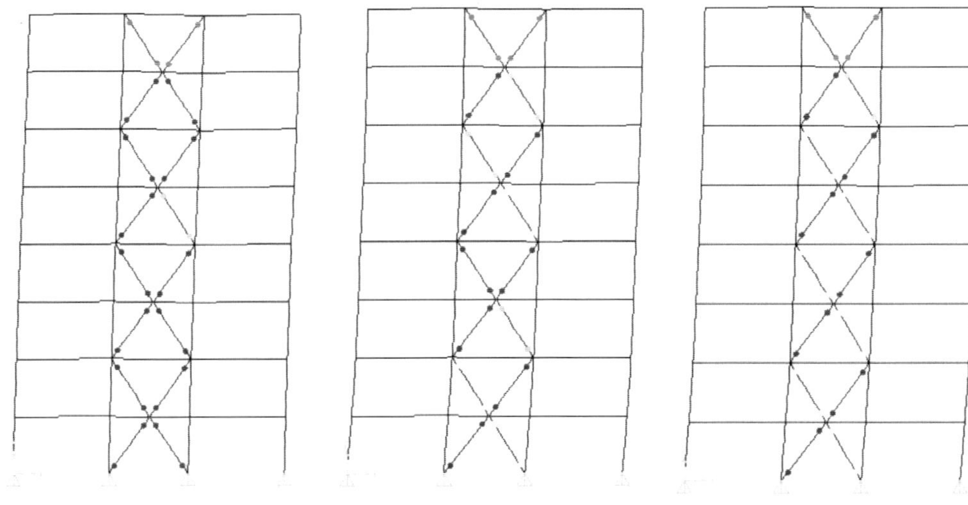

(a) (b) (c)

Figure 8. Plastic hinge pattern in pushover analysis (a) IO-level (b)LS-level(c) CP-level.

The spectral matching and the selection of ground motion records are done as per the guidelines in ASCE 7-16[3]. Two suites of records each consisting of seven numbers of ground motions spectrally matched to the conditional mean spectrums CMS (T_1) and CMS (T_2) are used for NLRHA. The recorded ground motions with site characteristics similar to the example site are obtained from PEER (Pacific Earthquake Engineering Research Centre) Strong Ground Motion Database [16]. The target conditional mean spectra and the spectra associated with the selected ground motions (GMs) along with their mean are shown in figure 9.

7. Results

The results obtained from NLRHA are presented in terms of structural response parameters such as story displacements, story drift ratio, story shear and moments. The responses are studied in two stages, first a comparison between CMS responses and second between CMS and UHS. Figure 10 shows that the CMS conditioned at larger period (fundamental period) i.e. CMS(T_1) provide a larger response for all parameters studied and govern the envelope of CMS(T_1) and CMS(T_2). However, this observation may not be valid for tall structures where the higher mode effects govern the response parameters at higher story levels. A comparison between CMS and UHS shown in figure 11 indicates that, the structural response parameters are almost the same in both cases except the story shears and drifts which is dominated by UHS in lower floors and by CMS in upper floors.

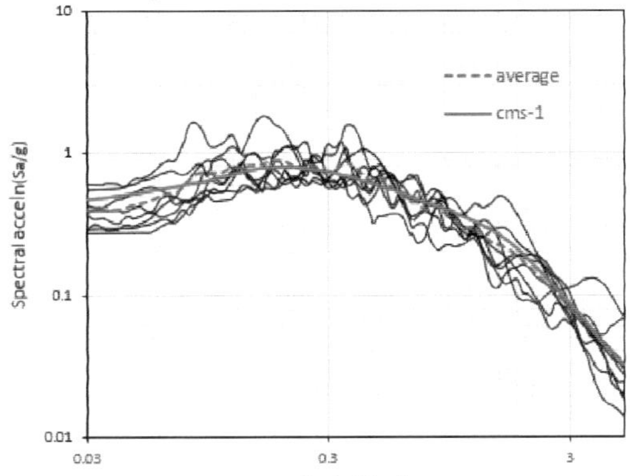

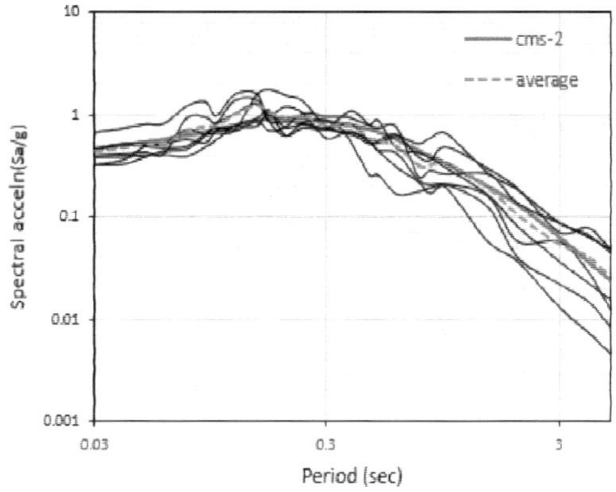

(a) GMs matched to CMS(T_1) (b) GMs matched to CMS(T_2)

Figure 9. Spectrally matched ground motions (GMs).

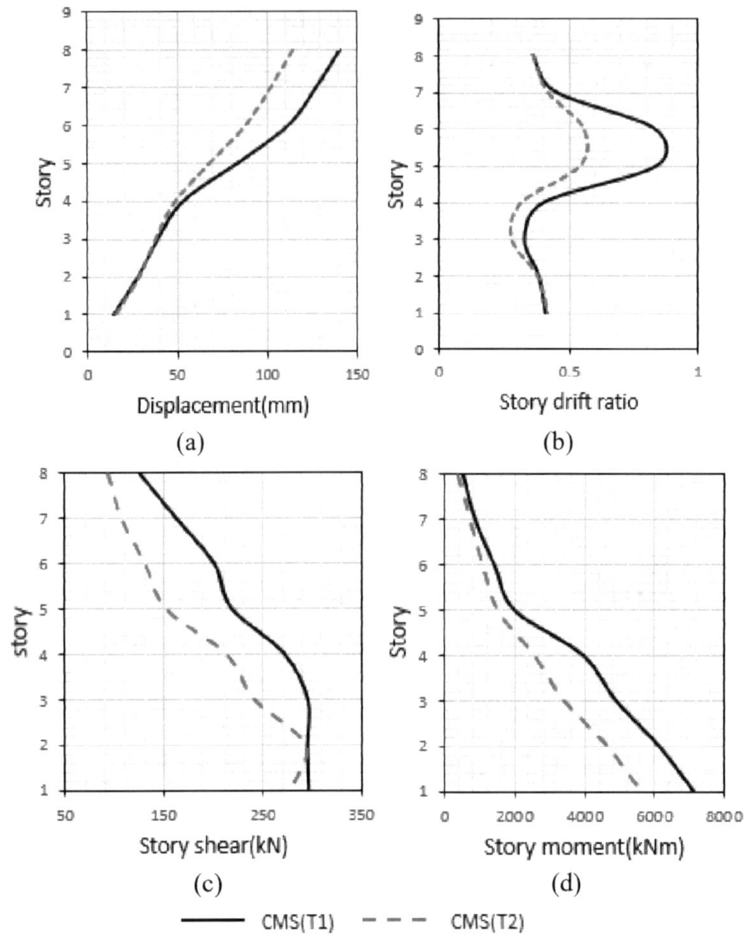

Figure 10. Structural response parameters for CMS at different conditioning periods.

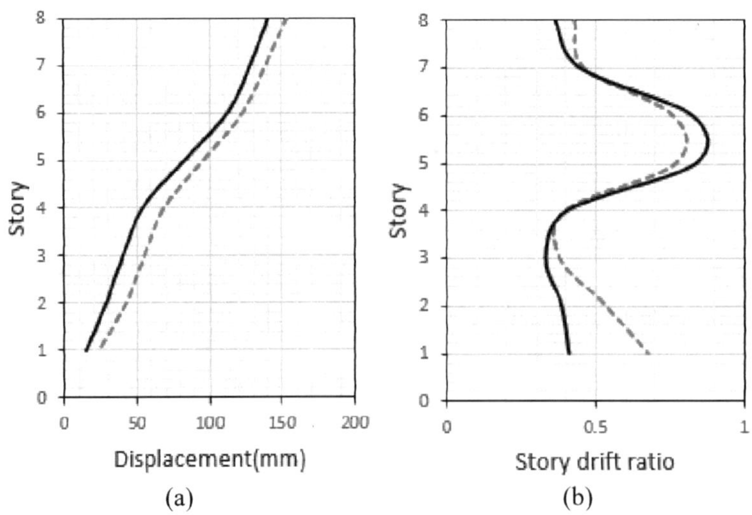

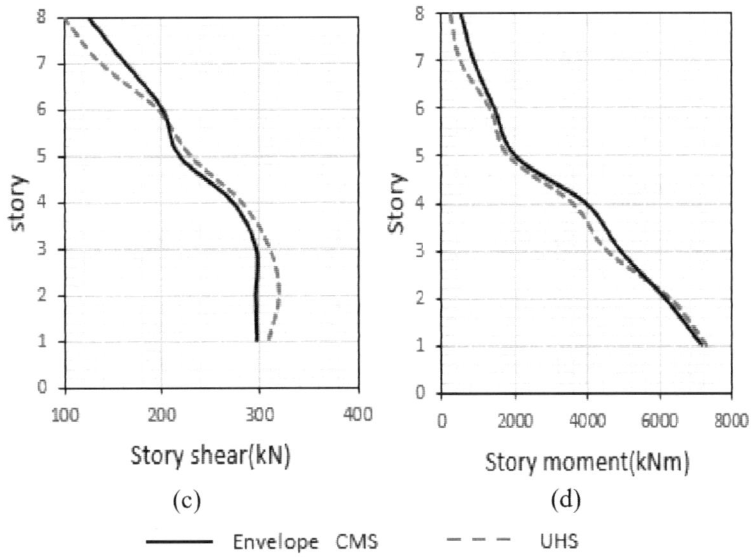

Figure 11. Comparison of responses obtained from CMS and UHS

8. Conclusions

The implementation of conditional mean spectrum (CMS) to estimate the seismic demands of an eight-story steel structure provided with ordinary concentric braced frame situated an example-site was investigated in the present study. The structural response parameters included storey displacement, inter-story drift, story shear and bending moment. The responses were compared between CMS conditioned at different natural periods of the subject structure. The seismic demands obtained through CMS and UHS were also compared. The study shows that the CMS conditioned at larger modal period (fundamental period) govern the structural response as compared to the one conditioned at higher modes. However, for tall structures, CMS conditioned at higher modes may govern certain structural response quantities. It is also observed that for the structure investigated, the UHS and CMS provide almost the same seismic demands except the story shears and drifts which is dominated by UHS in lower floors and by CMS in upper floors.

9. References

[1] ASCE/SEI 41-17 2017 *Seismic Evaluation and Retrofit of Existing Buildings* (American Society of Civil Engineers/Structural Engineering Institute)

[2] Haselton C B 2009 Evaluation of ground motion selection and modification methods: predicting median inter-story drift response of buildings *PEER Report 2009/01* (Pacific Earthquake Engineering Research Centre University of California Berkeley)

[3] ASCE/SEI 7-162016 *Minimum Design Loads for Buildings and Other Structures* (American Society of Civil Engineers/Structural Engineering Institute)

[4] Azarbakht A and Ghodrati A R 2018 The dependence of conditional spectra on the choice of target periods *Scientia Iranica Transaction A: Civil engineering* **25** 1–10

[5] Baker J W 2011 Ground motion selection for the performance-based engineering, and the conditional mean spectrum as a selection tool *Proc. of the 10thPacific Conf. on Earthquake Engineering-Building an Earthquake Resilient Pacific* (Sydney)

[6] EN 1998-12004 Eurocode 8 *Design of Structures for Earthquake Resistance – Part 1: General Rules, Seismic Actions, and Rules for Buildings* (European Committee for Standardization Brussels)

[7] Kramer S L 2003 *Geotechnical Earthquake Engineering* Prentice-Hall International Series in Civil Engineering and Engineering Mechanics (Pearson Education: India)

[8] National Research Council Report 1988 *Probabilistic Seismic Hazard Analysis* ed. Washington D.C. (National Academy Press)

[9] Douglas J 2011 *Ground-motion prediction equations 1964-2010 PEER Report 2011/102* Pacific Earthquake Engineering Research Center College of Engineering (University of California Berkeley)
[10] Abrahamson NA and Silva WJ 1997 Empirical response spectral attenuation relations for shallow crustal earthquakes *Seismological Research Letters* **68** 94–127
[11] Mathcad Professional The worldwide standard for technical calculations (https://www.ptc.com/en/products/mathcad)
[12] Baker J W 2011 Conditional mean spectrum: tool for ground motion selection *Journal of Structural Engineering* **137** 322–31
[13] FEMA 356 2000 *Prestandard and Commentary for the Seismic Rehabilitation of Buildings* (American Society of Civil Engineers Federal Emergency Management Agency: Washington, DC)
[14] ATC-19 1995 *Structural Response Modification Factors* (Applied Technology Council: Redwood City, California)
[15] SAP2000 version 20 Linear and nonlinear static and dynamic analysis of three-dimensional structures (Computer and Structures: Berkeley, CA)
[16] Peer Strong Motion Database2013/14 (https://peer.berkeley.edu/ngawest2 and https://peer.berkeley.edu/ngaeast)

Acknowledgements
The author would like to thank Pacific Earthquake Engineering Research Centre (PEER) for providing access to ground motion database and tools for searching, scaling and downloading ground motion data.

IOP Conf. Series: Materials Science and Engineering **936** (2020) 012031 doi:10.1088/1757-899X/936/1/012031

Strengthening of Confined Masonry Structures for In-plane Loads: a Review

Akshay Gupta[1*] **and Vaibhav Singhal**[2]

[1] Ph.D. Student, Department of Civil and Environmental Engineering, Indian Institute of Technology Patna, Bihta, 801106, India

[2] Assistant Professor, Department of Civil and Environmental Engineering, Indian Institute of Technology Patna, Bihta, 801106, India

*akshay_1921ce15@iitp.ac.in

Abstract. Confined masonry is a load-bearing masonry structure where walls are confined with nominally reinforced concrete elements at key locations. This building typology is popularly used in many countries as it generally performs better than the unreinforced masonry and non-ductile RC infill masonry under seismic loads. However, the confined masonry buildings have also experienced damage during past earthquakes. Poor seismic performance is generally seen due to construction errors, design flaws or material deficiencies in the building design and construction process. To overcome the structural deficiencies researchers have used different strengthening techniques for confined masonry walls such as application of ferrocement overlay, steel bands, and fiber-reinforced polymers. There are several inherent advantages and disadvantages of these strengthening techniques in terms of efficiency, bonding with the substrate material, durability, ease in construction and economy. This article will review and discuss the in-plane strengthening of confined masonry walls with conventional and fiber-reinforced materials and will highlight the gap areas and scope for future research.

1. Introduction

Confined masonry system was introduced in Italy as an alternative to unreinforced masonry (URM) buildings, which were almost completely destroyed in the 1908 Messina earthquake. Confined masonry (CM) structure has always gained vital importance in the construction practices, as its performance is much better than the unreinforced masonry and non-ductile RC infill structures during seismic loads. The CM structures are more popularly known for the low rise residential buildings of South America, Central America, Asian countries and European countries.

Past studies have revealed the satisfactory performance of confined masonry buildings during the 1939 Chile earthquake, 1985 Llolleo earthquake. However, few major damages were observed in these structures during the seismic event such as in M7.5 Llolleo earthquake, it was reported that 6% of the confined masonry buildings sustained repairable damage and 2% of the confined masonry buildings sustained irreparable damage [1]. Likewise, in the Tehuacan earthquake 1999 (Figure 1a), major damages were seen in the three or more storey CM buildings. Moreover, in the 2010 Haiti earthquake, inadequate performance of CM buildings was observed due to poor construction practices and the use of substandard materials (Figure 1b). The damages in the CM wall arise due to the poor design, detailing and construction. To improve the performance of CM buildings, many researchers have investigated different methods of strengthening. Generally, two types of techniques have been used for the strengthening of CM walls, which includes the retrofitting with conventional composites and retrofitting with reinforced fiber composites.

IOP Conf. Series: Materials Science and Engineering **936** (2020) 012031 doi:10.1088/1757-899X/936/1/012031

(a) (b)

Figure 1. Damages in confined masonry (a) Tehuacan earthquake; Alcocer et al. [2] (b) Haiti earthquake [3].

2. Common failures in confined masonry

Generally, three types of in-plane failure modes are seen in confined masonry walls, which include diagonal shear failure, sliding shear failure and in-plane bending failure, as shown in figure 2. The diagonal shear failure mechanism is the most common failure for confined masonry walls resulting from the combination of vertical and horizontal loads (seismic loads). The shear failure happens when the principal tensile stresses become more than the masonry tensile strength. The diagonal shear failure of the CM wall is shown in figure 2a [4]. The inadequate number of tie-columns and poor workmanship may result in sliding shear failure and subsequently large damage to confined masonry walls (Figure 2b). In some cases, the bond failure at the mortar-brick interface or tensile cracking at the masonry units was also found to be the major cause of the damage [5].

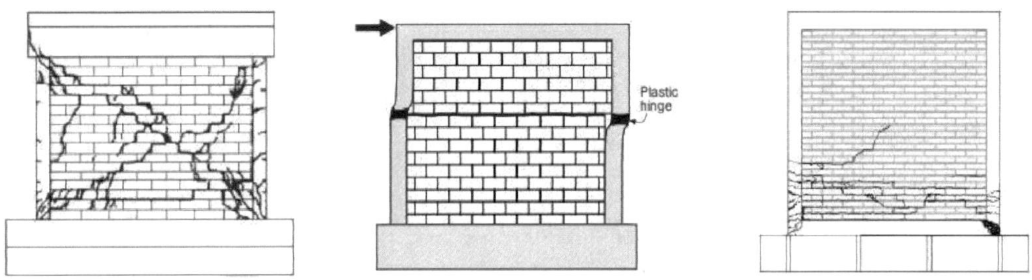

(a) (b) (c)

Figure 2. Common failures in confined masonry (a) Diagonal shear failure; Yoshimura et al. [4] (b) Sliding shear Failure; Singhal [6] (c) In-plane bending Failure; Zabala et al. [7].

When the CM walls are subjected to compression stresses at one end and tensile stresses at the other end, the type of failure is known as an in-plane bending failure (Figure 2c) [7]. The wall having panel height much larger than the panel length is more susceptible to a flexural mode of failure. To mitigate these failures, strengthening techniques have been developed using various composites.

3. Strengthening techniques

Many techniques have been used for the retrofitting of the CM walls. These techniques majorly use the application of composite materials on the surface of the walls. Broadly, it includes the retrofitting of CM structures by conventional composites or by fiber-reinforced composites. Retrofitting by conventional composites have been majorly used for strengthening, as these are readily available at a cheaper price. A conventional composite includes steel welded wire mesh, which is generally placed on the surface of the wall with the help of anchors/nails as shown in figure 3a. A layer of mortar (plaster) is then applied to the mesh. This strengthening composite is commonly known as ferrocement or wire reinforced cementitious matrix. On the other hand, the fiber-reinforced polymer composites in the form of carbon, glass, and basalt have also been widely used by several researchers for the retrofitting of CM walls. These composite materials can effectively strengthen the CM walls for better seismic performance. Earlier, fibers were applied on the surface of the CM wall with the help of chemical epoxy and cementitious matrix as shown in figures 3b and 3c. However, the organic binder or epoxy has now been replaced with the cementitious matrix due to its high durability, fire resistance and good compatibility with the substrate material (Figure 3c) [8], [9]. This type of strengthening technique is popularly known as fibre reinforced cementitious matrix (FRCM). Depending on the type of deficiencies, the fibers are applied either in a grid pattern or in a diagonal pattern. To overcome the failures like diagonal shear, flexure, interface, etc. these layouts are used. A typical example of the diagonal layout has been shown in figure 3c, in order to overcome the diagonal shear failure of CM wall specimen. Similar to solid CM walls, different techniques have been proposed to strengthen CM walls with door and window openings. In general, openings are confined by providing reinforcement along its periphery to mitigate failure of adjacent masonry piers and corners of opening.

Researchers have also explored the use of waste materials like plastic cement bag mesh for retrofitting of CM walls [10]. A mesh of waste plastic cement bag mesh was prepared and applied to the specimen. It was reported that the retrofitting through waste materials give a significant improvement in strength at a low cost. Further, modern materials such as engineered cementitious composite (ECC) was also used for the strengthening of confined masonry walls [11]. It was reported that the ECC overlay can significantly improve the in-plane strength of the CM wall. To evaluate different strengthening schemes, a repository of CM specimens available in the literature was created (Table 1). This repository consists of only single-story specimens strengthened for in-plane loads without any prior damage.

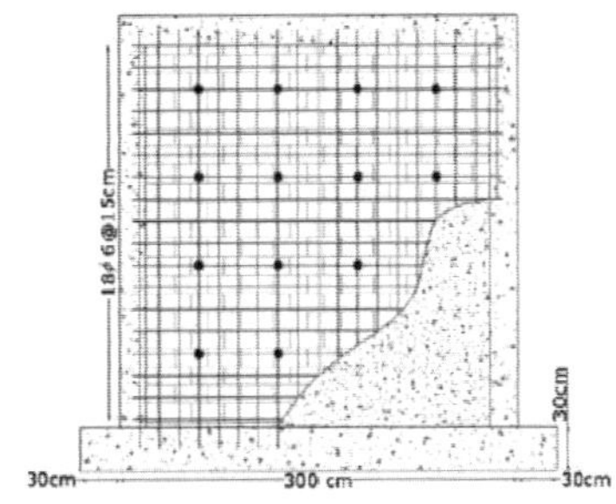

(a) WRCM [12]

(b) FRP applied using epoxy [13]

(c) FRP applied using cementitious matrix[14]

Figure 3. Commonly used strengthening techniques.

4. Performance of various strengthening techniques

The result of various experimental studies showed that there was a huge influence of the retrofitting agent on the response of the CM wall. Based on the details collected in the repository, the performance of different strengthening schemes was compared using cracking load β^c, cracking drift λ^c, peak load β^u, and ultimate drift λ^u for each strengthened and unstrengthened specimens (Table 2). The ultimate drift was considered at 80% of the peak load in the post-peak region. Further, the ultimate load and drift of strengthened specimens were compared with their respective control specimens by estimating ultimate load ratio β^u/β^{uc} and ultimate drift ratio λ^u/λ^{uc}(Table 2).

It was noted that the conventional strengthening techniques like wire reinforced cementitious matrix and the ferrocement were highly effective in enhancing the strength and deformation capacity of the CM wall. If these strengthening techniques are properly implemented along with the provision of suitable anchors then up to 150% improvement in the strength can be achieved (Table 2, Sl. No. 4). Further, it was reported that the larger diameter wire provides better deformation capacity and can sustain the peak load up to larger drifts. However, the over-reinforcement may also lead to brittle failure of the strengthened wall.

Table 1. Details of confined masonry specimens.

No.	Reference	A_s (mm × mm)	Composite	Φ (mm)	S_s	Remarks
1	Alcocer et al. [15]	2500 × 2500	C	-	-	-
2		2500 × 2500	WRCM	4.88	300	-
3		2500 × 2500	WRCM	6.35	300	-
4		2500 × 2500	WRCM	3.43	450	-
5	El-Diasity et al. [14]	2000 × 1820	C	-	-	-
6		2000 × 1820	C-W	-	-	Opening - 760 × 780mm
7		2000 × 1820	C-D	-	-	Opening - 680× 1410mm
8		2000 × 1820	FCM	1.6	15	Corner layout
9		2000 × 1820	FCM	1.6	15	Diagonal layout
10		2000 × 1820	FCM	1.6	15	Full wall cover
11		2000 × 1820	GFRP	0.013	-	At diagonals
12		2000 × 1820	GFRP	0.013	-	Full wall cover
13		2000 × 1820	FCM-W	1.6	15	Full wall cover
14		2000 × 1820	FCM-D	1.6	15	Full wall cover
15	Dimas et al. [16]	3000 × 2100	C-W	-	-	Opening -1200 × 1000mm
16		3000 × 2100	GFRP-W	1.0	-	Vertical cover
17	Zhou et al. [13]	1860 × 1500	C	-	-	-
18		1860 × 1500	C-W	-	-	Opening - 600 × 450 mm
19		1860 × 1500	BFRP	0.121	-	Solid wall
20		1860 × 1500	BFRP-W	0.121	-	Opening - 600 × 450 mm
21	Shakib et al. [12]	2580 × 2590	C	-	-	-
22		2580 × 2590	CFRP	-	-	Grid pattern
23		2580 × 2590	CFRP	-	-	Diagonal pattern
24		2580 × 2590	PFS	-	-	Fiber density 0.9 g/cm^3
25	Deng and Yang [11]	2370 × 1370	C-LMS	-	-	-
26		2370 × 1370	C-LMS	-	-	-
27		2370 × 1370	C-HMS	-	-	-
28		2370 × 1370	C-HMS	-	-	-
29		2370 × 1370	ECC-LMS	-	-	One side strengthening
30		2370 × 1370	ECC-LMS	-	-	One side strengthening
31		2370 × 1370	ECC-HMS	-	-	One side strengthening
32		2370 × 1370	ECC-HMS	-	-	One side strengthening
33		2370 × 1370	ECC-LMS	-	-	Both side strengthening
34		2370 × 1370	ECC-LMS	-	-	Both side strengthening
35		2370 × 1370	ECC-HMS	-	-	Both side strengthening
36		2370 × 1370	ECC-HMS	-	-	Both side strengthening

A_s – Area of specimen; Φ – Diameter of reinforcement; S_s – Reinforcement spacing; C – control specimen; W – Window opening; D – Door opening;;WRCM – strengthening with wire reinforced cementitious matrix; BFRP – strengthened with basalt fiber reinforced polymer; R – Repaired; GFRP – strengthened with glass fiber reinforced polymer; CFRP –retrofitting by carbon fiber reinforced polymer; PFS – retrofitted with poly-propylene fiber shotcrete; ECC –strengthening with engineered cementitious composites; LMS – low mortar strength; HMS – high mortar strength.

Similarly, the use of modern material for strengthening showed a significant increase in the strength of the CM wall specimens. The specimen strengthened with GFRP composite on full wall surface exhibited an approximately 80% increase in strength when compared to the control specimen. Moreover, for the ultimate drift, retrofitting with the CFRP composite was found highly effective. It was seen that an increment of more than 150% in the drift can be achieved when CFRP composite is used in a diagonal pattern (Table 2, Sl. No. 23). The strengthening with ECC exhibited a significant improvement in the strength, but there was no improvement in the deformation capacity. However, it is highly advantageous to use ECC as a strengthening composite, as it shows an approximately 250% rise in the strength of the CM structure when applied on both sides of the wall. Further, in comparison to other strengthening methods, the application of ECC for strengthening is relatively quicker and easier as it does not require a highly skilled manpower.

Table 2. Influence of composite material on CM wall specimen.

No.	β^c (kN)	β^u (kN)	λ^c (%)	λ^u (%)	β^u/β^{uc}	λ^u/λ^{uc}	No.	β^c (kN)	β^u (kN)	λ^c (%)	λ^u (%)	β^u/β^{uc}	λ^u/λ^{uc}
1.	82.5	112.5	0.42	0.76	-	-	21.	-	148.0	0.053	0.376	-	-
2.	160.0	237.5	0.60	1.28	2.11	1.68	22.	-	202.0	0.078	0.613	1.36	1.63
3.	167.5	215.0	0.40	1.39	1.91	1.83	23.	-	172.0	0.077	0.990	1.16	2.63
4.	207.5	292.5	0.58	0.79	2.60	1.04	24.	-	231.0	0.163	0.819	1.56	2.18
5.	210.0	290.0	0.109	0.550	-	-	25.	-	163.4	-	0.876	-	-
6.	152.5	254.0	0.109	0.896	-	-	26.	-	190.7	-	1.241	-	-
7.	130.0	207.0	0.055	1.420	-	-	27.	-	217.8	-	0.731	-	-
8.	345.0	365.0	0.329	0.760	1.26	1.38	28.	-	243.6	-	1.462	-	-
9.	335.0	378.5	0.329	0.790	1.31	1.44	29.	-	317.9	-	0.292	1.95	0.33
10.	376.0	376.0	0.219	0.640	1.30	1.16	30.	-	436.8	-	0.437	2.29	0.35
11.	315.0	373.0	0.219	0.745	1.29	1.35	31.	-	417.9	-	0.437	1.92	0.60
12.	525.0	525.0	0.659	0.796	1.81	1.45	32.	-	441.4	-	0.511	1.81	0.35
13.	227.5	325.0	0.055	1.049	1.28	1.17	33.	-	588.9	-	0.437	3.60	0.50
14.	177.5	272.5	0.164	1.145	1.32	0.81	34.	-	577.6	-	0.511	3.03	0.41
							35.	-	416.1	-	0.459	1.91	0.63
15.	58.9	82.2	0.084	0.695	-	-	36.	-	635.0	-	0.576	2.61	0.39
16.	62.5	110.0	0.122	0.793	1.34	1.14							
17.	192.0	234.0	0.087	0.660	-	-							
18.	184.0	217.0	0.073	0.356	-	-							
19.	218.0	322.0	0.146	1.359	1.38	2.06							
20.	204.0	266.0	0.126	0.800	1.23	2.25							

β^c – Cracking load; β^u – Ultimate load; λ^c – Cracking drift; λ^u – Ultimate drift (at 80% of peak load in the post-peak zone); β^u/β^{uc} – Ultimate load ratio; λ^u/λ^{uc} – Ultimate drift ratio.

5. Conclusions

This article provides preliminary review of various strengthening techniques used for the CM walls with and without openings. It was seen that strengthening techniques were significantly beneficial for the improvement of the overall strength and deformation capacity of the confined masonry walls. A comparative study was done to evaluate various techniques for the strengthening of CM walls. The specific increase of strength was particularly seen when conventional techniques like wire reinforced cementitious matrix and modern techniques like FRP, FRCM and ECC were used. It was seen that different layout of strengthening enhanced the performance differently and can mitigate different types of failures. The comparative analysis showed that the engineered cementitious composite (ECC) was most effective in improving the in-plane capacity of CM walls, whereas the carbon-fiber reinforcement laid in a diagonal pattern significantly improved the deformation capacity when compared with other strengthening schemes. The application of ECC overlay for strengthening is much simpler but the availability of material especially suitable polyvinyl alcohol (PVA) fibres is scarce in local market. Further, the use of ECC did not show any improvement in the deformation capacity. However, based on the test results, the strengthening using wire reinforced cementitious matrix (WRCM) is a better choice as it had shown improvement in both deformation and in-plane capacity and the required material is easily available.

The developed repository clearly indicates that the limited studies have been done till date on the strengthening and rehabilitation of the CM structures. A more detailed analytical and experimental investigation is required to understand the behaviour of strengthened CM walls and to develop the methodology for designing the strengthening scheme to achieve the desired performance level.

6. References

[1] Brzev S 2007 Earthquake resistant confined masonry construction *National Information centre of earthquake engineering, IIT Kanpur* ISBN: 81-904190-9-9

[2] Alcocer S M, Aguilar G, Flores L, Bitrán D, Durán R, López O A, Pacheco M A, Reyes C, Uribe C M and Mendoza M J 2001 The Tehuacán Earthquake of June 15, 1999 *Centro Nacional de Prevención de Desastres (SEGOB), Mexico City, Mexico* 198

[3] Haiti Earthquake Clearinghouse 2010 (http://eqclearinghouse.org /co/20100112-Haiti/)

[4] Yoshimura K, Kikuchi K, Kuroki M, Nonaka H, Tae Kim K, Wangdi R and Osikata A 2004 Experimental study for developing higher seismic performance of brick masonry walls *13th World Conference on Earthquake Engineering* (Vancouver, Canada) 1597

[5] Tomaževič M 1999 *Earthquake-Resistant Design of Masonry Buildings* (Imperial College Press: London)

[6] Singhal V 2014 Effect of toothing and openings on bi-directional seismic behaviour of confined masonry walls *Indian Institute of Technology Kanpur* 16

[7] Zabala F, Bustos J L, Masanet A and Santalucia J 2004 Experimental behaviour of masonry structural walls used in Argentina *Proc. 13th World Conference of Earthquake Engineering* (Canada) 1093

[8] American Concrete Institute 2013 Design and construction guide of externally bonded FRCM systems for concrete and masonry repair and strengthening *ACI 549.4R*

[9] Sagar S L, Singhal V and Rai D C 2019.In-plane and out of plane behaviour of masonry-infilled RC frames reinforced with fabric-reinforced cementitious matrix *Journal of composites for construction* **23(1)** 04018073–1

[10] Chaurasia A, Singhal S and Parashar J 2019 Experimental investigation of seismic strengthening technique for confined masonry buildings *Journal of Building Engineering* **25** 1–13

[11] Deng M and Yang S 2020 Experimental and numerical evaluation of confined masonry walls retrofitted with engineered cementitious composites *Engineering Structures* **207** 110249

[12] Shakib H, Dardaei S, Mousavi M and Rezaei M K 2016 Experimental and analytical evaluation of confined masonry walls retrofitted with CFRP strips and mesh-reinforced PF shotcrete *Journal of Performance of Constructed Facilities* **30(6)** 04016039

[13] Zhou D, Zhou S and Lei Z 2015 In-plane shear behaviours of constrained masonry walls externally retrofitted with BFRP *Journal of Composites for Construction* **20(2)** 04015059

[14] El-Diasity M, Okail H, Kamal O and Said M 2015 Structural performance of confined masonry walls retrofitted using ferrocement and GFRP under in-plane cyclic loading *Engineering Structures* **94** 54–69

[15] Alcocer S M, Ruiz J, Peneda J A and Sepeda J A 1996 Retrofitting of confined masonry walls with welded wire mesh *11th world conference on earthquake engineering* 1471

[16] Velázquez-Dimas J, Quiñonez-Esquivel B, Castorena-González J, Reyes-Salazar A, González-Cuevas J and López-López D 2012 In-plane behaviour of confined masonry walls with holes retrofitted with GFRP and subjected to lateral cyclic loading *15th world conference on earthquake engineering* (Portugal)

IOP Conf. Series: Materials Science and Engineering **936** (2020) 012032 doi:10.1088/1757-899X/936/1/012032

Investigation on Rehabilitation of Circular RCC Columns with FRP Composites

P Bindurani*, N Ganesan, P V Indira and V R Yadav

Department of Civil Engineering, National Institute of Technology Calicut, Kozhikode, Kerala, India - 673601

*binduranip@gmail.com

Abstract. A detailed experimental investigation was done on the effect of rehabilitation using (i) carbon fibre reinforced polymer (CFRP) and (ii) glass fibre reinforced polymer (GFRP) on reinforced cement concrete (RCC) circular columns in order to compare the rehabilitation potential of these selected composites. The CFRP and GFRP laminates respectively consist of unidirectionally woven carbon fibre mat (230g/sqm) and bidirectionally woven glass fibre mat (220g/sqm), both impregnated in the same epoxy resin matrix. Even though there are many studies reported in the literature, comparative studies among these composites are limited. In order to have a level playing field, tension tests have been conducted on test coupons of these composites, and it was found that one layer CFRP was equivalent to 3 layers GFRP. In this study, pre-distressed RCC circular columns of 150mm diameter and 850mm length were rehabilitated with these equivalent laminates and tested under uniaxial compression. For rehabilitation purposes, the levels of pre-distress induced were at 70%, 80% and 90% of the ultimate capacity of the specimen. The compression behaviour of the rehabilitated columns was compared in terms of strength and serviceability. It was noted that both the composites improve the strength significantly, and CFRP performed much better than GFRP in terms of strength and serviceability.

1. Introduction

Concrete deterioration in RCC structural members due to aging, unexpected overloading, unfavourable environmental conditions, corrosion of steel reinforcement etc. are very common and these may lead to a reduction in load-carrying capacity and the structural integrity of the RCC members. However, the replacement of such deteriorated elements requires a considerable amount of time and money. Structural rehabilitation will be one of the solutions, and it involves determining the origin and amount of distress, removing damaged materials and causes of distress, as well as selecting and applying appropriate repair materials that extend the life of the structure. In recent years, the technology of strengthening the distressed structural elements with fibre reinforced polymers (FRPs) gained more momentum over the technologies using traditional materials. The primary reason is its advantages like more strength to weight ratio, easy and rapid application, minimal change in structural geometry, etc.

Several studies were conducted on FRP rehabilitation [1-5] and confirm its effectiveness in enhancing the ductility and load-carrying capacity of RCC columns. Even though the studies comparing the confinement effectiveness of different FRPs on RCC columns, based on strength and serviceability are limited. This study deals with an experimental investigation on the enhancement of compression behaviour of preloaded RCC columns using (i) Carbon Fibre Reinforced Polymers (CFRP) and (ii) Glass Fibre Reinforced Polymers (GFRP). The parameters considered in this study are (i) the composite materials used for strengthening, and (ii) the amount of pre-distress induced to the columns.

2. Experimental Programme

The experimental work was conducted on eight circular columns having 150mm diameter and 850mm length. These columns were grouped into two with four columns each viz. Group-1 and Group-2. One column in each group was tested up to failure under uniaxial compression in a universal testing machine (UTM) of 300T capacity to find the ultimate capacity and kept as control specimens. The other columns in each group were preloaded to 70%, 80% and 90% of the ultimate capacity. The Group-1 and Group-2 preloaded columns were then rehabilitated with CFRP and GFRP respectively, by complete wrapping with these composites by wet lay-up procedure. The rehabilitated specimens were cured for two weeks at ambient temperature. These columns were then tested to failure and the behaviour was compared. In order to do the comparison study in a level playing field, initially, tension tests were conducted on CFRP and GFRP coupons to equalise their strengths.

2.1. Tension tests on FRP coupons

2.1.1. FRP composites used. CFRP consisted of unidirectional carbon fibre woven mat of weight 230g/sqm impregnated in epoxy resin. Similarly, GFRP consisted of bi-directionally woven glass fibre mat of weight 220g/sqm impregnated in the same resin material. The average thicknesses of the dry mat were 0.26mm and 0.18mm for carbon and glass respectively. The epoxy resin used for making both kinds of FRP was a two-part system of base and hardener(viz. CERA EPS) which was mixed in a proportion of 5:3 by weight. For preparing the FRP composites, the fibre and the resin were taken in a ratio of 1:1 by weight. The selected fibre mats and resin are shown in figure 1.

(a) Carbon fibre mat (b) Glass fibre mat (c) Epoxy resin

Figure 1. Constituents of composites selected for rehabilitation.

2.1.2. Test coupons. The tensile tests were conducted as per ASTM D3039/D3039M [6]. Test coupons were fabricated by following the wet lay-up procedure given in ASTM D7565/D7565M [7]. Coupons of width 25mm and length 280mm were cut from 300mm×300mm laminate sheets which were prepared as per the given procedure.Two flat rectangular aluminium tabs of thickness 1.5mm, width 30mm, and length 80mm were glued on the opposite faces of the test coupons at each end to prevent the gripping damage. This may ensure that the failure occurs within the gauge region. Typical test coupons are shown in figure 2.

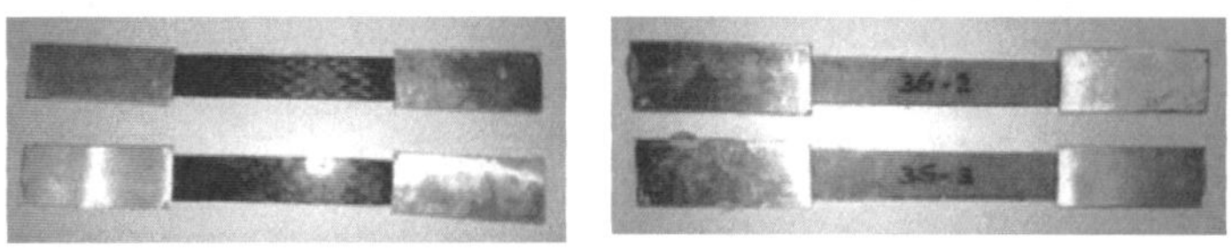

(a) CFRP (b) GFRP

Figure 2. Tension test coupons.

2.1.3. Test results. Coupons were prepared with different numbers of layers. CFRP coupons were prepared with one and two layers. GFRP coupons were prepared with two to five layers. Five samples were prepared with each layer group and tested to failure in a UTM of 200kN capacity. The rate of loading was set as constant with a standard head displacement of 2 mm/min, such that the failure occurs within 1 to 10 minutes as specified in ASTM D3039 [6]. The failure mode was noted and was found that

IOP Conf. Series: Materials Science and Engineering **936** (2020) 012032 doi:10.1088/1757-899X/936/1/012032

all the specimens failed in the recommended modes, which were either long splitting or lateral breaking of FRP laminate within the gauge region.The maximum tensile force per unit width for the specimens was calculated as per ASTM D7565 [7]. CFRP with 1 and 2 layers showed the tensile strength of 246.71N/mm and 541.24N/mm respectively. Similarly, GFRP coupons attained the tensile strengths of 118.43N/mm, 242.1N/mm, 312.46N/mm, 389.96N/mm respectively for 2 to 5 layers. It may be noted from these results that one layer of CFRP exhibits a tensile strength which is almost equal to the tensile strength given by three layers of GFRP. Hence CFRP with one layer and GFRP with 3 layers were selected as equivalent laminates for rehabilitating RCC columns for conducting the comparative study.

2.2. Preparation of column specimens
Eight circular columns of size 150mm diameter × 850mm length were cast using a design concrete mix of grade M35.The materials used for preparing the concrete mix were Portland Pozzolana cement, manufactured sand as fine aggregate, 12mm crushed stone as coarse aggregate and potable water. The mix obtained was 1:1.55:2.78 with a water-cement ratio of 0.5.Columns were axially reinforced with six numbers of 6mm diameter high yield strength deformed (HYSD) bars.Transverse reinforcement was of 6mm diameter HYSD bars at a centre to centre spacing of 85mm. The spacing of lateral ties was reduced to 50mm at both the ends of the column over a length of 100mm to avoid crushing at the ends due to stress concentration while loading. In order to ensure the occurrence of failure of columns within the mid-height region, internal confinement to the core concrete was provided using a 10mm square welded wire mesh, and external confinement to the cast column was provided using two layers of CFRP for a length of 50mm at both the ends.Reinforcement details of the column, an actual reinforcement cage with wire meshes at the ends, and an actual cast column with external CFRP confinement at the ends are shown in figure 3.

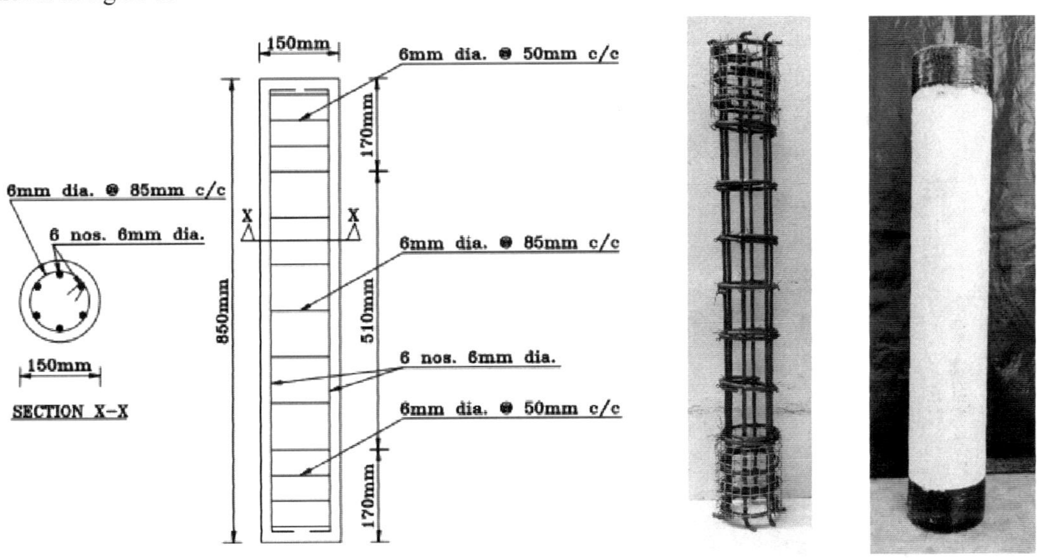

(a) Reinforcement details (b) Reinforcement cage (c) Actual specimen

Figure 3. Column specimen.

2.3. Preloading of columns
The Group-1 columns were meant for rehabilitation with CFRP and those in Group-2 with GFRP.The control specimens in both groups were tested to failure to find the ultimate load-carrying capacity.The remaining three columns in each group were tested up to 70%, 80% and 90% of the ultimate load for inducing pre-distress. The columns were designated with a letter and a number in which the letter indicates the composite (viz. C for CFRP and G for GFRP) and the number indicates the amount of pre-distress (viz. 100, 90, 80 and 70 respectively for the pre-distress levels of 100% (control specimen),

90%, 80%, and 70%). For example, the column specimen having a pre-distress of 80% of the ultimate load that rehabilitated with CFRP was designated as C80.

The specimens were preloaded under axial compression in a UTM with a capacity of 300T. The rate of loading was kept constant at 20kN/min. Two linear variable differential transformers (LVDT) with a least count of 0.01mm were placed longitudinally at mid-height to measure the axial deformation. The gauge length was kept as 100mm. Another two LVDTs were kept laterally at mid-height such that the tips touched two diametrically opposite points on the column surface, to measure the lateral deformation of the column. Two lateral extensometers were placed at a distance of 1/4thheight of the column from both the ends, to note the lateral deformations.The test set up is shown in figure 4.

2.4. Rehabilitation process

The Group-1 pre-loaded columns were rehabilitated with one layer of CFRP; and those in Group-2 with three layers of GFRP. The rehabilitation method adopted was full height wrapping of the column by using the equivalent laminate in order to ensure the complete confinement of the pre-distressed columns.Unidirectional CFRP was wrapped such that the fibre direction was along the circumference of the column. The surfaces of the pre-loaded columns were first applied with an epoxy primer (Cera EP18) to have a good bonding with the FRP laminate. The primer also was a two-part system of base and hardener that was mixed in a proportion of 2:1 by weight. The full height of the columns was applied with one coat of primer and cured at ambient temperature for 24 hours to get a smooth surface to bond with the composite. The application for FRP composite was made by wet lay-up procedure. The matrix which is the epoxy resin (EPS) was prepared by thoroughly mixing the base and hardener in the proportion 5:3 by weight. The FRP sheets were cut at the required sizes to cover the full height of the columns such that the CFRP was wrapped with one layer and three layers in the case of GFRP. The wrapping was done in such a way that there was no entrapped air and excess resin. The rehabilitated specimens were cured at ambient temperature for a period of 2 weeks to ensure the complete hardening of the laminate before final testing. The rehabilitated specimens are shown in figure 5.

Figure 4. Test set-up.

(a) Group-1 (CFRP) (b) Group-2 (GFRP)

Figure 5. Rehabilitated columns.

3. Test results and discussion

The rehabilitated specimens were tested under axial compression as explained in section 2.3. LVDT and extensometer readings were noted at 20kN intervals. The columns were tested until complete failure occurred, and the failure modes were noted. The post-peak behaviour was recorded until the load reduced below 50% of the ultimate load.

3.1. Ultimate load carrying capacity

The ultimate load of the rehabilitated columns is shown in table 1 along with the control columns (CC). It can be observed that the rehabilitation by both the FRPs increases the load-carrying capacity of the pre-distressed columns in which CFRP showed a better performance. Also, it may be observed that the load-carrying capacity increases with a decrease in the pre-distress level.

Table 1. The ultimate load of rehabilitated columns and control columns.

Specimen group no.	Beam designation	% Distress level	Ultimate load(kN)	% Increase in load carrying capacity	
				Absolute	Normalised
1 (CFRP rehabilitated)	C100 (CC)	100	602	-	1.00
	C90	90	960	59.47	1.59
	C80	80	1137	88.87	1.89
	C70	70	1185	96.84	1.97
2 (GFRP rehabilitated)	G100 (CC)	100	600	-	1.00
	G90	90	890	48.33	1.48
	G80	80	905	50.83	1.51
	G70	70	959	59.83	1.60

3.2. Failure modes

All the specimens failed due to FRP rupture and are shown in figure 6. The major rupture occurred around 1/3rd height from either the top or bottom ends. None of the specimens showed any cracks near the ends, which confirms the effectiveness of extra confinements provided at the ends both internally and externally. All the rehabilitated columns showed good ductility before the complete rupture of FRP laminates.

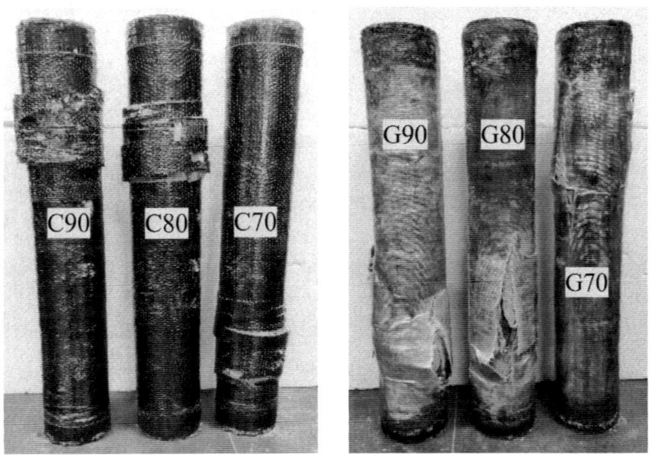

(a) Group-1 (CFRP) (b) Group-2 (GFRP)

Figure 6. Failure modes of rehabilitated columns with various pre-distress level.

3.3. Load-axial deformation behaviour

Compressive load versus axial deformation curves were plotted for both the groups of columns and are depicted in figure 7. It may be noted from the plots that the axial stiffness and the ductility are increased significantly due to rehabilitation. CFRP performed much better than GFRP in all levels of pre-distress. It is also noted that the increase in load-carrying capacity is more for lesser distressed columns, however these showed less ductility.

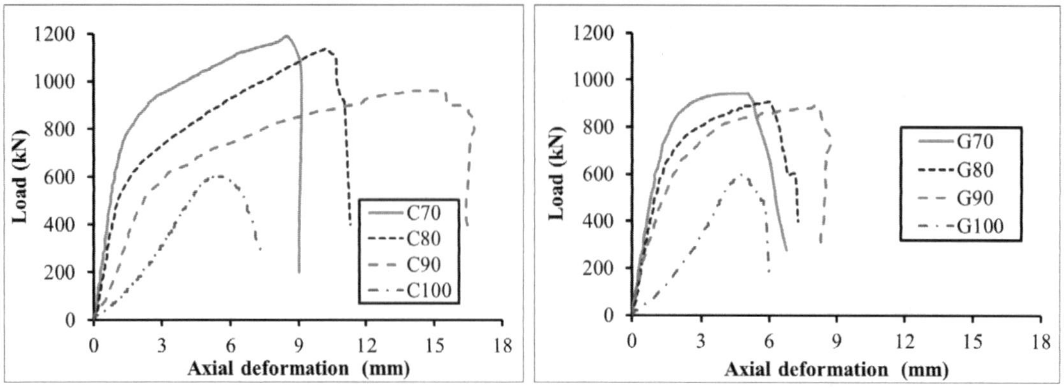

(a) Group-1 (CFRP) (b) Group-2 (GFRP)

Figure 7. Load-axial deformation behaviour of rehabilitated columns havingthe various pre-distress levels.

3.4. *Compressive stress versus axial strain behaviour*

The compressive stress-axial strain behaviour was compared between CFRP and GFRP rehabilitated columns at different levels of the pre-distress. The relationships are depicted in figure 8. It may be noted that there is a higher reduction in the stiffness of CFRP specimensas the pre distress increases than for the GFRP rehabilitated specimens. CFRP showed more ductility than GFRP at all levels of pre-distress.

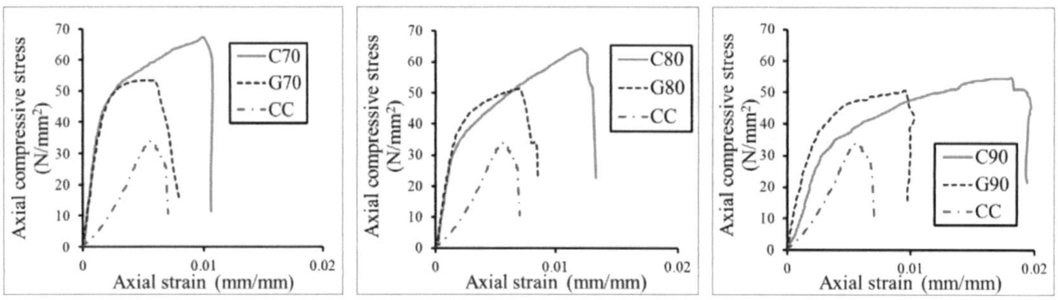

(a) 70% Pre-distress level (b) 80% Pre-distress level (c) 90% Pre-distress level

Figure 8. Compressive stress-axial strain behaviour of rehabilitated columns.

4. Conclusions

The following conclusions are arrived at based on the comparative study between CFRP and GFRP on their rehabilitation potential on RCC circular columns, pre-distressed to various levels of ultimate load.

- Both the FRP rehabilitation enhanced the load-carrying capacity significantly. The strength gain in CFRP rehabilitation was 59% to 97% over and above the control specimens, and that in GFRP, it was 48% to 60% with a decrease in pre-distress level from 90% to 70%.
- Axial deformation behaviour improved with a reduction in pre-distress level in both the FRPs, and CFRP showed a better bahviour than GFRP concerning stiffness and ductility.
- All the rehabilitated specimens showed a ductile failure, and at ultimate stage FRP rupture occurred away from the support region.
- CFRP outperformed GFRP in both the aspect of strength and serviceability in rehabilitating circular columns at all levels of pre-distress.

5. References

[1] Issa M A, Alrousan R Z and Issa M A 2009 Experimental and parametric study of circular short columns confined with CFRP composites *J. Compos.Constr.* **13(2)** 135-47

[2] Al Nimry H and Soman A 2018 On the slenderness and FRP confinement of eccentrically loaded circular RC columns *Eng. Struc.* **164** 92-108

[3] Widiarsa I B R and Hadi M N S 2013 Performance of CFRP wrapped squarereinforced columns subjected eccentric loading *Proc. Eng.* **54** 365-76
[4] Al-solloum Y A, Al-Amri G S, Siddiqui N A,Almusallam T H and Abbas H 2018 Effectiveness of CFRP strengthening in improving cyclic compression response of slender RC columns *J. Compos.Constr.* **22(3)** 04018009
[5] Benzaid R, Chikh N and Mesbah H 2009 Study of the compressive behavior of short concrete columns confined by fiber reinforced composite *Arab. J. Sci. Eng.* **34** 15-26
[6] ASTM D3039/D3039M2000 *Standard Test Method for Tensile Properties of Polymer Matrix Composite Materials* (ASTM International West Conshohocken USA)
[7] ASTM D7565/D7565M-10 2017*Standard Test Method for Determining Tensile Properties of Fiber Reinforced Polymer Matrix Composites used for Strengthening of Civil Structures* (ASTM International West Conshohocken USA)

Acknowledgments
The authors would like to acknowledge the National Institute of Technology Calicut, India for extending the facilities for the successful completion of this project.

Utilisation of Ultra-High Performance Fiber Reinforced Concrete(UHPFRC) for Retroffiting – a Review

Josniya Jose[1*], Praveen Nagarajan[2] and Mini Remanan[3]

[1] Research scholar, Department of Civil Engineering, NITC, Kerala, India

[2] Associate Professor, Department of Civil Engineering, NITC, Kerala, India

[3] Assistant Professor, Department of Civil Engineering, NITC, Kerala, India

*josniyaj@gmail.com

Abstract. Retrofitting has become a proven method for improving the load carrying capacity of existing concrete structural elements subjected to high mechanical loading, ageing or aggressive environments. Ultra-High Performance Fiber Reinforced Concrete (UHPFRC)is a relatively new generation of cement-based retrofitting material with a very high compressive strength and low permeability attributed to its tightly packed microstructure, also displaying excellent bonding ability with normal structural concrete. It has the potential to eliminate various problems caused by other retrofitting methods like FRPs or externally bonded steel plates as a result of the mismatch of stiffness with parent concrete. This study presents an overview of UHPFRC focusing on its properties, design, applications and challenges. The performance of UHPFRC concrete mix is greatly influenced by the type and degree of packing of its constituents in the cementitious matrix, necessitating the need for optimisation. Strength properties are very much higher when compared to that of conventional concrete alongside with increased durability aspects. The retrofitting technique showed an enhancement in the load carrying capacity of deteriorated structures. Therefore, UHPFRC can be considered as a practical solution to improve the sustainability of buildings and other infrastructural components. However, as a retrofit, there are some concerns regarding the loss of ductility. This broadens the need for methods to improve its ductility. This paper presents the state-of-art review of existing literature that deals with the retrofitting/rehabilitation of structures using UHPFRC. Altogether, the paper aims to expand the awareness regarding UHPFRC which was otherwise hindered, owing to limited design codes.

1. Introduction

A well-developed infrastructure reflects the growth and economic prosperity of a country. Most of the old reinforced concrete structures have undergone severe deteriorations over ages. This requisites the development of new, cost – effective strengthening and retrofitting strategies to extend the service life of these structures with a lower frequency of repair interventions. Growing demands in the construction industry necessitates materials of higher strength and durability properties. Some of the commonly available schemes of retrofitting include external prestressing, steel or concrete jacketing, fibre reinforced polymers (FRPs), near-surface mounting etc. Although these techniques are highly effective, certain drawbacks such as increase in dead load in case of concrete jacketing, corrosion of steel plates, debonding, long term durability, fire sensitivity and ageing of adhesion materials in case of FRPs are observed [1,2].

Over the past two decades an advanced cement-based material withimproved strength and durability characteristics, namely, Ultra-High Performance Fiber Reinforced Concrete (UHPFRC) is gaining interest worldwide. It defines a unique combination of low permeability, high strength and toughness.

2. UHPFRC

UHPFRC refers to a fiber reinforced superplasticised concrete, wherein coarse aggregates are replaced with very fine sand and highly active pozzolanic material. When compared to high performance concrete, ultra-high performance concrete has 2-3 times greater compressive strength (more than 150 MPa) and 2-6 times higher flexural strength [3]. It typically consists of cement, binders like GGBS, silica fume, quartz powder, fine sand, high range water reducing admixtures, fibers and low amount of water. UHPFRC combines the properties of self-compacting concrete, fiber reinforced concrete and high performance concrete as depicted in figure 1. Enhanced microstructure can be attained as a result of higher particle packing density, very low water content and proper mixing procedure [4]. This enables it to have extremely low permeability requiring less cover. Appropriate amount of well distributed fibers eliminates brittleness and enhances the mechanical properties.

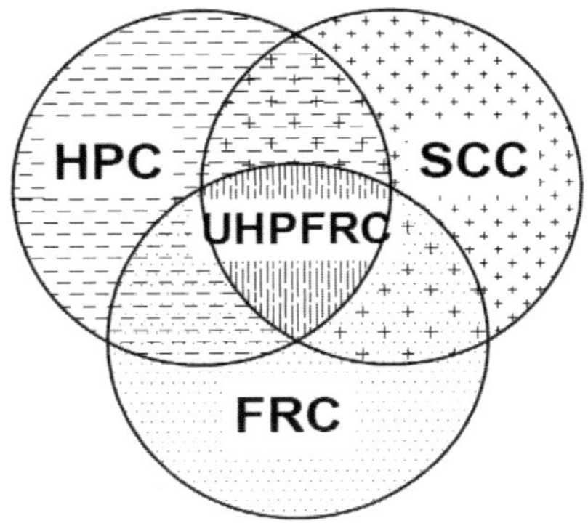

Figure 1. Different types of special concrete [5].

2.1. Development of UHPFRC

Immense research has been conducted over the past decades in the field of concrete technology to attain higher compressive strength.During 1980s, superplasticisers (SP) were introduced to reduce the water/binder (*w/b*) ratio up to 0.3. High dosage of silica fume and SP, further reduced the *w/b* ratio to 0.16. With optimal grain size distribution, compressive strength up to 280 MPa was achieved with reduction in porosity and enhancement of durability. These technological breakthroughs along with application of low porous materials led to the development of ultra-high performance materials. With the addition of steel fibers in the late 1980s, ductility of the matrix improved. Compact Reinforced Concrete (CRC) and Slurry Infiltrated Fiber Concrete (SIFCON) are two examples of this type.

In 1990s Richard et al. [6] developed Reactive Powder Concrete (RPC) with improved fineness and compressive strength in the range of 200-800 MPa. Ductal© was the first commercial UHPC developed through the RPC technology in the late 1990s. However, high production costs limited the use of this material [7]. Much progress was achieved in the 21st century with the introduction of UHPFRC [5,8]. To reduce the cement content, supplementary cementitious materials were used as substitutes. Furthermore, standard water curing in place of temperature curing was also reported [1,9-15]. The applications of ultra-high performance concrete are gaining ground as a result of emergence of cost-effective and environment-friendly UHPC.

2.2. Typical mix of UHPFRC

Optimum particle packing density and matrix homogeneity are the key factors determining the characteristics of UHPFRC. Hence, appropriate selection of raw materials plays an important role in mixture design of this cementitious composite in order to achieve good workability and strength. Basic principles identified by various researchers [16-19] in designing UHPC include (i) reducing *w/b* ratio and including a wide range of material classes reducing the matrix porosity, (ii) eliminating coarse aggregates thereby improving homogeneity, and (iii) incorporating steel fibers to improve ductility.

Various particle packing models have been reported for the mix design of UHPC. First one of these types was Linear Packing Density Model (LPDM) (1986) formulated by Larrard and Sedran [20]. But its linear nature failed to address the relations between packing density and material proportions. Further improvement led to Solid Suspension Model (SSM) (1994) and Compressible Packing Model (CPM)

(1999) based on virtual packing density and compaction index [21]. Aashay et al. [9] have successfully developed an economical UHPC based on CPM. Continuous models like Fuller Thompson model (1907) and Andreassen model (1930) assume presence of all possible particle sizes in the distribution system [21]. Yu et al. [15] used the modified Andreassen and Anderson model (1997) to develop UHPFRC with much lower binder content of 650 kg/m^3. This model recognised finite size of materials considering minimum particle size in distribution. Besides empirical methods, statistical methods based on neural networks were also adopted. Ghafari et al. [22] found the optimum contents of silica fume and cement to be 9% and 24% by volume of concrete respectively.

A typical UHPFRC mix contains Portland cement, fine sand, quartz powder, steel fibers, superplasticiser and water. The *w/b* ratio typically lies between 0.16-0.20 [23,24]. Supplementary cementitious materials and fillers like fly ash, GGBS, rice husk ash may be incorporated as a substitute to reduce large high cement contents. A GGBS replacement level up to 40% was found to be effective without sacrificing the strength properties [10].

Table 1. Typical mix components of UHPFRC [16].

Component	Weight range (kg/m^3)
Cement	610-1080
Fine sand	490-1390
Silica fume	50-334
Crushed quartz	0-410
Fibers	40-250
Superplasticiser	9-71
Water	126-261

2.3. Properties of UHPFRC

2.3.1. Workability. The relative slump flow ability linearly decreased with the addition of steel fibers as a result of high cohesive forces [15]. To have sufficient flow it is necessary to add sufficient quantity of high range water reducers into the mix, up to 5% mass of cement [23,24]. Improved workability can be achieved with the use of air-entraining agents as well as by appropriate incorporation of fillers to replace cement. Flow capacity is greatly influenced by the packing density of inclusions [25].

2.3.2. Compressive strength. Compressive strength increased with increase in fiber volume up to 3%, along with delayed crack propagation [13,14]. Main factors influencing compressive strength included the fiber content, curing conditions, loading rate and mix components. Elevated temperature cured specimens were found to have improved strength of 10-30% when compared to standard water cured specimens [1,11,26,27]. However, the lowered performance with standard water curing could be nearly compensated by increasing the fiber volume. Supplementary materials like silica fume and filler materials increased the strength as a result of pozzolanic action and intimate matrix packing [13,28-29].

2.3.3. Tensile and flexural strength. Tensile strength depends largely on the type, amount and orientation of fibers in the matrix. UHPFRC exhibits significantly higher strength compared to other cementitious materials as shown in figure 2. Studies have demonstrated a proportional increase in flexural and tensile strength in steel fiber ratios up to 3% in weight [2]. Tensile strength ranges from 7 to 15 MPa.

2.3.4. Durability. Low *w/b* ratio and enhanced microstructure have resulted in low porosity and permeability, resistance to medium erosion and better wear resistance. Micro-cracks may develop due to high rate of autogenousshrinkage[23]. However, UHPC has been found superior to conventional concrete in terms of wear resistance, carbonisation depth and chloride ion permeability [2,30,31]. Mixture design based on modified Andreassen model resulted in UHPCs with lower and comparable porosities [15].

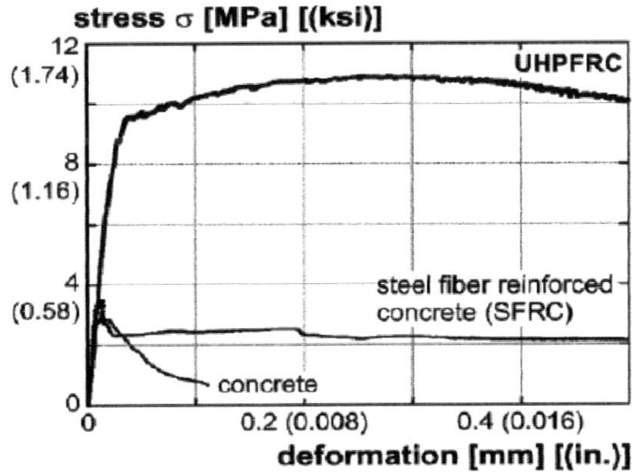

Figure 2. Comparison of uniaxial tensile behavior [32].

3. UHPFRC for retrofitting applications

The excellent performance of UHPFRC offers various applications in the field of building constructions, infrastructural strengthening and precast technology. Indeed, it is already being commercialised in many countries like Australia, Germany, France, New Zealand, etc. The first application of UHPC overlay was reported on a bridge over La Morge River, Switzerland [33]. One another example includes the retrofit of containment walls of a nuclear reactor in France [34].

3.1. Advantages over existing methods

One main concern for selection of a repair material is to have its mechanical properties close to that of its substrate. Most of the popular techniques of retrofitting using fiber-reinforced plastic laminates or externally bonded steel plates can result in brittle failure of the retrofitted structure due to the mismatch of tensile strength and stiffness [35,36]. Delamination failure of FRPs may limit the effective utilisation of the retrofitting material [37]. Higher tensile strength of these materials may possibly make the sections over-reinforced, thereby demanding alternate materials like UHPFRC of comparable strength parameters. Being a cement-based retrofitting material UHPFRC exhibits better compatibility and bonding ability with concrete. Prefabrication of strengthening strips can bring about standards in quality and the total cost might be much lower when compared to many other methods of strengthening [1]. Moreover, it can foster speedy construction with minimal interventions [38].

3.2. UHPFRC – NC bond strength

Studies reported a 40% increase in bond strength of UHPC overlay when compared to normal strength concrete. Wet surface displayed higher bonding than dry surface [39]. Specimens jacketed using sandblasting technique performed better under combined shear and compression [3,40]. Epoxy bonded specimens exhibited better performance in peeling tension attributed to its higher tensile strength when compared to Normal Concrete (NC) and UHPFRC. In essence, the UHPFRC-NC bond quality is excellent regardless of surface preparation [3,41].

3.3. Flexural strengthening

Flexural strength of UHPC could be 2-6 times greater than High Performance Concrete (HPC) [42]. Addition of UHPFRC layer to the tension side of beams showed improved stiffness, increased load carrying capacity and delayed flexure and shear crack propagation. The strengthened beams behaved monolithically [3,12,43,44]. However, beams strengthened on its three faces as in figure 3 displayed high capacity with significant loss in ductility when compared to the former configuration [11,43,45]. UHPFRC retrofitted beam-column joints also exhibited increased load carrying capacity, ductility and energy dissipation characteristics. Additional confinement lowered the rate of stiffness degradation which is significant in structures subjected to seismic loading [37,46].

Epoxy bonded prefabricated UHPFRC strips on the tension side increased the cracking load when compared to jacketed specimens due to the high tensile strength of epoxy [3,35]. The strips along with the parent beam behaved as an integral unit with no de-lamination[14,47]. Original load carrying capacity was regained after strengthening even in beams subjected to 90% preloading [47]. The bending performance of laminates was found to improve with the addition of reinforcing bars into the laminate increasing the ultimate load carrying capacity by 35% [14]. Shrinkage strain was found to reduce with the addition of steel fibers [11]. Different bonding techniques were employed for attaching the precast strips on to the concrete surface, viz., anchoraging and epoxy gluing. The glued specimens showed a 5% increase in ultimate load carrying capacity than anchoraged ones [14]. Provision of anchorages further reduced the effective area, thereby weakening the UHPFRC layer [1]. UHPC strengthening in slabs increased the overall stiffness and excellent energy absorption capacity resulting in a ductile flexural failure. However, not much improvement in ultimate strength was noticed [43].

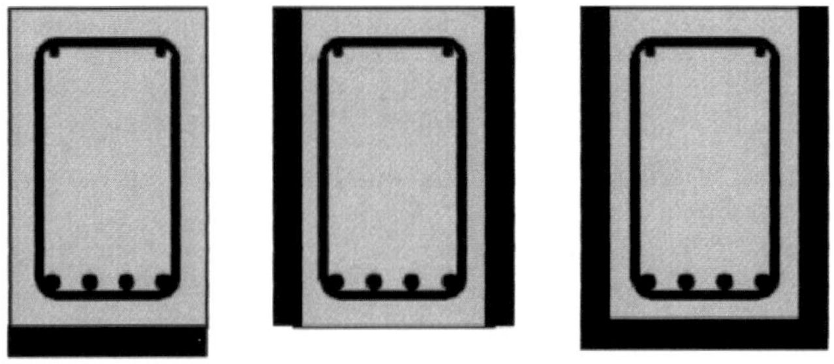

Figure 3. Different strengthening configurations in a beam [38].

3.4. Shear strengthening

Alaee et al. [45], conducted study using short strips adhered onto the tensile and vertical faces of beam specimens along with long continuous strips near to the support region. Flexural mode of failure was noticed with increased load carrying capacities. Punching shear capacity increased around 70% when UHPC was provided within the critical punching shear area of RC slabs [43,48]. Retrofitting of seismically deficit beam-column joints enhanced the shear capacity and stiffness characteristics. Prefabricated strips showed sufficient increase in strength compared to jacketing technique. However, the strength degradation using strips was observed to be 30% higher than jacketed members as a result of delamination [46]. Diagonal tension shear strength and ductilty under lateral loading in an RC shear wall were found to increase due to the confinement provided by UHPFRC jacket [49].

3.5. Torsional strengthening

Limited works have been carried out in the field of torsional strengthening using UHPFRC. Torsional capacity and cracking strength was found to improve in retrofitted beam members. A minimum increase of 65% in side wrapped specimens and a maximum of 260% in fully wrapped specimens were achieved proving the effectiveness of using UHPFRC in torsional strengthening [50].

3.6. Limitations and future recommendations

Preparation of UHPC with comparable performance under strict quality control and limited design guidelines restricts its widespread usage. Application method and type of retrofitting was found to influence the behaviour and ultimate strength of the parent member [14]. The peak load and ultimate displacement of epoxy bonded specimens were 15–20% lower when compared to jacketed specimens [3]. Embedding of reinforcing bars in the laminate finally resulted in sudden delamination of UHPFRC strips from the substrate surface [14]. Research can be extended to investigate retrofitting RC members using hybrid FRP wrapping along with strengthening strips. Although the stiffness is improved, 25–30% reduction in ductility was noticed in strengthened members when compared to unstrengthened members [35,47]. Enhancement in ductility in case of prefabricated strips by incorporating different strip configurations could be studied. Further, efforts to reduce carbon footprint by exploiting alternate raw material replacement combinations in UHPC mix could be carried out.

4. Conclusions

Matrix homogeneity and improved packing density make UHPC an excellent material with higher durability aspects and mechanical properties. Incorporation of suitable quantity of supplementary cementitious materials could sufficiently lower the high cement content in UHPFRC. Elevated temperature curing improves the mechanical properties when compared to standard water curing. Being a cement based retrofitting material, UHPFRC displays outstanding bond quality with parent concrete. As a retrofit, UHPFRC effectively complemented the flexural and shear strength of RC members. However, there are some issues regarding the loss of ductility and strength degradation. If the above limitations are rectified, UHPFRC proves to be a promising solution to significantly improve the durability and structural reliability of deteriorated concrete structures.

5. References

[1] Tanarslan H M, Alver N, Jahangiri R, Yalçınkaya Ç and Yazıcı H 2017 Flexural strengthening of RC beams using UHPFRC laminates: Bonding techniques and rebar addition *Constr. and Building Mat.* **155** 45–55

[2] Zhou Mi, Lu Wei, Song Jianwei and Lee George C 2018 Application of Ultra-High Performance Concrete in bridge engineering *Constr. and Building Mat.* **186** 1256–67

[3] Al-Osta M A, Isa M N, Baluch M H and M K Rahman 2017 Flexural behavior of reinforced concrete beams strengthened with ultra-high performance fiber reinforced concrete *Constr. and Building Mat.* **134** 279–96

[4] Spyridon A, Paschalis, Andreas P, Lampropoulos and Ourania Tsioulou 2018 Experimental and numerical study of the performance of ultra high performance fiber reinforced concrete for the flexural strengthening of full scale reinforced concrete members *Constr. and Building Mat.* **186** 351–66

[5] Azmee N M and Shafiq N 2018 Case study Ultra-high performance concrete: From fundamental to applications *Case Studies in Constr. Mat.* **9** e00197

[6] Richard P and Cheyrezy M 1995 Composition of reactive powder concretes *Cem. Concr. Res.* **25** 1501–11.

[7] Acker P and Behloul M 2004 Ductal© technology: a large spectrum of properties, a wide range of applications *Proc. of the Int. Symp. on UHPC Kassel* (Germany) pp 11–23

[8] Hussein Luaay and Amleh Lamya 2015 Structural behavior of ultra-high performance fiber reinforced concrete-normal strength concrete or high strength concrete composite members *Constr. and Building Mat.* **93** 1105–16 http://dx.doi.org/10.1016/j.conbuildmat.2015.05.0300950-0618

[9] Arora Aashay, Almujaddidi Asim, Kianmofrad Farrokh, Mobasher Barzin and Neithalath Narayanan 2019 Material design of economical ultra-high performance concrete (UHPC) and evaluation of their properties *Cem. and Conc. Compos.* **104** 103346

[10] Ganesh P and Murthy Ramachandra 2019 Tensile behaviour and durability aspects of sustainable ultra-high performance concrete incorporated with GGBS as cementitious material *Const. and Building Mat.* **197** 667–80

[11] Lampropoulos A P, Paschalis S A, Tsioulou O T and Dritsos S E 2016 Strengthening of reinforced concrete beams using ultra highperformance fibre reinforced concrete (UHPFRC) *Eng. Struct.* **106** 370–84

[12] Paschalis S A, Lampropoulos A P and Tsioulou Ourania 2018 Experimental and numerical study of the performance of ultra high performance fiber reinforced concrete for the flexural strengthening of full scale reinforced concrete members *Constr. and Building Mat.* **186** 351–66

[13] Raheem Ahmed H A, Mahdy M and Mashaly Asmaa A 2019 Mechanical and fracture mechanics properties of ultra-high-performance concrete *Constr. And Building Mat.* **213** 561–6

[14] Tanarslan H M 2017 Flexural strengthening of RC beams with prefabricated ultra high performance fibre reinforced concrete laminates *Eng. Struct.* **151** 337–48

[15] Yu R, Spiesz P and Brouwers H J H 2014 Mix design and properties assessment of ultra-high performance fibre reinforced concrete (UHPFRC) *Cem. and Conc. Research* **56** 29–39

[16] Bassam A T, Abu Bakar B H, Megat Johari M A and Yen Lei Voo 2013 utilisation of ultra-high performance fibre concrete (UHPFC) for rehabilitation: a review *Procedia Engineering* **54** 525–38

[17] Hassan, Jones S W and Mahmud G H 2012 Experimental test methods to determine the uniaxial tensile and compressive behaviour of ultra high performance fibre reinforced concrete (UHPFRC) *Constr. and Building Mat.***37** 874–82
[18] Ma J and Schneider H 2002 Properties of ultra-high-performance concrete *Leipzig Annual Civil Engineering Report (LACER)* **7** pp 25–32
[19] Naaman A E and Wille K 2012 The path to ultra-high performance fiber reinforced concrete (UHP-FRC):Five decades of progress *Proc. of Hipermat 3rd Int. Symp. on UHPC and Nanotechnology for High Performance Constr. Mat.* pp 3–16
[20] Larrad F and Sedran T 1994 Optimisation of ultra-high-performance concrete by the use of a packing model *Cem and Conc. Res.* **24** 997-1009
[21] Kumar Senthil V and Santhanam Manu 2003 Particle packing theories and their application in concrete mixture proportioning: A review *The Indian Con. J.* 1324–31
[22] EhsanGhafari, Hugo Costa and Eduardo Júlio 2015 Statistical mixture design approach for eco-efficient UHPC *Cem. and Conc. Compos.* **55** 17–25
[23] Eide Mari B and Hisdal J M 2012 Ultra high performance fibre reinforced concrete (UHPFRC) – State of the art *COIN Project report* p 44
[24] Shaikh Faiz U A, Luhar Salmabanu, Arel Hasan S, Luha Ismail 2020 Performance evaluation of Ultra high performance fibre reinforced concrete – A review *Constr. and Building Mat.* **232** 117152–79
[25] Amanjean E Nguyen, Mouret M and Vidal T 2019 Effect of design parameters on the properties of ultra-high performance fibre-reinforced concrete in the fresh state *Constr. and Building Mat.***224** 1007–17
[26] Songlin Yang and Bo Diao 2009 influence of curing regime on the ductility of ultra-high performance fibre reinforced concrete (UHPFRC) *ICCTP 2009: Critical Issues in Transportation Systems Planning, Development, and Management* pp 1–7
[27] Zong-cai D, Xiao Rui and Shen Chen-liang 2013 A review on preparation and properties of ultra-high performance concrete *Mater.* **9** 66–9
[28] Abdulkareem O M, Ben Fraj Amor, Bouasker Marwen and Khelidj Abdelhafid 2018 Mixture design and early age investigations of more sustainable UHPC *Constr. and Building Mat.* **163** 235–46
[29] Khan M I, Abbas Y M and Fares G 2017 Review of high and ultrahigh performance cementitious composites incorporating various combinations of fibers and ultrafines *J. of King Saud Univ.-Eng. Sci.* **29(4)** 339–47 (doi:http://dx.doi.org/10.1016/j.jksues.2017.03.006)
[30] Liu S G, Xiang Z, Huang R H, Wang D H and Ju Y Z 2019 The durability of reactive powder concrete: a review *The 7th Global Conf. on Materials Science and Engineering* p 11
[31] Pimienta P 2005 Durability of UHPFRC specimens kept in various aggressive environments *10DBMC Int. Conf. On Durability of Building Materials and Components* (Lyon) pp 17–20
[32] Habel K 2004 Structural behaviour of composite UHPFRC-concrete elements Doctoral Thesis, Swiss Federal Institute of Technology, Lausanne, Switzerland (doi:10.5075/epfl-thesis-3036)
[33] SAMARIS Management Group 2015 Full Scale Application of UHPFRC for the Rehabilitation of Bridges–From the Lab to the Field Document number: SAM_GE_DE22v03_01
[34] Corvez D and Masson B 2013 UHPFRC solutions for the retrofit of nuclear reactor containment walls *Proc. of Int. Symp. on Ultra High Performance Fiber-Reinforced Concrete* pp 147–156
[35] Aldahdooh M A A, Bunnori N M, Johari M A, Jamrah A and Alnuaimi Ali 2016 Retrofitting of damaged reinforced concrete beams with a new green cementitious composites material *Compos. Struct.* doi:http://dx.doi.org/10.1016/j.compstruct.2016.01.067
[36] Murthy Ramachandra, Rani P V and Priya S D 2018 Fatigue behaviour of damaged RC beams Strengthened with Ultra high performance fibre Reinforced Concrete *Int. J. Of Fatigue* doi:https://doi.org/10.1016/j.ijfatigue.2018.06.046
[37] Sharma Raju and Bansal Prem P 2019 Behavior of RC exterior beam column joint retrofitted using UHP-HFRC *Constr. and Building Mat.* **195** 376–89
[38] Zhu Yanping T, Zhang Yang, Hussein Husam H and Chen Genda 2020 Flexural strengthening of reinforced concrete beams or slabs using ultra-high performance concrete (UHPC): A state of the art review *Eng. Struc.* **205** 110035

[39] Farzad M T, Shafieifar Mohamadreza and Azizinamini Atorod 2019 Experimental and numerical study on bond strength between conventional concrete and ultra high - performance concrete (UHPC) *Eng. Struct.* **186** 297–305

[40] Jafarinejad Soheil, Rabiee Amirmahdi and Shekarchi Mohammad 2019 Experimental investigation on the bond strength between ultra high strength fiber reinforced cementitious mortar & conventional concrete *Constr. and Building Mat.* **229** 116814

[41] Valikhani Alireza, Jahromi Azadeh J, Mantawy Islam M and Azizinamini Atorod 2020 Experimental evaluation of concrete-to-UHPC bond strength with correlation to surface roughness for repair application *Constr. and Building Mat.* **238** 117753

[42] Lubbers A 2003 Bond Performance of Ultra-High-Performance Concrete and Prestressing Strands *MS Thesis Ohio University*

[43] Hor Yin, Teo Weeb and Kazutaka Shirai 2017 Experimental investigation on the behaviour of reinforced concrete slabsstrengthened with ultra-high performance concrete *Constr. and Building Mat.* **155** 463–74

[44] Safdar M, Matsumoto T and Kakuma K 2016 Flexural behavior of reinforced concrete beams repaired with ultra-high performance fiber reinforced concrete (UHPFRC) *Compos. Struct.* doi:http:// dx.doi.org/10.1016/j.compstruct.2016.09.010

[45] Alaee Farshid J and Bhushan Lal Karihaloo 2003 Retrofitting of reinforced concrete beams with CARDIFRC *J. of compos. for constr.* **7** 174–186

[46] Khan M I, Al-Osta M A, Ahmad S and Rahman M K 2018 Seismic Behavior of Beam-Column Joints Strengthened with Ultra-High Performance Fiber Reinforced Concrete *Compos. Struc.* (doi:https://doi.org/10.1016/j.compstruct.2018.05.080)

[47] Murthy Ramachandra, Karihaloo B L and Priya S D 2018 Flexural behavior of RC beams retrofitted with ultra-high strength concrete *Constr. and Building Mat.* **175** 815–824

[48] Zohrevand Pedram, Yang Xiong, Jiao Xin and Mirmiran Amir 2015 Punching Shear Enhancement of Flat Slabs with Partial Use of Ultra high-Performance Concrete *J. of Mat. in Civil Eng.* **27**04014255:1–10

[49] Sakr Mohammed A, El-khoriby Saher R, Khalifa Tarek M and Nagi Mohammed T 2019 Modeling of RC shear walls strengthened with ultra-high performance fibre reinforced concrete (UHPFRC) jackets *Eng. Struc.* **200** 0141–296

[50] Mohammed Thaer J, Bakar Abu B H and Bunnori Muhamad N 2016 Torsional improvement of reinforced concrete beams using ultra high-performance fiber reinforced concrete (UHPFC) jackets – experimental study *Constr. and Building Mat.* **106** 533–42

A Review on Fractal Analysis and its Applications in Structural Engineering

Ayisha Anwar [1*] and S Adarsh[2]

[1]Post Graduate Student, Department of Civil Engineering, TKM College of Engineering, Kollam, India

[2]Associate Professor, Department of Civil Engineering, TKM College of Engineering, Kollam, India

[*]ayisha.civil17@gmail.com

Abstract. Fractals refer to rough or fragmented geometric shape in which each sub-section is a reduced-size copy of the whole. Its potential to describe the complex and irregular phenomena leads to wider applicability, and its enormous unexplored potential in the field of structural engineering is to be attended by the research community. This paper first presents a brief theoretical background of the most popular algorithms, such as the box-counting method, Multifractal Detrended Fluctuation Analysis (MFDFA), and Multifractal Detrended Cross-Correlation Analysis (MFDCCA). This paper further presents a comprehensive review of the potential of fractals and multifractals in four specific domains, such as crack identification, pore structure analysis and chloride resistance, and structural damage detection. The review of literature further reveals that 2 dimensional and multi-dimensional extension of fractal theory can be effectively coupled with image analysis for micro-level examination of pore structures, which opens the scope for abundant applications in research in the field of structural materials. The applications of MFDFA and cross-correlation analysis structural health monitoring are very recent research contributions in the field, and they are in the infant stage of applications. The review of research works performed in this study shows that the complete potential of fractals in the structural engineering field is not yet explored by the researchers.

1. Introduction

A fractal is a rough or fragmented geometric shape that can be split into parts, each of which is a reduced-size copy of the whole [1]. Fractals are distinguished over traditional Euclidean geometry as they define irregular and complex phenomena and paved a new research path on fractal applications [2]. Most of the complex time series exhibits fractal behaviour including medical and physiological time series [3], geophysics [2], financial time series [4], and many more. If the scaling behaviour is described by a single exponent, it is referred to as monofractals, and if scaling behaviour is described by multiple scaling exponents in different parts of the series, it is referred to as multifractals. Fractal objects are commonly characterised by fractal dimension and express the degree of complexity involved. The various approaches for computing fractal dimensions include the box-counting method, power spectrum method, and fracture island method [3]. Among the various methods for estimation of fractal dimension estimation, the box-counting method is the most popular one [5]. This method outperforms the other methods due to its simplicity in the calculation. The scaling behaviour of many time series records are complicated and do not exhibit monofractal behaviour. Thus, different scale exponents are required to explain different parts of the series. Hence, the multifractal analysis should be opted in such situations. Multifractal detrended fluctuation analysis (MFDFA) proposed by Kantelhardt et al. [6] is the most popular algorithm to describe it. Even though fractality and multifractality were analysed successfully in many scientific disciplines, the applications of the fractal theory are not much explored in the field of structural engineering except few studies in the domains of pore structure analysis, crack identification and structural health monitoring.

The main objective of this paper is to give an insight into the present researches and applications of fractal and multifractal analysis in structural engineering, thereby, provide assistance to future researches in this domain. First, the algorithms of most commonly used methods of fractal/multifractal analysis are presented briefly, then the detailed review on applications of fractal/multifractal analysis in structural engineering is presented by categorising into four parts. The conclusions drawn from the review are presented in the final section.

2. Algorithms of popular methods for fractal and multifractal analysis

2.1. Box-counting method

In this method, a grid of squared boxes is superimposed on the object of interest, and the number of boxes that cover the object is counted. The procedure is repeated for different grid sizes. The fractal dimension is obtained from the following relation:

$$N(r) \propto \left(\frac{1}{r}\right)^D \tag{1}$$

where D represents the fractal dimension. The slope of the log-log plot $\left(\frac{1}{r}\right)$ against log $N(r)$ gives the fractal dimension D. The concept of the box-counting process for the determination of crack pattern is illustrated in figure 1.

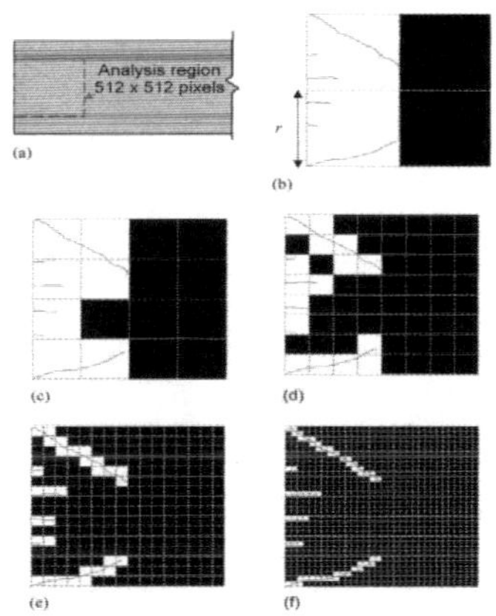

Figure 1. Box counting method applied for the analysis of crack pattern [20]

2.2. Multifractal detrended fluctuation analysis

The multifractal detrended fluctuation analysis (MF-DFA) proposed by Kantelhardt et al. [6] is one of the most popular tools for the detection of the scaling behaviours and multifractal properties of non-linear and nonstationary time series. The procedure for MF-DFA is described below:

1. Consider a time series X, having length N. The time series is converted into a random walk by subtracting the mean value of the time series.

$$Y(i) = \sum_{k=1}^{i} \left[x_k - \bar{x}\right] \tag{2}$$

   where i =1, 2,...., N
2. The profile $Y(i)$ is divided into N_s non-overlapping segments, where $N_S = \text{int}(N/s)$ and s stands for scale. N may not always be a multiple of s, so the procedure is done for a total of 2 Ns segments.
3. In order to find the local trend in each segment, least-square fitting is done for each segment for variance

$$\upsilon = 1, 2, \ldots, N_s \tag{3}$$

$$F^2(s,v) = \frac{1}{s}\sum_{i=1}^{s}\{Y[N-(v-N_s)s+i]-y_v(i)\}^2 \tag{4}$$

for $\upsilon = N_s+1, \ldots, 2N_s$

Here $y_\upsilon(i)$ is the fitting polynomial in segment υ.

4. Fluctuation function is found out by averaging over all segments

$$F_q(s) = \left\{\frac{1}{2N_s}\sum_{v=1}^{2N_s}\left[F^2(s,v)\right]^{q/2}\right\}^{1/q} \tag{5}$$

Fluctuation function depends on scale s and q.

5. The scaling behaviour of the fluctuation functions is obtained from the slope of log-log plots of fluctuation functions and scale s and is referred to as Hurst exponent $h(q)$.

Hurst exponent is an indication of long-range memory of time series. When $h(q)$=0.5, the series is uncorrelated. $h(q)$<0.5 indicates the series is negatively correlated and $h(q)$>0.5, the series is positively correlated. The scale to be chosen is of critical importance while computing this method. The minimum scale should at least greater than (k+2), where k is the order of the polynomial chosen, and the maximum scale (s_{max}) should be chosen such that a sufficient number of segments is produced (s_{max}<=L/4, where L is the data length). Some guidelines, along with the selection of MFDFA control parameters, are available at [7].

2.3. Multifractal detrended cross-correlation analysis (MF-DCCA)

In order to quantify the presence of correlation between different signals, various statistical measures were developed, out of which the most popular and simplest one is the Pearson correlation coefficient. For the real field data (which often possess non-linear and non-stationarity characteristics) spurious correlation may be reported as they are always characterised by trend. Because of this, an estimation measure supported by the detrending operation is advisable. Podobnik and Stanley [9] proposed the detrended cross-correlation analysis (DCCA), to investigate power-law cross-correlations between two candidate non-stationarity time series. Multifractal Detrended Cross-Correlation Analysis (MF-DCCA) can be used to investigate the long-term cross-correlations between two nonstationary time series.

The procedure for MF-DCCA is described below:

1. For two time series x_i and y_i having equal length N., The time series are converted to random walk by subtracting their respective mean values.

$$x(k) = \sum_{i=1}^{k}[x_i - \bar{x}] \text{ and } y(k) = \sum_{i=1}^{k}[y_i - \bar{y}] \tag{6}$$

where, $i= 1,2,3\ldots\ldots, N$

2. Both the profiles are divided into N_s non-overlapping segments, where N_S =int(N/s)and s stands for scale. N may not always be a multiple of s, so the procedure is done for a total of 2 Ns segments.
3. In order to find the local trend for both series in each segment, least-square fitting is done for each segment and determine the covariance:

$$f_{DCCA}{}^2(s,i) = \left\{\frac{1}{s}\sum_{k=i}^{i+n}\left[(x_k-\bar{x})(y_k-\bar{y})\right]\right\} \tag{7}$$

Repeat the procedure for various scales s and q values.

4. Detrended covariance is obtained by summing over all overlapping all segments of length N

$$F_{xy}(s) = \left\{\frac{1}{2N_s}\sum_{i=1}^{2N_s}\left[f_{DCCA}{}^2(s,i)\right]\right\}^{1/q} \tag{8}$$

The scaling behaviour of the detrended covariance is obtained from the slope of log-log plots of $F_{xy}(s)$ and scale s and is referred to as cross-correlation exponent $\lambda(q)$.

3. Applications of fractal and multifractal analysis in structural engineering

In this section, the review of applications of fractal/multifractal theory in four major fields namely, pore structure analysis, crack detection, damage identification, and chloride analysis, are presented.

3.1. Pore Structure Analysis

The Fractal behaviour is well exhibited by the pore structures of many natural and artificial materials. Three types of fractality are observed in pore structures, namely, pore surface fractal, solid surface fractal [10], and pore surface fractal [11]. Surface fractal characteristics were used to find out ionic diffusivity [10]. Gao et al. [12] proposed a generalised fractal approach that can be used to examine all the three types of fractality present in pore structures. Intermingled Fractal Units (IFU) model has been adopted to analyse the microstructure by many researchers as they could reproduce fractal as well as non-fractal behaviour of pore structures. The virtual representation of pore-size distribution in the IFU model increases the efficiency in predicting sorption characteristics [11].

Bednarska et al. [13] analysed the effect of freeze-thaw cycles on pore structure based on fractal dimension. The experiment was conducted on a spliced cement mortar specimen inside PVC tubes. The surface of the specimen is polished and then painted with a black stain. The image of the painted surface was analysed, and the fractal dimension was calculated by the box-counting method. As the number of freeze-thaw cycles increased, there was a rise in the fractal dimension. The permeability at each stage was executed from corresponding to the fractal dimension value. On determining the surface fractality of pore structure in high strength ground granulated blast-furnace slag (GGBS), increased fractal dimension showed an increase in pore volume [14]. Another study conducted by Yang et al. [15] on GGBS cement pastes proved that the fractal dimension had been improved on blending the cement paste with GGBS. By evaluating the fractal dimension of individual pores, the accuracy of pore structure analysis can be increased [16].

3.2. Crack Identification

Fracture energy is a representation of crack growth. Won et al. [17] related fractal dimension with fracture energy, and they exhibited a proportional relationship. The size and location of cracks in beams were identified using fractal dimension crack detector (FDCD) and proved its potential in practical applications [18]. The beneficial performance of FDCD was extended from 1-D to 2-D structures [19]. The main attraction of the method is that it gives insight into all crack characteristics and requires only a small portion of the signal for analysis. The use of FDCD can also be extended to the crack detection of composite plates. Multifractal analysis of 2D images taken in the visible spectrum provided a quantitative measure of damage in reinforced concrete shear walls (RCSW), by retrieving surface defect patterns [20]. Flowchart showing the procedure of surface crack detection using multifractal analysis is provided in figure 2.

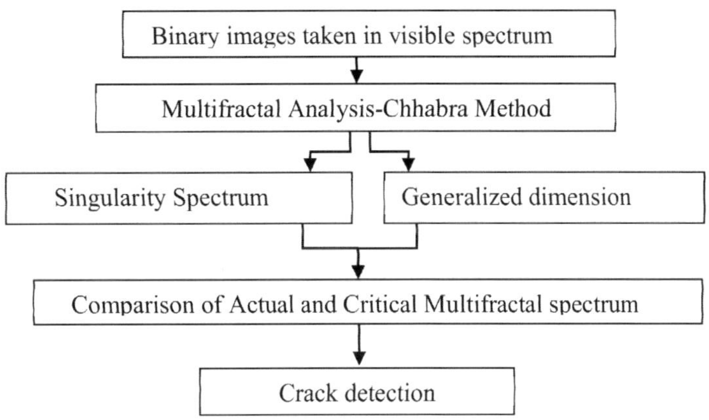

Figure 2. The procedure of surface crack detection using multifractal analysis [20].

For this experiment, four synthetic crack patterns were considered and referred to as CASE0, CASE1, CASE2, and CASE3, as shown in figure 3. CASE0 is the reference pattern. Singularity spectrum curves were plotted for the determination of crack, as shown in figure 4. CASE1 curves almost coincided with a positive range of q with CASE0 curves since the overall shape was maintained. For CASE2, the overall shape was altered by flipping the crack patterns. Hence, they show deviation from the base curve for $q<0$. A clear distinction of deviation was exhibited by CASE3 curves on both sides due to crack pattern growth.

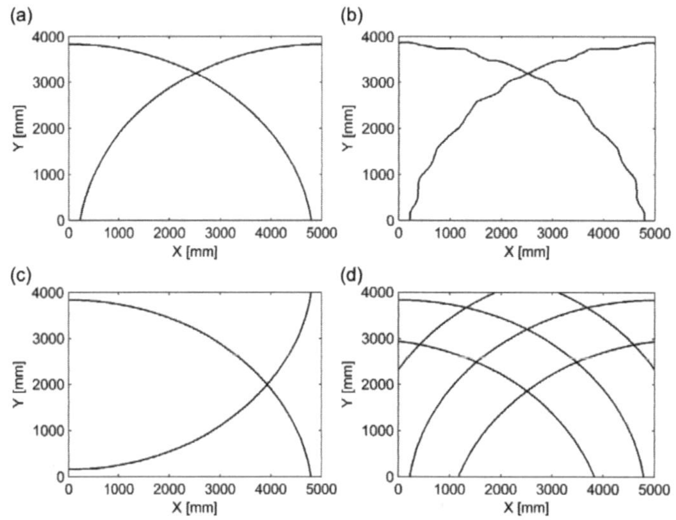

Figure 3. Synthetic crack patterns (a) CASE0 (b) CASE1 (c) CASE2 (d) CASE3.

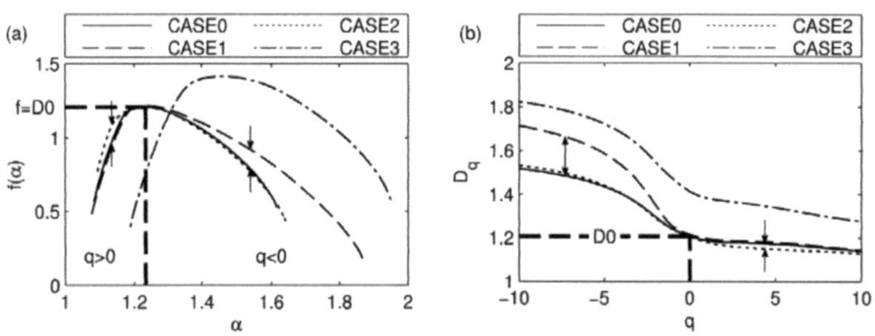

Figure 4. Multifractal description of images of crack patterns (a) singularity spectrum; (b)generalised fractal dimension plot [20].

Furthermore, the fractal and multifractal analysis were employed to detect surface cracking patterns of prestressed concrete girders [5]. Developing an automatic alarming system for crack identification makes the solution using fractal analysis more realistic.

3.3. Damage Detection

Fractal dimension based damage identification is effective even for strong noise-contaminated signals [21]. Moustafa and Salamone [22] proposed a Relative Fractal dimension (RFD) approach to determine the fractality of Lamb waves. They validated the same by recognising surface damage in isotropic structures. The dimension change in displacement data can be used to monitor the state of buildings [23]. Lin et al. [24] proposed a structural health monitoring system based on the multifractal analysis. The flowchart showing the procedure of Structural health monitoring using multifractal analysis is given in figure 5. A four-storey steel structure was simulated in SAP2000 software. A mass of 120 kg was assigned to each storey. The damage was simulated by the removal of bracings in respective floors. A 1MW noise signal was used to excite the four-storey steel structure. The time history response from each storey was recorded and then, analysed for damage condition and damage location using DFA and DCCA, respectively. Compared to reference curve, figure 6(a), figure 6(b) showed the third and fourth curves have a large area in between from their previous curves. This indicates the damage in the third and fourth floors.

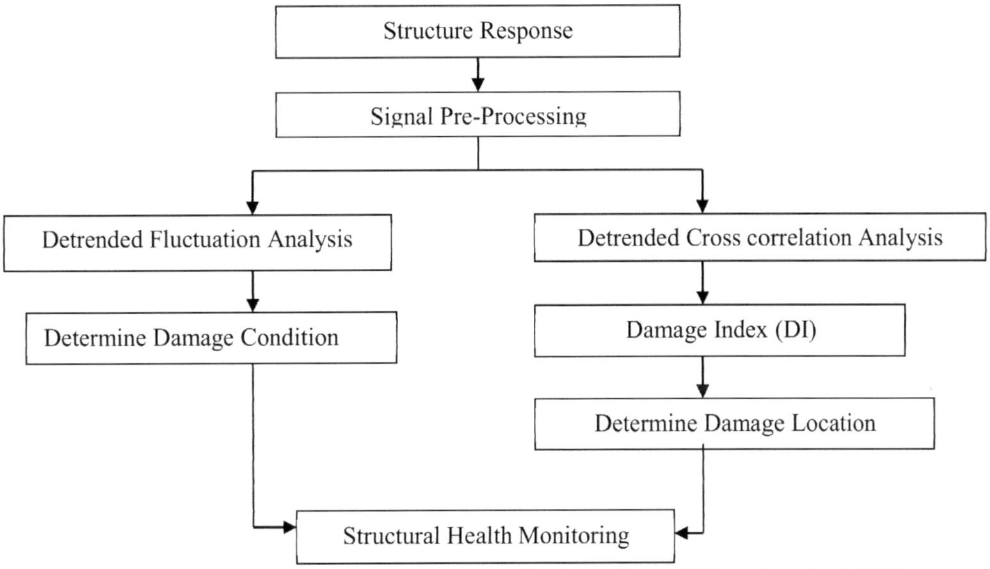

Figure 5. Procedure for Structural health monitoring using multifractal analysis [24].

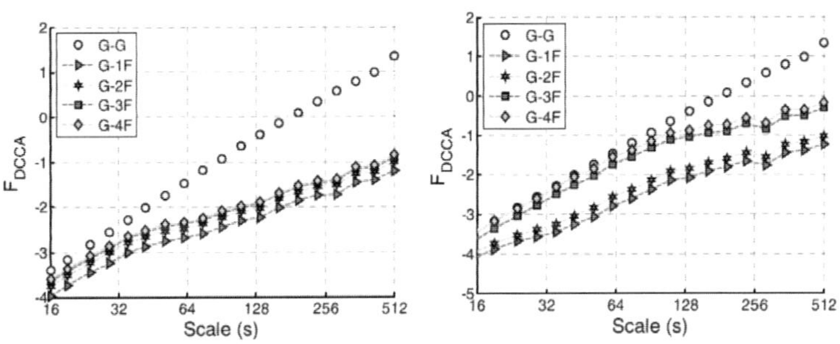

Figure 6. Fluctuation function plots obtained by DCCA between ground displacement and floor displacement of example structure (a) healthy condition; (b) damage in the third and fourth floor [24].

Fractal analysis integrated with Bayesian learning was proposed by Huang et al. [25] for damage detection. They proved the real-life potential of the proposed method by successfully detecting the damage in the Yonghe bridge. There is ample scope for utilising the potential of advanced variants of cross-correlation studies such as Multifractal Cross-Correlation analysis (MFCCA) [4] and conjunctive use of statistical evidence theory and multifractal analysis [26] for damage identification of structural systems.

As there exist enormous opportunities for multifractal analysis in developing a well defined structural monitoring, our experiments were focussed in this area. A seven-storey structure was developed in SAP2000 finite element software. The damage on each floor was simulated by the removal of bracings in each floor. The structure was excited using a 1 MW noise signal, and the corresponding displacement response of each floor was recorded. A total of 16 damage cases were considered. DFA was employed to assess the damage condition. From figure 7, it is clear that the Hurst exponent value increases as the severity of damage increases. The Hurst value due to damage in the first floor is higher compared to the damage caused by the second, third, fourth, fifth, sixth, and seventh floors. This indicates the greater potential of the first floor in contributing to global damage.

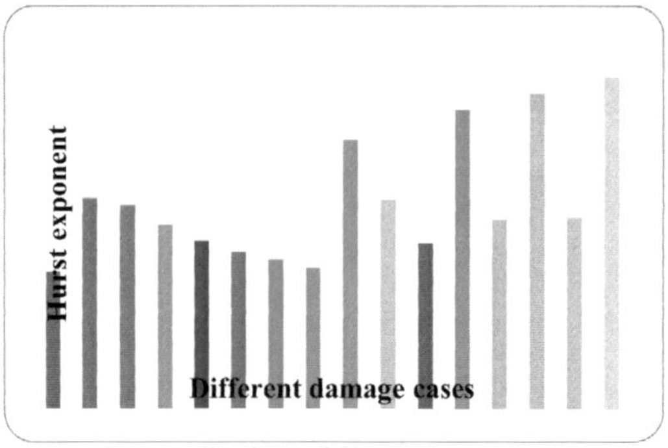

Figure 7. Distribution of Hurst exponent value for different damage cases on the first floor.

3.4. Chloride Resistance

The fractal dimension theory was utilised by Xue et al. [27] to establish the impact of fine aggregate on chloride resistance of concrete and proved its proportional relationship. The equations relating the void volume, chloride resistance, and volume fractal dimension in the case of mortar and concrete had been established by using fractal theory [28].

4. Conclusions

This study performed a review of research works on applications of fractal and multifractal analysis in the field of structural engineering. Even though the characterisation of signals from seismic ground motion and dynamical systems using fractals were explored by the researchers, the complete potential of the fractal and multifractal theory in this field is not fully explored yet. The detection of fractal properties like fractal dimension has enormous potential in micro-level examinations like crack detection, chloride resistance, and pore-structure analysis. The multifractal analysis can be effectively used for signal characterisation, and the Detrended Cross-Correlation Analysis (DCCA) and its multifractal extension enhances its applicability in structural health monitoring. The 2D variant of MFDFA, coupled with image analysis, is believed to be a potential tool in micro-level investigations in structural engineering.

5. References

[1] B B Mandelbrot 1983The fractal geometry of nature (New York: WH freeman) **173** 51

[2] Kantelhardt J W 2009 Fractal and multifractal time series *Encyclopedia of Complexity and Systems Science* 3754–79

[3] Lopes R and Betrouni N 2009 Fractal and multifractal analysis: A review *Medical Image Analysis* **13** 634–49

[4] Watorek M, Drozdz S, Oswiecimka P and Stanuszek M 2018 Multifractal cross-correlations between the world oil and other financial markets in 2012- 2017 *Energy Economics* **81** 874–85

[5] Ebrahimkhanlou A, Athanasiou A, Hrynyk T D, Bayrak O and Salamone S 2019 Fractal and multifractal analysis of crack patterns in prestressed concrete girders *J. Bridge Eng.* **24** 04019059 1–13 https://doi.org/10.1061/(ASCE)BE.1943-5592.0001427

[6] Kantelhardt J W, Zschiegner S A, Koscielny-Bunde E, Havlin S, Bunde A and Stanley H E 2002 Multifractal detrended fluctuation analysis of nonstationary time series *Physica A: Statistical Mechanics and its Applications* **316** 87–114

[7] Ihlen E A F E 2012 Introduction to multifractal detrended fluctuation analysis in MATLAB *Frontiers in physiology* **3** 141

[8] Piao L and Fu 2016 Quantifying distinct associations on different temporal scales: comparison of DCCA and Pearson methods *Scientific Reports* **6** 36759

[9] Podobnik B and Stanley H E 2008 Detrended cross-correlation analysis: a new method for analysing two nonstationary time series *Physical review letters* **100** 084102

[10] Gao Y, Jiang J and Wu K 2016 Fractal modeling of pore structure and ionic diffusivity for cement

paste *Advances in Material Science and Engg.* **124** 237–43 http://dx.doi.org/10.1155/2016/7645954
[11] Pia G, Siligardi C, Casnedi L and Sanna U 2016 Pore size distribution and porosity influence on Sorptivity of ceramic tiles: From experimental data to fractal modelling *Ceramics International* **42** 9583–90
[12] Gao Y, Wu K and Jiang J 2016 Examination and modeling of fractality for pore-solid structure in cement paste: Starting from the mercury intrusion porosimetry test *Construction and Building Materials* **124** 237–43
[13] Bednarska D, Wieczorek A and Koniorczyk M 2018 *AIP Conf. Proc.* 1988 020005 https://doi.org/10.1063/1.5047599
[14] Kim J, Choi Y C and Choi S 2017 Fractal Characteristics of Pore Structures in GGBFS-based Cement Pastes *Applied Surface Science* **428** 304–14
[15] Yang J, Haung J, He X, Su Y, Tan H, Chen W, Wang X and Strnadel B 2019 Segmented fractal pore structure covering nano- and micro-ranges in cementing composites produced with GGBS *Construction and Building Materials* **225** 1170–82
[16] Gao Y, Jing H and Zhou Z 2019 Fractal analysis of pore structures in graphene oxide-carbon nanotube based cementitious pastes under different ultrasonication *Nanotechnol Rev* **8** 107–115
[17] Won J P and Kim S A 2002 Crack growth behaviours of cement composites by fractal analysis *KCI Concrete Journal* **14** 30–5
[18] Hadjileontiadis L J, Douka E and Trochidis A 2005 Fractal dimension analysis for crack identification in beam structures *Mechanical Systems and Signal Processing* **19** 659–74
[19] Hadjileontiadis L J, and Douka E 2007 *Engg.* Crack detection in plates using fractal dimension *Engineering Structures* **29** 1612–25
[20] Ebrahimkhanlou A, Farhidzadeh A and Salamone S 2016 Multifractal analysis of crack patterns in reinforced concrete shear walls *Structural Health Monitoring* **15** 81–92 https://doi.org/10.1177/1475921715624502 1-12.
[21] Li H and Huang Y 2011 Fractal Dimension-Based Damage Detection Method for Beams with a Uniform Cross-Section *Computer-Aided Civil and Infrastructure Engineering* **26** 190–206
[22] Moustafa A and Salamone S 2012 Fractal dimension–based Lamb wave tomography algorithm for damage detection in plate-like structures *J. of Intelligent Material Systems and Structures* **23** 1269–76
[23] Yang C, Zhao X, Yao Y and Zhang Z 2016 Application of fractal theory in brick-concrete structural health monitoring *Engineering* **8** 646–56
[24] Lin T K and Fajri H 2017 Damage detection of structures with detrended fluctuation and detrended cross-correlation analyses *Smart Materials and Structures* **26** 19
[25] Huang Y, Li H, Wu S and Yang Y 2018 Fractal dimension based damage identification incorporating multi-task sparse Bayesian learning *Smart Materials and Structures* **27** 18
[26] Zhu Y, Wang K, Lin Y, Yin H, Hou D, Yu J, Huang P and Zhang G 2020 An online contaminant classification method based on mf-dcca using conventional water quality indicators *Processes* **8** 178
[27] Xue W, Zhang L, Chen J and Jin W 2011 Relationship between fractal dimension of fine aggregates and chloride resistance of concrete *International Conf. on Durability of Building Materials and Components* pp 1–8
[28] Xue W and Chen J 2016 Effect of Fractal Dimension of Fine Aggregates on the Concrete Chloride Resistance *5th International Conf. on Durability of Concrete Structures* (Shenzhen University, Shenzhen, Guangdong Province, P.R.China) pp 89-93

IOP Conf. Series: Materials Science and Engineering **936** (2020) 012035 doi:10.1088/1757-899X/936/1/012035

Production and Utilisation of Artificial Coarse Aggregate in Concrete - a Review

George K George[1*] and P Revathi[2]

[1] Ph.D. Research Scholar, Department of Civil Engineering, Pondicherry Engineering College, Pillaichavady, Puducherry, India

[2] Assistant Professor, Department of Civil Engineering, Pondicherry Engineering College, Pillaichavady, Puducherry, India

*gkgpec@gmail.com

Abstract. This present work provides a study of several successful approaches in producing artificial aggregates using a pelletiser machine and the parameters affecting efficiency. These aggregates were contrived from substances of nature such as fly-ash, GGBS, Rice husk ash, volcanic ash, etc. Different sets of artificial aggregates are to be made with different molarity. The common methods of developing them include cold bonding, autoclaving, and sintering. The aggregate pellets (with lightweight and possessing less specific gravity and high impact value than gravel) were batch manufactured by a cold bonded technique using disc pelletiser (used to perform agglomeration). The prepared aggregates are characterised based on their crushing and impact strength rate and also moisture absorption. This paper explicitly reviews various effective and eco-friendly artificial aggregate and their potential usage in the construction industry.

1. Introduction

In day-to-day life, enormously booming construction field demands mass production of Portland cement and aggregates. The tremendous growth in the infrastructural sector around the world and the demand for a huge quantity of cement and aggregates in construction have occurred. Aggregates are obtained from natural resources such as granite and other rocks, especially the ones which are fine grained find their source enormously from natural river sand. Due to the enormous demand for these aggregates, natural resources are depleting faster leading to demand for alternate mode of aggregates. The scarcity of river sand has increased the burden on coarse aggregates. So crushed granite is replaced by river sand in almost all construction activities. Hence there is no scarcity of fine aggregate due to the availability of M sand. It is produced from hard granite stone by crushing. The enormous production of M-sand leads to the depletion of natural resources. This leads to serious issues of the unavailability of coarse aggregates also. So in-order to avoid the depletion of natural aggregates, coarse aggregates are replaced by recycled aggregates or artificial aggregates. Recycled aggregates are nothing but released aggregates from different concrete waste products in construction. The major limitation with respect to the utilisation of recycled aggregates is the quality of recycled aggregates, such as high water absorption and lower strength. So there is a major necessity of utilisation of artificial coarse aggregates for construction purposes. They were obtained from the engineered co-products such as ashes of fly, husk or volcanic type and GGBS. In India, in most of the industries (like the thermal power plant, iron industry) coal is burnt in large numbers. Thus in the above process, a significant amount of fly types of ashes were produced in our country, but they turned as a menace to our society owing to their health hazard. This burning issue was overcome by stacking them on low-lands, producing bricks, and proportioning concrete mix [1]. The fly-ash replaced coarse aggregate in construction with a mix of cement as a binder gave good cementation quality and also provided an efficient solution by depressing the environment concern. For purpose of sustainable city with green purpose, fly-ash aggregates are also critical as it is manufactured by recycling of fly-ash from industries. Like fly-ash, several raw materials can be used for the manufacture of artificial aggregates. Identical to waste materials like fly-ash, several others can also be used effectively for the production of aggregates. Their utilisation as by-products of artificial aggregates production also found to address the disposal issue in these industrial waste along with concern for healthy environment [2]. This paper reviews the techniques, materials, and methodologies for producing and utilising artificial aggregates in the construction industry.

2. Artificial aggregates

2.1. Materials

The fractional binding alternative by FA for the creation of artificial aggregate improves their cementitious properties. These artificially fabricated ones decoded the major environmental issue of pollution and natural resource depletion. The parent materials of concrete production plays the main governing factor in the behaviour of their properties [3]. By the process of geopolymerisation, full replacement of cement by other materials is possible. Geopolymers, with an amorphous microstructure and chemical composition of zeolites, is a co-polymerisation product of high pH and single alumino and silicate species (composing silicon and aluminium in dissolved form) with soluble alkali metal silicates. The unused alumino-silicate materials in surroundings were utilised as building and mining materials, as they possessed satisfactory chemical and physical properties [4]. By pelletisation technique, artificial geopolymer aggregates can be made from volcano ash, which reacts with the geopolymer binder and alkaline activator (sodium silicate(Na_2SiO_3) and sodium hydroxide (NaOH) solutions). These aggregates sized with 14-20 mm are sintered at temperature 500 °C, 600 °C and 800 °C for 1 hour [5]. Various waste materials like FA, Volcanic ash, etc. can be used for the production of artificial aggregates. Upon close examination of these materials and their properties, minimal change in their nature and strength was noticed. The physical properties of different materials for the production of aggregates are shown in table 1.

Table 1. Physical properties of raw materials [6].

Raw materials	Physical properties		
	Specific gravity	Mean particle size	Surface area (m^2/kg)
Cement	3.15	–	346
FA	2.20	23.4	311
GGBS	2.86	9.2	435
RHA	2.06	25.3	–

By and large, their creation is observed primarily to occur when the ground discarded materials as new pellets are blended in their predetermined dimension. The diverse varieties of raw materials, binders and additives used in the manufacture of artificial aggregates with reference to the literatures are particulared below in table 2.

Table 2. The different types of raw materials, binders and additives used for the manufacture of artificial aggregates.

Sl No	Raw material	Binder	Mol-ality	Silica Modulus /Alkali ratio	Inferences	Further Research	Reference
1	Class F fly ash, GGBFS, RHA	Na_2SiO_3, NaOH	10 M	1.5	Various factors affecting the production efficiency of the cold bonding method are binder	Properties of cold-bonded aggregates can be improved by treating their surface and reduce water absorption.	[7]

					content, pellatisation, curing etc.		
2	FA, GGBS and RHA	Na_2SiO_3, NaOH	10 M	1.5	Improved crushing strength of Light Weight Aggregate (LWA).	Study can be extended with various raw materials.	[6]
3	Volcanic Ash	sodium hydroxide and sodium silicate	14 M	2.5	Geopolymer volcano ash aggregate has specific gravity of 1.1 to 1.8 and water absorption of 12 to 16%.	Find the best mix design, curing temperature and fire resistance with arrived product thereby expanding the knowledge on their performance and properties.	[8]
4	Fly Ash	NaOH and Na_2SiO_3	10-14 M	2.5	12M of NaOH concentration shows good result.	Efficiency of aggregate in different temperature.	[9]

2.2. Methods of manufacturing

Production of different artificial aggregates from various sources involves various methods. The primary focus is integration of the reduced residue to any definite predetermined size for their targeted production. After the process of agglomeration through pelletisation, the new post-processed pellets achieved forte as aggregate replacement in concrete. Sintering, autoclaving, and cold bonding are the commonly available techniques used for the post-processing of fly-ash aggregates.

2.2.1. Sintering. This technique involves the fusion of green pellets (GP) at a temperature of more than 1200°C [10]. For that, a large quantity of coal requirement helped this course in producing the aggregates. But the huge energy requirement for making aggregates makes the process undesirable. Better durability properties like corrosion resistance and permeability exhibited by them, showed this process advantageous. In this process, the fused FA particles at the contact points form good strength pellets. For this process, spot welding with very high temperature is preferred additionally with maintained minimal temperature for overcoming enormous fusion. Through the sintering surface, the 200-300 mm thick GP is conveyed on the endless belt. The high volume of coal content, causing higher sintering temperature, leads to enormous energy usage and money for the preparation of aggregates [11].

2.2.2. Autoclaving. This method mainly consists of chemical accumulation like cement, lime, or gypsum in the agglomeration phase. The formed material binding property produced good aggregate strength. At 1400°C temperature, the new pellets are healed in coerced drenched vapour. This method gave fast aggregate production with a minimal binding material and curing time[12].

2.2.3. Cold bonding. In this method, normal curing water at room temperature was used. At ordinary room temperatures, this process leads to FA reaction with calcium hydroxide resulting in a moisture-proof binder. By using compaction agglomeration techniques, drying shrinkage and the creep of the bonding material can be overwhelmed. Agglomeration stands for the conversion of the complete consolidation of solid particles into larger shapes. This can be achieved by agitation granulation by drum granulation, disc granulation, etc. [11]. This method helps to prevent energy utilisation compared to other methods of manufacturing. The increment of curing time provided the comparable aggregate properties as on autoclaving and steam curing and gave the maximum strength to the aggregates. [13].

2.3. Properties of artificial aggregates

The improved moisture absorption and lesser density are the fundamental assets of the artificial fly ash aggregates. Their mechanical properties rest on the binder kind and measure, curing temperature, and its duration. In contrast, the moisture content influences the size, shape, and texture of the aggregates and angle of pelletisation used. Mechanical properties like crushing strength and specific gravity are the significant properties of artificial aggregates. The average value of crushing strength is generally expressed in terms of their 'aggregate crushing value'. Mainly artificial aggregates were synthesised using the pelletisation technique of various fraction of waste material and cement. Numerous mix proportions with replacement of 20, 40, 60, 80, and 100 % of natural replacing artificial aggregate in concrete were found out with their characteristics [1]. Mechanical Properties of artificial geopolymer aggregates are mainly specific gravity, water absorption, etc. The Specific Gravity of artificial geopolymer aggregate produced by the sintering process is less compared to natural aggregates. For the production of lightweight concrete, these types of aggregates are beneficial. The water absorption for artificial geopolymer aggregate with volcano ash is more compared to natural aggregates. The water absorption can be controlled by adjusting the expansion sintering temperature [5]. Pelletisation based on aggregate size obtained from the various works of literature is listed in Table 3. The aggregate strength based on various experimental methods is tabulated below in table 4.

Table 3. Pelletisation based on aggregate size.

Sl. No.	Size of Aggregate (mm)	Pelletiser dimensions		Mechanical parameters of pelletiser		Pelletisation duration (min)	Reference
		Diameter (mm)	Depth (mm)	Angle (°)	Speed (rpm)		
1	16-20	800	300	53	35	-	[6]
2	4–8	400	100	35 & 40	20,30,40	15	[7]
3	10	400	150	43	45	20	[14]
4	2-8	1000	150	45	15	15	[15]
5	10- 12	500	270	36	55	15	[16]
6	10- 20	560	250	55	50	7- 14	[17]
7	4- 14	800	350	30 -92	54	20	[18]

Table 4. The strength of the aggregate based on experimental approaches.

Sl. No.	Raw materials	Crushing Strength of LWA (MPa)	Water Absorption (%)	Specific gravity (g/cm^3)	Reference
1	FA + GGBS + RHA	8.1–8.8	9.8–10.1%	-	[6]
2	Fly Ash	22.7	13.23	2.12	[19]
3	FA	5.7	10.6%	-	[6]
4	Volcanic Ash	5.1—8.6	12-16%	1.1 to 1.8	[8]
5	FA + GGBS	15.5–15.7	7.8–8.3%	-	[6]
6	FA + RHA	6.0–8.1	10.8–20.5%	-	[6]

2.4. Properties of concrete produced from artificial aggregates

Artificial aggregates comprise of the waste products like fly-ash, GGBS whose properties depend mainly on the native material, cementing material, the structure to develop them, and other factors, which primarily affect the strength and durability of concrete.

2.4.1. Properties of fresh concrete. The fresh concrete is controlled by the texture and shape of aggregate, notably the smooth-edged ones stimulate the workability, and the pointed nature helped with better workability. The cost reduction in concrete by well-graded aggregate consumed lesser and denser cement paste, by reducing the void ratio which further lead to the decreased amount of paste for void filling. This reduction in void reduced the workability of the concrete as the latter depends on the volume fraction of cold bonded aggregate, which can be overcomed by proper gradation of aggregate[12].

2.4.2. Hardened properties of concrete. In the case of hardened properties, such as compressive strength, flexural strength, tensile strength, etc.; they change based on the aggregate properties. The strength also varies with the volume fraction of cold bonded aggregates. The novel geopolymer artificial aggregate concrete showed an average compressive strength of 37 MPa, which is the vital element of the final mix strength[20]. The high strength artificial aggregates (fly ash) concrete produced high strength than the conventional aggregate concrete [2].

2.4.3. Durability of concrete produced from artificial aggregates. Concrete durability depends upon the degree of exposure, the concrete grade (or strength), and the cement content. Since concrete is a porous material, reinforcement bars within the concrete will be subject to possible corrosion. So permeability is an important parameter in the durability of concrete. The permeability of artificial aggregate concrete was lower than the conventional concrete, whereas when cold-bonded fly ash was used for lightweight concrete, the permeability value increases [21]. Their properties such as the resistance to acid attack, corrosion, etc., are primary factors affecting the durability of the concrete. Also, the durability of concrete is determined by the pore structure. The value of durability is measured by the water penetration depth. Its value was found to be less for concrete. The water depth should be measured after the water penetration test on the hardened concrete cubes [15]. Water and chloride ion permeability of fly ash aggregate lightweight concrete were slightly lower than normal-weight concrete. This transportation by the water within the unsaturated concrete specimens can be determined from the sorptivity. The Nano-silica particles usage in concrete reduces the water sorptivity and improve the durability of concrete [22].

3. Conclusions

The main observations which can be drawn from the study are as follows:

- The outbreak of construction activities and the fast depletion of natural resources demands the use of artificial aggregate.
- The specific gravity of artificial aggregate is 30 to 50 % less than natural aggregates favouring the lightweight concrete structures.
- Water absorption of artificial aggregates is 2 to 5 times higher when compared with normal aggregate.
- The compressive strength of the mix with artificial aggregates was found to be almost 5- 15 % less than conventional concrete.
- The durability varies according to different methods in manufacturing of artificial aggregates and mix design of concrete and it is less than conventional concrete.

4. References

[1] Manju K, B Dhanush Kumar, S Suresh Kumar and S Nirmal Kumar 2018 Behaviour of concrete by using artificial aggregates a review *Engineering and Technolog* **7** 255–8

[2] Harilal B and Thomas J 2013 Concrete made using cold bonded artificial aggregate *American Journal of Engineering Research* **1** 20–5

[3] Santhi A S, P Priyadharshini and G M Ganesh 2012 Effect of cold bonded Fly ash aggregates on strength and restrined shrinkage properties of concrete *Science and Management* **3** 160–4

[4] Van Deventer and Hau Xu 2000 The Geopolymerisation of alumino- silicate minarals *Miner. Process* **59** 247–66

[5] Rafiza A R, A M Mustafa Al Bakri, H Kamarudin, I KhairulNizar, D Hardjito and Y Zarina 2013 Evaluation of pelletized artificial geopolymer aggregate manufactured from volcano ash *Basic and Applied Sciences* **7** 15–20

[6] Bui L A, Hwang C, Chen C, Lin K and Hsieh M 2012 Manufacture and performance of cold bonded lightweight aggregate using alkaline activators for high performance concrete *Construction and Building Materials* 35 1056–62

[7] Tajra F, M AbdElrahman and Stephan D 2019 The production and properties of cold-bonded aggregate and its applications in concrete: A review *Construction and Building Materials* **225** 29–43

[8] Razak R A, Abdullah A M, Hussain K and Ismail K N 2014 Assessment on the potential of volcano ash as artificial lightweight aggregates using geopolymerisation method *Material science and engineering* 7 65–73

[9] Abdullah A, Razak R A, Hussain K and Ismail K N 2014 Compressive strength and morphology of fly ash based geopolymer as artificial aggregate with different curing temperature *Engineering Materials* **3** 151–5

[10] Terzic A, Pezo L, Mitic V and Radojevic Z 2015 Artificial fly ash based aggregates properties influence on lightweight concrete performances *Ceramics International* 41 2714–26
[11] Bijen J M 1986 Manufacturing processes of artificial lightweight aggregates from fly ash *The International Journal of Cement Composites and Lightweight Concrete* **8(3)** 191–9
[12] Manikandan R and Ramamurthy K 2008 Effect of curing method on characteristics of cold bonded fly ash aggregates *Cement and Concrete Composites* **30(9)** 848–53
[13] Geetha S and Ramamurthy K 2010 Environmental friendly technology of cold-bonded bottom ash aggregate manufacture through chemical activation *Journal of Clean Production* **18(15)** 1563–9
[14] Kockal N U and Ozturan T 2011 Characteristics of lightweight fly ash aggregates produced with different binders and heat treatments *Cement and Concrete Composites* **33** 61–7
[15] Tang P, Florea M V A and Brouwers H J H 2017 Employing cold bonded pelletisation to produce lightweight aggregates from incineration fine bottom ash *J. Cleaner Prod.* **165** 1371–84
[16] Gomathi P and A Sivakumar 2015 Accelerated curing effects on the mechanical performance of cold bonded and sintered fly ash aggregate concrete *Construction and Building Materials* **77** 276–87
[17] Geetha S and K Ramamurthy 2015 Reuse potential of low-calcium bottom ash as aggregate through pelletisation *Waste Manage.* 30 1528–35
[18] Gesoğlu M, E Güneyisi, H Ö Öz 2012 Properties of lightweight aggregates produced with cold-bonding pelletisation of fly ash and ground granulated blast furnace slag *Mater. Struct.* **45** 1535–46
[19] Priyadharshini P, Ganesh G M and Santhi A S 2011 Experimental study on cold bonded fly ash aggregates *International Journal of Civil and Structural Enggineering* **2** 493–501
[20] Gunasekara C, Sujeeva S, David W L, Nick W and Trevor B 2018 Engineering properties of Geopolmer aggregate concrete *Materials in Civil Engineering* **30** 11–22
[21] Kockal N U and Ozturan T 2010 Effects of lightweight fly ash aggregate properties on the behavior of lightweight concretes *Journal of Hazardous Materials* **179** 954–65
[22] Their J M and Ozakca M 2018 Devoloping geopolymer concrete by using cold- bonded fly ash aggregate, nano silica and steel fiber *Construction and Building Materials* **180** 12–22

Design of Flexural Members Reinforced with GFRP Bars

Abinash Kumar Sethi[1*], Trupti A Kinjawadekar[2], Praveen Nagarajan[1] and A P Shashikala[1]

[1] Department of Civil Engineering NIT, Calicut, India

[2] Manipal School of Architecture and Planning, MAHE, India

[*]abhinas.sethi@gmail.com

Abstract. Reinforced concrete has emerged as a primary construction material since the nineteenth century. A passive combination of concrete and steel reinforcing bar (rebar), this composite material is widely used in various types of structures. However, corrosion of steel reinforcement has been identified as a significant problem affecting the structural integrity of such concrete structures. In this context, the use of GFRP bars is gaining prominence due to their non-corrosive nature. This paper presents behaviour of flexural members reinforced with GFRP bars. The strength reduction factors and guidelines available with different codes for flexural members are studied. The design of flexural members reinforced with GFRP bars has been done using Indian Standard code parameters. The flexural capacity of the beams is estimated from the strain compatibility and equilibrium of forces. The theoretical and experimental results are compared. It is seen that designing the beams as an over-reinforced section will increase the stiffness of the beams.

1. Introduction

In reinforced cement (RC) concrete design of structures quality and strength of structures are central points under thought. The interior steel gets corroded over time resulting in reduced durability of structures. Ports, coastal structures and various industry offices are structures, which are subjected to the aggressive environment causing corrosion problem in the steel reinforcement [1]. Research has been done to recommend a substitute in the form of a glass fiber reinforced polymer (GFRP) bar for regular steel rebar. It is a composite bar made up of glass fiber and epoxy resin.[1]

Figure 1. GFRP bars [2].

1.1. Advantages and disadvantages of GFRP bar

Favourable aspects of utilising GFRP bar is that it has less weight compare to steel rebar, it has more strength and is non-corrosive in nature. It can be viewed as the best alternative to use GFRP bars in construction particularly in the marine environment [2]. GFRP is a ribbed composite bar in which glass strands are impregnated and bound by amasingly strong polymeric epoxy resin. GFRP rebars have different specific properties, for instance, high quality to weight proportion, magnificent behaviour under fatigue, non-conductivity, and high rigidity, additionally the thermal expansion value of the GFRP is close to that of cement. When it is used as rebar in RCC bending elements, the modulus of elasticity, bond stress and behaviour under tension is the main mechanical properties to be considered in studying the structural behaviour of components [1]. The behaviour of GFRP bars under tension is given by a linear stress-strain relationship to failure. The sections reinforced with GFRP bars are generally over-reinforced sections, and they exhibit brittle failure without notification.

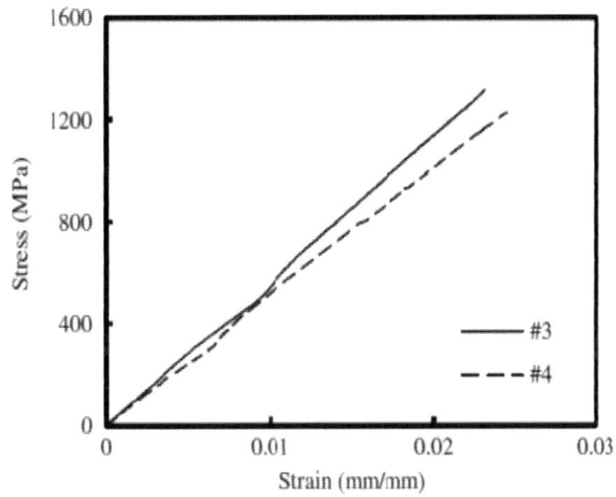

Figure 2. Stress- Strain behaviour: GFRP bars [2].

2. Literature survey

Harris et al. (1998) [3] experimented and studied concrete structures using hybrid ductile reinforced bars, and discovered that the unique technique of reinforcement yields bilinear stress-strain characteristics so as to implement in the distressed and new concrete structures. In addition hybrid reinforced bars possess comparatively superior properties like lower weight, corrosion resistance and excellent strength against the traditional reinforcement. Also, the tensile experimentation yields linear stress-strain characteristics. After experimentation, it is observed that the beam exhibits the capacity to withstand larger deformation. The ductility index is assessed considering the energy, curvatures, and deflections, considering the FRP reinforced beam (three in numbers) is establish to behave similar to that of a steel rebar [3]. Joseph R. Yost et al. (2001) presented studies over shear strength [4]in the beam reinforced using GFRP bars and also observed that the longitudinal reinforcement has not much effect over shear capacity in the beams reinforced using GFRP bars with numerous ratios ranging from 2.10 - 4.32 ρ_b. So, those standard pragmatic equations can be used for the approximation of shear strength. The behavior under shear failure in the beams reinforced with the GFRP is found comparable to that of concrete beams reinforced using the steel. However, shear strength is found to be of lower value in beams reinforced using GFRP when compared to beams reinforced using steel. The shear strength equation is developed considering the data obtained from experimentation over steel-reinforced member which is found to be unconservative in the beams reinforced using GFRP[5]. H.Y. Leung et al. (2003) experimented on seven rectangular RC flexural members strengthened with steel and GFRP rebars (hybrid reinforcement).The length of the member was 2.5m, depth 200 mm width 150 mm. Different types of rebars were used wherein they were utilised in two separate layers along the tension side of the specimen. Reinforcing bars with a high yield strength and a diameter of 10 mm which displayed the value of elastic modulus and yield strength as 200 GPa and 460MPa, respectively were used. Also, the two designed concrete mixes provided strengths of 30MPa and 50MPa each. Four point bending experimentation was conducted over beams as depicted in figure 1 [6].

M.W. Goldston et al. (2017) experimented and studied Flexural behaviour in the concrete beam with GFRP reinforced ultra-high and high strength concrete [7]. Observations showed that using Ultra High Strength Concrete with 117 MPa is more advisable for enhancing the load carrying capacity in over-reinforced beams (GFRP-RC) comparing with the High Strength Concrete with 95 MPa [7]. Also as the strength of concrete is enhanced to 117MPa UHSC from 95 MPa, load carrying capacity is elevated by 13% and 27% for the reinforcement ratio $\rho_f = 1.0\%$ and 2.0%, respectively [7]. The mid-span deviation is observed to elevate with strength of the concrete 117 MPa from 95 MPa [7].

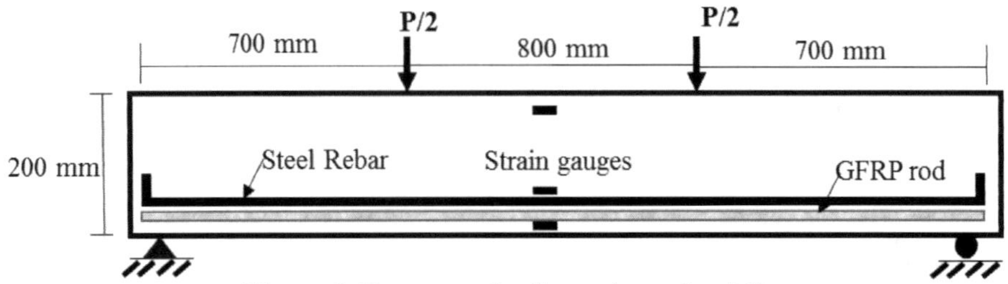

Figure 3. Test setup for flexural member [6]

3. Design of flexural member

3.1. Design modulus of elasticity
The modulus of elasticity of FRP (E_f) is the slope of the stress-strain curve within the elastic range

3.2. Ultimate rupture strain
The ultimate rupture strain is taken as 0.012 to 0.015 for GFRP bars

3.3. Ultimate tensile strength
The tensile strength of FRP (F_{fu}) is the maximum tensile load that can be applied while stretching.

3.4. Positioning of neutral axis
From strain compatibility diagram,

$$X_{u,lim} = \left(\frac{0.0035}{0.0035+0.0128}\right) d = 0.214\, d \quad (1)$$

where,

$X_{u,lim}$ = Maximum depth of Neutral Axis,
d = effective depth of beam

Total compressive and tensile force can be written as,

$$C = 0.542\, f_{ck} x_u b \quad (2)$$

$$T = A_f E_f\, \varepsilon_f \quad (3)$$

where,

x_u = Actual depth of Neutral Axis
b = width of beam
E_f = Modulus of elasticity of GFRP bars
A_f = Area of GFRP reinforcement
f_{ck} = Characteristic Compressive strength of concrete

Equating both the forces we get,

$$x_u = \frac{A_f E_f \varepsilon_f}{0.542\, f_{ck} b} \quad (4)$$

4. Theoretical analysis vs experimental studies
For verification of the above equation, experimental results obtained by VG Kalpana and K Subramanian [9], Maher A et al. [10] have been taken into consideration.

Table 1. Comparison of ultimate load carrying capacity.

Beam ID	Grade of Concrete(f_{ck}) (MPa)	Theoretical Ultimate Load (kN)	Experimental Ultimate Load (kN)	Variation %	Average Variation %
M20 D16	20	62.41	81.5	+23.42	
M20 D20	20	79.76	105	+24.03	24.8
M20 D24	20	83.98	115	+26.97	
M40 D16	40	86	115	+25.21	
M40 D20	40	102.71	135	+23.91	19.45
M40 D24	40	131.56	145	+9.26	
M60 D16	60	95	125	+24	
M60 D20	60	110.36	150	+26.42	21.23
M60 D24	60	147.41	170	+13.28	

It is observed that, maximum capacity to carry the load is increased considerably using high strength concrete. Table 1 shows ultimate load capacity based on change in reinforcement ratio and variation in concrete grade. The percentage increase of 0.50 in reinforcement ratio in high strength and regular concrete results in load carrying capacity about 20% to 30%, respectively.

Table 2. Comparison of ultimate load carrying capacity.

Beam ID	Grade of Concrete(f_{ck}) (MPa)		Theoretical Ultimate Load (kN)	Experimental Ultimate Load (kN)	Variation %	Average Variation %
	Target	Actual				
A25-1	25	24.5	35	45.9	+23.74	
A25-2	25	24.5	30.91	40.7	+24.05	20.3
A25-3	25	24.5	65.32	75.2	+13.13	
A45-1	45	48	43	55.8	+22.93	
A45-2	45	48	66	81.9	+19.41	22.65
A45-3	45	48	81.66	109.8	+25.62	
A70-1	70	74.4	70	84.6	+17.25	
A70-2	70	74.4	100.22	132.7	+24.47	18.95
A70-3	70	74.4	123.132	145.1	+15.13	

It is found that when the grade of concrete is increased, the ultimate load carrying capacity of the specimen also increaseed.The GFRP RC members are showing better performance with High strength concrete.

5. Experimental investigation

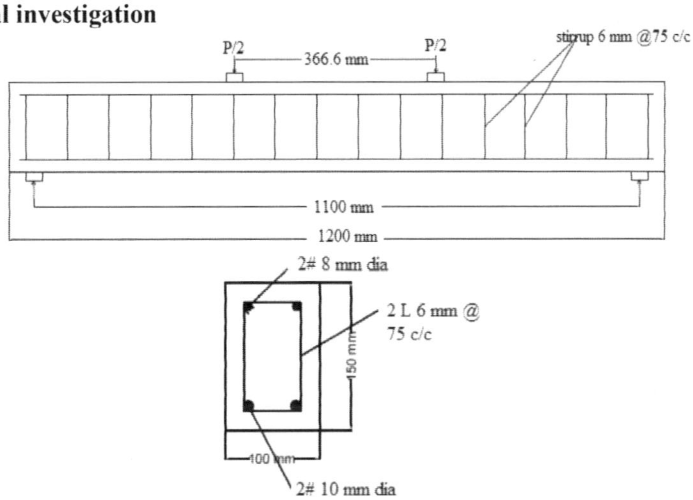

Figure 4. Cross-section and reinforcement details.

IOP Conf. Series: Materials Science and Engineering **936** (2020) 012036 doi:10.1088/1757-899X/936/1/012036

Table 3. Behaviour of GFRP RC Flexural member.

Beam ID	$b \times D$ (in mm)	Grade of concrete(f_{ck}) (MPa)	Theoretical Ultimate Load (kN)	Experimental Ultimate Load (kN)	Type of Failure	% Variation
C-T08 B10	100 × 150	30	26.88	35.216	Concrete crushing	+ 23.67

The experimental results show that the specimen was found flexure critical and crushing of concrete at the top occurred due to over-reinforcing the bottom layer. As we increase the percentage of reinforcement, the load carrying capacity also increased. The experimental ultimate load is found to be 35.216 kN. The experimental load varies 23.67 % with the Theoretical ultimate load.

Figure 5. Testing of Specimen.

The specimen was tested under Two-point loading and for the measurement of strain, LVDTs were attached at the top and bottom of the specimen and to measure the displacement at the mid span one dial gauge was attached below the specimen, to measure the total deflection.

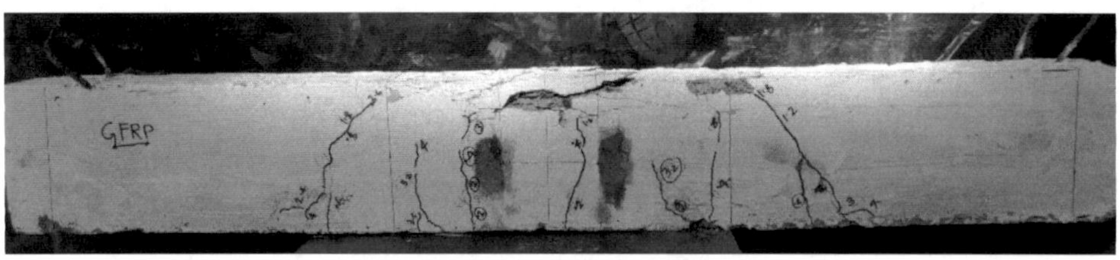

Figure 6. Failure of C-T08 B10.

The cracks in the vicinity of the tension zone within and near the constant moment region are found to be vertical. Those vertical cracks start propagating towards the compression zone as the load increases and small branches appear near lower tension surface up to approximately 60% of the maximum load.

6. Summary and conclusions

- From the Analytical analysis, it is concluded that increase in reinforcement ratio in the flexural member increases the load carrying capacity of the member and it is recommended to design the GFRP RC structure as over-reinforced.

- The variation in theoretical and experimental results range between 18% to 25 %.
- The GFRP reinforced concrete structure gives better performance with the high strength and ultra-high strength concrete. If the grade of concrete is increased then the ultimate load carrying capacity is also increased
- From the literature study, it is observed that hybrid reinforcement plays a significant role to enhance the bending strength in the concrete beams. It improves flexural strength.

7. References

[1] Tavares D H, Giongo J S and Paultre P 2008 Behavior of reinforced concrete beams reinforced with GFRP bars *Rev. IBRACON Estruturas e Mater.* **1(3)** 285–95

[2] Nanni A, De Luca A and Zadeh H J 2014 *Reinforced Concrete with FRP Bars: Mechanics and Design* (CRC Press)

[3] Harris H G, Somboonsong W and Ko F K 1998 New ductile hybrid FRP reinforcing bar for concrete structures *Journal of Composites for Construction* **2(1)** 28–37

[4] Yoo D Y, Banthia N and Yoon Y S 2016 Flexural behavior of ultra-high-performance fiber-reinforced concrete beams reinforced with GFRP and steel rebars *Engineering Structures* **111** 246–62

[5] Yost J R, Gross S P and Dinehart D W 2001 Shear strength of normal strength concrete beams reinforced with deformed GFRP bars *Journal of composites for construction* **5(4)** 268–75

[6] Leung H Y and Balendran R V 2003 Flexural behaviour of concrete beams internally reinforced with GFRP rods and steel rebars *Structural Survey* **21(4)** 146–57

[7] Goldston M W, Remennikov A and Sheikh M N 2017 Flexural behaviour of GFRP reinforced high strength and ultra high strength concrete beams *Construction and Building Materials* **131** 606–17

[8] Bank L C, Campbell T I and Dolan C W 2003 Guide for the design and construction of concrete reinforced with FRP bars reported by ACI Committee 440 *Concrete* 1–42

[9] Varghees K and Subramanian K 2011 Behavior of concrete beams reinforced with GFRP bars *Journals of Reinforced Plastics and Composites* 30(23) 1915–22

[10] Adam M A, Said M, Mahmoud A and Shanour A 2015 Analytical and experimental flexural behavior of concrete beams reinforced with glass fibre reinforced polymers bars *Construction and Building material* 84(13) 354–66

IOP Conf. Series: Materials Science and Engineering **936** (2020) 012037 doi:10.1088/1757-899X/936/1/012037

A Review on Ambient Cured Geopolymer Concrete-Sustainable Concrete for the Future

A V Nithin[1*], Mini Soman[2] and S Deepa Raj[3]

[1]Research Scholar, Department of Civil Engineering, College of Engineering, Thiruvanathapuram, 695017, India

[2]Department of Civil Engineering, College of Engineering, Thiruvanathapuram, 695017, India

[3]Department of Civil Engineering, Government College of Engineering Bartonhill, 695035, India

[*]nithinav2312@gmail.com

Abstract. The cement industry is one of the major contributors to global warming due to the release of greenhouse gases. An alternative low emission binding agent is needed to reduce the environmental impact caused by cement production. Geopolymer binder is an ideal material to substitute cement binder. Geopolymer Concrete (GPC) is formed by the polymerisation of aluminates and silicates formed by the reaction of solid aluminosilicates with alkali activators. The source materials studied were waste by-products from industries. This paper reviews the strength and durability characteristics of Geopolymer Concrete under ambient curing conditions. The comparison and study of the morphology of different aluminosilicate source materials through X-ray diffraction (XRD) pattern and Scanning electron microscopy (SEM) analysis under different curing temperatures and the study on the chemical composition and its effect on alkali activators and aggregates helped to understand the nature and type of source materials and activators most desirable to develop GPC in ambient conditions. GPC formed in ambient curing conditions have considerable strength and durability characteristics and can be adopted as a sustainable replacement for conventional concrete.

1. Introduction

Concrete is the most widely used man-made substance. Ordinary Portland Cement (OPC) is the most commonly used primary binder in the conventional concrete. The production of OPC is an energy-intensive process and it releases huge amount of greenhouse gases to the atmosphere. For every tonne of cement produced, about one tonne of CO_2 is released into the atmosphere [1]. The annual global production of cement is about 4 billion tonnes [2]. The global demand for concrete is increasing rapidly and its production will increase to about 18 billion tonnes per annum by 2050 [3]. An alternative environment-friendly sustainable and low emission primary binding material needs to be widely adopted to replace OPC based concrete. For a developing country like India, there is a high demand for cement to meet the ongoing infrastructural activities. India is in the second position globally, in both production and consumption of cement with an annual production of 502 million tonnes per annum in 2018. Alternative materials rich in aluminates and silicates like Fly ash (FA), ground granulated blast furnace slag (GGBFS) etc. can be utilised to develop cement-like binders without compromising efficiency and quality of concrete. Geopolymer technology replaces OPC binder completely with an aluminosilicate binder material and about 80% of greenhouse gas emitted by the cement industries [4]. 'ENVIS centre on fly ash', estimated the production of fly ash in India to be around 600 million tons by 2030 from about 160 million tons in the year 2010. 70% of electricity demand in India is met by coal-based thermal power plants [4]. India and China are the major producers but have less than 50% utilisation rate [5]. According to Central Electricity Authority's report 2018, the highest utilisation of fly ash was in cement sector with 25.6 per cent of the total fly ash used but the concrete had only 0.66 percent of the total fly ash used. Ground granulated slag is the waste generated during the steel manufacturing process and about 17.236 million tons were produced for the year 2017-18 in India [6]. The government of India has estimated steel production to hit 128.6 million tonnes by 2021. The Blast furnace slag specific waste generation is 320 kg/ton of steel. 40% of this is utilised in cement industry [7]. An average of 30000 tons per year of silica fume is produced by Steel Authority of India [8]. An average of 40 kg of Rice husk ash (RHA) can be produced for every tonne of rice [9]. According to the report by the United States Department of Agriculture (USDA), 114 MT of rice production in India was forecasted during the year 2019-20. There is an adequate generation of aluminosilicate industrial by-products in India. The utilisation to generation ratio of these materials is very less. Adopting GPC as a viable alternative to OPC concrete enables to minimise the impact of cement production-related environmental problems and creates a sustainable utilisation of these harmful industrial by-products.

2. Geopolymer concrete

During the history of the concrete, there was first-generation lime concrete, second-generation ordinary Portland and special cement-based concrete. Geopolymer concrete (GPC) can be considered as a third-generation type of concrete [10]. Geopolymer technology was first developed by Joseph Davidovits in 1972 by polycondensation reaction of kaolinite with NaOH [11]. In Geopolymer concrete, OPC binder is totally replaced by a polymerised aluminosilicate binder formed by the reaction of an alkaline solution with aluminosilicate rich source materials. Geopolymer compounds are crystalline in hydrothermal setting conditions and amorphous in ambient setting conditions [12]. The geopolymers are composed of source materials rich in silicates and aluminates which when dissolved in an alkaline activating solution, polymerises and forms the binder [13].

In Geopolymer concrete (GPC), these binders (geopolymer) binds the aggregates. The behaviour of geopolymer concrete and conventional Portland concrete are different due to the variation in the microstructural behaviour of source materials and the process of formation of binders [14]. The studies on geopolymer concrete were mostly based heat cured fly ash blended types. Elevated temperatures were required for fast geopolymerisation reaction [15]. Heat-cured geopolymer concrete possesses high compressive strength, low drying shrinkage and creep, and good resistance to sulphate and acid environments. The heat curing mechanism is difficult and costly. Thus, there occurred a need to research and develop ambient cured geopolymer concrete. In this review study, an attempt has been done to study the properties of different aluminosilicate sources materials and its effect on the properties of GPC under ambient curing and are detailed in the subsequent sections.

3. Geopolymer source materials

The geopolymer source materials are either natural minerals like kaolinite, clay or industrial by-products like fly ash, GGBFS, Silica fume, Rice husk Ash, etc. The reaction between the source materials and alkaline activators and their interaction affects the properties and behaviour of Geopolymer concrete [16]. The main factors affecting the properties of geopolymer are the chemical composition and fineness of various source materials. The generally adopted tests done to examine the microstructure and chemical composition are X-ray diffraction (XRD) pattern and Scanning electron microscopy (SEM) analysis. The chemical composition of various source materials and its specific surface area is given in table 1 [17]. A description of source and microstructure properties of different aluminosilicates sources and their behaviour are discussed below.

3.1. Fly ash

Fly ash is the most commonly used binder because of its pozzolanic character, less cost, availability, richness in Silicon and Aluminium oxides, low water demand and high workability. Class F fly ash is obtained by burning of pulverised coal extracted from anthracite and bituminous coal reserves [25]. Fly ash (Class C) is obtained on the burning of lignite and has Calcium more than 10% [26]. Class F fly ash showed better performance than class C fly ash when subjected to heat curing [27]. The SEM and XRD of various source materials are given in figure 1 and figure 2. From figure 1 (a) it can be seen than Class

F fly ash consisted mostly of hollow spherical particles (cenospheres) of size ranging from 1 to 300μm [28]. The SEM Analysis shows solid spheres with irregular shape particles [29]. The spherical shape imparts ball-bearing effect [30]. The spherical shape of fly ash particles will impart easy blending and free-flowing nature to the binder. From figure 2 (a), the XRD pattern of Class F Fly ash had sharp peaks of quartz, mullite, and mellite which indicate crystalline phase and broad humps showing amorphous phase. Class F Fly ash had a higher mullite content. This indicates less abundance of amorphous phase as there is an inverse relationship between mullite and amorphous phase. [31]. The presence of the crystalline phase in Class F Fly ash resulted to use heat curing as an effective method to accelerate geopolymerisation reaction. Figure 1 (b) shows the SEM analysis of Class C Fly ash, which consists of sphere, agglomerate and angular shape particles and figure 2 (b) shows XRD pattern of class C fly ash with smooth and less steep humps. This indicates the presence of both crystalline and amorphous phase [32]. The Fly ash-based GPC had considerable workability compared to GPC formed with other aluminosilicate-based materials.

3.2. Ground Granulated Blast furnace slag (GGBFS)

GGBFS is obtained during the manufacture of pig iron for the steel industry [33]. GGBFS is hydraulic in nature due to the presence of oxides of Calcium. When alkali-activated it resulted in the C-S-H microstructure formation compared to an Al-Si network typically formed during geopolymerisation [20]. But a faster rate of strength development and ambient temperature curing are the advantages of using GGBFS as aluminosilicate binder. Figure 1 (c) shows SEM images of GGBFS indicating very fine glassy and granular shaped particles and these particles were finer than 30μm [34] [36]. The granular shaped particles of GGBFS resulted in the formation of geopolymer binder with less workability and making the slag-based GPC difficult to work with. GGBFS based GPC had lower setting time for geopolymer binders. As in figure 2 (c), smooth peaks were observed in XRD patterns of GGBFS [35]. The smooth hump indicates the amorphous phase. C-S-H formation helps to accelerate geopolymerisation reaction at ambient temperature. Due to these properties, Fly Ash based GPC partially replaced with GGBFS had better strength compared to Fly Ash based GPC at ambient temperature.

3.3. Rice husk ash

Rice husk ash (RHA) is obtained by burning rice husk [37]. It contains SiO_2 greater than 90%. Since it is obtained from the agricultural industry, it is important to maintain efficient burning at a controlled temperature of 500 to 600°C and is then ground in jar mill to obtain required fineness. From figure 1 (d), the average size of RHA was found nearly three times smaller than fly ash particles and are irregular in shape [20]. Due to which GPC having RHA is found to have less setting time. RHA gets polymerised faster due to less size compared to Fly ash. From figure 2 (d), the majority of composition was found to be in the amorphous phase due to smooth humps [20]. The amorphous phase of silica in RHA reacts with an alkaline solution to produce binders at ambient temperature conditions.

3.4. Silica fume

Silica fume (SF) is obtained from the ferrosilicon industry [38]. Silica fume contains about 85-97% SiO_2. It is a highly reactive material suitable for geopolymerisation. Figure 1 (e) shows SEM analysis of silica fume and particle size were extremely small particles sizes [31]. The reactivity was higher due to a larger surface area to mass ratio. GPC with SF has a lower setting time. The SEM analysis showed dense microstructure. The dense microstructure improved the durability properties in Fly ash based GPC partially replaced by SF. Figure 2 (e) shows XRD pattern of SF as amorphous structured due to its board smooth humps and presence of a high amount of amorphous silicon dioxide which provide a highly reactive pozzolanic material [20][39]. SF reacts with an alkaline solution to produce binders at ambient curing conditions.

3.5. Alccofine

Alccofine is a low calcium micro-fine silicate slag. It is a fine powder obtained by powdering the dried granular material produced by quenching molten iron slag in a blast furnace in water or steam [40]. Figure 1 (f) of alccofine shows SEM image, which reveals ultra-fine and angular particles for alccofine

[40]. Fly ash-based GPC with partial replacement of Alccofine had an early setting and dense microstructure was observed. The dense microstructure improved the durability properties in Fly ash-based GPC partially replaced by Alccofine. Figure 2 (f) shows XRD pattern of Alccofine as amorphous since it shows the Calcite phase with broad smooth humps [41]. Alccofine reacts with an alkaline solution to produce binders at ambient curing conditions.

3.6. Red mud

Red Mud is a silt muddy alkaline solid industrial by-product obtained during the extraction of alumina from bauxite ores by the Bayer process. The Presence of Alumina and high alkalinity makes them useful in geopolymer. Fe_2O_3, SiO_2 and Al_2O_3 constitute 75% of total composition [42]. Figure 1 (g) shows the SEM image of Red Mud. It contains superfine amorphous aragonite and a few calcite micrites and the particle sizes of the grinded Red Mud are less than 45µ [43]. The fineness imparts quick setting properties to the binder. Figure 2 (g) shows the XRD pattern of Red Mud. The presence of sharp peaks indicated the crystalline phases and smooth humps indicated amorphous phase [44].

Table 1. The chemical composition (% mass) of Class F Fly ash (FFA) [11],Class C Fly ash (CFA), GGBFS [19], Rice Husk Ash(RHA) [20], Nano silica(NS) [21], Alccofine [22] , Silica fume(SF) [23] and Red Mud (RM) [24]

Composition (% mass)	FFA	CFA	GGBFS	RHA	NS	Alccofine	SF	RM
SiO_2	59.7	38.4	34.06	91	99.8	35.3	92	8.88
Al_2O_3	30.2	18.7	20	0.35	-	21.4	0.46	25.4
CaO	0.7	24.6	32.6	-	-	32.2	0.29	0.92
SO_3	0.02	1.4	0.9	1.21	-	0.13	0.19	0.22
Fe_2O_3	2.8	5.1	0.8	0.41	-	1.2	1.60	40.03
MgO	0.8	5.1	7.89	0.81	-	8.2	0.28	-
Na_2O	-	-	-	-	-	-	-	8.75
LOI	0.8	0.3	0.9	<1	0.2	-	1	8.7
Surface area(m^2/kg)	1280	900	375	20000	90000	12000	8900	655

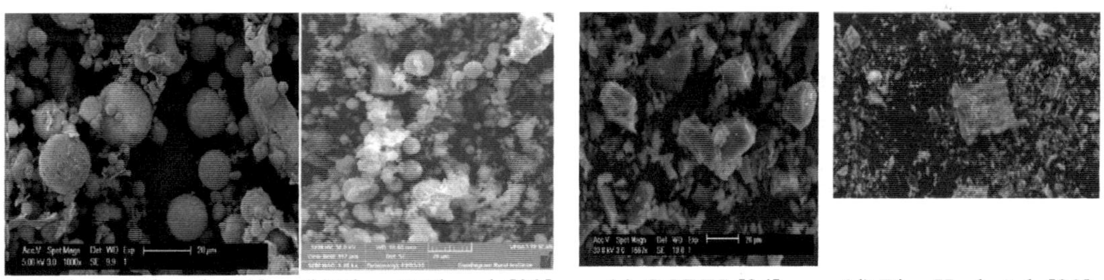

(a) Class F Fly ash [29] (b)Class C Fly ash [32] (c) GGBFS [34] (d) Rice Husk Ash [20]

(e) Silica Fume [31] (f) Alccofine [40] (g) Red Mud [43]

Figure 1. SEM analysis of aluminosilicate sources.

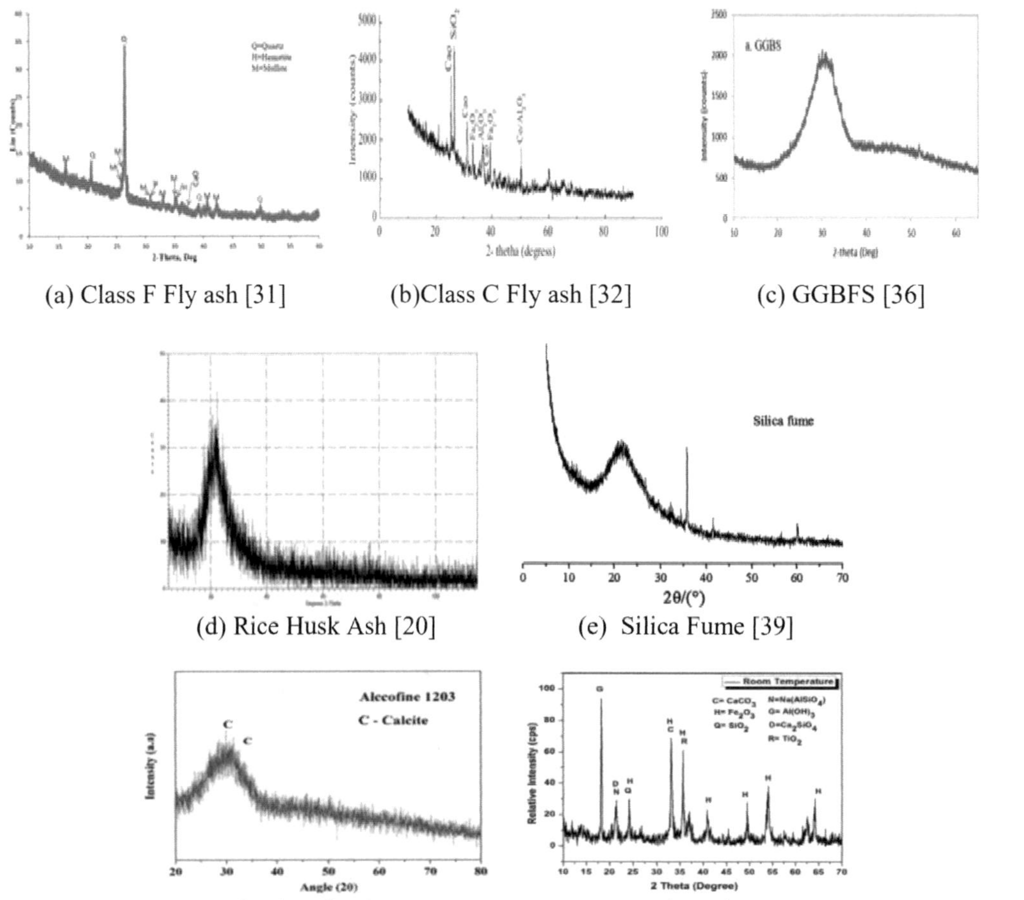

(a) Class F Fly ash [31] (b)Class C Fly ash [32] (c) GGBFS [36]

(d) Rice Husk Ash [20] (e) Silica Fume [39]

(f) Alccofine [41] (g) Red Mud [45]

Figure 2 XRD patterns of aluminosilicate sources.

4. Previous Studies on Geopolymer concrete cured at ambient temperature

To enhance the strength of GPC in ambient curing conditions, the geopolymerisation process needs to be accelerated at the early stages. For this, amorphous structured aluminosilicate source materials need to be effectively used. A Literature survey was conducted to understand the factors affecting the strength, workability and durability of ambient cured geopolymer concrete. These factors are summarised on the following sections.

4.1. Effect of Aluminosilicate Sources

Geopolymer binder with fly ash as the source material required higher temperature for initiating the geopolymerisation reaction. The heat-cured GPC had high early-age strength gains and has better durability properties compared to conventional concrete. [46][47][48]. Partial replacement of fly ash with amorphous structured aluminosilicates or those binders having calcite enabled to develop GPC at ambient curing. Majority of earlier studies were done on heat-cured low calcium fly ash. This limited the applications to precast members and as a solution, further research was carried in developing cost effective ambient cured GPC using two or more aluminosilicate source materials. Class C fly ash showed better strength and performance at ambient cured conditions [35]. Class F fly ash geopolymer concrete (GPC) partially replaced by GGBFS, OPC or Calcium Hydroxide under ambient curing had setting time lower to a value similar to that of OPC based concrete. It had better early strength compared to fly ash GPC cured at ambient conditions but late age strength at 28 and 90 days showed not much significant difference [49].This behaviour is due to the amorphous phase in GGBFS activated by the Calcium Silicate Hydrates (C-S-H) when OPC reacts with water during geopolymerisation at ambient temperature. The addition of GGBFS in fly ash GPC increased the strength and decreased the

workability of GPC [19][33][35][36][50]. The properties of ambient cured GPC were enhanced by the addition of GGBFS but the rate at which strength improved slowed down after 28 days [51]. GGBFS blended ambient cured GPC has strength comparable to that of OPC at the ambient cured condition. The setting time reduced while replacing more percent of fly ash with GGBFS [49]. This is mainly due to the variation in particle size and shape at the microstructural level. The spherical Fly ash particles were replaced by the angular and irregular shaped GGBFS particle, resulting in reduced workability. The addition of OPC accelerated the geopolymerisation reaction as well as affected the workability and setting time [49][52]. The strength of ambient cured fly ash GPC was also improved by the small percentage replacement by OPC. The micro cracks reduced as the hydration utilised expelled water in the geopolymerisation process [53]. Fly ash-based GPC cured at ambient temperature can also be blended with ordinary Portland Cement (OPC) and micro-silica (MS) and GPC with sufficient strength can be attained at ambient curing conditions [54]. Early strength and durability were achieved in ambient cured geopolymer concrete blended with red mud, GGBFS and fly ash [42]. Red Mud contains alkalies that help to reduce the molarity of NaOH activator solution and compressive strength comparable to OPC was obtained by partial replacement [42]. Fly ash geopolymer concrete blended with alccofine improved compressive strength and durability at ambient curing conditions [22]. The fly ash nano-silica modified GPC cured at ambient temperature showed higher compressive strength and enhanced durability properties compared to heat-cured GPC and OPC [55]. GGBS based fly ash had dense microstructure. It performed well under sulphate attack [56]. Aluminosilicate source materials having fineness in micro units formed very dense structured GPC. The compact binder formed by alkali activation improved durability properties of GPC. Using aluminosilicate source materials with amorphous structured microstructure and accelerating the reaction by adding calcite sources or very fine silicate sources, GPC with strength comparable to conventional OPC concrete were developed.
Another ongoing research was to modify the microstructure of fly ash before mixing during geopolymerisation. The compressive strength of geopolymer concrete was enhanced by the mechanical activation of fly ash material [30]. Mechanical activation results in increased rate of geopolymerisation by reducing particle size and increasing the amorphous phase [57]. The heat activation of fly ash was another method adopted and used for preparing ambient cured GPC. It improved the early strength properties of GPC. Here early activation of fly ash helped in early polymerisation and less amount of crystalline compound. About 9% less energy was required in heat-activated fly ash material blended geopolymer concrete than in heat-cured GPC for heat activation [55].

4.2. Effect of alkaline activators

The alkaline activators used to activate aluminosilicates are Sodium or Potassium based hydroxides or silicates. Sodium hydroxide (NaOH), potassium hydroxide (KOH), sodium silicate (Na_2SiO_3) and potassium silicate (K_2SiO_3), and NaOH are more commonly used as it has the capacity to release more silicate and aluminate monomers [58]. The Sodium Silicate to Sodium Hydroxide ratio (SS/SH) and molarity of Sodium Hydroxide affects the properties of geopolymer concrete cured at ambient temperature. The slump value and flow value are reduced when SS/SH ratio is increased [49] and lower value of SS/SH ratio showed better later age strength in fly ash GPC blended with GGBFS. Increasing the molarity of NaOH from 4.0 to 14.5 increased the strength of geopolymer paste but further increase causes early precipitation of aluminosilicates, which hindered the development of strength [51]. For obtaining higher compressive strength, the molarity of NaOH must be very high [59]-[61]. Less concentration of NaOH is required when geopolymerisation was done on elevated temperatures [62]. The compressive strength was increased and the setting time was decreased when molarity of NaOH was increased subject to a subsequent increase in slag content [63]. Higher molarity for NaOH is required for faster geopolymerisation reaction for Ambient cured GPC compared to heat-cured GPC.

The current research on activators is focussed on developing GPC by using single activator. The behaviour of Geopolymer concrete using single activator NaOH showed that replacing binder with 100% GGBS gained sufficient strength in 28 days. Fly ash used as a sole aluminosilicate source in ambient cured geopolymer concrete resulted in very low strength [64].

4.3. Effect of aggregates

The strength of geopolymer binders is less affected by the interaction of aggregates and more on the chemical behaviour of the geopolymer binder. Compared to OPC, less treatment of C&DW is required for geopolymer based concrete. Thus, geopolymer technology was effective in developing concrete using recycled industrial construction waste [65]. The strength of GPC improved with the increase in the mean size of the coarse aggregates. There was a decrease when the mean size was increased after the mean size of 18.75 mm [66]. The aggregate–paste interface was bigger for large size aggregates. This resulted in wider cracks, which interacted with cracks in the binder as well as in other interfacial zones resulting in loss of the strength. When internal bleeding occurs, water gets trapped in between the bigger aggregates and when this water evaporates, voids were formed which reduced the strength of the concrete. The maximum value for compressive strength was obtained for aggregates of size range 12.5–25 mm [66]. Thus, the strength of fly ash-based geopolymer concrete increased with an increase in total aggregate size and proportioning of total aggregate content and ratio of fine aggregate to total aggregate affected the behaviour of GPC [67].

5. Conclusions

The paper reviewed the strength and durability studies on ambient cured geopolymer concrete formed from different aluminosilicate sources. Ambient cured GPC can be used as an alternative for the conventional OPC concrete as it had better durability and comparable strength parameters. The early strength is very less when compared heat-cured Fly ash GPC. India has an abundant quantity of aluminosilicate materials as industrial by-products. The source materials used in geopolymer binders must be pozzolanic in nature. The microstructure and chemical composition of aluminosilicate source materials played a major role in the determination the strength and durability behaviour of GPC. It also helped to understand the nature in which source materials behaves during geopolymerisation reaction. The oxides of silica, alumina and calcium were the critical elements affecting the behaviour of geopolymer binder. The crystalline phase in aluminosilicate source materials required a higher temperature to initiate geopolymerisaton reaction compared to the amorphous phase in which geopolymerisation occurs at ambient temperature. The particle size and presence of amorphous phases determines the rate of geopolymerisation reaction. GGBFS, RHA and Alccofine blended with Fly ash GPC had sufficient strength at ambient curing condition. The particle shapes and size study were important to compare the reactivity and workability of the binders. The workability was reduced in angular shaped aluminosilicate source materials and often very fine sizes attributed to the early setting of geopolymer binder. This widened the research area on the effect of superplasticiser on the geopolymer binders. The addition of the calcium-based aluminosilicates enhanced the properties of the fly ash-based geopolymers cured at ambient temperature. The current research on ambient curing is mainly focused on low-cost blended geopolymer concrete using tertiary binders. Another area of research is on developing activated geopolymer source materials. Ambient GPC achieved by mechanical activation of aluminosilicate source materials or by developing ambient cured geopolymer concrete using aluminosilicate source materials having calcite, appropriate molarity for alkali activator and alkali to binder ratio is found to be the current area of interest. By avoiding heat curing and reducing the concentration of alkali activator, the cost of GPC can be reduced considerably. The current researches need to be focused on developing low-cost alkali activators.

6. Reference

[1] Khalil M G, Elgabbas F, El-Feky M S and El-Shafie H 2020 *Construction and Building Materials* **242** 118090
[2] U S Geological Survey 2018 *Mineral Commodity Summaries*
[3] Babu K G, Rao H S and Amarnath Y (Eds.) 2020 *Lecture Notes in Civil Engineering*
[4] Shiv Pratap Raghuvanshi, Avinash Chandra and Ashok Kumar Raghav 2006 *Energy Conversion and Management* **47(4)** 427–41
[5] Arpita Bhatt, Sharon Priyadarshini, Aiswarya Acharath Mohanakrishnan, Arash Abri, Melanie Sattler and Sorakrich Techapaphawit 2019 *Case Studies in Construction Materials* **11** e00263
[6] Indian Minerals Yearbook 2018, 57th edition, *Part -II: Metals and Alloys, Slag-Iron and steel*

[7] Neelam Sharma, Viswanathan N Nurni, Vilas Tathavadkar and Somnath Basu 2017 *Mineral Processing and Extractive Metallurgy* (Taylor & Francis)

[8] Ram Babu Nimma 2019 *Indian Conference on Geotechnical and Geo-Environmental Engineering (ICGGE-2019)*

[9] Zerbino R, Giaccio B and Isaia G C 2011 *Construction and Building Materials* **25** 371–8

[10] Chakradhara R Meesala, Nikhil K Verma and Shailendra Kumar 2019 *Structural Concrete* 1–16

[11] Joseph Davidovits 1994 *Alkaline Cements and Concrete* (KIEV, Ukraines)

[12] J Davidovits 1991 *Journal of thermal analysis* **37**1633-56

[13] Panias D and Giannopoulou I 2004 *1st International Conference on Advances in Mineral Resources Management and Environmental Geotechnology AMIREG* **1** p 40712

[14] Daniel L Y Kong ,Jay G Sanjayan 2010 *Cement and Concrete Research* **40(2)** 334–39

[15] Shaikh F U A and Vimonsatit V 2014 *Fire and Materials* **39(2)** 174–88

[16] Mohammed Haloob Al-Majidi , Andreas Lampropoulos , Andrew Cundy and Steve Meikle 2016 *Construction and Building Materials* **120** 198–211

[17] Karen L. Scrivener, André Nonat 2011 *Cement and Concrete Research* **41** 651–65

[18] Scott Z Jones, Dale P Bentz, Kenneth A Synder, Nicos S Martys, Daniel S Hussey and David L Jacobson 2015 *International Concrete Sustainability Conference* (Miami, Florida)

[19] Mallikarjuna Rao Goriparthi and Gunneswara Rao T D 2017 *Advances in Concrete Construction* **5(4)** 313–30

[20] Srinivasreddy A, McCarthy T J and Lume E 2013 *26th Biennial Concrete Institute of Australia's National Conference (Concrete 2013)* pp 1–10

[21] Haruehansapong S, Pulngern T and Chucheepsakul S 2014 *Construction and Building Materials* **50** 471–7

[22] Bharat Bhushan Jindal, Dhirendra Singhal, Sanjay K Sharma, and Parveen 2017 *Journal of Engineering and Applied Sciences* **12(9)** 3022-31

[23] Debabrata Dutta and Somnath Ghosh 2018 *Rasayan J. Chem.* **11(3)** 1018–33

[24] Nevin Koshy , Kunga Dondrob , Liming Hu , Qingbo Wen and Jay N Meegoda 2019 *Crystals* **9** 572

[25] Barry E Scheetz, Russell Earle 1998 *Current Opinion in Solid State and Materials Science* **3(15)** 510–20.

[26] Xueying Li , Xinwei Ma, Shoujie Zhang and Enzu Zheng 2013 *Materials* **6(4)** 1485–95

[27] Wardhono A 2018 *Conf. Ser.: Mater. Sci. Eng.* **288** 012019

[28] Mustafa Al Bakri A M, Abdulkareem O A, Kamarudin H, Khairul Nizar I, Rafiza A R, Zarina Y and Alida A 2013 *Applied Mechanics and Materials* **421** 342–8

[29] Bijoy Krishna Halder, Vivek Tandon, Anthony Tarquin and Ramana V Chintalapalle 2013 *World of Coal Ash Conference*

[30] Temuujin J, Williams R P and Van Riessen A 2009 *Journal of Materials Processing Technology* **209** 5276–80

[31] Ruben Snellings, Gilles Mertens and Jan Elsen 2012 *Reviews in Mineralogy & Geochemistry* **74** 211–78

[32] Venkata Suresh Gand Karthikeyan J 2016 *Journal of Structural Engineering* **43(4)** 381–9

[33] Suman Saha and Rajasekaran C 2017 *Construction and Building Materials* **146** 615–20

[34] Wan H, Shui Z, Lin Z 2004 *Cem. Concr. Res.* **34(1)** 133–7

[35] Ravikumar D, Peethamparan S and Neithalath N 2010 *Cement Concrete Compos.* **32(6)** 399–410

[36] Apriany Saludung, Yuko Ogawa, and Kenji Kawai 2018 *MATEC Web of Conferences* **195** pp 01013

[37] Tambichik M A, Samad A A A, Mohamad N, Bosro M Z M and Iman M A 2020 *Materials Science and Engineering* **713**

[38] Venkatraman S and Vinothkumar S 2018 *International Journal of Pure and Applied Mathematics* **119(12)** 8981–9

[39] Xin Wang, Jian Huang , Baoguo Mac, Shaobin Dai , Qi Jiang and Hongbo Tan 2019 *Construction and Building Materials* **195** 537–46

[40] Narasimha Prasad K N, Ganesh N , Rakesh S, Radhakrishna and Venkatesh S V 2015 *I manager's Journal on Structural Engineering*

[41] Parveen, Dhirendra Singhal , Talha Junaid M , Bharat Bhushan Jindal and Ankur Mehta, 2018 *Construction and Building Materials* **180** 298–307
[42] Anuj Kumar and Sanjay Kumar 2013 *Construction and Building Materials* **38** 865–71
[43] Yongshuang Zhang, Yongxin Qu and Shuren Wu 2001 *Environmental Geology* **41** 249–56
[44] Guoping Zhang, Jian He and Robert P. Gambrell 2010 *Transportation Research Record Journal of the Transportation Research Board*
[45] Harjeet Nath, and Abanti Sahoo 2014 *International Journal on Applied Bioengineering* **8(1)**
[46] Prabir K Sarker , Rashedul Haque and Karamchand V Ramgolam 2013 *Materials and Design* **44** 580–6
[47] Venkateswara Rao, Srinivasa Rao and Rambabu 2017 *Proceedings of the Institution of Civil Engineers* Construction Materials 1–9
[48] Yacob N S, ElGawady M A, Sneed L H and Said A 2019 *Engineering Structures* **196** 109298
[49] Pradip Nath and Prabir Kumar Sarker 2014 *Construction and Building Materials* **66** 163–171
[50] Mallikarjuna Rao G and Gunneswara Rao T D 2014 *Arab. J. Sci. Eng.*
[51] Kiatsuda Somna, Chai Jaturapitakul, Puangrat Kajitvichyanukulu and Prinya Chindaprasirt 2011 *Fuel* **90** 2118–24
[52] Mahya Askarian , Zhong Tao, Georgius Adam and Bijan Samali 2018 *Construction and Building Materials* **186** 330–7
[53] Lateef Assi, SeyedAli Ghahari, Edward (Eddie) Deaver, Davis Leaphart and Paul Ziehl 2016 *Construction and Building Materials* **123** 806–13
[54] Prinya Chindaprasirt, Pornnapa Kasemsiri, Suttipong Poomsrisa-ard and Patcharapol Posi 2019 *International Journal of GEOMATE* **16(54)** 29–35
[55] Dibyendu Adak and Saroj Mandal 2019 *Journal of Materials in Civil Engineering* **31** 9
[56] Gopalakrishnan R, and Chinnaraju K 2019 *Construction and Building Materials* **204** 70–83
[57] Kamal Neupane 2016 *Mechanics of Materials*
[58] Duxson P, Fernández-Jiménez A, Provis J. L, Lukey G. C, Palomo A and Van Deventer J. S. J 2007 *J Mater Sci* **42** 2917–33
[59] Sathish Kumar V, Ganesan N and Indira P V 2017 *Earth and Environmental Science* **80** 012011
[60] Solomon Oyebisi, Anthony Ede, Festus and Adeyemi Olutoge 2018 *International Journal of Civil Engineering and Technology* **9(8)** 734–43
[61] Yun Yong Kim, Byung-Jae Lee, Velu Saraswathy and Seung-Jun Kwon 2014 *Hindawi Publishing Corporation Scientific World Journal* **1** 10
[62] Gökhan Görhan and Gökhan Kürklü 2014 *Composites: Part* B **58** 371–7
[63] Guohao Fang, Wing Kei Ho, Wenlin Tu, Mingzhong Zhang 2018 *Construction and Building Materials* **172** 476–87
[64] Ghina M Zannerni, Kazi P Fattah and Adil K Al-Tamimi 2020 *Sustainable Materials and Technologies* **23** e00131
[65] Abbas Mohajerani, David Suter, Tristan Jeffrey-Bailey, Tianyang Song, Arul Arulrajah, Suksun Horpibulsuk and David Law 2019 *Clean Technologies and Environmental Policy* **21** 493–515
[66] Ernesto J Guades 2017 *European Journal of Environmental and Civil Engineering* 2116-7214
[67] Benny Joseph and George Mathew 2012 *Scientia Iranica A* **19(5)** 1188–94

IOP Conf. Series: Materials Science and Engineering **936** (2020) 012038 doi:10.1088/1757-899X/936/1/012038

Effect of Adhesive Debonding on Static and Dynamic response of PZT Bonded Aluminium Beam

Mohammed Aslam[*], Praveen Nagarajan and Mini Remanan

Department of Civil Engineering, National Institute of Technology Calicut, Kerala, India

[*]pcaslam4@gmail.com

Abstract. Piezoelectric actuators and sensors have been extensively used in smart structures. A structure becomes smart when piezoelectric patches are either surface bonded or embedded into the host structure. Under the application of electric potential piezoelectric materials may expand/contract, which serves them as an actuator. Similarly, when it gets strained, an electric potential is induced, which serves them as sensors. Piezoelectric materials (PZT) are usually surface bonded to the structures using suitable bonding agents. This bonding agent plays a crucial role in transferring the forces from the piezoelectric layer to the host layer. This paper presents the effect of adhesive debonding on static as well as dynamic response of PZT bonded aluminium beam. The results show that debonding failure adversely affects the actuating capabilities of the PZT transducer. It is seen that the tip displacement decreases with increase in debonding size. The results also highlight that neglecting the adhesive layer in the finite element (FE) models overestimates the tip displacement as well as its natural frequencies.

1. Introduction

The ability of piezoelectric transducers to convert electrical loads to mechanical effects and vice versa has made them applicable for various engineering applications such as vibration control, energy harvesting, shape control, structural health monitoring (SHM) etc. Piezoelectric patches (PZT) are superior compared to other transducers due to distinct features like high force-to-weight ratio, fast response, low power consumption and high sensitivity. Moreover, they can be quickly bonded/embedded into structures, and they possess broad operating bandwidths [1,2]. Piezoelectric sheets are generally attached to the host structure using appropriate bonding agents, which provides a medium for strain transfer between the PZT layer and the host structure. The effect of the adhesive layer on the interaction between the PZT layer and the host structure has been studied recently by various researchers using experimental and numerical investigations [3-6]. During the service life of any PZT bonded structures, piezoelectric transducers get detached from the host structure due to prolonged use, temperature change, high peeling stress concentration at the adhesive edge, poor bonding agents etc. As a result of debonding, the sensing and actuation abilities of PZT transducers may get affected. Hence it is essential to investigate the consequence of adhesive debonding on the static and dynamic response of PZT bonded structures. Recently, the effect of debonding on stress transfer between PZT and host structure is studied by Yu and Wang [7]. Raja et al. [8] studied the effect of edge debonded actuator on the transverse deflection of the piezoelectric beam under different boundary conditions. Liu et al. [9] experimentally investigated the effect of the actuator as well as sensor debonding in SHM systems. A comprehensive analysis tool is developed by Hosking and Sotoudeh [10] to study the dynamic and aeroelastic behaviour of PZT attached beam. However, here the effect of the adhesive layer is not considered. In most of the past studies, the effect of the bonding layer is neglected, or its debonding effects are studied either by 1D or 2D analytical or numerical models. Since most of the piezoelectric materials are orthotropic, a three-dimensional analysis is necessary to evaluate the performance of

IOP Conf. Series: Materials Science and Engineering **936** (2020) 012038 doi:10.1088/1757-899X/936/1/012038
debonded smart beams. This paper presents a three-dimensional numerical simulation of a smart aluminium beam with partially debonded PZT actuator. Static analysis is carried out to examine the effect of partial debonding of the piezoelectric patch on tip deflection. Nine different debond configurations are considered for the study. Additionally, the effect of the engineering properties of the bonding layer and the effect of its thickness were also evaluated. Eigenvalue analysis is also performed to extract the natural frequencies for a fully bonded as well as partially debonded smart cantilever beam.

2. Finite element model

The PZT bonded aluminium beam illustrated in figure 1 is used for the current study. The beam is 152.4 mm long, 25.4 mm wide and is fixed at its left end. The aluminium beam is 15.24mm thick, while the bonding layer and PZT layer thicknesses are 0.254 mm and 1.524 mm, respectively. For details regarding the material properties of smart beams, refer to [11]. The top surface of the PZT layer is imposed with a voltage of 12.5 kV while the bottom surface is grounded.

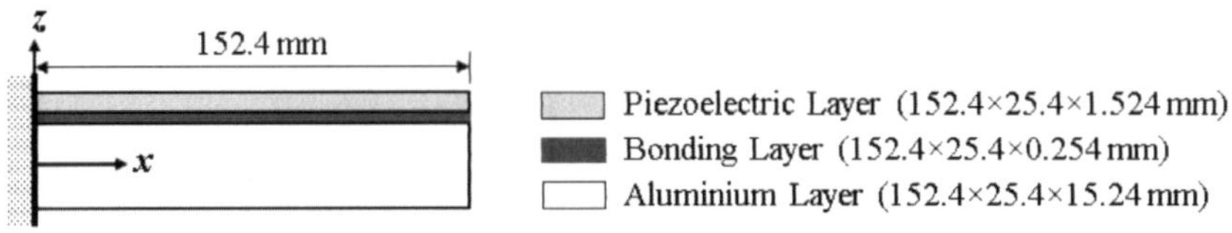

Figure 1. The geometry of the smart cantilever beam.

The behaviour of the PZT bonded smart beam is studied using commercially available FE software Abaqus CAE. The 3D numerical model of the smart beam is shown in figure 2. A linear hexahedral element is utilised to discretise the aluminium and bonding layer, while a linear piezoelectric hexahedral element is used to discretise the piezoelectric layer. A mesh refinement study was conducted initially to arrive at an optimum number of finite elements. A total of 3660 elements is chosen for the PZT layer and adhesive layer, while 8700 elements were used for the aluminium layer. To avoid any relative motion between the layers, tie constraints are assigned at the interface.

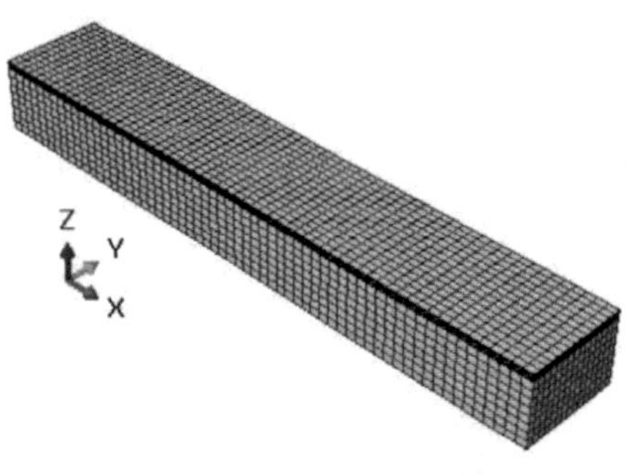

Figure 2. 3D finite element model of the PZT bonded smart beam.

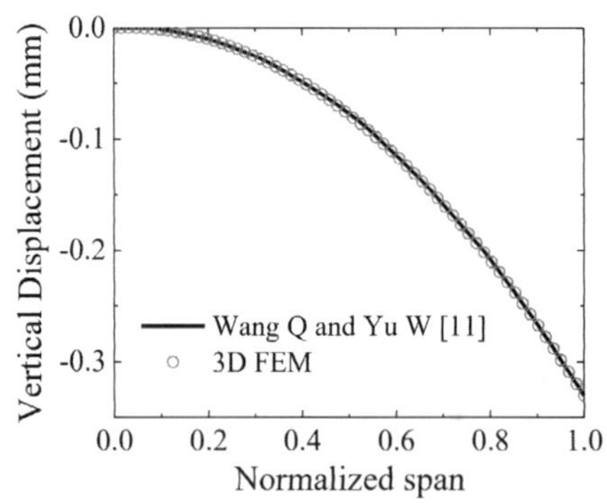

Figure 3. Comparison of the transverse displacement along the length of the beam.

For validating the numerical model, the transverse displacement along the length of the beam obtained by finite element analysis is compared with the solutions attained by Wang and Yu [11]. The comparison of transverse deflection is shown in figure 3. It is noticed that excellent agreement exists between the results. Hence the current FE model is used for additional studies. Various debond configurations are considered to study the effect of debonding. Case 1 – debond along the length of PZT, case 2 – debond along the width of PZT and case 3 – Debond along both length and width of PZT. For each case, the percentage of debond area is taken as 0%, 25% and 75%. The different partial debond configurations with 50% debond area is displayed in figure 4.

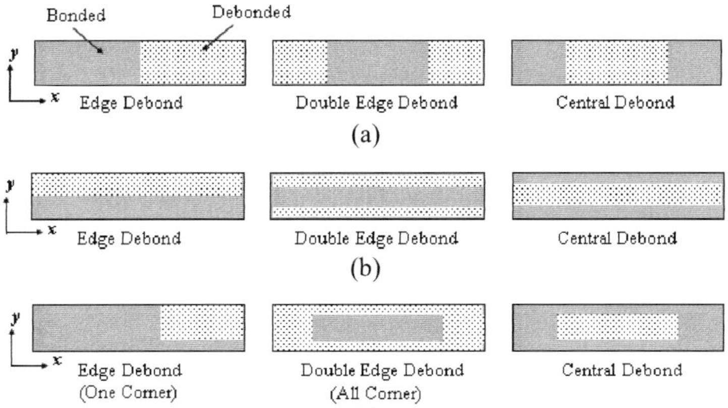

Figure 4. Different partial debond configurations (a) Debond case 1 (b) Debond case 2 (c) Debond case 3.

3. Results and Discussion

3.1. Influence of partial debond on tip deflection

Figure 5 shows the variation in tip displacement for different debond cases. It is noted that for all the cases, the tip displacement decreases with increase in percentage debond area. However, the percentage reduction in tip displacement for central debond is less compared to the edge, and double edge debond. The maximum percentage reduction for central debond is obtained as 12% for debond case 2 with 75% debond area. This indicates that the central debond, which is challenging to visualise and identify, does not cause many effects in the static response. The worst situation under all the cases considered is for debond case 1 with double edge debond. Here the percentage reduction in tip displacement is obtained as 81% for 75% debond area. This decrease is due to the reduction in the stiffness of the adhesive layer. When the adhesive gets degraded, the efficiency of the bonding layer in transmitting the forces to the host layer from the piezoelectric layer reduces.

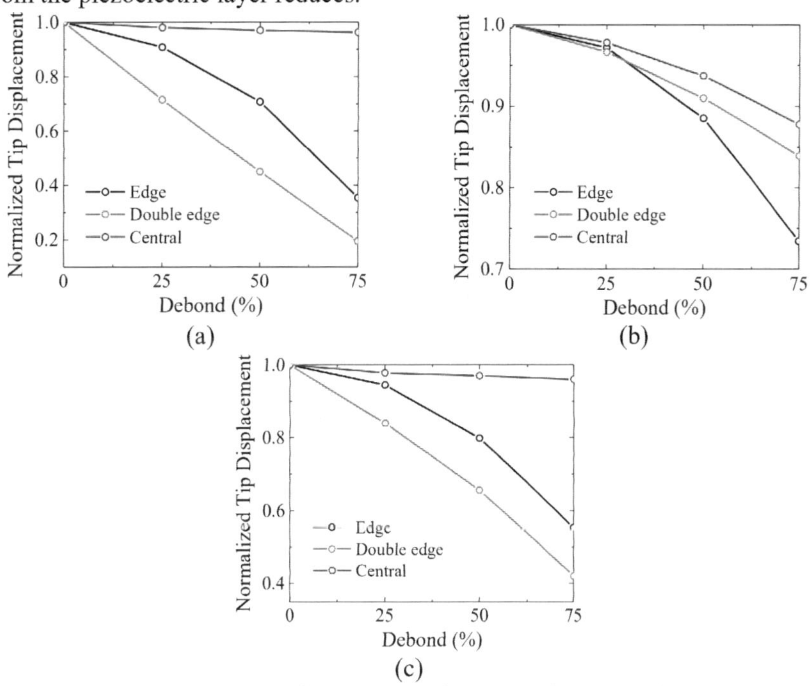

Figure 5. Variation in normalised tip displacement for (a) Debond case 1 (b) Debond case 2 and (c) Debond case 3.

3.2. Influence of partial debond on natural frequencies

The system's natural frequencies and their corresponding mode shapes are also studied for the healthy as well as for different debonding cases. It is found that the presence of debonding decreases the natural frequencies, and by increasing the debonding size, the natural frequencies decrease for all higher modes. Figures 6 and 7 show the variation in first and twentieth natural frequencies, respectively, for different debond cases. It is observed that for debond case 2, the percentage reduction in first natural frequency is less compared to the other two cases. The maximum percentage reduction when debond is considered along the width of PZT (debond case 2) is 5% for 75% debond area. It is also noted that the edge debond is most critical under all debond cases. A maximum percentage reduction of 85% is obtained for debond case 1 with edge debond (75% debond area). The results also highlight that the debond cases with central debond do not have much effect on the first natural frequency. A maximum percentage reduction of 11% is obtained for debond case 1. To study the effect of debonding at high frequencies, the variation in 20th natural frequency is plotted with different debonding size. Figure 6 shows the change in 20th natural frequency for different debond cases. It is seen that for almost all debond cases, there is a considerable reduction in natural frequency. A maximum percentage reduction of 78%is obtained for debond case 3 with 75% debond area.

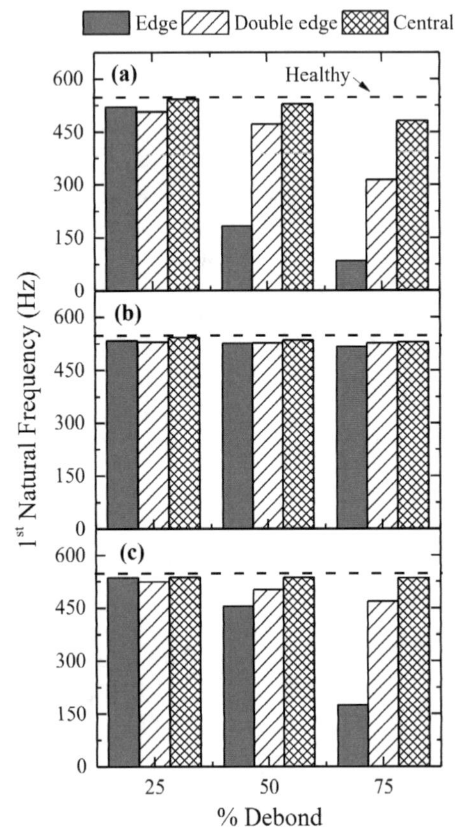

Figure 6. Variation in 1stnatural frequency for (a) Debond case 1 (b) Debond case 2 and (c) Debond case 3.

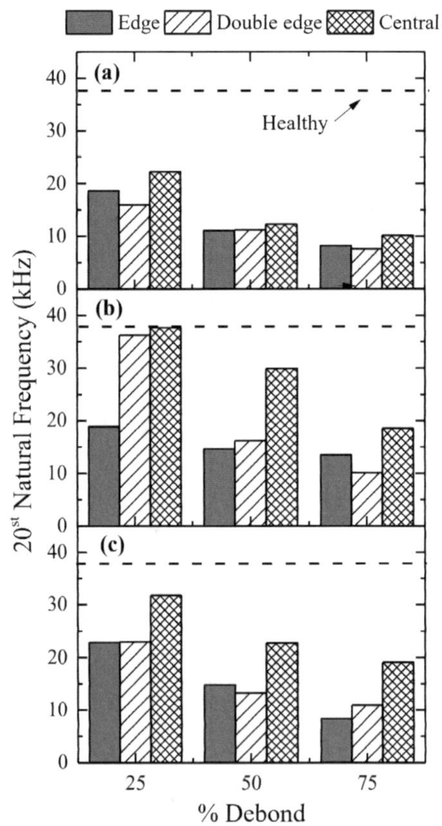

Figure7. Variation in 20th natural frequency for (a) Debond case 1 (b) Debond case 2 and (c) Debond case 3.

3.3. Influence of engineering properties of bonding agent

Piezoelectric materials are attached to the host structure using various kinds of bonding agents viz. thermoplastic tape, silver epoxy, Loctite adhesive, instant glue and Araldite epoxy [12]. The selection of a specific bonding agent is vital for achieving an ideal coupling between the host layer and the piezoelectric layer. The coupling governs the capability of the piezoelectric patches in actuating and receiving voltage signals. The structural adhesives available in the market can broadly be classified into

IOP Conf. Series: Materials Science and Engineering **936** (2020) 012038 doi:10.1088/1757-899X/936/1/012038

three types A1, A2 and A3 [13]. Adhesives - A1is classified under the rubbery region. These have a higher Poisson's ratio of $\nu = 0.49999$and a lower Young's modulus of $E = 0.001$ GPa. Adhesives- A2 are classified under rubber-to-glass transition region which has an intermediate value of $\nu = 0.4$ and $E = 1.0$ GPa whereas adhesives – A3 in the glassy region have a lower Poisson's ratio of $\nu = 0.3$and higher Young's modulus of $E = 10$ GPa. Figure 8(a) shows the variation in tip displacement for different adhesive types. It is observed that maximum tip displacement is obtained for adhesive type A3 whereas for adhesives A1 and A2 the percentage reduction in tip displacement is obtained as 97% and 9% respectively. The change in first and fiftieth natural frequency for different types of adhesives is plotted in figure 8 (b) and (c) respectively. The natural frequency obtained for adhesive A1, A2 and A3 is 465.8 Hz, 532.18 Hz and 545.3 Hz, respectively.

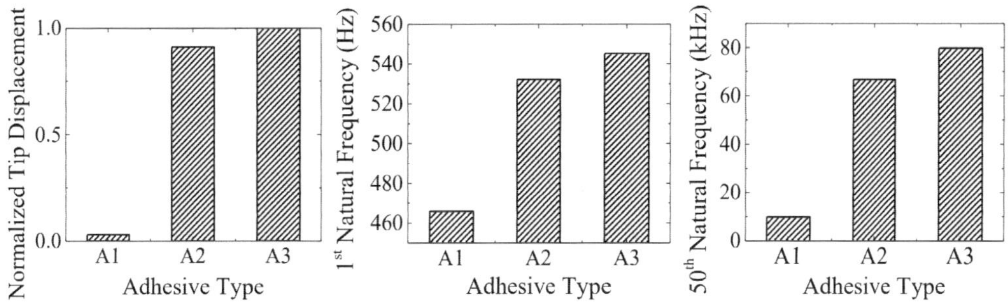

Figure 8. Effect of adhesive types on (a) tip deflection (b) first and (c) fiftieth natural frequency.

3.4. *Effect of thickness of the adhesive*

Shear lag is one of the unavoidable phenomena that occurs due to the presence of the bonding layer in smart structures. The previous studies indicate that as the bonding layer thickness increases, the effect of shear lag becomes more dominant [14]. The transfer of shear force between the host layer and the piezoelectric layer in such cases becomes less efficient. Figure 9 (a) shows the variation in tip displacement for different bonding layer thickness. It is seen that the magnitude of tip displacement decreases with an increase in adhesive thickness. From figure 9 (b) and (c) it is observed that the natural frequency also gets affected due to increased adhesive thickness. It is also realised that without considering the adhesive layer, the system overestimates its natural frequencies.

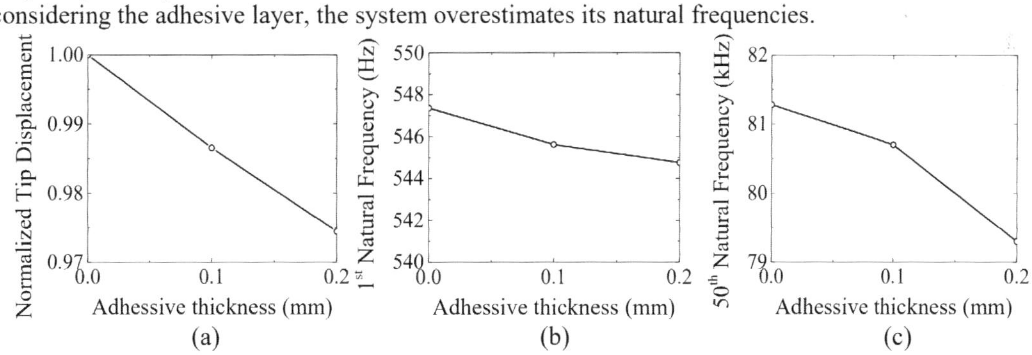

Figure 9. Effect of thickness of bonding agent on (a) tip displacement (b) first and (c) fiftieth natural frequency.

4. Conclusions

This paper presents the effect of adhesive debonding on the static and dynamic response of PZT bonded aluminium beam. A three-dimensional numerical study is done to investigate the effect of partial debonding on tip deflection and the natural frequencies of the smart cantilever beam. It is noted that the efficacy of the bonding layer on stress transfer between the host layer and piezoelectric layer depends on adhesive layer stiffness. Hence, partial debonding significantly alter the actuating capabilities of PZT actuator as well as its natural frequencies. The partial debonding effects are also substantial at high frequencies. The numerical results also emphasise the significance of considering the bonding layer in the finite element model. Also, it is observed that the system overestimates its natural frequencies if the bonding layer is neglected.

5. References

[1] Duan W H, Wang Q and Quek S T 2010 Applications of piezoelectric materials in structural health monitoring and repair: Selected research examples *Materials* **3(12)** 5169–94
[2] Aslam M, Nagarajan P and Remanan M 2019 Numerical studies on pzt-bonded aluminium beam *Journal of The Institution of Engineers (India): Series A* **100(1)** 117–30
[3] Altammar H, Dhingra A and Salowitz N 2018 Ultrasonic sensing and actuation in laminate structures using bondline-embedded d35 piezoelectric sensors *Sensors* **18(11)** 3885.
[4] Chen N, Yan P and Ouyang J 2019 A generalised approach on bending and stress analysis of beams with piezoelectric material bonded *Sensors and Actuators A: Physical* **290** 54–61
[5] Kirillova E V, Seemann W and Shevtsova M S 2019 The Influence of an adhesive layer on the interaction between a piezo-actuator and an elastic 3d-layer and on the excited wave fields *Materials Physics and Mechanics* **42(1)**
[6] Aslam M, Nagarajan P and Remanan M 2020 Dynamic response of piezoelectric smart beam with adhesive debonding *Mat Design Process Comm.* 1–6 (https://doi.org/10.1002/mdp2.159)
[7] Yu H and Wang X 2017 Modelling and simulation of surface-bonded piezoelectric actuators with bending effects *Journal of Intelligent Material Systems and Structures* **28(4)** 507–20
[8] Raja S, Rao K V and Gowda T M 2017 Improved finite element modeling of piezoelectric beam with edge debonded actuator for actuation authority and vibration behaviour *International Journal of Mechanics and Materials in Design* **13(1)** 25–41
[9] Liu X, Xu Y, Li N, Wang X and Zhang W 2019 Effect of adhesive debonding on the performance of piezoelectric sensors in structural health monitoring systems *Sensors* **19(23)** 5070
[10] Hosking N and Sotoudeh Z 2019 Comprehensive analysis of smart beams *AIAA Journal* **57(1)** 408–20
[11] Wang Q and Yu W 2012 Asymptotic multiphysics modeling of composite slender structures *Smart Materials and Structures* **21(3)** 035002
[12] Mustapha S and Ye L 2015 Bonding piezoelectric wafers for application in structural health monitoring–adhesive selection *Research in Nondestructive Evaluation* **26(1)** 23–42
[13] He X and Oyadiji S O 2001 Influence of adhesive characteristics on the transverse free vibration of single lap-jointed cantilevered beams *Journal of Materials Processing Technology* **119(1-3)** 366–73
[14] Moharana S and Bhalla S 2012 Numerical investigations of shear lag effect on PZT-structure interaction: review and application *Current Science* **25** 685–96

IOP Conf. Series: Materials Science and Engineering **936** (2020) 012039 doi:10.1088/1757-899X/936/1/012039

Hysteresis Behaviour of Precast Shear Wall – Slab Connection under Reverse Cyclic Loading

S Arthi[1*] and K P Jaya[2]

[1] Faculty, Department of Civil Engineering, National Institute of Technology-Puducherry, India

[2] Professor, Division of Structural Engineering, Department of Civil Engineering, Anna University, Chennai, Tamilnadu, India

*arthisubramanian.s@gmail.com

Abstract. An experimental study focused on precast shear wall-slab joint failure is presented in this paper. For the precast specimen, the shear wall and slab are precast members and were connected with cast-in situ concrete using dowel bars. The specimen was tested under reverse cyclic loading and the structural behaviour of the connection was evaluated. The parameters considered are cracking pattern, hysteresis behaviour, energy dissipation and ductility. This study also aimed to develop 3-D finite element (FE) model of precast shear wall – slab dowel connection using ABAQUS software in order to investigate the joint failure of the specimen such as damage and hysteresis behaviour. The non-linear behaviour of concrete was defined using concrete damaged plasticity (CDP) model and the interactions between the precast members were modelled using cohesive element. FE model was then validated against the experimental results of precast dowel connection which are vulnerable to wall-slab joint shear failure. The failure pattern of the tested specimen showed the "Strong joint-Weak member" concept. The comparison between experimental and numerical results showed that the FE model developed was capable of simulating the seismic performance and the joint failure of precast shear wall-slab connection using dowel bars.

1. Introduction

Precast construction technology has gained consistent growth in recent years. The prefabrication comprises of different members cast in a factory and erected in the site to create a structural system. The crucial region in precast structures was the connection between the members. This was due to the important role they play in controlling the behaviour of the building subjected to seismic loading. The connection between shear wall and slab was a critical region in structures to resist lateral load. The intense stress occurred at the joint between the shear wall and slab when earthquakes occurred [1]. Zenunovic and Folic [2] have studied the connections between the cast-in situ wall with the precast slab. The energy dissipation of the connection was observed to be similar to the cast-in-situ connection. Arthi S and Jaya KP [3] studied the seismic response of shear wall and diaphragm connection using dowel bars. The precast connection showed superior behaviour with respect to ultimate load and ductility compared with monolithic connection. Therefore, it was imperative to analyse the behaviour of the precast wall and diaphragm connection under seismic loading. The investigation of such performance needs computation of structures, materials and the loading effects. The FE method will be the best solution to include the variation in these parameters. Due to the advancements in the computation, the efficiency of FE programs to predict the performance of RC structures has already been established [4], [5], [6].Generally, the different types of connections in precast structures are emulative, dry, welded and bolted connection. In case of wet connection, the horizontal forces between precast members are transferred by dowel bars, shear keys and interface friction. The performance of connection between beam and column using dowels was investigated [7].Yuksel et al. [8] have discussed different types of connection between beam and column. . The authors have proposed one of the connection detailing with the presence of corbel. In the proposed connection, half depth of the beam was precast and the longitudinal reinforcement was tied with the stirrups protruding from the precast beam. Finally second concreting was done. Finite element analysis of precast connection using dowels was done using the ABAQUS software to reproduce the cyclic performance of the connection. The dowel bar from the column are connected to the beam, and then the gap was filled with mortar [9], [10].It was found that defining the material properties and the contact surface between the precast members was important for FE analysis. [11]. Soudki et al. [12] have studied the precast connections between walls. The precast connection was proposed by using reinforcement, shear keys, and tendons between walls. Feng et al. [13] investigated the shear performance of precast column using grouted sleeve connected rebar subjected to seismic loading and developed the non-linear FE model using ABAQUS software. The specimens were tested under cyclic loading. The authors observed that the grouted sleeve connection ensured the transfer of shear force and showed similar behaviour as compared to the cast-in situ connection. The connection between the precast walls using dowel bars and bolted steel channels was studied by Taheri et al. [5]. The interface shear strength at the contact surfaces was significant for the modelling of precast structures. [14], [15]. Some research work explained the capability of the Concrete Damaged Plasticity model which exactly simulates the behaviour of the structures [16], [6]. From the literature survey, it was understood that, the connection between precast members was important and only minimum works was carried out related to wall and slab connection. Therefore, the objective of this work is to assess the structural performance of exterior precast shear wall and slab connection by conducting an experimental testing and finite element modelling subjected to seismic loading.

2. Experimental work

2.1. Detailing of precast specimen

The critical forces of the structural members were obtained by analysing the -8 storey structure using staad pro software. From the analysis, the shear force at the joint between wall and slab was found to be 963.04 kN. The wall and slab was designed as per IS 456-2000 [17] and detailed as per IS 13920-1993[18]. The dowel reinforcement between precast wall and slab was designed to resist the shear force acting at the connection region [19]. The reinforcement detailing is shown in figure 1.The connection detailing is shown in table1 and figure 2.

Table 1. Reinforcement detailing of precast connection between shear wall-slab.

Sl. No	Connection between precast structural member	Reinforcement detailing	Development length
1	shear wall-slab	4 nos of 6 mm diameter @ 200 mm c/c	270 mm
2	shear wall - wall	5 nos of 10 mm diameter @ 175 mm c/c	450 mm

2.2. Assembly of precast specimen

One-third scaled-down specimen was cast using M-30 grade of concrete for testing. The cubes were tested on 28th day and the compressive strength (f_{ck}) obtained was 39.2 N/mm^2. The construction sequence of the specimen was as follows:

1. The precast slab was provided with a 20 mm diameter circular duct for connecting the dowel bars protruding from the precast shear wall and the space was filled with high strength grout.
2. Dowel bars from the precast slab was tied with the reinforcement which is placed above the precast slab. Screed concreting was done.
3. The upper wall panel was erected and the space was filled by grouting. The construction stage of the specimen is shown in figure 3.

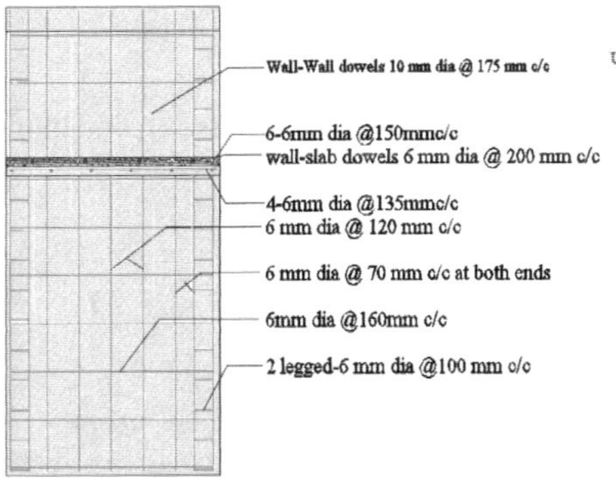

Figure 1. Reinforcement Detailing.

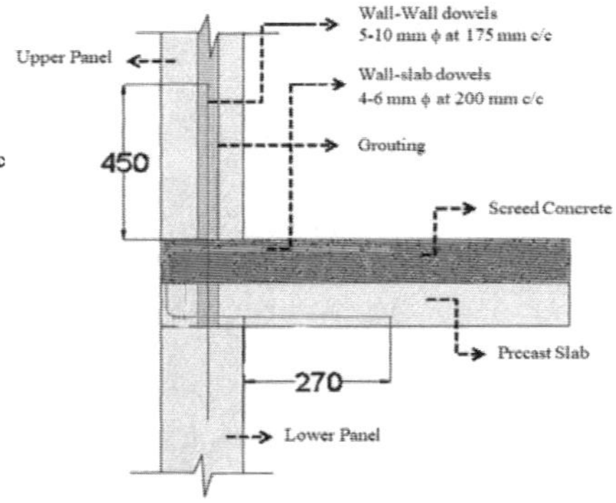

Figure 2. Precast connection Detaling.

Figure 3. Assembly of precast connection.

2.3. Test set-up

The axial load was kept constant throughout the test. During an earthquake, the wall and slab joint would be subjected to an in-plane moment. This was simulated by applying forces at the ends of the slab as shown in figure 4. The displacement and load were monitored using LVDT and load cells which were placed at the slab ends (Figure 5).

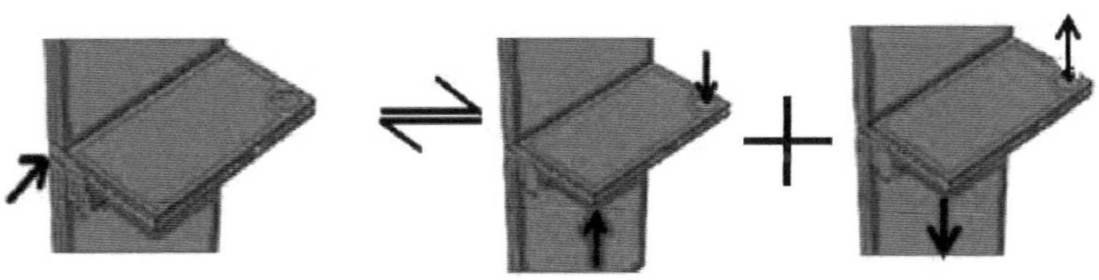

Figure 4. Simulation of reverse cyclic loading.

2.4. Loading sequence

The sub-assemblage of the specimen was tested under displacement controlled loading protocol. Three cycles for each displacement level was applied at the slab ends. The loading protocol for the tested precast specimen consisted of displacement of ±1 mm, ±2mm, ±3 mm, ±5 mm, ±7 mm, ±10 mm, ±13 mm, ±16 mm, ±20 mm, and ±27 mm. The loading protocol used for testing is shown in figure 6.

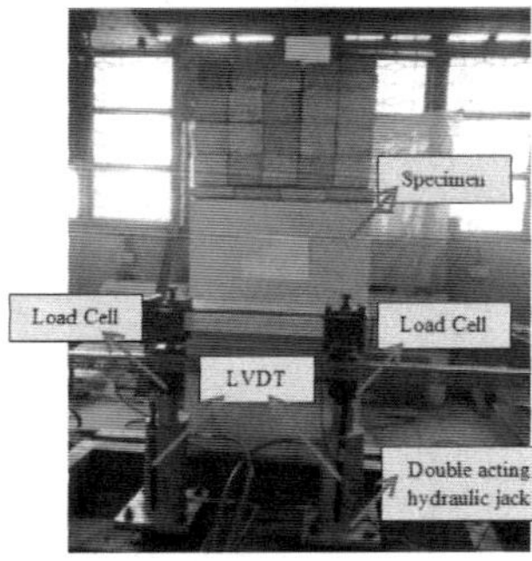

Figure 5. Test set- up.

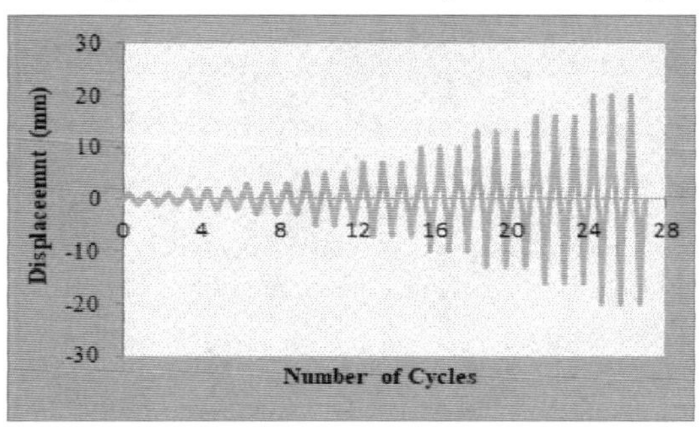

Figure 6. Loading protocol.

3. Testing results

3.1. Failure mode

The failure of the precast specimen is shown in figure 7. The damage of precast specimen tested under reverse cyclic loading is as follows:

Precast Specimen	❖ The initial crack started from the left side of the slab. ❖ The cracks started diagonally from the loading point at 3mm (5.25kN) positive displacement cycle. At 3mm (5.50kN) negative displacement cycle, the crack extended and developed nearer to the joint region. ❖ Crack at the joint region started at 10mm (7.95kN) positive displacement cycle and gets widened at 13mm (7.97kN) negative displacement cycle. ❖ At 16mm (7.50kN) negative displacement cycle, the crack widened at the right side of the precast slab and concrete crushed at the joint region. ❖ The crack width at the joint region on the left side of the slab was 4mm and at the top face of the slab was 2mm. There was a debonding between the lower panel and slab at 20mm (6.22kN) positive displacement cycle. ❖ The tested specimen reached a maximum displacement of about 27 mm.

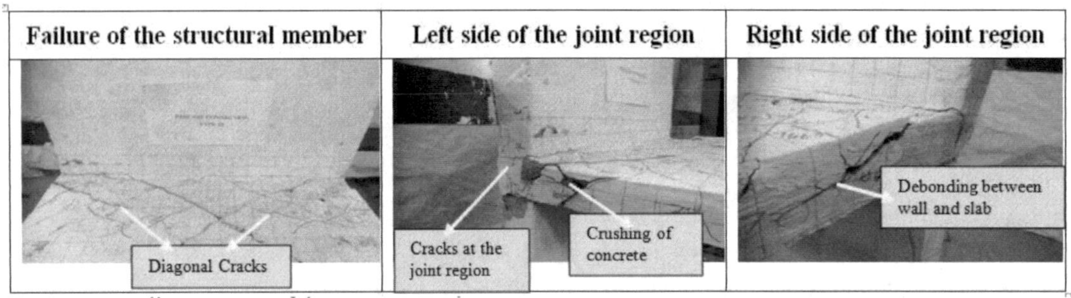

Figure 7. Failure pattern of the precast specimen.

3.2. Hysteresis behaviour

The hysteresis behaviour of the precast dowel connection regarding the load-displacement relationship has been discussed in this section. The experimentally tested specimen exhibited load-carrying capacity of 8.35 kN at 3.2 % drift (13 mm) and 7.97 kN at 3.2 % drift (13mm) in the positive (push) and negative (pull) direction loading, respectively. From the load-displacement graph (Figure 8), it was observed that there was linear hysteresis behaviour at the initial loading stage and later pinching effect was observed as displacement increased. After reached the ultimate load, there was a drop in load-carrying capacity of the specimen.

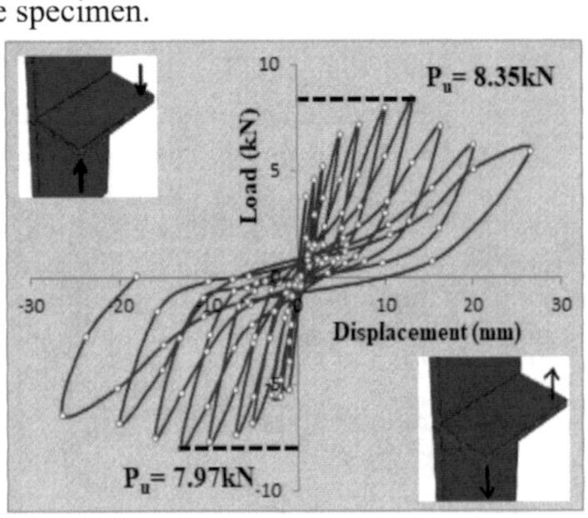

Figure 8. Load-displacement curve.

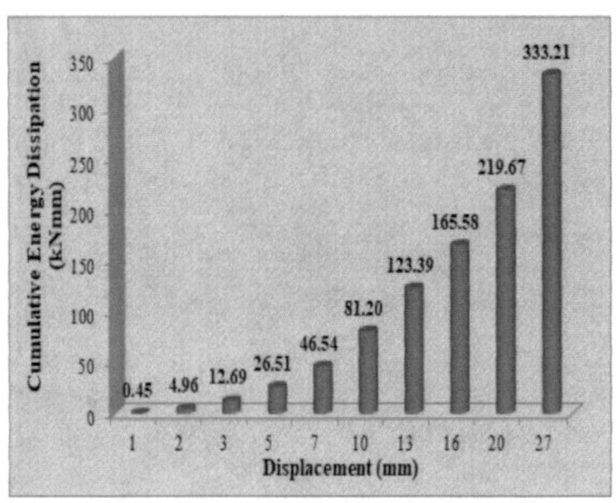

Figure 9. Energy dissipation capacity.

3.3. Energy dissipation

The total energy dissipated by the specimen was found by the area under the load-displacement (P-Δ) relationship. The cumulative value was found by adding the energy dissipated by the specimen in the successive cycles of displacement. The area under the P-Δ curve in each displacement became larger as displacement increased as shown in figure 9. The cumulative energy dissipation of the tested specimen was found to be 333.21 kNmm.

3.4. Ductility
Ductility was the crucial index in defining the seismic performance of precast specimen. It was obtained by the dividing the ultimate by yield displacement of the specimen. The calculated ductility factor of the tested specimen is shown in table 2.

Table 2. Ductility of precast specimen.

S.No	Specimen	Δ_y (mm)		Δ_u (mm)		μ		μ (Average)
		Push	Pull	Push	Pull	Push	Pull	
1	Precast	5.73	5.02	23.2	26.5	4.05	5.28	4.66

*Where Δ_y – yield displacement, Δ_u – ultimate displacement, μ - ductility.

4. Finite Element Modelling
One – third scaled-down specimen was modelled using FE software ABAQUS as similar to the experimental programme. The modelling stages in FE analysis are as follows:

4.1. Modelling
In the present work, the concrete and grout parts are modelled using a three dimensional solid element and the assigned concrete region was meshed using C3D8R element with hourglass control. The reinforcement and the dowel bars were modelled using 3D wire planar elements and meshed by using a 2-noded B31 beam element. The modelled specimen is shown in figure 10.

4.2. Material property
Concrete Damaged Plasticity (CDP) model was used for defining the nonlinear behaviour of concrete. The Young's modulus of concrete defined in the FE analysis was 32725.49 N//mm^2 and poisson ratio was 0.2. The value of eccentricity, dilation angle, initial biaxial to uniaxial ratio, viscosity , *k*, defined in this study was 0.1, 38^0, 1.12, 0.666 and 1 and The performance of concrete to dynamic loads in the CDP model is shown in figure 11. The modulus of elasticity of steel was 200000 N/mm^2 and Poisson ratio was and 0.3.

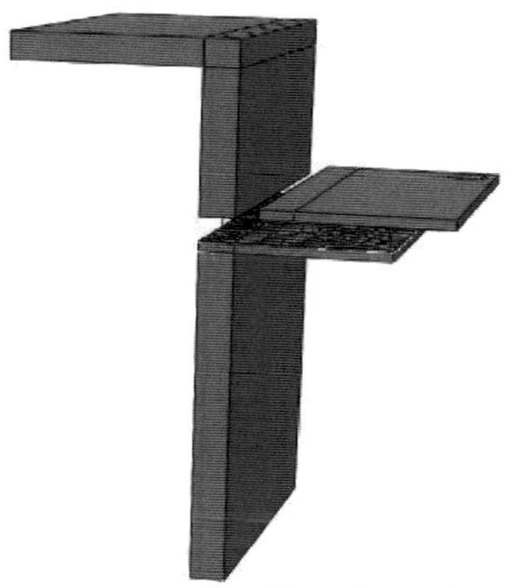

Figure 10. Assembly of Specimen.

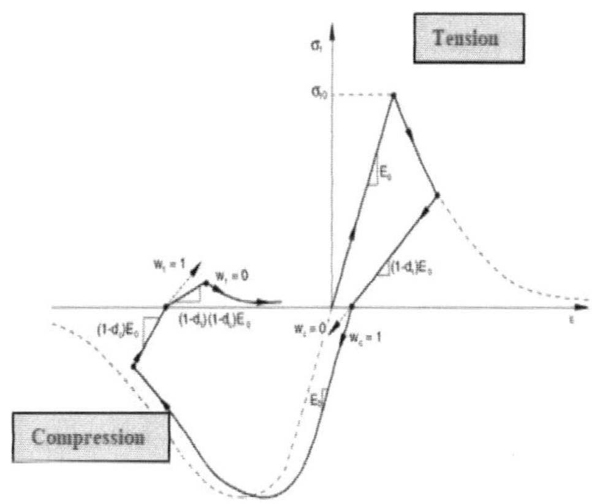

Figure 11. Behaviour of concrete in CDP model.

4.3. Interactions
The contact surface and the interaction between the precast members play an essential role in the finite element modelling for nonlinear analysis. In this analysis, five contact surfaces were defined between the structural members and shown in table 3.

Table 3. Interaction Property.

S.No	Contact Surface	Interaction Property
1	Precast lower shear wall-precast slab contact surface	Cohesive Element (eight noded, three-dimensional cohesive elements COH3D8)
2	Precast slab-screed concrete contact surface	Cohesive Element
3	Screed concrete- upper shear wall	Tie Constraint
4	Interaction between dowels-concrete/grout	Embedded Constraint
5	Reinforcement and concrete	Embedded Constraint

5. Finite element results

5.1. Hysteresis behaviour

From figure 12, the precast specimen showed spindle-shaped hysteresis behaviour with low pinching due to the bond between the reinforcement and concrete in the joint region. The maximum drift ratio (%) undergone by the precast specimen was 6.8% (27mm). The specimen exhibited load-carrying capacity of 9.35 kN at 3.2% drift (13 mm) and 8.82 kN at 3.2% drift (13mm) in the push and pull loading, respectively. The failure of the precast specimen from the FE analysis is shown in figure 13.

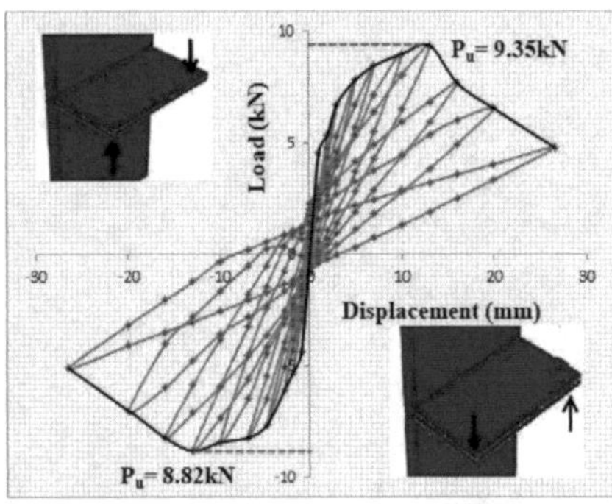

Figure 12. Load-Displacement curve.

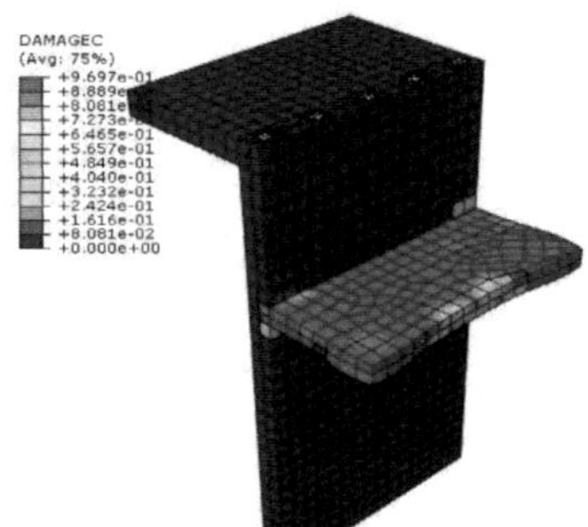

Figure 13. Damage of the specimen.

5.2. Ductility

The ductility factor of the modelled specimen was calculated and shown in table 4. Although ABAQUS can estimate the capacity and behaviour of specimens well, it cannot account for pinching effects on cyclic behaviour of RC structures [4], [20]. The ductility obtained by finite element analysis was 10.9% higher than the experimental result.

Table 4. Ductility of precast specimen.

S.No	Specimen	Δ_y (mm)		Δ_u (mm)		μ		μ (Average)
		Push	Pull	Push	Pull	Push	Pull	
1	Precast	3.52	3.84	17.99	20.05	5.11	5.22	5.17

*Where Δ_y – yield displacement, Δ_u – ultimate displacement, μ - ductility.

6. Conclusions

In this study, a one-third scaled-down exterior precast shear wall – slab specimen was subjected to reverse cyclic loading. The tested specimen was also modelled using finite element software as similar to the experimental programme. The conclusions made from this study are as follows:

1. The precast specimen exhibited an ultimate load-carrying capacity of 8.35 kN in the positive direction and 7.97 kN in the negative direction.

2. It was observed that there was linear hysteresis behaviour at the initial stage and the area under the hysteresis loop becomes larger as displacement increases which showed good energy dissipation capacity of the dowel connection.
3. The proposed precast dowel connection showed ductile behaviour and the ductility was found to be 4.66.
4. The crack propagation was more in the precast slab than the joint region which satisfied the "strong joint – weak member" concept.
5. The finite element results were found to be 11% greater than the test results and also exactly predicted similar behaviour as experimental testing.

7. References

[1] Kaushika S and Dasgupta K 2015 Seismic damage in shear wall-slab junction in rc buildings *Procedia Engineering* **144** 1332–9

[2] Zenunovic D and Folic R 2012 Models for behaviour analysis of monolithic wall and precast or monolithic floor slab connections *Eng. Struct.* **40** 466–78

[3] Arthi S and Jaya KP 2020 Seismic performance of precast shear wall–diaphragm connection: a comparative study with monolithic connection *International Journal of Civil Engineering* **18** 9–17

[4] Surumi RS, Jaya KP and Greeshma S 2015 Modelling and assessment of shear wall–flat slab joint region in tall structures *Arab J Sci Eng.* **40** 2201–17

[5] Taheri H, Hejazi F, Vaghei R, Jaafar M S and Aabang Ali A A 2016 New precast wall connection subjected to rotational loading *Periodica Polytechnica Civil Engineering* **60(4)** 547–60

[6] Dere Y and Koroglu MA 2017 Nonlinear fe modeling of reinforced concrete *International Journal of Structural and Civil Engineering Research* **6** (1)

[7] Zoubek B, Fischinger M and Isakovic 2016 Seismic response of dowel connections in RC structures *NZSEE Conference* (New Zealand)

[8] Yuksel E, FarukKaradogan H, EnginBal I, Iiki A, Bal A and Inci P 2015 Seismic behavior of two exterior beam–column connections made of normal-strength concrete developed for precast construction *Engineering Structures* ***99*** 157–72

[9] Magliulo G, Ercolino M, Cimmino M, Capozzi V and Manfredi G 2014 Seismic behavior of beam-to-column dowel connections: numerical analysis vs experimental test *Second European conference on Earthquake Engineering and Seismology* (Istanbul) p 1–11

[10] Kremmyda G D, Fahjan Y M and Psycharis I N 2013 Analytical prediction of the shear resistance of precast rc pinned beam-to-column connections *4th ECCOMAS Thematic Conference on Computational Methods in Structural Dynamics and Earthquake Engineering* (Kos Island, Greece)

[11] Zoubek B, Fahjan Y, Fischinger M and Isokovic T 2014 Nonlinear finite element modelling of centric dowel connections in precast buildings *Computers and Concrete* **14** 463–77

[12] Soudki K A, West J S, Rizkalla S H and Blackett B 1996 Horizontal connections for precast concrete shear wall panels under cyclic shear loading *PCI Journal* 64–80

[13] Feng B, Xiong F, Liu B, Chen J and Zhang Y 2016 Shear performance of horizontal joints in short precast concrete columns with sleeve grouted connections under cyclic loading, *Plos One* **11(11)** e0165988

[14] Mohammad M E and Ibrahim I S 2015 Interface shear strength of concrete-to-concrete bond with and without projecting steel reinforcement *Jurnal Teknologi (Sciences&Engineering)* **75(1)** 169–72

[15] Papanikolaou V K and Thermou G E 2015 Concrete-to-concrete interfaces under cyclic loading - Finite element analysis towards experimental verification *5th ECCOMAS Thematic Conference on Computational Methods in Structural Dynamics and Earthquake Engineering* (Crete Island, Greece)

[16] Alfarah B, López-Almansa F and Oller S 2017 New methodology for calculating damage variables evolution in Plastic Damage Model for RC structures *Engineering Structures* **132** 70–86

[17] IS 456:2000 *Indian Standard Plain and Reinforced Concrete Code of practice* (Bureau of Indian Standards: New Delhi, India)
[18] IS 13920:1993 *Indian Standard Ductile Detailing of Reinforced Concrete Structures Subjected to Seismic forces* (Bureau of Indian Standards: New Delhi, India)
[19] Elliot KS 2017 *Precast Concrete Structure,* Second Edition, CRC press (Taylor & Francis: NewYork)
[20] Wan S, Loh C H and Peng S Y 2001 Experimental and theoretical study on softening and pinching effects of bridge column *Soil Dynamics and Earthquake Engineering* **21** 75–81

Acknowledgement
This research was supported by the Council of Scientific & Industrial Research (CSIR). The authors are grateful to the funding agency for their support.

Drift Response Evaluation of Buckling-Restrained Braced Frames (BRBFs) under Sequential Seismic Disturbances

Muhamed Safeer Pandikkadavath[1*], Sujith Mangalathu[2], Romanbabu M Oinam[3], P C Ashwin Kumar[4], Dipti Ranjan Sahoo[5] and N M Anoop Krishnan[5]

[1] Department of Civil Engineering, National Institute of Technology Calicut, Kerala, India

[2] Department of Civil and Environmental Engineering, University of California, Los Angeles, USA

[3] Department of Civil and Environmental Engineering, Indian Institute of Technology Tirupati, India

[4] Department of Earthquake Engineering, Indian Institute of Technology Roorkee, Uttarakhand, India

[5] Department of Civil Engineering, Indian Institute of Technology Delhi, New Delhi, India

[*]msafeerpk@nitc.ac.in

Abstract. Buckling Restrained Braced Frames (BRBFs) are reliable lateral load resisting systems featured with Buckling Restrained Braces (BRBs) to safeguard building frames against earthquake-induced damages. BRBs are axial yielding type metallic dampers characterised with stable and balanced hysteretic behavior along with excellent low-cycle fatigue capacity. However, BRBFs are susceptible to relatively large drift responses mainly due to low axial stiffness of BRBs resulted from the higher seismic response modification factor, *R*. In the event of sequential seismic actions immediately after mainshock may induce additional drift demands which may adversely affect the reusability and retrofitting or in the worst case, it may lead to premature collapse of the BRBFs. The present study encompasses on drift response demand evaluation of BRBFs under the combined action of the main shock and sequential secondary ground shocks. For the study, a medium-rise BRBF used in the past investigation is selected and re-designed as per the recommended prevailing seismic provisions. Nonlinear time history analysis of BRBFs under the action of a suit of sequential ground disturbances is carried out and the results are analysed in terms of drift responses.

1. Introduction

Structural systems located in seismic regions are vulnerable to sequential tremors followed by Mainshocks. The Secondaryshocks or Aftershocks soon after Mainshocks may cause extended damages to the already yielded (under Mainshocks) structural components [1]. Normally the seismic analysis and design provisions account for the effect due to Mainshocks or Primaryshocks only. Investigations showed that the Secondaryshocks vary in terms of magnitude, space, frequency content, and duration compared to Primaryshocks. Many times the variation of these parameters under Aftershocks is more detrimental compared to the Mainshocks[2].This, in turn, poses serious concern over the adequacy of post Mainshock-Aftershock performance of the Structures. The effect of combined Primaryshock-Secondaryshock action on Buckling Restrained Braced Framed structural systems also not different from the above-discussed threat.

Buckling-Restrained Braced Frames (BRBFs) are ductile and stable lateral load resisting system which can be used to shield the structures against seismic induced hazards. BRBFs are characterised by axial yielding of Buckling Restrained Braces (BRBs). BRBs include a metallic core plate placed inside a mortar filled steel encasement, a de-bonding agent, that is embedded at the inter-face of metallic core plate and the surrounding mortar filled steel casings. The de-bonding agent will ensure the free axial movement of BRBs under inelastic reversed cyclic loading conditions and the surrounding mortar filled steel/metallic casing can act as a lower mode buckling-restrainer and to guide the metallic centrally placed metallic (mostly steel) core plate to yield in compression as well [3]. This makes the BRBs to be capable of yielding in both compression and tension and providing a balanced and stable hysteretic response under the inelastic cyclic loading conditions [4]. Figure 1 shows the schematic longitudinal as well as cross-sectional details of BRBs. The BRB metallic core plate has relatively longer restrained (using mortar filled steel encasing) yielding zone at the middle portion, followed by restrained transition zones and un-restrained connection zones towards either end successively.

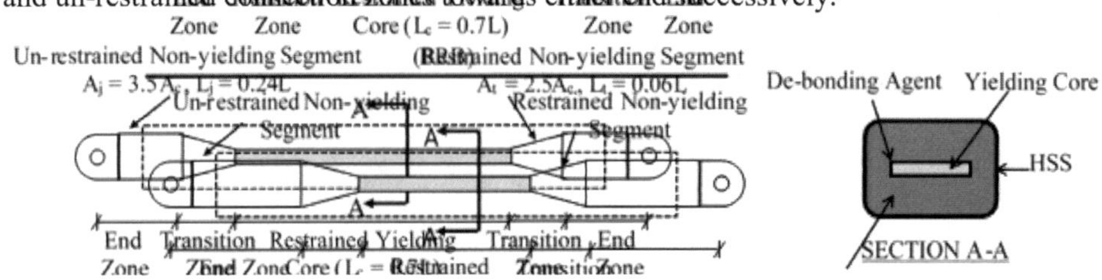

Figure 1. Schematic representation of BRB along with longitudinal and cross-sectional directions respectively.

However, past studies on Mainshock response evaluation showed that BRBFs are vulnerable to large lateral deformations and resulting drift responses under moderate to high seismic disturbances [5-7]. The relatively large response modification factor value, the capability of BRBs to yield in both tension and compression, and the resulting lower axial stiffness (elastic and post yield) may cause this unintended drift response of BRBFs [7-9]. In the event of sequential ground disturbances, this effect may become even worst and may adversely affect the reliability of the structure in the post-earthquake regime, so that the reusability and retrofitting of BRBFs may become unattainable. Further, the previous investigation reports on seismic response of BRBF under the sequential ground shakings are very limited and hence it warrants further investigations [10, 11]. In the present analytical study, the effect of sequential seismic disturbances, which is Mainshock or Primaryshock followed by Aftershock or Secondaryshock on a six-storey BRBF designed as per current seismic provisions is focussed. For the analysis a suite of ground motion records which comes under Maximum Considered Earthquake (MCE) hazard level is selected and a set of non-linear time-history analysis is performed. Successively, the seismic response evaluation of BRBF is carried out based on the residual (permanent) drift ratio and maximum inter-story drift ratio of the six-story BRB frame.

2. Study building frame and modeling

To carry out the present analytical investigation, a six-story building frame located in Los Angeles-USA is selected [7-9, 12] and the building is assumed to be situated in soil site class D. The building has an equal bay length of 9.14 m each along with both horizontal directions in the plan. The bottom storey-height of the building is 5.49 m and all other top floors have a storey height of 3.96 m each. Along any horizontal direction, the building has 5 bays and six symmetrically placed BRBF along its periphery as shown in figure 2.Two storey-X configurations are adopted for all the BRBs along with the height. The selected BRBF (marked in figure 2) is designed as per current seismic provisions by taking a response modification factor, R of value 8 [13]. Accordingly, the building frame is identified as 6XBRBFR8. The seismic design weight of the building is obtained as 61080 kN. The design spectral acceleration corresponding to a long-period (1.0 s) and short period (0.2 s) is found to be 0.77$g$and 1.39g respectively, where g is the acceleration due to gravity. The seismic design response co-efficient and the approximate time period of the designed building frame are found to be 0.17 and 0.82 s respectively. To analyse, design, and model the beams and columns of study frame, nominal material yield strength f_y of 345 MPa, the corresponding material overstrength (corresponding to yielding) factor R_y is taken as 1.1. Similarly, for BRBs these (f_y and R_y) values are taken as 248 MPa and 1.3 respectively [9]. The entire storey shear due to lateral loads are assumed to be resisted by respective BRBs and the corresponding BRB axial forces are obtained from the critical load combinations as per the recommended provisions. The strength adjustment factors corresponding to compression (β) and tension (ω) are adopted as 1.1 and 1.4 respectively based on past reports [14]. All the BRBs are assumed with pinned end connections and all the column-beam joints are taken as moment free type for the study frame. For the first three floors, beams and columns are *W14x120* and *W14x211* and for the remaining top floors, the beams and columns are *W14x82* and *W14x145* respectively. Figure 2 gives the BRB cross-sectional details on respective floors obtained from the design.

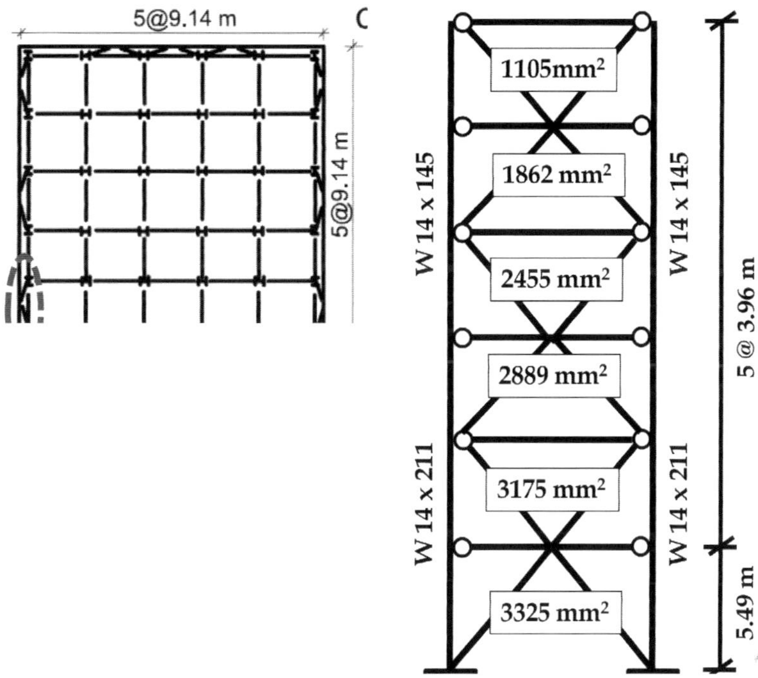

Figure 2. Building plan, elevation details of study frame 6XBRBFR8 along with BRB cross-sectional areas and Tri-linear plastic hinge model for BRB.

The building frame is modelled in CSI SAP 2000 platform with applicable provisions [15]. The columns and beams are incorporated as all beam-column elements (from the in-built library) by applying axial load-moment (*P-M*) and bending moment-rotation (*M-θ*) interaction hinges at a critical location of beams and columns [16]. End length off-sets for all the beams and columns are taken as 10% each at both ends. Similarly, for BRBs, a yielding zone length of 70% is assigned and applied with a tri-linear plastic hinge at mid-location of respective BRBs to include inelastic response shown in figure 2. The post-yield/elastic stiffness of BRBs is taken as 2% of its un-yielded (elastic) stiffness [7, 8] and the combined hardening model is considered for catching the hysteretic response of BRBs. In order to account for the *P*-Delta effect of loads, a continuous leaning column at a distance of one bay width from the study frame is modelled and these are connected with rigid elastic beams at respective floor heights. For all the non-linear time-history analysis a 2% Rayleigh damping is considered.

For the non-linear time-history analysis, a suite of 7 ground-motion records which comes under 2% exceeding probability in 50 years (MCE level) is considered [17]. All the selected ground motions have a time duration of 60*s*. The PGA of the selected ground motion records vary between 0.5*g* to 1.33*g* and the magnitude varies from 6.7 to 7.1 respectively. Figure 3 shows the response spectra (*g*) of adopted ground motion records along with details such as Peak ground shaking Acceleration, PGA. For one set of Mainshock-Secondaryshcok analysis, the same ground motion record is selected for both cases referring to past investigations [18, 19].

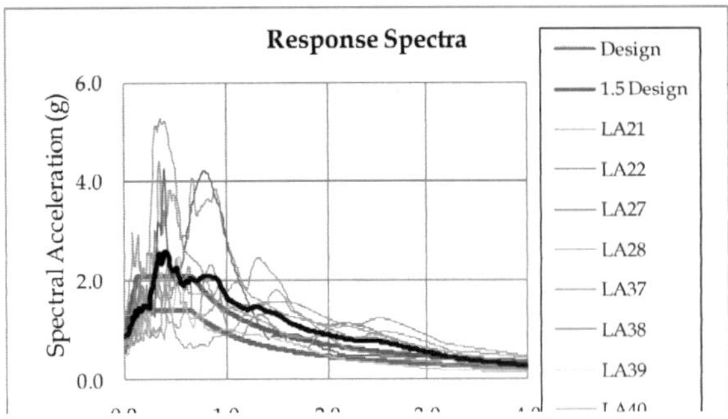

Record	Magnitude	PGA (g)
LA21	6.9	1.28
LA22	6.9	0.92
LA27	6.7	0.93
LA28	6.7	1.33
LA37	7.1	0.71
LA39	7.1	0.50
LA40	7.1	0.63

Figure 3. Response spectra, magnitude and PGA of selected ground motions along with design spectra of study frame.

3. Analysis Results and Discussions

From the analysis, the first mode time period of the study frame is obtained as 1.05 *s*. The nonlinear time-history analysis of 6XBRBFR8 under the selected set of Mainshock-Secondaryshock is carried out. The analysis results are presented in terms of the Residual Drift Ratio (RDR) and Inter-Storey Drift Ratio (ISDR) of the frame. The ratio of a maximum of the relative storey-displacement (lateral) to the corresponding storey height at any time during the seismic action is the maximum ISDR and the permanent drift left with the storey after the completion of seismic action is termed as the RDR. Both ISDR and RDR can be taken as an index to the extent of damage experienced to the frame system.

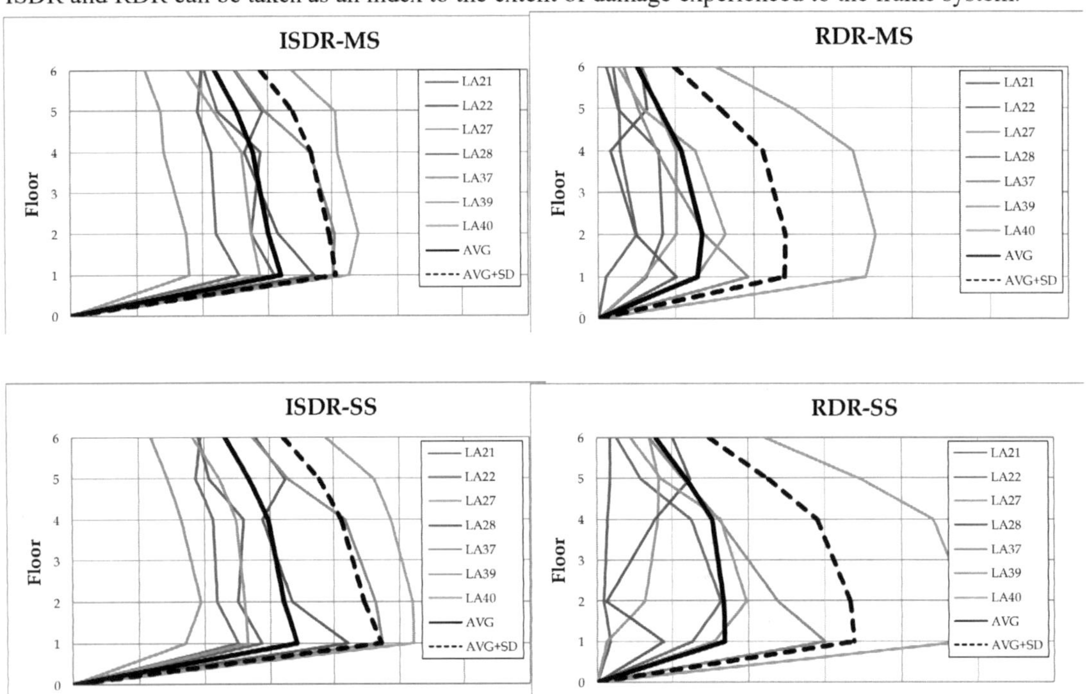

Figure 4. Variation of Inter-Storey Drift Ratio and Residual Drift Ratio under the selected set of Mainshock (MS) and Secondaryshock (SS).

Figure 4 shows the variation of RDR and ISDR of 6XBRBFR8 under the action of selected Mainshock-Secondaryshock seismic simulations. The maximum ISDR of 4.40% (under Mainshock) is experienced due to LA40 on the second floor of the study frame. Similarly, the average maximum ISDR under Mainshock ($ISDR_{avg\text{-}MS}$) is found to be 3.00% at the second floor itself with a standard deviation ($SD_{ISDR\text{-}MS}$) of 0.95%. A similar trend is obtained for RDR response as well. Due to the action of

Mainshock the frame experienced a maximum RDR of 1.77% on the second floor of the frame. The corresponding average maximum RDR (RDR_{avg-MS}) and standard deviation (SD_{RDR-MS}) are 0.67% and 0.52% respectively. Secondarshock action effect on the frame also showed a comparable trend observed in the Mainshock results. Both the maximum ISDR and average maximum ISDR ($ISDR_{avg-SS}$) and the standard deviation corresponding to $ISDR_{avg-SS}$ ($SD_{ISDR-SS}$) are obtained as 5.20%, 3.20%, and 1.26%, respectively. Whereas the maximum RDR, average maximum RDR (RDR_{avg-SS}), and the standard deviation corresponding to RDR_{avg-SS} (SD_{RDR-SS}) under Secondaryshocks are obtained as 2.42%, 0.84%, and 0.84%, respectively. The results showed the increase of ISDRand RDR is in the range of 5-10% and 20-25%, respectively.

Based on the non-linear time-history analysis, fragility curves are developed to understand the approximate seismic hazard levels of the frame considered. ISDR and RDR under Mainshock and Secondaryshock are considered as the Engineering Demand Parameter (EDP) for the analysis and fragility curves as per Gaussian distribution are plotted [20]. Provided, the EDP related mean as μ and standard deviation as *SD*, the cumulative normal distribution function is given as

$$p[EDP > x] = 1 - \phi\left(\frac{x - \mu}{SD}\right) \tag{1}$$

Figure 5 shows the fragility analysis curves (based on the normal distribution) of the study frame under Mainshock-Secondaryshock considering ISDR and RDR are Engineering Demand Parameters. The exceeding probability of ISDR over 4% is found to be 19% and 33%, respectively under Mainshock and Secondrayshock seismic disturbance level. Whereas the exceeding probability of RDR by 1% is obtained as 31% and 48%, respectively under the selected Mainshock and Secondrayshock ground disturbance levels.

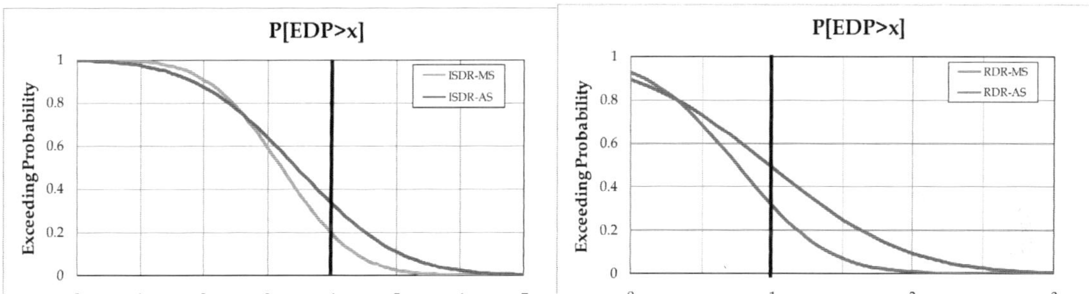

Figure 5. Fragility curves of 6XBRBFR8 considering ISDR and RDR.

4. Conclusions

The nonlinear time-history analysis under the selected suit of Mainshock-Secondaryshock seismic records is carried out on a six-storey Buckling Restrained Braced Frame. The investigation results are presented in terms of maximum Residual Drift Ratio (RDR) and Inter-Storey Drift Ratio (ISDR). Subsequently, fragility curves based on normal distribution are plotted taking Engineering Demand Parameters as ISDR and RDR. The Drift response evaluation showed that the increase of ISDR from Mainshock to Secondaryshock disturbances is in the range of 5-10%, whereas the increase of RDR is in the range of 20-25%. The resulting fragility curved showed that the exceeding probability of ISDR over 4% (from Mainshock to Secondaryshock hazard level) is in the range of 19-33% whereas the exceeding probability of RDR over 1% in the range of 31-48%. That is RDR response of 6XBRBFR8 showed higher variation in magnitude compared to ISDR.

5. References

[1] Li Q, and Ellingwood B R 2007 Performance evaluation and damage assessment of steel frame buildings under main shock–aftershock earthquake sequences *Earthquake Engineering and Structural Dynamics* **36** 405–27

[2] Aljawhari K, Freddi F and Galasso C 2019 State-dependent vulnerability of case-study reinforced concrete frames *Computational Methods in Structural Dynamics and Earthquake Engineering (COMPDYN) 2019* (Crete, Greece)

[3] Watanabe A, Hitomi Y, Saeki E, Wada A and Fujimoto M 1988 Properties of brace encased in buckling-restrained concrete and steel tube *Proceedings 9th World Conference on Earthquake Engineering* (Tokyo, Japan)

[4] Black C J, Makris N and Aiken I D 2004 Component testing seismic evaluation and characterisation of Buckling Restrained Braces *ASCE Journal of Structural Engineering* **131(6)** 880–94

[5] Fahnestock L A, Sause R and Ricles J M 2007 Seismic response and performance of Buckling Restrained Braced Frames *ASCE Journal of Structural Engineering* **133(9)** 1195–204

[6] Erochko J, Christopoulos C, Tremblay R and Choi H 2011 Residual drift response of SMRF and BRB frames in steel buildings designed according to ASCE 7-05 *ASCE Journal of Structural Engineering* **137(5)** 589–99

[7] Pandikkadavath M S and Sahoo D R 2016 Analytical investigation on cyclic performance of buckling restrained braced frames *International Journal of Steel Structures* **16(4)** 1273–85

[8] Pandikkadavath M S and Sahoo D R 2017 Mitigation of seismic drift response of braced frames using short yielding-core BRBs *Steel and Composite Structures* **23(3**) 285–302

[9] Chao S H, Karki N B and Sahoo D R 2013 Seismic behavior of steel buildings with hybrid braced frames *ASCE Journal of Structural Engineering* **139(6)** 1019–32

[10] Veismoradi S, Cheraghi A and Darvishan E 2019 Probabilistic mainshock-aftershock collapse risk assessment of buckling restrained braced frames *Soil Dynamics and Earthquake Engineering* **115** 205–16

[11] Morfuni F, Freddi F and Galasso C 2019 Seismic performance of dual systems with BRBS under mainshock-aftershock sequences *13th International Conference on Applications of Statistics and Probability in Civil Engineering ICASP13* (Seoul, South Korea)

[12] Sabelli R, Mahin S A and Chang C 2003 Seismic demands on steel braced frame building with buckling restrained braces *Engineering Structures*,**32(9)** 2950–8

[13] ANSI/ASCE 7-16 2016 Minimum design load for buildings and other structures *American Society of Civil Engineers* Reston (VA)

[14] Sahoo D R and Chao S H 2010 Performance based plastic design method for buckling-restrained braces *Engineering Structures* **32 (9)** 2950–8

[15] CSI 2013 SAP 2000 user guide *Computers and Structures Inc.(*Berkeley, CA, USA)

[16] FEMA 356 2000 Pre-standard and commentary for the seismic rehabilitation of buildings *Federal Emergency Management Agency* (Washington DC)

[17] Somerville P G, Smith M, Punyamurthula S and Sunj J 1997 Development of ground motion time histories for phase 2 of the FEMA/SAC steel project. Report no. SAC/BD-97/04 (Sacramento (CA), SAC Joint Venture)

[18] Jalayer F and Ebrahimian H 2017 Seismic risk assessment considering cumulative damage due to aftershocks *Earthquake Engineering and Structural Dynamics* **46(3)** 369–89

[19] Mirtaheri M, Amini M and Rad M D 2017 The effect of mainshock- aftershock on the residual displacement of buildings equipped with cylindrical frictional damper *Earthquakes and Structures* **12(5)** 515–27

[20] Ghowsi A F and Sahoo D R 2013 Seismic performance of buckling-restrained braced frames with varying beam column connections *International Journal of Steel Structures* **19(1)**173–90

IOP Conf. Series: Materials Science and Engineering **936** (2020) 012041 doi:10.1088/1757-899X/936/1/012041

Probabilistic Assessment of Torsional Buildings

Philip Luke Karuthedath[1*], Robin Davis[2] and Pradip Sarkar[3]

[1] PG Scholar, National Institute of Technology Rourkela, India

[2] Assistant Professor, National Institute of Technology Calicut, India

[3] Professor, National Institute of Technology Rourkela, India

[*]k.philip.luke@gmail.com

Abstract. Asymmetry/irregularity is an undesired building characteristic that nevertheless exists commonly due to modern-day structural definitions. Asymmetry in plan, due to the uneven distribution of mass or stiffness can cause torsional coupling under seismic loads. A well-defined parameter that can sufficiently represent the damage of torsional buildings is non-existent, which may be why probabilistic seismic assessment methods such as fragility analyses have not been done for such buildings. The present study proposes 'Resultant Drift' as a parameter that can define the failure behaviour of asymmetric buildings and utilises the 2000 SAC FEMA method to analyse the seismic risk and develop fragility curves. The study observes the dependency of the extent of asymmetry and storey height on the failure probability. Investigations on the maximum storey rotation for the selected parameters show the necessity of international standard definitions in the performance levels of rotation behaviour.

1. Introduction

Plan asymmetry in building structures is said to exist when the Centre of Mass (CM) and Centre of Stiffness (CS) are not co-incident. When hazardous loading such as earthquakes occur, such buildings possess rotational motion in addition to lateral movement. This torsional-translational coupling poses a serious threat to the buildings regarding their stability. The vulnerability of such plan asymmetric structures has been of interest to many researchers. Tso and Dempsey [1] studied the effect of the ratio of torsional frequency on lateral frequency and observed the variation of ductility demand. A study on the time-history response of asymmetric buildings [2] showed the inadequacy of building codes. Tso and Ying [3] divided the structural elements into flexible and rigid elements based upon their distance from asymmetry and observed their varying ductility demands. Samali et al. [4] investigated the effectiveness of base isolators on asymmetric buildings, while Vial et al. [5] observed the decreasing seismic demands when a frictional damper is incorporated. Numerous other studies [6–11] investigate the behaviour of torsional buildings and their seismic demands and attempts to understand or to provide a solution to this statement.

However, a proper index that can adequately represent the damage and quantify the vulnerability in an asymmetric building has not been accounted for and is virtually non-existent. Probabilistic methods such as fragility curves [12] provide an excellent base for risk as well as uncertainty quantification and have been generally used for various structural systems [13–19] but not for complex torsional buildings. Fragility curves represent the cumulative probability distributions that indicate whether a component/system will be damaged to a given damage state or a more severe one, as a function of a particular demand, such as the intensity of ground motion. The present study thus attempts to develop fragility curves for torsional buildings. Firstly, the study attempts to propose an engineering damage

IOP Conf. Series: Materials Science and Engineering **936** (2020) 012041 doi:10.1088/1757-899X/936/1/012041

parameter that can adequately represent the damage of an asymmetric building. Using this parameter, the seismic vulnerability of the torsional buildings is quantified using fragility curves.

2. Modelling of the structure

Since the investigation is on torsional buildings, the study considers 3 dimensional RC frames having 5 bays in both the plan directions. The number of storeys for the frames vary evenly from 4 to 10. They are designed according to relevant Indian standard IS 456: 2000 with the grades of concrete and steel as 25 MPa and 415 MPa, respectively. A storey height of 3.2 m and a bay width of 5 m is given for each of the frames considered. The base of the frames is assumed to be fixed. Elevation, plan, and section details of all the frames are given in figures 1-3. Rigid diaphragms are provided at each floor level to model the slab stiffness.

To account for the asymmetry in the models, mass eccentricity is induced by lumping heavy masses along the selected axis for each of the floors such that eccentricity to breadth ratio (*e/b*) of the models are varied, thereby obtaining various asymmetric levels for each of the building models. A symmetric counterpart is also considered as a reference. The study limits itself to uniaxial mass eccentricity that remains equal for every floor level. The terms eccentricity and asymmetry will be used interchangeably from now on to denote the level of difference between CS and CM.

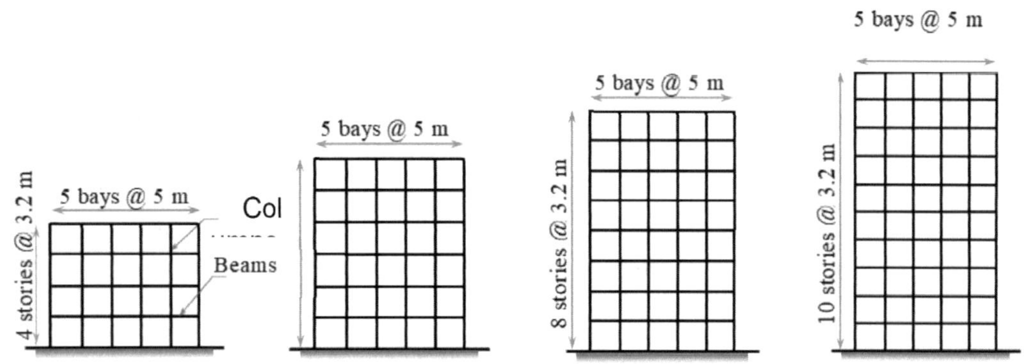

Figure 1. Elevation of the building frames.

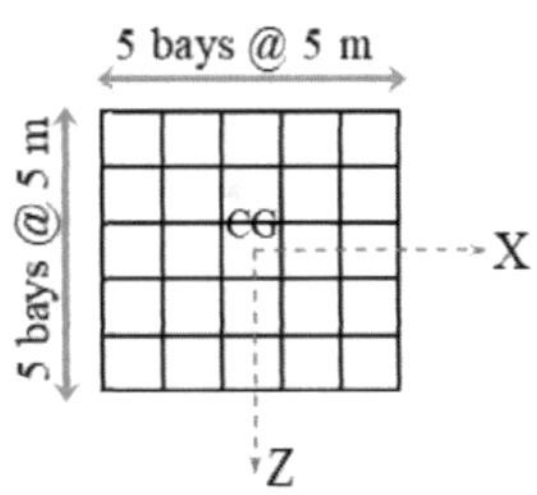

Figure 2. Plan view of the building frames.

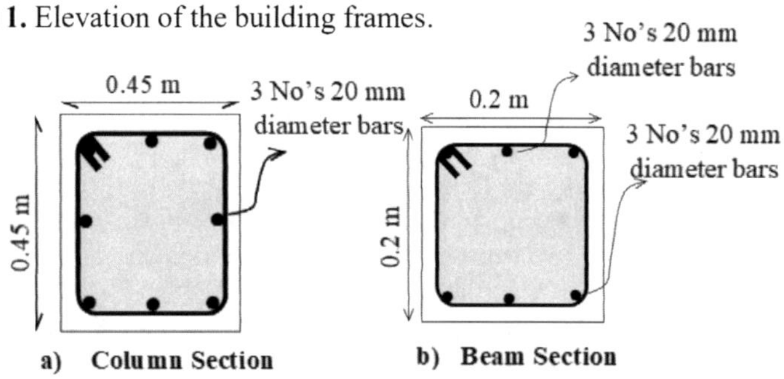

Figure 3. Section detailing for the models.

Selected buildings are modelled for nonlinear time history analysis using the Open System for Earthquake Engineering Simulation (OpenSees) framework [20]. Concrete modelling is done using the modified Mander's model [21]. Reinforcement steel behaviour is accounted for using a Menegotto-Pinto model [22] and the elements are modelled as fiber section [23].

3. Modelling of uncertainties & ground motion

Because of the error and randomness that can occur in the design, construction, and functioning; uncertainties are unavoidable in a real time building structure. They can occur as material, geometric properties. and loading conditions of the building. Significant material properties that can affect the nonlinear response of buildings are obtained from previous studies and are taken as random variables for the study. Table 1 presents the details of the selected random variables used in the study. Randomness in the seismic load is considered by using a suite of 22 pairs of far-field earthquake ground motions [24] and converting them according to Indian standards.

Table 1. Statistical details of random variables used for fragility analysis.

Material/Property	Variable	Mean	COV (%)	Distribution	Remarks
Concrete	f_{ck}	30.28 MPa	21	Normal	Uncorrelated
Steel	f_y	468.90 MPa	10	Normal	Uncorrelated
Global Damping ratio	ξ	5%	40	Normal	Uncorrelated

4. Resultant drift

The failure of a structure can be quantitatively related to certain standard parameters such as deflection, inter-storey drift (ISD), base shear etc.; exceedance of which indicates failure. Such parameters are known as Engineering Damage Parameters (EDP). For a symmetric building, the unidirectional translation of a building can sufficiently represent its inter-storey drift and thus the failure. However, for an asymmetric building, bidirectional translational is expected due to rotation even under uniaxial loading, and some alternate criteria that can represent its failure are requisite. Investigations on the nonlinear response of asymmetric buildings have shown that the resultant of lateral displacement in both the plan directions may adequately characterise its torsional behaviour. The idea behind this methodology is illustrated in figure 4 and formulated in Equation (1) and Equation (2).

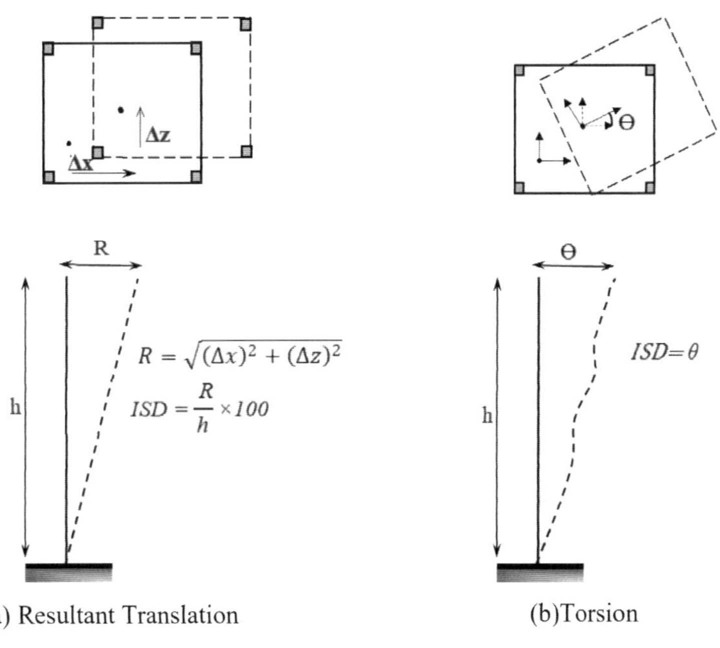

Figure 4. Deflection behaviour of asymmetric buildings.

$$R = \sqrt{(\Delta x)^2 + (\Delta z)^2} \tag{1}$$

$$ISD_r = \frac{R}{h} \times 100 \tag{2}$$

where Δx and Δz represents the translation of the floor in the plan directions, and h represents the storey height. The advantage behind this parameter is that it can satisfactorily explain the torsional behaviour while associating with the international standards of drift. Likewise, rotation of the building is another parameter that can represent the torsional behaviour, still the lack of existing standards of performance levels make it an unworthy option to obtain the seismic fragility.

5. Performance level

Non-linear time history analysis done on selected buildings give data on the selected damage parameters, which are compared with limit states or performance levels. The performance level stands for the structural and non-structural damage that may be allowed during earthquakes. In the present study, for the selected damage parameters, namely resultant translation and element strain, performance levels are specified from existing standards.

As for the resultant translation, maximum ISD limit states are taken from ASCE standards [25], which gives the failure criterion/performance levels for RC buildings. Table 2 summarises these limit states.

Table 2. Limit states for performance levels in ISD.

Performance Level	Permissible ISD
Immediate Occupancy (IO)	1%
Life Safety (LS)	2%
Collapse Prevention (CP)	4%

6. Results and discussion

Asymmetry divides the plan into flexible and rigid sides. For the uniaxial mass eccentric system considered, the edge closer to the CM will act as the flexible side while the edge closer to the CS will be the rigid side. A schematic diagram representing the deflection of the torsional building in the plan is shown in figure 5.

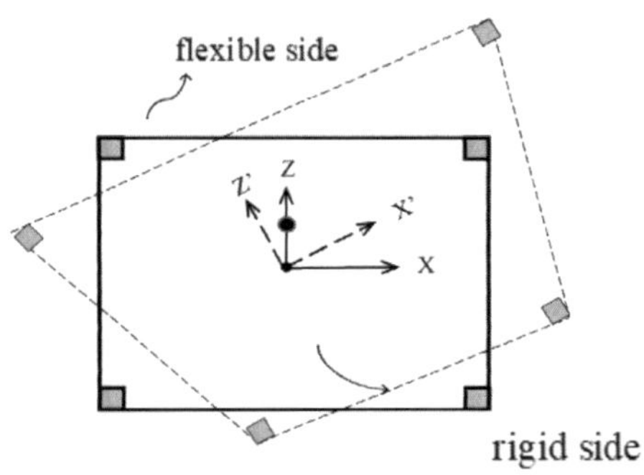

Figure 5. Deflection in plan due to eccentricity.

For the present study, the ground motion is applied in a direction perpendicular to the axis of asymmetry, and the deflection behaviour of the building will depend upon the intensity of eccentricity. Eccentricity is varied for each of the building models considered, and a comparative study is done on the building fragility using the resultant translation as the damage parameter.

Fragility curves for the four-storey frame having various levels of asymmetry for different performance levels are shown in figures 6 – 8. It is found that with each increase in eccentricity, the flexible side becomes more fragile while the rigid side becomes more strong while the fragility of the symmetric system remains the same irrespective of the side whose deflection is measured. This can be explained from the idea that, with each increment in eccentricity, the CM comes closer to the flexible side and farther to the rigid side. Thus when inertia force acts through CM, the nearness of the sides to the CM affects their damage behaviour. It can also be inferred that, the effective probability of exceedance of limit state in a building become higher with subsequent increase in eccentricity/asymmetry. Another observation that can be made is that, for a particular level of asymmetry, the flexible side is more vulnerable to earthquakes, which is illustrated in figure 9.

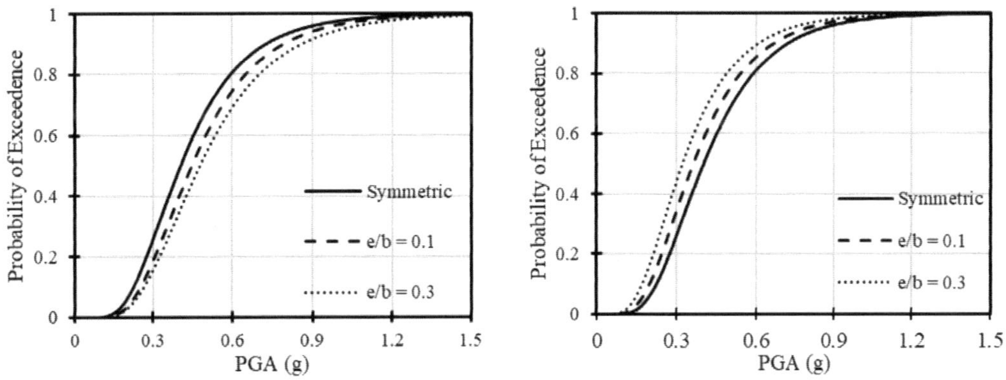

(a) Rigid Side (b) Flexible Side

Figure 6. Fragility curves for 4 storied frame – IO performance level.

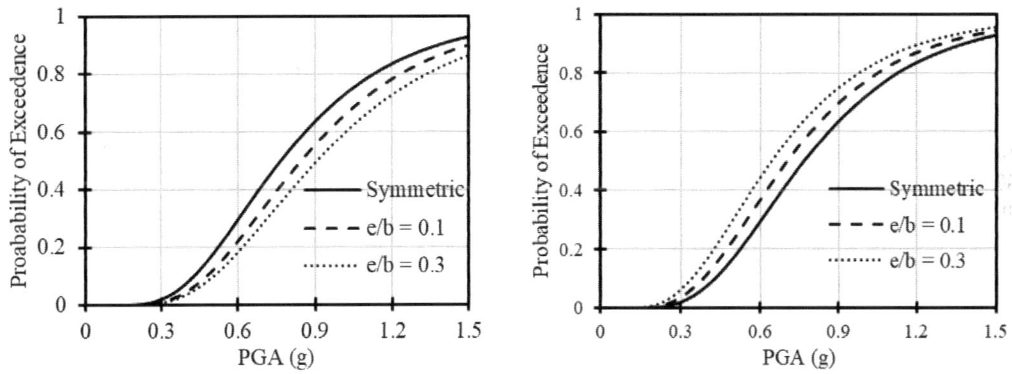

(a) Rigid Side (b) Flexible Side

Figure 7. Fragility curves for 4 storied frame – LS performance level.

The 6, 8, and 10 storeyed frames were also observed to show similar behaviour with the variation of asymmetry. Because of space constraints, the results are not included in this paper. A storey-wise comparison of all the models considered for each level of eccentricity is as given in figure 10 and figure 11. The graphs show that no matter what the level of asymmetry is; with subsequent increase in the building height, the seismic fragility of buildings increases. Moreover, the stiffness of the sides doesn't matter in this regard. However, the variation becomes negligible in higher storeyed buildings. The same trend is observed for all levels of asymmetry for all performance levels. The rest of the results are not included for the brevity of the manuscript.

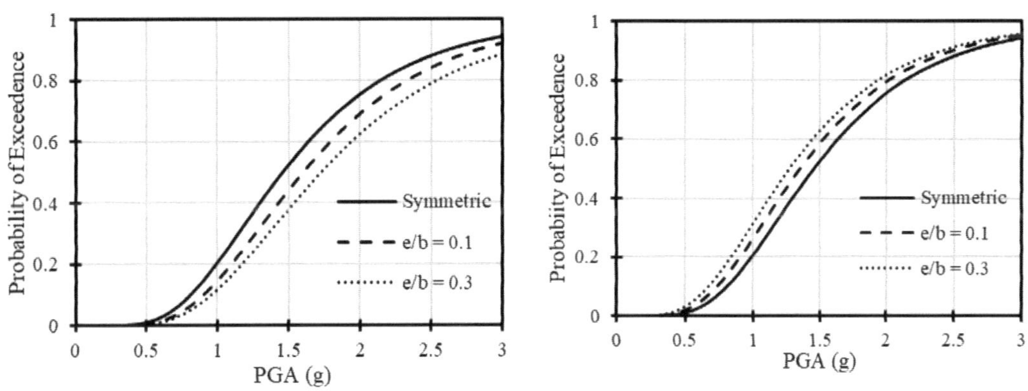

(a) Rigid Side (b) Flexible Side

Figure 8. Fragility curves for 4 storied frame – CP performance level.

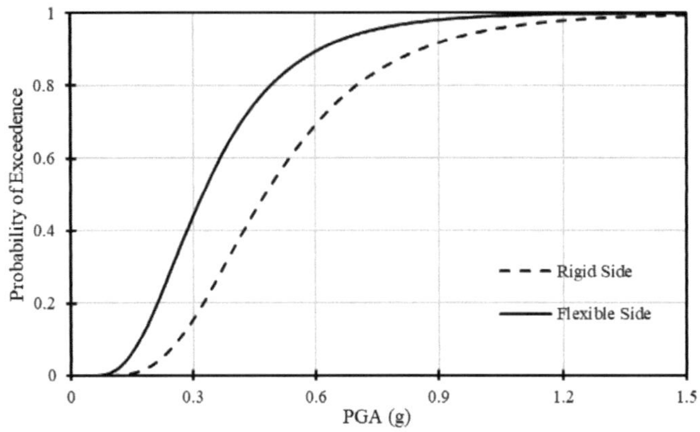

Figure 9. Four Storied building with eccentricity /breadth = 0.3, IO performance level.

Asymmetry in the plan is defined by the presence of rotational movement in buildings in addition to translational movement. This causes a torsional-lateral coupling, thereby making the structure weak under ground motion. It is, therefore, necessary to investigate the effect of eccentricity and storey height on the rotational behaviour of buildings. It should be noted that the amount of rotation is independent of the relative edge stiffness and is equivalent to all the points of a particular floor. Figure 12 plots the magnitude of maximum rotation for the building frames concerning the intensity of the ground motion. Various levels of asymmetry are considered as before.

For the symmetric system, the rotation is in the order of zero as expected. Still, with each increase in the eccentricity between the CS and CM, the amount of rotation is observed to be increasing and reaches the peak when *e/b* =0.5, the maximum possible eccentricity in a building. Figure 13 represents the storey wise variation in rotation. For a symmetric case, irrespective of the storey height, rotation value approximates to zero. In contrast, for an eccentric case, the rotation is observed to be a direct function of the building height.

Figures 12 and 13 show the presence and effect of rotation in buildings due to asymmetry as a result of the uneven distribution of mass/stiffness. Even though the present chapter deals with developing fragility curves of asymmetric building, the study deliberately avoids plotting fragility curves by considering the rotation as a damage parameter. This is because of the unavailability of international standards that define performance levels/limit states with which permissible rotation in buildings is quantified.

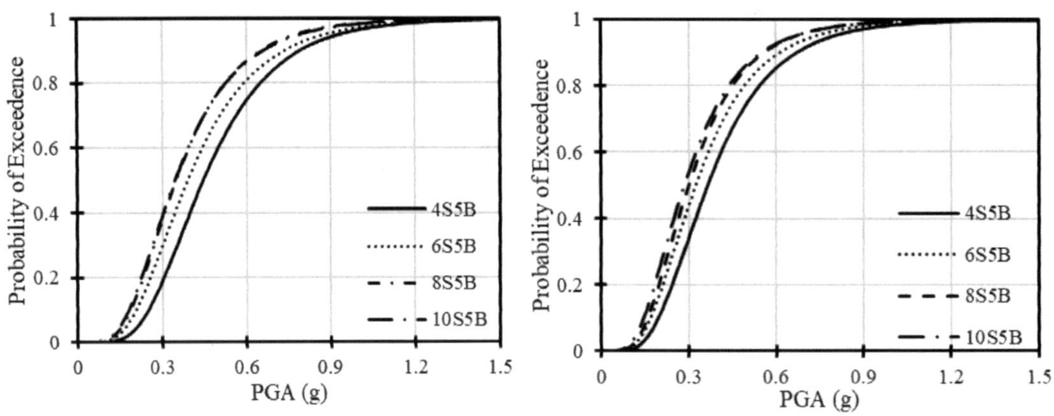

(a) Rigid Side (b) Flexible Side

Figure 10. Fragility curves for *e/b* = 0.1.

IOP Conf. Series: Materials Science and Engineering **936** (2020) 012041 doi:10.1088/1757-899X/936/1/012041

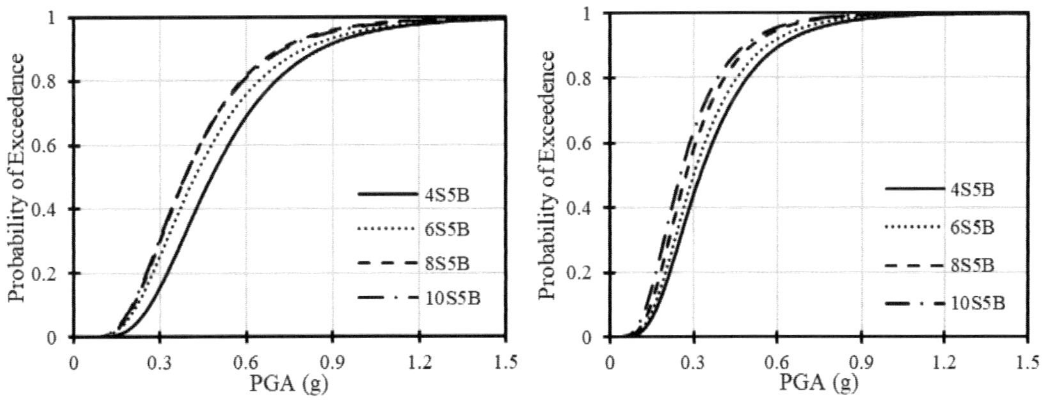

(a) Rigid Side (b) Flexible Side

Figure 11. Fragility curves for $e/b = 0.3$.

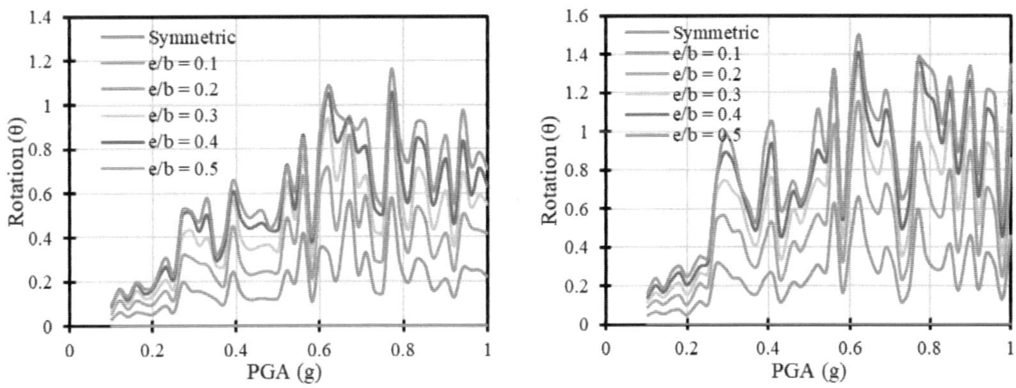

(a) 4 storied model (b) 6 storied model

Figure 12. Maximum rotation for the earthquakes considered.

The investigation thus proposes the necessity of developing adequate performance levels that can account for the rotational behaviour in buildings.

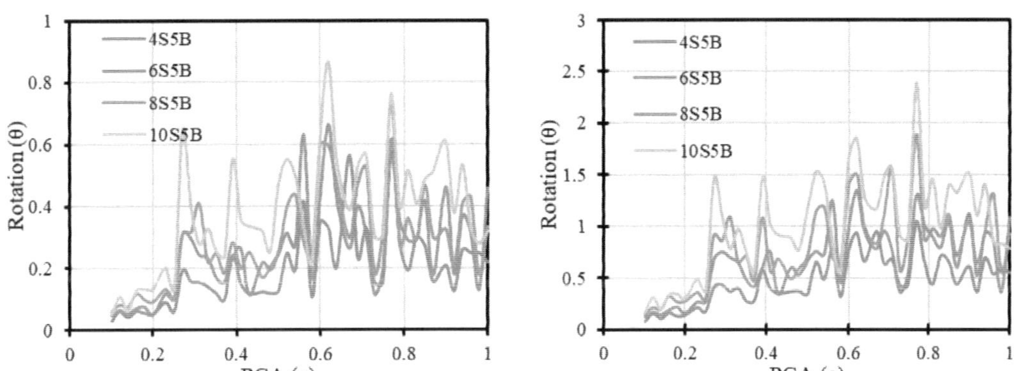

(a) $e/b = 0.1$ (b) $e/b = 0.3$

Figure 13. Height wise distribution of maximum rotation.

7. Conclusions

The study investigates the seismic performance of asymmetric buildings by considering their seismic vulnerability as the performance index. Resultant drift is proposed as an alternate damage parameter that can define the damage of asymmetric buildings. Fragility curves using this parameter are developed. The study proves that the presence of asymmetry increases the probability of failure in a building and that this condition further increases with an increase in asymmetry. The fragility also depends directly on the building height. Further studies were carried out to evaluate the rotational behaviour as a function of eccentricity and building height and the increasing rate of rotation with an increase in eccentricity and storey height is observed. The study proposes the need for well-defined standards that can assess the performance level in terms of rotation in buildings.

8. References

[1] Tso W K and Dempsey K M 1980 Seismic torsional provisions for dynamic eccentricity *Earthq. Eng. Struct. Dyn.* **8** 275–89
[2] Chandler A M and Hutchinson G L 1987 Evaluation of code torsional provisions by a time history approach *Earthq. Eng. Struct. Dyn.* **15** 491–516
[3] Tso W K and Ying H 1990 Additional seismic inelastic deformation caused by structural asymmetry *Earthq. Eng. Struct. Dyn.* **19** 243–58
[4] Samali B, Wu Y M and Li J 2003 Shake table tests on a mass eccentric model with base isolation *Earthq. Eng. Struct. Dyn.* **32** 1353–72
[5] Vial I J, Llera J C D la, Almazan J L and Ceballos V 2006 Torsional balance of plan-asymmetric structures with frictional dampers: Experimental resu *Earthq. Eng. Struct. Dyn.* **35** 1875–98
[6] Annigeri S, Mittal A K and Jain A K 1996 Uncoupled frequency ratio in asymmetric buildings *Earthq. Eng. Struct. Dyn.* **25** 871–81
[7] Anagnostopoulos S and Stathopoulos K 2000 Inelastic earthquake response of buildings subjected to torsion *Proc. 12th World Conf. Earthq. Eng.* (Auckland, New Zealand) pp 1–8
[8] Bagheri B, Firoozabad E S and Yahyaei M 2012 Comparative study of the static and dynamic analysis of multi-storey irregular building *World Acad. Sci. Eng. Technol.* **6** 1847–51
[9] Chopra A K and Goel R K 2004 A modal pushover analysis procedure to estimate seismic demands for unsymmetric-plan buildings *Earthq. Eng. Struct. Dyn.* **33** 903–27
[10] Georgoussis G, Tsompanos A and Makarios T 2015 Approximate seismic analysis of multi-story buildings with mass and stiffness irregularities *Procedia Engineering* vol 125 (Elsevier B.V.) pp 959–66
[11] Hejal R and Chopra A K 1989 Lateral-torsional coupling in earthquake response of frame buildings *J. Struct. Eng.* **115** 852–67
[12] Hwang H H M and Jaw J 1990 Probabilistic damage analysis of structures *J. Struct. Eng.* **116** 1992–2007
[13] Singhal A and Kiremidjian A S 1996 Method for probabilistic evaluation of seismic structural damage *J Struct. Eng.* **122** 1459–67
[14] Cornell C A, Jalayer F, Hamburger R O and Foutch D A 2002 Probabilistic basis for 2000 SAC federal emergency management agency steel moment frame guidelines *J. Struct. Eng.* **128** 526–33
[15] Ellingwood B R 2001 Earthquake risk assessment of building structures *Reliab. Eng. Syst. Saf.* **74** 251–62
[16] Moehle J and Deierlein G G 2004 A framework methodology for performance-based earthquake engineering *13th World Conf. Earthq. Eng.* pp 3812–4
[17] Ramamoorthy S K, Gardoni P and Bracci J M 2006 Probabilistic demand models and fragility curves for reinforced concrete frames *J. Struct. Eng.* **132** 1563–72
[18] Shiravand M R, Nashtaee M A and Veismoradi S 2017 Seismic assessment of concrete buildings reinforced with shape memory alloy materials in different stories *Struct. Des. Tall Spec. Build.* e1384
[19] Zentner I, Gündel M and Bonfils N 2017 Fragility analysis methods: Review of existing approaches and application *Nucl. Eng. Des.* **323** 245–58

[20] McKenna F, Scott M H and Fenves G L 2010 Nonlinear finite-element analysis software architecture using object composition *J. Comput. Civ. Eng.* **24** 95–107

[21] Mander J B, Priestley M J N and Park R 1988 Theoretical stress-strain model for confined concrete *J. Struct. Eng.* **114** 1804–26

[22] Menegotto M and Pinto P E 1973 Method of analysis for cyclically loaded R C plane frames including changes in geometry and non-elastic behavior of elements under combined normal force and bending *Proc. IABSE Symp. Resist. Ultim. Deform. Struct. Acted by Well Defin. Loads* pp 15–22

[23] Lee T H and Mosalam K M 2004 Probabilistic fiber element modeling of reinforced concrete structures *Comput. Struct.* **82** 2285–99

[24] Whittaker A S, Atkinson G M, Baker J W, Bray J D, Grant D N, Hamburger R, Haselton C and Somerville P G 2012 Selecting and scaling earthquake ground motions for performing analyses *15th World Conf. Earthq. Eng.* pp 1–256

[25] FEMA 2000 Prestandard and Commentary for the Seismic Rehabilitation of Buildings *Rehabil. Requir.* 1–518

IOP Conf. Series: Materials Science and Engineering **936** (2020) 012042 doi:10.1088/1757-899X/936/1/012042

Deep Learning-based Metamodeling Technique for Nonlinear Seismic Response Quantification

Anirban Kundu* **and Subrata Chakraborty**

Department of Civil Engineering, Indian Institute of Engineering Science and Technology Shibpur, West Bengal-711103, India

*anirbankundu1997@gmail.com

Abstract. Classical physics-based numerical techniques such as finite element method (FEM) usually takes a huge computational resource and time for simulation-based uncertainties in structural analysis. Especially, it is expensive when nonlinear time history analysis is involved. The metamodeling technique becomes an alternative with an ability to predict the time history response of both elastic and inelastic structural system in a data-driven fashion. Various machine learning and deep learning-based techniques have been attempted to have their limitations in handling huge and sequential data while it comes to predict the whole response time history given stochastic ground motion acceleration. In this regard, Long Short Term Memory (LSTM) approach is found to be useful. The present study used LSTM deep learning network based metamodeling approach for nonlinear seismic response history approximation. The novelty of the approach lies in its capability of capturing record to record variability even for an inelastic structure. This metamodel can work with the desired level of accuracy with very limited data. The proposed approach has shown satisfactory results to approximate seismic response of a nonlinear single degree of freedom system.

1. Introduction

The finite element method is more efficient for structural dynamic analysis but demands huge computational effort. It is a well known fact that the finite element method based vibration response analysis of nonlinear systems demands a small time step and considerably small mesh size to provide an accurate and numerically stable solution. Monte Carlo simulation is the best way to evaluate the reliability when uncertainty in external load as well as internal structural parameters is involved. This method requires a huge number of simulations. Also, incremental dynamic analysis while doing the structural fragility assessment demands numerous simulations. In these cases, the finite element method fails in terms of time and computational demand. This limitation demands an alternative approach that is metamodel i.e. a fast running surrogate model, which is mathematically trained and learns the response prediction laws from very few available inputs and output data and reproduces it for unknown input data. Various emerging adaptive machine learning-based approaches e.g. artificial neural network (ANN) [1,2], Kriging [3], polynomial chaos expansion (PCE) [4,5], and support vector machines (SVM) [6] are notable in this regard. But such metamodeling approaches have some limitation such as selection and extraction of feature requires human intervention. These are not suitable for very large and structured data such as sequential data. Recent developments in the deep neural networks (DNNs) allow learning sequential models for forecasting and predicting in the field of nonlinear structural analysis. DNN includes Multi-layer perceptron (MLP), convolutional neural network (CNN), recurrent neural network (RNN), auto-encoder decoder, etc. Multi-layer perceptron (MLP), a classical ANN-based network has shown state-of-the-art performance in the field of structural damage diagnosis [7, 8]. Though, MLP is limited to the dynamic analysis of the linear system and fails to model a very complex nonlinear system. CNN is capable of handling spatial data with grid-like topology though a modified version i.e. 1-

dimensional CNN has been used in the field of seismic response prediction but while large plastic deformation is involved, the accuracy of 1D CNN drastically falls in predicting the response [9, 10]. Zhang et al. [11] have developed a data-driven physics-guided CNN model for seismic response prediction. It is trained with limited data gained from simulation; and the network output is constrained by available laws of physics and dynamics to alleviate the overfitting issue. The recent development of RNN in handling sequential data and huge success in the field of speech and handwriting recognition [12, 13] motivate to apply this in the field of nonlinear response history prediction of structure. As discussed, this study focuses on quantifying uncertainty in the record to record variability of ground motion acceleration by establishing a surrogate model that can predict the whole response time history. The RNN is found most suitable for this. However, the major two drawbacks of the RNN are vanishing gradient and exploding gradient problems which prevents the model to update the weights at the beginning of the sequence resulting in a zero-learning situation [14]. For this RNN is not capable of predicting response involving long term dependency and complex nonlinearity in time-varying data.

The most robust solution is to use a gated recurrent unit that enables us to learn and forget information selectively through the gates and keep a long track of memory even in case complex nonlinearity is involved. Such a gated cell called long short term memory (LSTM) is explored little in nonlinear stochastic dynamic response approximation. Using the LSTM network Yang et al. [15] have developed a deformation pattern of a landslide along with time, which is complex, nonlinear, and periodically dependent on external factors such as rainfall, fluctuation of reservoir-water-level, etc. To make a well-informed decision-making system for an emergency responder to attend the most critically affected area after severe earthquake, Mangalathu et al. [16] have developed an LSTM based classification model based on textual description gathered from the social media platform. Using semantic analysis i.e. converting language to information, an indexed library is made that classifies description of the severity of building damage. Yin et al. [17] have developed a model to detect anomaly of earthquake precursor data using LSTM. Zhang et al. [18] have developed a seismic response prediction model with a stacked LSTM model. The present study used LSTM deep learning network based metamodeling approach for nonlinear seismic response history approximation. The novelty of the approach lies in its capability of capturing record to record variability even for an inelastic structure. The capability of the proposed metamodel to approximate nonlinear seismic response with the desired level of accuracy with very limited data is numerically demonstrated.

2. Long short-term memory (LSTM) network

The network architecture of LSTM is made of LSTM and other hidden layers having input and output layers at the two surfaces. The LSTM layers are made of several LSTM cells which are the constitutive component of the LSTM architecture.

2.1. The LSTM cell

LSTM cells are equivalent to the unrolling of the recurrent operation by a standard RNN network but with some additional gated operation. These gates keep the relevant information memorised throughout a long sequence. Each LSTM cell has an independent set of weights and biases that is shared across the entire temporal space within that specific layer and get updated through training. This works through three elementary operations (i) selective writing in the memory (ii) selective reading from the memory and (iii) selectively forgetting the irrelevant part from the memory. The selectively written hidden state from the previous cell and new input to the current cell, collectively generate a temporary cell state. The flow of input activation is controlled by the input gate. Selective forget operation is conducted by the forget gate which produces the cell state for the next time step combining with previous cell state. Then the output get helps in producing the hidden state which is served as the output from that particular time step. This way the LSTM cell forget irrelevant information adaptively and store the relevant information for long term dependencies [19].

2.2. The architecture

An LSTM network is composed of one or more LSTM layers followed by fully connected (FC) or dense layers. Each LSTM layer is composed of multiple LSTM cells or nodes. Each node in the layer is analogous to a mini-network which is exposed to the input sequence and produces an output. The number of cells in a layer represents the number of times the recurrent operation is repeated. The input sequence is processed one time step at a time. Each step of input for a node results in output and an internal state. Both are used in the processing of the subsequent time step. Next, the FC layer which is provided to obtain the necessary number of output features available at the output layer. FC layers have full connections to all activation nodes in the previous layer. Adding dropout layers [20] is a good way to prevent overfitting which is a very common problem in deep learning. Typically, it is applied before the FC layer due to having a large number of learnable parameters which make this layer prone to overfitting.

2.3. Sequence to sequence modelling

In this study an LSTM network is trained where a stochastically varied ground motion signature is taken as input and the response time history obtained from nonlinear time history analysis (NLTHA) is taken as the target. Input ground motion records are denoted as $X = \{x_1, x_2, x_3 .....x_n\}^T \in \mathbb{R}^{n\times p}$ and the output structural response $Y = \{y_1, y_2, y_3 ......y_n\}^T \in \mathbb{R}^{n\times q}$. Here n is the number of time steps, p and q are the numbers of input and output features respectively. In the current study p and q, both are considered to be one i.e. one set of ground motion acceleration corresponds to a set of displacement time history which made both X and Y of dimension $n\times 1$. LSTM architecture with multiple datasets prescribes three-dimensional structuring of the input and output layer which is discussed in detail in the numerical study section.

3. Numerical study: A non-linear single degree of freedom (SDOF) system

The nonlinear spring-mass SDOF system is considered to elucidate the effectiveness of the present LSTM network. A nonlinear spring is connected to the ground having a lumped mass on its free end. The nonlinear spring behaviour is described in figure2 (b) [21]. Rayleigh type damping is assumed where it is proportional to the mass and stiffness.

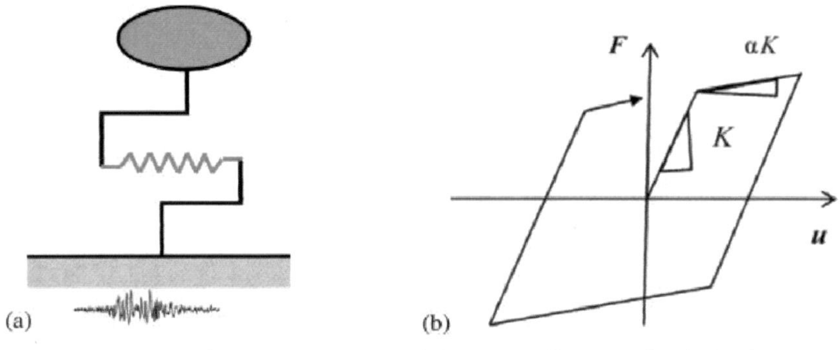

Figure 1. (a) The SDOF system and (b) the nonlinear spring behaviour.

The nonlinear seismic response of the SDOF system is obtained by numerical integration in the OpenSees platform which is taken from the study by Ghosh et al.[22]. All the structural parameters are assumed to be deterministic. The various parameters characterising the nonlinear SDOF system is described in table 1.

In the current study, a suite of synthetically generated accelerograms with the most vulnerable surface wave magnitude (M) and epicentral distance (R) combination are used whose properties are determined by probabilistic seismic hazard analysis (PSHA).More details on this may be seen in Ghosh et al. [22]. For the current study, 100 numbers of synthetic accelerograms are generated with different combinations of M ranging from 6.0 to 8.0 and R ranging from100 to 300 km for rock site. The stochastic method as proposed by Boore [23] is considered for the generation of synthetic accelerograms. The formulation, methodology, and MATLAB codes are adapted from the study of Ghosh et al. [22]. The ground motion generated represents the earthquake nature of the Guwahati region of north-east India.

Table 1. Values of the SDOF system parameters.

System parameters	Values
Natural frequency (*rad/sec*),ω	6.28
Damping ratio, ζ	0.02
Yield force (N), F_y	2.5
The ratio of post-yield to elastic stiffness, α	0.1

First, synthetically generated ground motions are normalised with their peak ground acceleration (PGA) values. The duration of each ground motion is 40 seconds having 2000 data points. A set of randomly generated PGA values lying between 0.1 and 1.0 are multiplied with all the normalised ground motions. The set is divided into two parts each having 50 ground motion accelerations. The first one is used in training while the other one is kept for the prediction purpose. Both training and prediction data set is formatted and prepared in three-dimensional tensor as suggested earlier. The training input (ground motion acceleration) format is [50, 2000, 1] whereas the training output (response time history) format is [50, 2000, 1]. Here, the first dimension depicts the number of events or samples, the second dimension denotes the total number of time steps, and the third dimension denotes the number of input or output features which is one for both the cases in the present study. As the equal number of samples is preserved for the prediction; the data format will be the same as of training data. Two LSTM layers each consist of 100 nodes are provided. An FC layer with 100 nodes followed by a dropout layer with 20 percent is provided. Dense layer with node number equal to the number of features i.e. one (displacement time history of the SDOF system) in the current study connects the previous FC layer. The data are normalised between 0 to 1 using *MinMaxScaler*. From the training set, the ground motion (as input) and the displacement response (as output) obtained from NLTHA (as target) are fed into the model for training.

During the training process, at each epoch, the training set is shuffled to increase the speed of convergence and 15 samples are kept for validation while the rest of 35 samples are used in training. Unlike the feed-forward neural network, the biases and weights are updated after each epoch using the backpropagation through time (BPTT) algorithm that results in reducing both the training and validation loss as shown in figure 2. In this study, it is observed that approximately after 10,000 epochs of training, validation loss decreased to a much lower value which indicates satisfactory performance. The network has a validation error of 2.8532×10^{-5} is saved as the best model. Total training time is 3.7 hours i.e. 1 sec/epoch is required. A batch size of five is used. Adam (Adaptive Momentum Estimation) optimiser is used with a learning rate of 0.002 and a decay rate of 0.00001 which is decided by the grid search method. The loss function is taken as a mean squared error (MSE) for this regression problem.

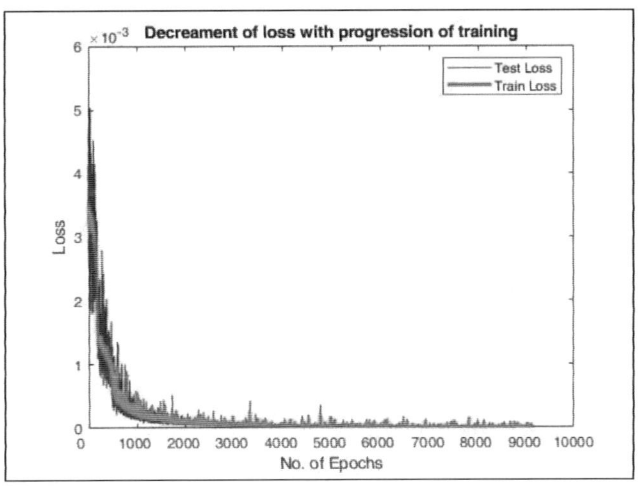

Figure 2. Decrement of 'train' and 'test' loss with progression of training.

The best model is saved, fed with the prediction-ground motion set and response time history is predicted. Along with this, the response is calculated with NLTHA from the prediction set of ground motion and compared with the response predicted from LSTM based metamodel. Such compared response for the two typical ground motions is shown in figure 3. It is seen that the predicted response is quite the same as the true response obtained from NLTHA.

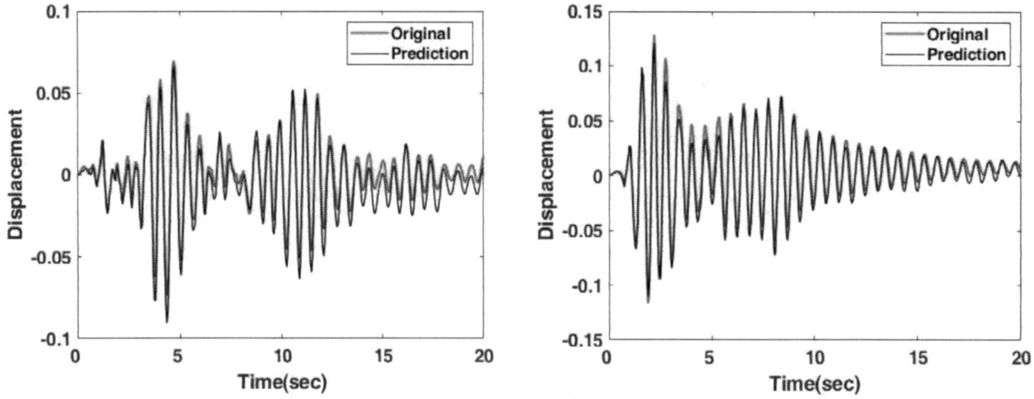

Figure 3. The response of the nonlinear SDOF system at unknown ground motion.

To further study the accuracy of the prediction, regression analysis is conducted. Let, $\overline{y}_t^{(j)}$ is denoted as the drift of the jth sample from the test set at the tth time step of the response time history predicted by the LSTM model. Similarly, $y_t^{(j)}$ denotes the response time history generated from NLTHA. The regression analysis is performed on the prediction data set (50 samples) by plotting the histogram normalised by the probability density of the Pearson's coefficient of correlation obtained for each time step, with the help of 50 pairs of data points. Pearson's correlation coefficient (*r*) is calculated as $r_{XY} = \dfrac{COV(X,Y)}{\sigma_X \sigma_Y}$ where, $X = \{y_t^{(1)}, y_t^{(2)} ......y_t^{(50)}\}$, $Y = \{\overline{y}_t^{(1)}, \overline{y}_t^{(2)} ......\overline{y}_t^{(50)}\}$ and $t \in \{1, 2....2000\}$ all the r_{XY} is prepared for all $t \in \{1, 2....2000\}$, and the probability density function is plotted with 2000 generated Pearson's coefficient values. As the correlation coefficient is calculated over the sample points at each time step, it truly reflects the temporal distribution of the error. A near to 1 correlation coefficient depicts a superior performance while a close to zero value indicates poor performance. Figure 4 represents the probability distribution of the Pearson's correlation coefficient. It can be seen that the probability of correlation coefficient being near 1 is very high. Notably, the majority of the correlation coefficients are over 0.9 and very few are below 0.7 or less than that which indicates a high prediction accuracy of the model.

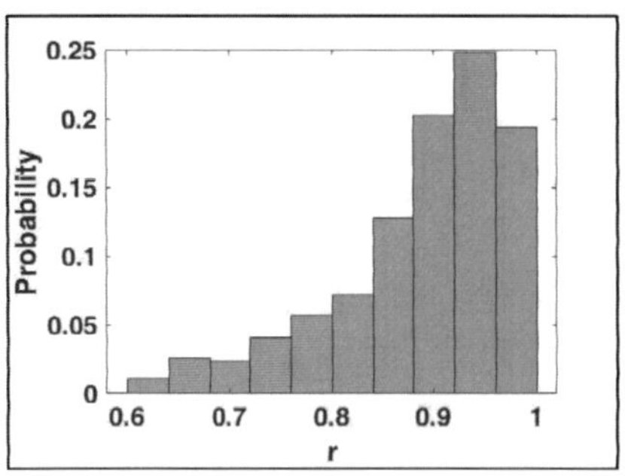

Figure 4. *Regression analysis*: Probability distribution of the Pearson's correlation coefficient calculated with 50 pairs of true and predicted response values at each time step showing the temporal distribution of error.

To illustrate the range of error for each sample from the prediction set, a box-whisker plot of normalised error percentage is presented in figure 5. The normalised error is defined as $\left| \dfrac{y_i^{true} - y_i^{pred}}{\max(y_i^{true})} \right| \times 100$. It is a series of errors at each time step for a particular sample. 1st and 3rd quartile values are shown by the blue box where the red mark in middle shows the median in figure 5. The majority of the sample shows a median of error below 5 and very few shows above 10. Red '+' marks in the figure indicate the outliers. The figure also indicates the maximum and minimum error for each sample.

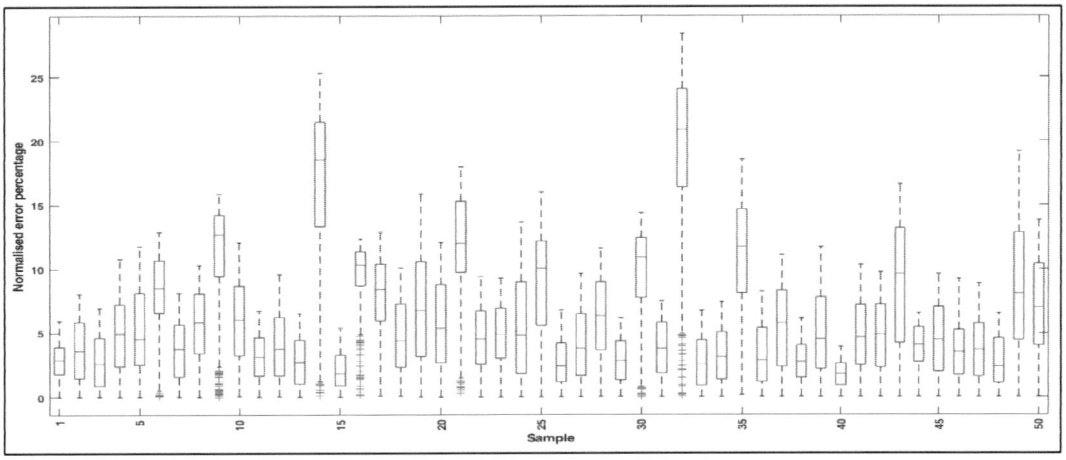

Figure 5. Box-whisker plot of normalised error percentage calculated for each sample from the prediction data set.

4. Summary and Conclusions

An LSTM based metamodeling approach is applied in seismic response approximation which is flexible in handling long term dependency in sequential data. The present algorithm is demonstrated for seismic response approximation of a nonlinear SDOF system subjected to ground motion generated synthetically. A satisfactory result is found through regression analysis and a box-whisker plot of normalised error. Outliers can be treated as a good sign as it is found below the lowest error which implies better accuracy of prediction. However, the accuracy could not be achieved without optimising the hyperparameters involved in the training phase. It is important to choose these juristically and optimise using the grid search method. From the result obtained, it is shown that the LSTM algorithm proposed is potentially having an ample scope to explore further in the seismic safety assessment of structures. It is reliable and robust. The proposed algorithm is generic and can be readily extended to a multi-degree of freedom system considering system parameter uncertainty which is under consideration at this stage.

5. References

[1] Papadrakakis M, Papadopoulos V and Lagaros N D 1996 Structural reliability of elastoplastic structures using neural networks and Monte Carlo simulation *Comp. Methods Appl. Mech. Eng.* **136** 145–63

[2] Chojaczyk A A, Teixeira A P, Neves L C, Cardoso J B and Soares C G 2015 Review and application of Artificial Neural Networks models in reliability analysis of steel structures *Struct. Saf.* **52** 78–89

[3] Kaymaz I 2005 Application of Kriging method to structural reliability problems *Struct. Saf.* **27(2)** 133–51

[4] Kersaudy P, Sudret B, Varsier N, Picon O and Wiart J 2015 A new surrogate modeling technique combining Kriging and polynomial chaos expansions-Application to uncertainty analysis in computational dosimetry *J. Comp. Phy.* **286** 103–17

[5] Sudret B and Blatman G 2010 An adaptive algorithm to build up sparse polynomial chaos expansions for stochastic finite element analysis *Probab. Eng. Mech.* **25** 183–97

[6] Ghosh S, Roy A and Chakraborty S 2018 Support vector regression based metamodeling for seismic reliability analysis of structures *Applied Mathematical Modelling* **64** 584–602

[7] Lightbody G and Irwin George W 1996 Multi-layer perceptron based modelling of nonlinear systems *Fuzzy Sets Syst* **79(1)** 93–112

[8] Huang C S, Hung S L, Wen C M and Tu T T 2003 A neural network approach for structural identification and diagnosis of a building from seismic response data *Earthquake Eng. Struct. Dyn.* **32(2)** 187–206

IOP Conf. Series: Materials Science and Engineering **936** (2020) 012042 doi:10.1088/1757-899X/936/1/012042

[9] Sun Shan-Bin, He Yuan-Yuan, Zhou Si-Da and Yue Zhen-Jiang 2017 A data-driven response virtual sensor technique with partial vibration measurements using convolutional neural network *Sensors* **17(12)** 2888

[10] Wu Rih-Teng and Jahanshahi Mohammad R 2018 Deep convolutional neural network for structural dynamic response estimation and system identification *J. Eng. Mech.* **145(1)** 04018125

[11] Zhang, Ruiyang, Yang Liu, and Hao Sun 2020 Physics-guided convolutional neural network (PhyCNN) for data-driven seismic response modeling *Engineering Structures* **215** 110704

[12] Graves, Alex, Abdel-rahman Mohamed, and Geoffrey Hinton 2013 Speech recognition with deep recurrent neural networks *2013 IEEE International Conference on Acoustics, Speech and Signal Processing*

[13] Doetsch, Patrick, Michal Kozielski, and Hermann Ney 2014 Fast and robust training of recurrent neural networks for offline handwriting recognition *2014 14th International Conference on Frontiers in Handwriting Recognition*

[14] Bengio Yoshua, Simard Patrice and Frasconi Paolo 1994 Learning long-term dependencies with gradient descent is difficult *IEEE Trans. Neural Netw.* **5**(2) 157–66

[15] Yang Beibei et al. 2019 Time series analysis and long short-term memory neural network to predict landslide displacement *Landslides* **16(4)** 677–694

[16] Mangalathu Sujith and Henry V Burton 2019 Deep learning-based classification of earthquake-impacted buildings using textual damage descriptions *International Journal of Disaster Risk Reduction* **36** 101111

[17] Cai Yin et al. 2019 Anomaly detection of earthquake precursor data using long short-term memory networks *Applied Geophysics* 1–10

[18] Zhang Ruiyang, Chen Z, Chen S, Zheng J, Büyüköztürk O and Sun H 2019 Deep long short-term memory networks for nonlinear structural seismic response prediction *Computers and Structures* **220** 55–68

[19] Hochreiter Sepp, and Jürgen Schmidhuber 1997 Long short-term memory *Neural computation* **9(8)** 1735–80

[20] Srivastava N, Hinton G, Krizhevsky A, Sutskever I and Salakhutdinov R 2014 Dropout: a simple way to prevent neural networks from overfitting *J. Mach. Learn. Res.* **15(1)** 1929–58

[21] Ghosh S, Roy A, and Chakraborty S 2018 Support vector regression based metamodeling for seismic reliability analysis of structures *Applied Mathematical Modelling* **64** 584–602

[22] Ghosh S and Chakraborty S 2017 Seismic performance of reinforced concrete building in Guwahati city, northeast India *Scientia Iranica* **24(4)** 1821–33

[23] Boore D M 2003 Simulation of ground motion using the stochastic method *Pure Appl. Geophys.* **160** 635–76

IOP Conf. Series: Materials Science and Engineering **936** (2020) 012043 doi:10.1088/1757-899X/936/1/012043

Seismic Fragility Analysis of Vertically Irregular Steel Framed Buildings

Brij M Shah[1*], Robin Davis[2] and C G Nanda Kumar[3]

[1] M.Tech, National Institute of Technology Calicut, India

[2] Assistant Professor, National Institute of Technology Calicut, India

[3] Faculty, National Institute of Technology Calicut, India

*brij.shah35@gmail.com

Abstract: Disaster mitigation of the country requires the data of fragility curves for various typical buildings. As the number of steel buildings in the cities is increasing, development of seismic fragility curves of steel buildings are also the need of the hour. This study discusses the fragility curves and damage index functions of a 7-storey regular and irregular steel-framed buildings subjected to different earthquake ground motions. Each building was designed based on the codes, IS 800 (2007) and IS 1893 part 1 (2016). Time History Analysis of various vertically irregular framed buildings was performed and displacements of each floor were obtained. Inter-storey drift ratio and spectral acceleration at fundamental period are taken as damage parameter and intensity measure for deriving the fragility curves. The probabilistic seismic demand models for typical stiffness and mass vertical irregular buildings were developed. Steel framed buildings with ground storey open were found to be the vertically irregular building which is the most vulnerable out of all the selected vertical buildings.

1. Introduction

With the increase in demand for different types of usages of a building, it is no longer possible to design a regular building with equal distribution of strength, stiffness, and mass always. Changes in the parameters of different storeys can lead to the occurrence of irregularity in strength, stiffness or mass in a structure. The Indian standard earthquake code [1] defines two types of irregularities, vertical and plan irregularity. Many studies [2] & [3] have been done on the seismic performance of vertical irregular steel buildings. The seismic performance of various types of vertically irregular RC buildings was evaluated by [4]. Irregularity indicators were proposed [5] [6] [7] for the quantification of vertical irregularity in the buildings. [8] studied seismic performance of vertically irregular setback buildings and compared the fragility curves of irregular and regular steel frames by considering different limit states.

Although there are some studies on the seismic performance of vertical irregular buildings in general, the study on the seismic fragility of vertically irregular steel-framed buildings is very limited. Hence the present study is an evaluation of the various type of vertically irregular buildings for their seismic performance using fragility curves and damage index functions.

2. Details of the selected buildings

Regular steel-framed buildings (*R*) having 7 storeys with a storey height of 3.3 m is considered. The plan and elevation of the building considered are shown in figure 1a and 1b respectively. The building is assumed to be in seismic zone III (PGA=0.16g) and accordingly seismic design base shear is calculated as per [1]. Members are designed according to [9]. Details of typical beams and columns are provided in table 1. The slab is considered as a reinforced concrete slab supported on steel beams. Slab thickness is taken as 120 mm [10]. Two types of mass irregular buildings are considered by assuming a swimming pool in the top storey and at the intermediate storey. Mass irregular building with a swimming pool at the top storey is assumed to have a mass of 500% compared to other typical storeys. This building is designated as MT-5. Other mass irregular buildings having a swimming pool at the intermediate storey are designated as MB-5. The performance of mass irregular buildings is evaluated regarding a regular building (MR) having uniform masses in all storeys.

To study the performance of various stiffness irregular buildings a reference regular building (SI-R) having cross bracings in all the storeys is considered. Various stiffness irregular (SI) buildings are considered by keeping a particular storey open (without bracings). The building having ground storey free of any infill walls is designated as (SI-G-O). Similarly, buildings having an intermediate open storey is designated as (SI-M-O). Details of all irregular buildings considered in the study are illustrated schematically in figure 2.

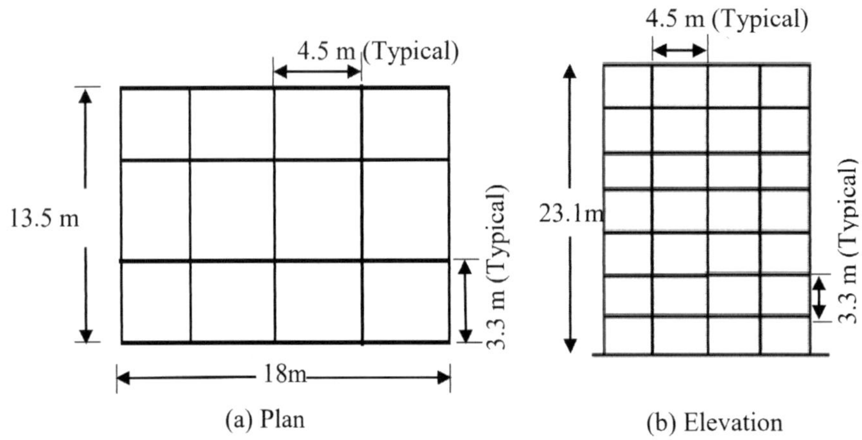

Figure 1. (a) Plan of building considered (b) Elevation of building considered.

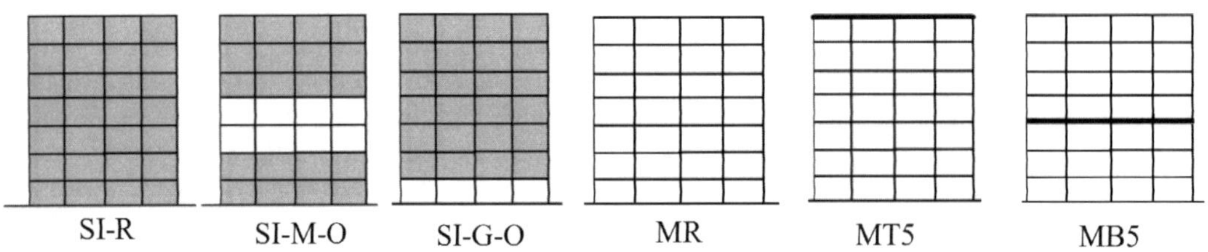

Figure 2. Different cases of stiffness (SI) and mass irregular (M) structures.

Table 1. Different type of sections.

Type	Column Section	Beam Section
MR	ISHB 350	ISMB 350
SI-R	ISHB 350	ISMB 350
MT5	ISHB 400	ISMB 300
MB5	ISHB 400	ISMB 300
SI-M-O	ISHB 250	ISMB 300
SI-G-O	ISHB 300	ISMB 300

3. Methodology

Seismic fragility curves of steel building frames in the present study are developed as per as an accepted methodology proposed by [11]. It uses probabilistic seismic demand models (PSDM) which is the relationship between S_a and inter-storey drift (ISD) of the building for various ground motions. From the PSDM models, median Drift Demand D is found which can be represented as:

$$D = a(S_a)^b \tag{1}$$

Where 'a' and 'b' are taken as constant coefficients. [12] defined $\beta_{D|S_a}$ as a measure of dispersion. Regression analysis of the inter-storey drifts from non-linear dynamic analysis gives the values of three constants a, b and $\beta_{D|S_a}$ required for development of PSDM.

The probability of exceeding a certain limit state for a specific pseudo-spectral acceleration (S_a) is defined as seismic fragility. The fragility function is defined as a probability distribution function that demonstrates the probability of a structure being damaged at a given damage state for a particular ground motion intensity. Fragility curves can be developed for any particular limit state by using the expression:

$$\mathrm{P}\,(C - D \leq 0|S_a) = \emptyset\left(\frac{ln(^D/_C)}{\sqrt{\beta^2_{D|S_a} + \beta^2_c + \beta^2_m}}\right) \tag{2}$$

Where D, C is the median of demand and median of chosen performance level respectively. β_c is defined as a dispersion in the capacity which is taken as 0.25.

4. Modelling for non-linear dynamic analysis

OpenSees laboratory tool [13] was used for non-linear time history analysis of the models of selected buildings. Beams and columns were modelled using non-linear beam-column fiber section elements for non-linear time history analysis. Seismic masses were assigned as lumped masses in each floor.

The uncertainty in the earthquake was modelled by a set of 44 ground motions with PGA increasing gradually from 0.1g to 1.0g. Global damping ratio and yield strength of steel are taken as random variables for considering uncertainty. 44 sets of random variables are generated using Latin hypercube sampling (LHS) method. The mean and coefficient of variation for the random variables considered are given in table 2.

Table 2. Random variables considered.

Material	Mean	COV (%)	Source
Yield Strength of Steel	250 MPa	10	Ranganathan (1999)
Global Damping Ratio	5%	40	Davenport and Carroll (1986)

5. Probabilistic Seismic Demand Models

A total of 44 computational models were subjected to 44 ground motions with a particular PGA [14]. The range of each PGA is taken from 0.1g to 1.0g. Maximum ISD values obtained in each of the building model from non-linear time history analysis were recorded. The maximum inter-storey drift values and corresponding S_a values were plotted to obtain the PSDMs of each frames as shown in figure 3. A power-law relationship was fitted using regression analysis to obtain the constant 'a', 'b' and $\beta_{D|S_a}$. The PSDM obtained from all the frames are shown in figure 3. Table 3 shows the equations of the PSDMs for the different models considered. Figure 3a depicts that inter-storey drift is highest for SI-G-O frame and lowest for SI-R. Figure 3b portrays that MB5 frame is having the highest inter-storey drift whereas it is lowest for MR frame.

Table 3. PSDM model and measure of dispersion.

Frame	PSDM	R^2	β_{D/S_a}
SI-R	$0.9161(S_a)^{0.8929}$	0.8630	0.1752
SI-M-O	$2.1976(S_a)^{1.0817}$	0.8351	0.2369
SI-G-O	$2.5431(S_a)^{1.2995}$	0.8153	0.2891
MR	$3.8993(S_a)^{1.3736}$	0.7101	0.4003
MB5	$5.3982(S_a)^{1.3962}$	0.6314	0.4213
MT5	$4.7215(S_a)^{1.4294}$	0.5501	0.4831

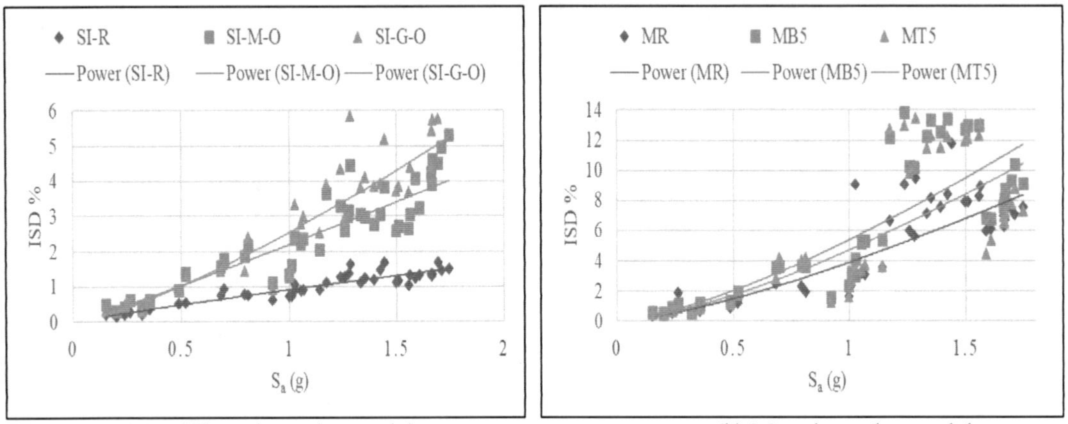

(a) Stiffness irregular models (b) Mass irregular models

Figure 3. PSDMs for stiffness and mass irregular models.

6. Comparison of fragility curves

The fragility curves for the stiffness and mass irregular building frames in comparison with a regular building are shown in figure 4a and 4b respectively. The figure depicts the fragility curve at the IO limit state level. The vulnerability of SI-M-O frame is more than that of SI-R frame. The frame, SI-G-O is more vulnerable than the frame SI-M-O. In the case of mass irregular buildings, the vulnerability of MB5 frame is more than that of MR frame. MB5 frame is more vulnerable than MT5 frame. This is due the fact that when heavier mass is introduced at the intermediate storey, it is more vulnerable than the building having heavier mass at the top storey.

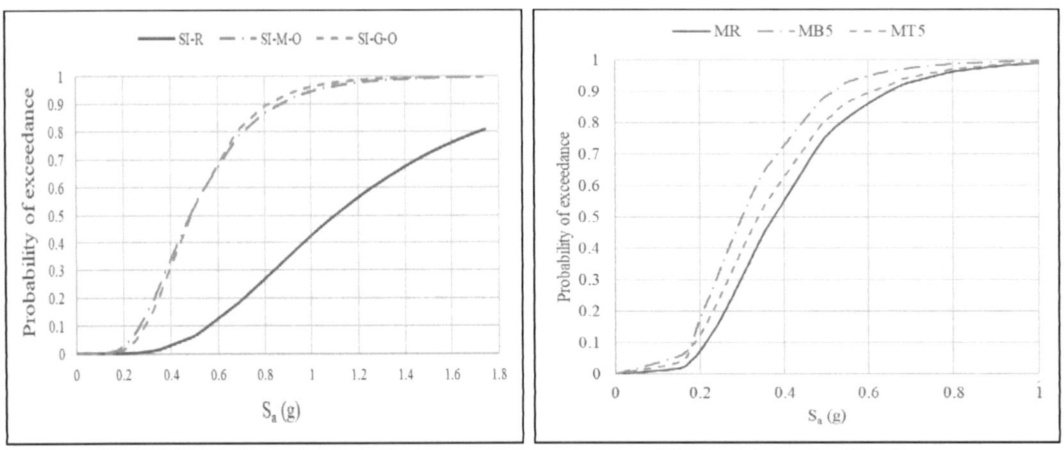

(a) Stiffness irregular buildings (b) Mass irregular buildings

Figure 4. Fragility curves for stiffness and mass irregular buildings at IO level.

7. Damage Index Function

A damage index function is used to determine the damage/loss of the structure at different damage states. In this paper, a standard probability of exceedance of 50% is used to convert a fragility function to a damage index format. Equation 3 is considered to determine the damage index where:

s represents S_a, α, β represents the Weibull distribution factors and F(S_a) represents damage index function.

$$F(S_a) = 1 - e^{-(\frac{S_a}{\alpha})^{\beta}} \tag{3}$$

Figure 5 represents the graph between damage index and S_a. It can be seen that the stiffness irregular building with an open ground storey (SI-G-O) is having the highest damage index as compared to SI-R and SI-M-O. It is also justified from the fragility curve that SI-G-O is the most vulnerable among all other buildings.

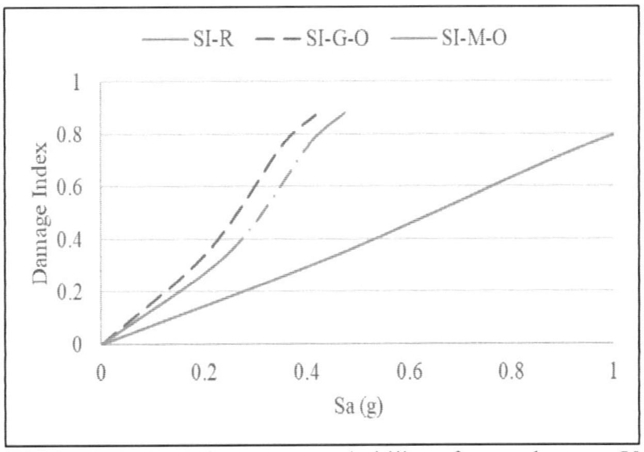

Figure 5. Damage Index vs S_a (probability of exceedance = 50%)

8. Conclusions

This paper focusses on svulnerability of various vertical irregular steel framed buildings. An accepted seismic risk assessment was used to evaluate the performance. The following conclusions can be made from this study.

- The effect of stiffness irregularity along the height of the structure influences the seismic response of the structure. When an irregularity is introduced at the open ground storey, the inter-storey drift and probability of exceedance were highest compared to a regular structure.
- The location of heavier mass has a significant effect on the seismic response of the building. When a heavier mass is positioned at the intermediate floor, the probability of exceedance is increased compared to a regular structure.

9. References

[1] IS 1893 (part 1) 2016 *Criteria for Earthquake Resistant Design of Structures* (Bureau of Indian Standartds: New Delhi)

[2] Le-Trung K, Lee K, Lee N and Lee D H 2012 Evaluation of seismic behaviour of steel special moment frame buildings with vertical irregularities *Struct. Design Tall Spec. Build.* **21** 215–32

[3] Bohlouli and Poursha 2016 Seismic evaluation of geometrically irregular steel moment resisting frames with setbacks considering their dynamic characteristics *Bull Earthquake Eng.* **14** 2757–77

[4] Bhosale A S, Davis R and Sarkar P 2017 Vertical Irregularity of buildings: regularity index versus seismic risk *J. Risk Uncertainty Eng. Syst. Part A: Civ.Eng.* **3(3)** 04017001

[5] Sarkar P, Prasad A M and Menon D 2010 Vertical geometric irregularity in stepped building frames *Eng. Struct.* **32(8)** 2175–82

[6] Karavasilis T L, Bazeos N and Beskos D E 2008 Seismic response of plane steel MRF with setbacks: Estimation of inelastic deformation demands *Journal of Constructional Steel Research* **64** 644–54

[7] Bhosale A S, Davis R and Sarkar P 2018 New seismic vulnerability index for vertically irregular buildings *J. Risk Uncertainty Eng. Syst. Part A: Civ.Eng.* **4(3)** 04018022

[8] Fadzli Mohamed Nazri, Chee Ghuan Tan and Siti Nur Aqilah Saruddin 2017 Fragility curves of regular and irregular moment-resisting concrete and steel frames *Int. J. Civ. Eng.* **16(8)** 917–27

[9] IS 800 2007 *General Construction in Steel – Code of Practice* (Bureau of Indian Standartds: New Delhi)

[10] IS 456 2000 *Plain and Reinforced Concrete – Code of Practice* (Bureau of Indian Standartds: New Delhi)

IOP Conf. Series: Materials Science and Engineering **936** (2020) 012043 doi:10.1088/1757-899X/936/1/012043

[11] Cornell C A, Jalayer F, Hamburger R O and Foutch D A 2002 The probabilistic basis for the 2000 SAC/FEMA steel moment frame guidelines *J. Struct. Eng.* **128(4)** 526-33
[12] Shome N and Cornell C A 1999 Probabilistic seismic demand analysis of nonlinear structures Rep. No. RMS-35 (Dept. of Civil and Environmental Engineering, Stanford Univ.: Stanford, CA)
[13] McKenna F, McGann C, Arduino P and Harmon J A 2014 OpenSees laboratory (https://nees.org/resources/openseeslab)
[14] FEMA 2012 *Quantification of Building Seismic Performance Factors* (FEMA P695)

IOP Conf. Series: Materials Science and Engineering **936** (2020) 012044 doi:10.1088/1757-899X/936/1/012044

Failure Mode Recognition of Columns Using Artificial Neural Network

Chippy Edward[1] and A S Balu[2*]

[1] PG student, Department of Civil Engineering, National Institute of Technology Karnataka, Surathkal, Mangalore- 575025, India

[2] Associate Professor, Department of Civil Engineering, National Institute of Technology Karnataka, Surathkal, Mangalore- 575025, India

[*]asbalu@nitk.edu.in

Abstract. Columns are one of the most vital segments in bridgessince its post-seismic behaviour is of much importance. The retrofitting methods and rehabilitation strategies of bridges mainly rely on the identification of the failure mode of columns.It has been witnessed in various studies on columns that the mode of failure highly depends on section and material properties and there is no specific boundary between the modes, which makes their identification more sophisticated. This paper uses an artificial neural network to predict the modes of failure by analysing the effects of such soft computing methods. In this study, machine- learning models were generated from the experimental data of 253 columns of rectangular cross-section and its accuracy of failure mode prediction was evaluated by considering failure modes mainly flexure, flexure-shear, and shear. The optimal input parameters have also been evaluated for the machine-learning algorithm that enhances the efficiency of failure mode prediction.

1. Introduction

Bridge columns are the most vital part regarding the load-carrying capacities and its post-seismic behaviour. Rectangular columns are more preferred considering its ease of construction and better strength capabilities. Different failure patterns can be exhibited by the columns depending upon their geometric and design details. The column behaviour is highly influenced by the failure modes;a brittle failure is observed in shear failure mode, flexural failure mode indicates a gradual failure by exhibitinglarge deformation capacity, whereas the flexure-shear columns undergo a ductile shear failure. Various researches have been carried out to explore the post-earthquake performance of RC columns. The identification of flexure and shear failure modes was done by analysing geometric and reinforcement details, as well as empirical models, developed based on shear strength of the columns [1][2][3]. However, the flexure-shear interaction makes the study of flexural-shear failure difficult. Thus, identifying the failure modes, especially the flexure-shear failure mode is of greater importance since the rehabilitation approach and the retrofitting methods to be adopted in the bridges are highly dependent on the column failure modes [4].

2. Experimental database

The study used an experimental database of 253 columns with rectangular cross-sections provided with lateral ties around the perimeter. Among this, 199 columns exhibit flexural failure, 18 columns undergo shear failure and 36 columns have a flexural-shear failure. The identification of these failure modes was from various experimental observations. The various parameters evaluated include the following, and they are falling within the given ranges.

- Compressive strength of concrete:16 MPa $\leq f_c \leq$ 118 MPa

- Yield strength of longitudinal reinforcement: 318 MPa $\leq f_y \leq$ 586.1 MPa
- Yield strength of transverse reinforcement: 249MPa $\leq f_{yh} \leq$ 1424 MPa
- Transverse reinforcement ratio: 0.002$\leq \rho s \leq$ 0.067
- Axial load ratio: 0.027 $\leq P/(f_cA_g) \leq$ 0.9
- Shear span to depth ratio: 1$\leq l/d \leq$ 7.64
- Longitudinal reinforcement ratio: 0.0068 $\leq \rho l \leq$ 0.067

where P is the axial load on the column. The different properties of columns including breadth, length, depth, splice length, hoop diameter, longitudinal bar diameter, the total number of bars, number of hoop sets, etc., were examined in detail for the current study.The input parameters used for the prediction are shown in table 1.

Table 1. Input parameters

Input parameter	Notation
Shear span to depth ratio	l/d
Axial load ratio	P/f_cA_g
Main steel index	$\rho l f_y/f_c$
Lateral ties index	$\rho s f_{yh}/f_t$

These input parameters were particularly chosen after thorough investigations on the column database [5]. Column failure modes can be predicted as a function of these four variables, eventhough they cannot be specifically correlated. In this study, the tensile strength(f_t) is calculated as $0.625\sqrt{f_c}$ based on recommendations as per ACI. To analyse the relations between different input parameters in failure mode prediction, data visualisation was carried out and presented in figure 1, where two variables are considered for each mode, noticed the following.

- Columns with low values of l/d exhibit shear failure whereas higher values show flexural failure. For an l/d ratio around 2, columns exhibit all the three failure modes varying with other parameters. Thus, l/d ratio alone cannot be relied on for the identification of column failure modes.
- With the increase in l/d ratio, mode switches from shear to flexure-shear and shear.
- The boundaries which differentiate the three modes are irregular and nonlinear.
- The design variables cannot be specifically correlated as seen in figure 1.

3. Overview of Artificial neural network

An artificial neural network (ANN) is a computational network developed based on the system of biological neurons [6]. ANN consists of an input layer,several hidden layers, and an output layer, as shown in figure 2. The input layer corresponds to the input parameters which can be fed into ANN to generate a predictive model. This layer does not take part in actual output modification but transports the data to the hidden layer. The hidden layer has an arbitrary number of layers with an arbitrary number of neurons in it. The nodes in the hidden layer are known as active since they take part in actual output modification.The number of neurons can be varied for improving the classification accuracy in identifying column failure modes. The output values for the neural network are obtained in the output layer. The nodes in this layer are active ones. ANN works through optimised weight values.

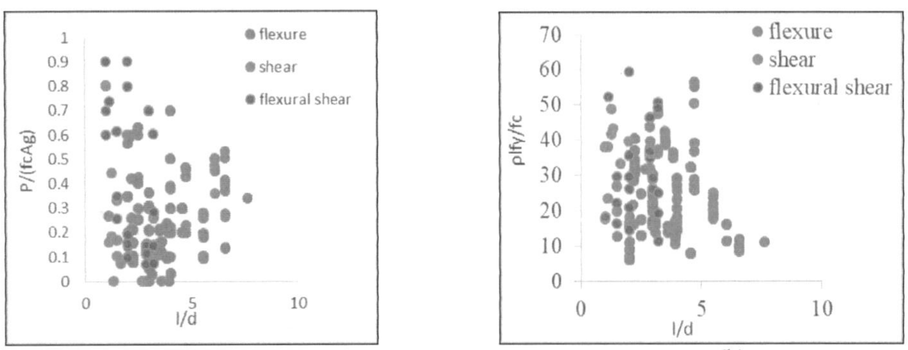

(a) (b)

IOP Conf. Series: Materials Science and Engineering **936** (2020) 012044 doi:10.1088/1757-899X/936/1/012044

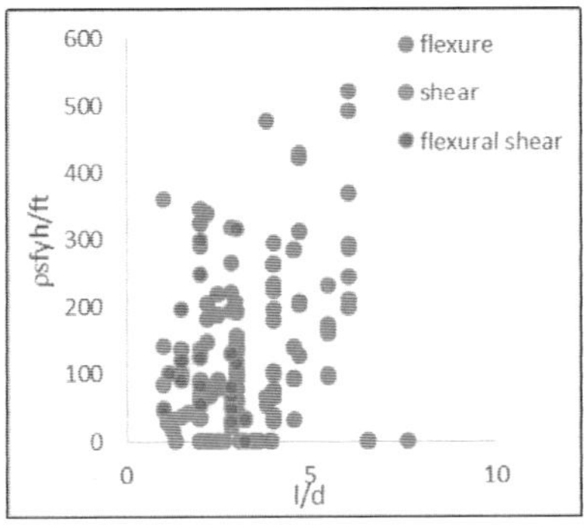

(c)

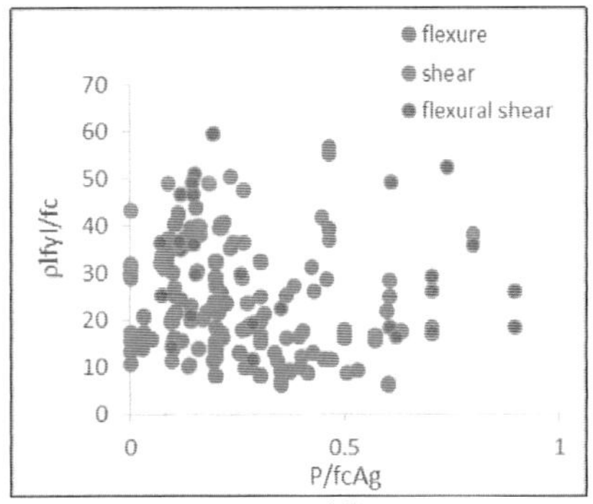

(d)

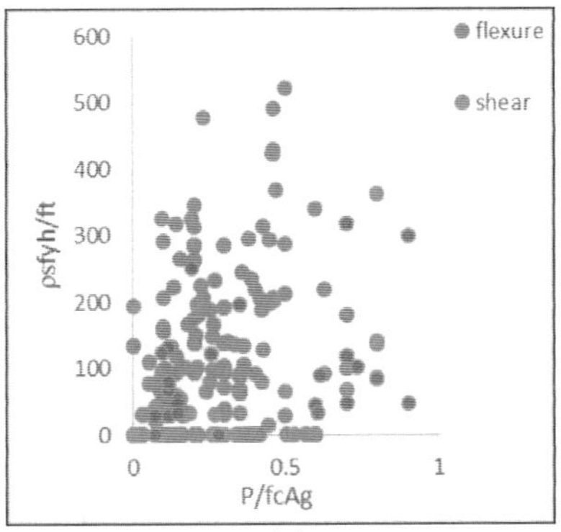

(e)

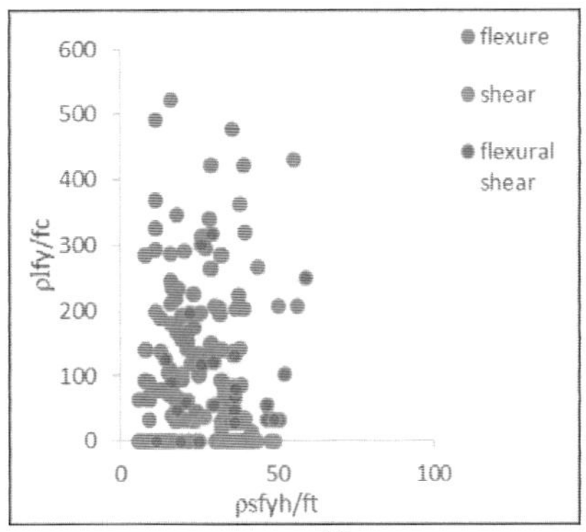

(f)

Figure 1. Plots of parameters for rectangular columns:

(a) $\frac{l}{d}$ Vs $\frac{P}{fc\,Ag}$; (b) $\frac{l}{d}$Vs $\rho l\frac{fy}{fc}$; (c) $\frac{l}{d}$ Vs $\rho s\frac{fyh}{ft}$

(d) $\frac{P}{fc\,Ag}$ Vs $\rho l\frac{fy}{fc}$; (e) $\frac{P}{fc\,Ag}$ Vs $\rho s\frac{fyh}{ft}$; (f) $\rho s\frac{fyh}{ft}$ Vs $\rho l\frac{fy}{fc}$.

For the prediction or classification of three failure modes such as shear, flexure, and flexure-shear, machine learning algorithm such as Artificial Neural Network (ANN) is particularly used in this study. In this study, $Y= (Y_1=\frac{l}{d}$, $Y_2=\frac{P}{fc\,Ag}$, $Y_3=\rho l\frac{fy}{fc}$, $Y_4=\rho s\frac{fyh}{ft})$are used as input variables [7][8].The network is trained many times with the training data to minimise the error in predicting the model by varying the weights allotted to the neurons and the efficiency is checked with the test data set, which remains separate from the dataset used for training purposes.

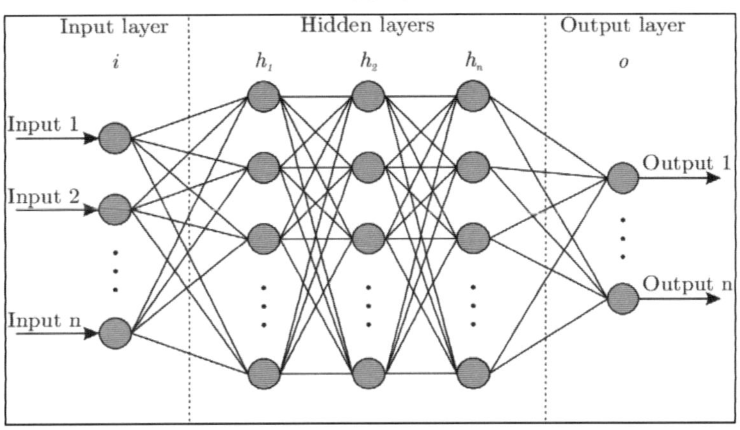

Figure 2. Artificial Neural Network

3.1. Performance of ANN in failure mode identification

To identify the failure modes, the whole data set was divided into two; one for the training purpose while the other is for testing purposes. The predictive model was generated using the training dataset and its efficiency was analysed with the test dataset. The network is trained several times by changing the number of neurons to improve the efficiency of the prediction model [9]. The entire dataset was divided into 70% and 30% for training and test respectively and these values were randomly assigned. The regression plots obtained after training the entire dataset are shown in figure 3.

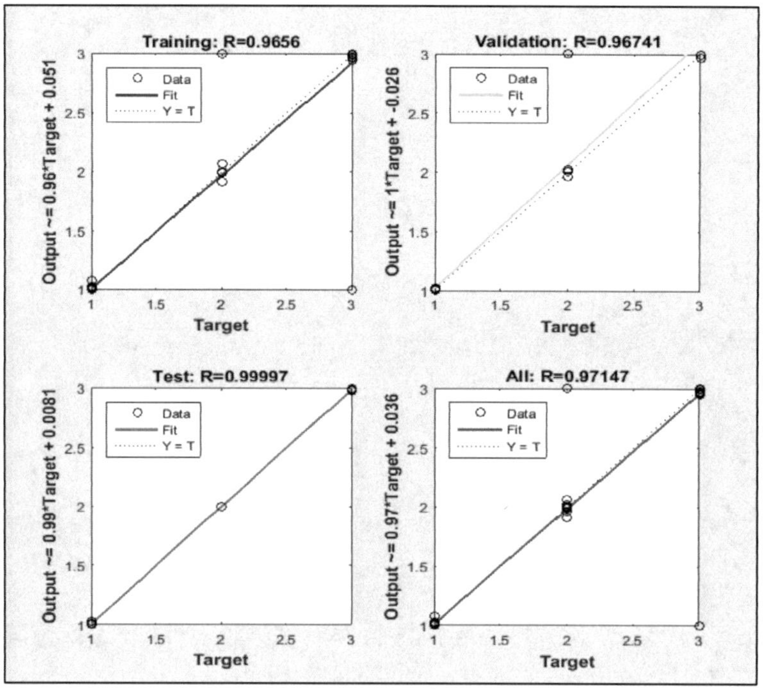

Figure 3. Regression plot

4. Predicted outputs and error variation

It is noted that the classification accuracy in identifying the failure mode increases with the increase in the number of neurons until an optimal value of 15 neurons. With an increase in the neurons beyond 15, the accuracy of the test set and the entire set decreases. The percentage error value has decreased from 21.35% to 19.73% while training the entire dataset by changing the number of neurons from 10 to 15. Out of 61 total specimens, 52 specimens are correctly predicted regarding their failure modes. The prediction accuracy is above 85%, which is far better than any other existing classification methods. Out of 51 specimens, which exhibits flexural failure mode, 48 have been predicted accurately. 3 out of 6 specimens and 1 out of 4 specimens had exact predictions in flexural-shear and shear failure mode respectively. Flexural-shear failure mode which remains as a grey area in most of the existing methods, has more than 50% prediction accuracy using ANN. Since most of the data fed for training involves specimens exhibiting flexural failure mode, it has a high percentage of accurate predictions. Finding out and training more database of columns exhibiting shear and flexural-shear failure modes may improve the prediction efficiency above 90% regarding these failure modes as well as for the whole dataset.

5. Conclusion

The identification of the failure mode of columns is important as it influences the methods being adopted for repairs. Existing methods of classification are based on the aspect ratio [10], displacement ductility, empirical models developed based on the shear strength of columns [11] etc. An artificial neural network was used here. A dataset of 253 test results on rectangular columns including the failure modes, various geometric properties, material properties, reinforcement ratios, and axial load ratios was assembled by carrying out a thorough study of existing literature. It was clear from the preliminary data study that the column failure modes depend on the sectional and reinforcement details, and the boundaries which differentiate the three failure modes are complex and irregular. Also, the design variables cannot be specifically correlated. The regression plots obtained after training of the dataset shows satisfactory results.The error histogram also showed satisfactory results, since the error function was minimised by training the dataset many times by changing the number of neurons. The ANN-based model showed greater efficiency compared to all the other theoretical and experimental methods. ANN has improved prediction accuracy over the flexural-shear failure mode identification than all other methods.

6. References

[1] Kowalsky M J, and Priestley M J N 2000 Improved analytical model for shear strength of circular reinforced concrete columns in seismic regions *ACI Struct. J.* **97(3)** 388–96

[2] Priestley M J N, Verma R, and Xiao Y 1994 Seismic shear strength of reinforced concrete columns *J. Struct. Eng.* **120(8)** 2310–29

[3] Sezen H and Moehle J P 2004 Shear strength model for lightly reinforced concrete columns *J. Struct. Eng.* **130(11)** 1692–703

[4] Zhu L, Elwood K and Haukaas T 2007 Classification and seismic safety evaluation of existing reinforced concrete columns *J. Struct Eng.* **133(9)** 1316–30

[5] Ma Y and Gong J X 2018 Probability identification of seismic failure modes of reinforced concrete columns based on experimental observations *J. Earthquake Eng.* **22(10)** 1881–99

[6] Zhang Y, Burton H V, Sun H and Shokrabadi M 2018 A machine learning framework for assessing post-earthquake structural safety *Struct. Saf.* **72** 1–16

[7] Mangalathu S, Heo G and Jeon J S 2018 Artificial neural network based multi-dimensional fragility development of skewed concrete bridge classes *Eng. Struct.* **162** 166–76

[8] Mangalathu S and Jeon J S 2018 Classification of failure mode and prediction of shear strength for reinforced concrete beam-column joints using machine learning techniques *Eng. Struct.* **160** 85–94

[9] Wang Z, Pedroni N, Zentner I and Zio E 2018 Seismic fragility analysis with artificial neural networks: Application to nuclear power plant equipment *Eng. Struct.* **162** 213–25

[10] Qi Y L, Han X I and Ji J 2013 Failure mode classification of reinforced concrete column using Fisher method *J. Central South Univ.* **20(10)** 2863–9

[11] Ghee A B, Priestley M J N and Paulay T 1989 Seismic shear strengthof circular reinforced concrete columns *ACI Struct. J.* **86 (1)** 45–59

IOP Conf. Series: Materials Science and Engineering **936** (2020) 012045 doi:10.1088/1757-899X/936/1/012045

ANN Based Design Parameter Estimation for Structural Systems

M Nagesh[1] and A S Balu[2*]

[1] Ph.D. Scholar, Department of Civil Engineering, National Institute of Technology Karnataka, Surathkal, India

[2] Associate Professor, Department of Civil Engineering, National Institute of Technology Karnataka, Surathkal, India

[*]asbalu@nitk.edu.in

Abstract. Estimation of the probability of failure of multi-dimensional structural systems is expensive from the computation perspective. To decrease the burden of computation, one can use simple approximation methods like Surrogate models, Kriging model, Support vector machine, Artificial neural network, and more based on the suitability for the problems. In Surrogate or Response surface modeling, the limit state function of any system is suitably approximated by making use of known mathematical models like polynomials, exponentials, etc. During the construction of surrogates, variables in the model should be well known prior to the approximation. In practical consideration, the design parameters are the unknowns that need to be evaluated before reliability-based design. Inverse Response surface procedure is proposed in the paper to address the above-mentioned issue. The procedure developed is the combination of adaptive Response surface method with appropriate experimental design i.e. Halton low discrepancy sequence sampling technique for evaluating the probability of failure or reliability index and an Artificial neural network is utilised as an inverse reliability procedure for design optimisation. The method gives an accurate result and the efficiency is increased for the same number of iterations in comparison to the work of David Lehky and Martina Somodikova [1] with Latin hypercube sampling as experimental design.

1. Introduction

To reach the expected level of probability of failure during any design is a challenging task. Uncertainties arise at every stage of the process in the design stages. For example live load, dead load, material properties, and many more in case of the structural design. Consideration of these uncertainties during reliability evaluation or engineering design is very essential. Reliability evaluation needs the help of forwarding approaches for evaluating the probability of failure or reliability index. In contrast, during an engineering design inverse approach is required for estimating the design variables or parameters to arrive at target reliability levels.

For any structure, the probability of failure (P_f) or reliability index (β) is estimated at a particular limit state. In complex systems, the evaluation of P_f or β becomes an expensive task given computation. To reduce the burden due to computation approximate metamodels are utilised without accuracy compromise. These approximate methods use a simpler function in place of Limit State Function (LSF) which consumes less time for computation. Estimation of P_f or β is carried out with the help of classical simulation methods like the Monte-Carlo technique. Some of the popular metamodeling techniques are the Response surface method (RSM) [1], Support vector machine [2], Kriging metamodel [3], and Artificial neural network (ANN) [4]. The inverse reliability approach for design parameter estimation has been carried out by many researchers by different techniques like Newton-Raphson iterative algorithm [5][6], decomposition technique [7], ANN [8][9][10]. To increase the accuracy for estimating the design parameter for required P_f or β with reducing the computation time, the inverse metamodeling approach is proposed by using RSM with a polynomial model for a forward approach and ANN for inverse approach. Sampling procedure based on Halton Low discrepancy sequence (LDS) is utilised here due to its less discrepancy between samples generated for any parameter.

2. Response surface approximation

Let **Y** denote the vector of input variables to represent the limit state function in equation (1).

$$h(\mathbf{Y}) \le 0 \tag{1}$$

The LSF in equation (1) is replaced with equation (2) a polynomial function of second-order as a surrogate model

$$\tilde{h}(\mathbf{Y}) = a + \sum_{i=1}^{n} b_i Y_i + \sum_{i=1}^{n} \sum_{j=1}^{n} c_{ij} Y_i Y_j \tag{2}$$

where a, b_i, c_{ij} are the unknown coefficients with $i, j = 1, 2, ..., n$. Number of numerical experiments are performed by utilising appropriate experimental design for the input variables to evaluate the regression coefficients. The total number of experiments essential for evaluation of coefficients in equation (2) is $1+n+n(n+1)/2$. The process can be further simplified by utilising polynomial function without considering mixed terms $Y_i Y_j$ as stated in equation (3) shown by Bucher and Bourgund [11] and also shown that accuracy is not much affected by neglecting the mixed terms.

$$\tilde{h}(\mathbf{Y}) = a + \sum_{i=1}^{n} b_i Y_i + \sum_{i=1}^{n} c_i Y_i^2 \tag{3}$$

By neglecting mixed terms the number of evaluations of experiments decreases to $2n+1$. Assuming uncorrelated normal variables $\tilde{h}(\mathbf{Y})$ is used to get Y_D and the middle point Y_M is obtained as follows.

$$Y_M = \bar{Y} + (Y_D - \bar{Y}) \frac{\tilde{h}(\bar{Y})}{\tilde{h}(\bar{Y}) - \tilde{h}(Y_D)} \tag{4}$$

where $\bar{Y}$ denotes the mean point and Y_D is the design point evaluated for the surface $\tilde{h}(\mathbf{Y}) = 0$. The new center point Y_M is utilised to update the model in equation (3) and $4n+3$ the number of evaluations has to be carried out in each iteration. After the number of iterations, an accurate response surface will be obtained which can be replaced with LSF of the given problem.

3. Artificial neural network-based inverse reliability method

Inverse reliability analysis deals with the evaluation of design parameters for required reliability. A reliability-based design optimisation problem can be generally stated as in equation (5).

$$\begin{aligned} &\text{Given: } P_{f,j} \text{ or } \beta_j \\ &\text{Find: } \mathbf{d} \text{ and/or } \mathbf{r} \\ &\text{Subject to: } Z_j = h(\mathbf{Y}, \mathbf{d}, \mathbf{r})_j = 0 \text{ for } j = 1, 2, \ldots, m \end{aligned} \tag{5}$$

Z_j indicates safety margins, $P_{f,j}$ represents target failure probabilities or β_j depicts reliability indices, where $j = 1, 2, ..., m$ is the number of LSFs. $\mathbf{d}$ & $\mathbf{r}$ represents deterministic and random variables parameter vector respectively. Also $\mathbf{Y} = \{Y_1, Y_2, ..., Y_n\}$ represents the vector of basic random variables. ANN is utilised in the present work as a surrogate model for the unknown inverse function. ANN builds relation between reliability indicators $\mathbf{I} = \boldsymbol{\beta}$ (or $\mathbf{I} = \mathbf{P}_f$) and the design parameters $\mathbf{P} = (\mathbf{d} \cup \mathbf{r})$ as in equation (6) based on preliminary data generated from suitable experimental design [12].

$$\mathbf{P} = f_{ANN}^{-1}(\mathbf{I}) \tag{6}$$

In the present work, the Halton LDS sampling technique has been considered for the preparation of the training set. LDS technique is chosen here since it generates values in the required region with a low discrepancy between the nearest values in the domain in comparison with other experimental designs [13]. The values generated from sampling are used for ANN training, testing, and validation. ANN architecture as shown in figure 1 consists of non-linear hidden layers and linear output layers. The design parameters corresponding to the required reliability level can be evaluated using a trained ANN model.

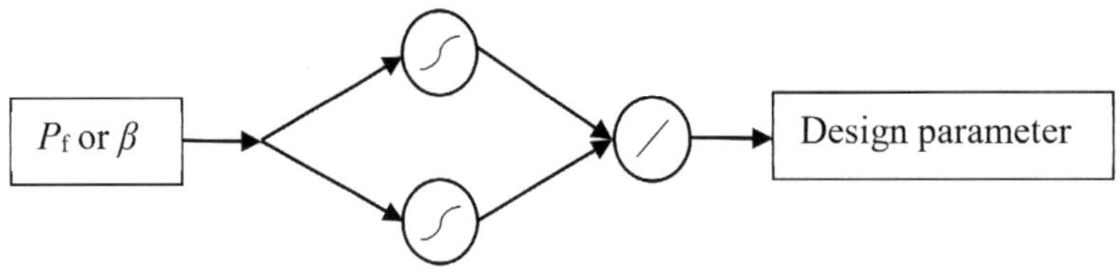

Figure 1. Artificial neural network architecture.

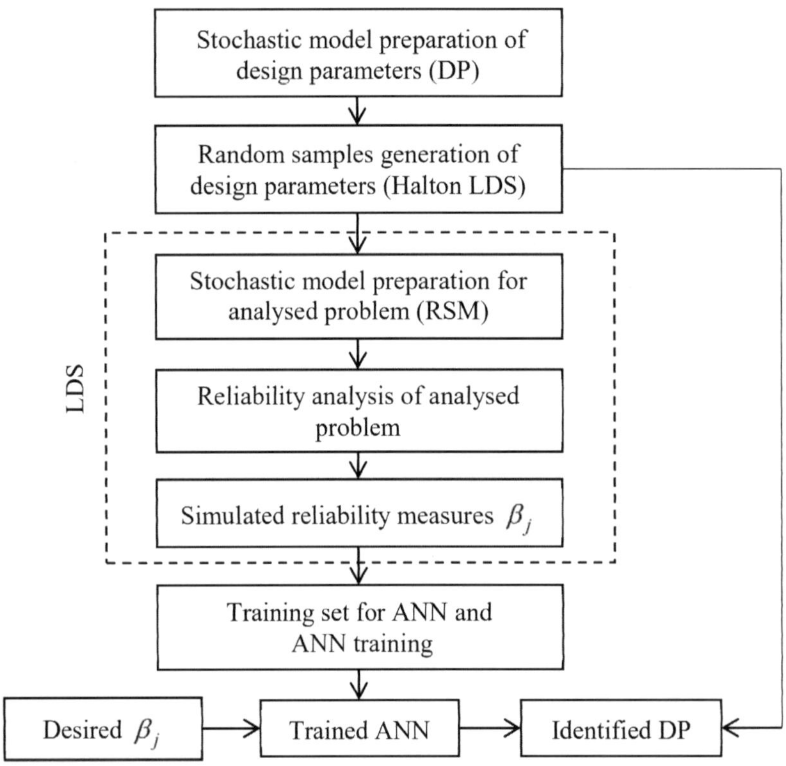

Figure 2. Flowchart representing the iterative procedure of the methodology.

4. Inverse response surface method

The response surface approach is used to replace the original LSF by a much simpler mathematical model as detailed in Section 2. Construction of the Response surface requires to know the design variables prior to the process. To get these values, the inverse approach i.e. ANN is utilised to generate an unknown surrogate model. The proposed method uses ANN with the Halton LDS sampling technique for better efficiency. Figure 2 represents the iterative scheme followed in the present work for upgrading the Response surface along with the inverse reliability analysis [6, 14].

a) The approximation by polynomial function is constructed in the direct procedure using the initial design parameter values.

b) By making use of the Response surface constructed in (a) the reliability indicators for a set of design parameters (formed from Halton LDS) are generated.
c) ANN-based inverse reliability analysis is carried out. The set of design parameters and reliability indicators are used for training ANN and an ANN surrogate model is obtained.
d) A new estimate of the design point is obtained from the ANN model in the inverse approach.
e) A new point is estimated from the design point by utilising equation (4). The new point is used to update the Response surface.
f) The process from (a) to (e) is repeated until the desired or acceptable level of design parameter is obtained.

5. Numerical example

Let the explicit nonlinear function [15] in equation (7) represent an LSF.

$$h(\mathbf{Y}) = \exp[0.4(Y_1 + 2) + 6.2] - \exp(d) - 200 \tag{7}$$

where - Y_1 is a random variable and d is a certain unknown number.

Numerical simulation is performed on equation (7) for a target value of $d = 5.163$ 10 million samples. Reliability Index (β) is defined as the shortest distance of the origin to the failure region and can be obtained as the ratio of mean to standard deviation. The reliability index value obtained is $\beta = 2.688$. Table 1 shows the parameter randomisation of d. The values of mean, standard deviation, and boundary values are used in the RSM construction and to conduct the inverse approach.

Table 1. Randomisation of design parameter.

Variable	μ	σ	Range
d	6.0	1.155	4.0-8.0

The design parameter d in equation (7) is estimated by the proposed methodology. The ANN architecture in this case consists of two non-linear neurons in the hidden layer and a linear output layer associated with d. The values of β and d obtained in the iterative procedure are tabulated in table 2. The methodology proposed utilising the Halton LDS method provides efficient results nearer to the target values of d in comparison to the use of the LHS method by Lehky and Martina Somodikova [1] as depicted in figure 3.

Table 2. Results after the iterative procedure.

Parameter	Identification			Target value
	Trial 1	Trial 2	Trial 3	
d	5.471	5.199	5.160	5.163
β	2.660	2.674	2.687	2.688

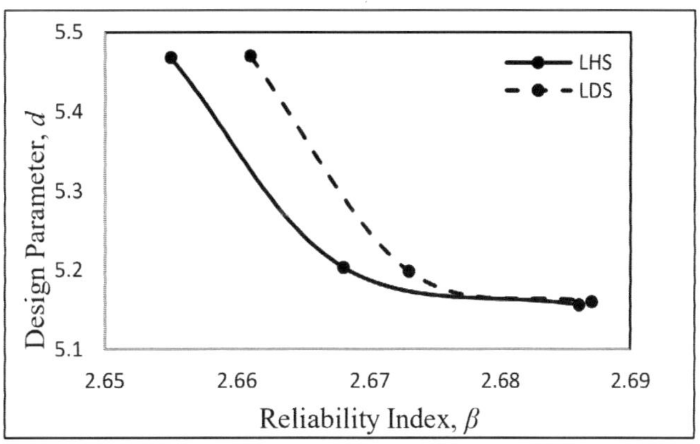

Figure 3. Results showing comparison between LHS and LDS.

6. Conclusion

The iterative procedure adopted in the work along with the Halton LDS sampling technique as an experimental design increase the quality of response surface in few iterations, resulting in improving the accuracy of estimation of design parameter in comparison with LHS sampling technique. A direct approach coupled with inverse analysis cannot yield good results without the iterative procedure. Original LSF and initial values of the design parameter range selected govern the efficiency of the results from the procedure. For non-linear and more complex problems the second-order polynomial may not be sufficient, higher-order polynomials can be utilised in such cases. The response surface approach is one of the suitable ways in comparison to other methods in terms of computation effort acceptability. For accuracy assurance, one should make use of the iterative procedure while utilising inverse response surface approaches like ANN.

7. References

[1] Lehky D and Somodikova M 2017 Inverse response surface method in reliability-based design *Transactions of the VŠB - Technical University of Ostrava Civil Engineering Series* **17(2)** 37–42

[2] Hurtado J E 2004 An examination of methods for approximating implicit limit state functions from the viewpoint of statistical learning theory *Structural Safety* **26(3)** 271–93

[3] Kaymaz I 2005 Application of Kriging method to structural reliability problems *Structural Safety* **27(2)** 133–51

[4] Lehky D and Somodíkova M 2017 Reliability calculation of time-consuming problems using a small-sample artificial neural network-based response surface method *Neural Computing and Applications* **28** 1249–63

[5] Li H and Foschi RO 1990 An inverse reliability method and its application *Structural Safety* 257–70

[6] Lehky D, Slowik O and Novak D 2018 Reliability-based design: Artificial neural networks and double-loop reliability-based optimisation approaches *Advances in Engineering Software* 123–35

[7] Minguez R, Castillo E and Hadi A S 2005 Solving the inverse reliability problem using decomposition techniques *Structural Safety* 1–23

[8] Shayanfar M A, Massah S R and Rahami H 2007 An inverse reliability method using neural networks and genetic algorithm *World Applied Science Journal* 594–601

[9] Cheng J and Li Q S 2009 A hybrid artificial neural network method with uniform design for structural optimisation *Computational Mechanics* 451–9

[10] Antonio C and Hoffbauer L 2010 Uncertainty propagation in inverse reliability-based design of composite structures *International Journal of Mechanics and Materials in Design* 89–102

[11] Bucher C G and Bourgund U 1990 A fast and efficient response surface approach for structural reliability problems *Structural Safety* 57–66

[12] Lehky D and Novak D 2012 Solving inverse structural reliability problem using artificial neural networks and small-sample simulation *Advances in Structural Engineering* 1911–20

[13] Dai H and Wang W 2009 Application of low-discrepancy sampling method in structural reliability analysis *Structural Safety* 55–64

[14] Balu A S and Rao B N 2012 Inverse structural reliability analysis under mixed uncertainties using high dimensional model representation and fast Fourier transform *Engineering Structures* 224–34

[15] Kim S H and Na S W 1997 Response surface method using vector projected sampling points *Structural Safety* 3–19

IOP Conf. Series: Materials Science and Engineering **936** (2020) 012046 doi:10.1088/1757-899X/936/1/012046

Modelling and Optimal Design of Thick-Walled Composite Pipes under In-Service Conditions

T Wang, O Menshykov*, M Menshykova and I A Guz

School of Engineering, University of Aberdeen, AB24 3UE, Scotland, UK

*o.menshykov@abdn.ac.uk

Abstract. The current study is focused on the stress-strain analysis of multi-layered thick-walled fibre reinforced composite pipes under torsional load. Three-dimensional elasticity solution has been obtained and Finite Element Method model has been developed to predict the stress-strain distribution and compute the Tsai-Hill failure coefficients. The models have been validated by comparing with the results available in the literature. As a numerical example the composite pipe subjected to pure torsion load has been considered. The effects of the torque magnitude and layers' thickness for specific pipe lay-ups have been investigated, and the appropriate engineering conclusions and design suggestions have been reported.

1. Introduction

Composite materials were introduced to a range of technical applications over the last decades. Due to their outstanding specific stiffness, strength, lightweight, and chemical resistance, the composite products are applied in aerospace, automotive, military, energy, civil and subsea industries as replacements of the traditional engineering materials (such as metals and alloys). With the increasing customer demands, the composite materials science and manufacturing techniques have been continuously developing and improving [1, 2].

Currently the modelling and design of reinforced composite structures is one of the most dynamically developing branches of structural engineering and materials science because of the variety of in-service loading conditions, component shapes and material combinations (in particular, fibre volume ratios, angles of reinforcement and the appropriate stacking sequences, etc.). The special attention is paid to composite pipes, rods and liners under pressure, bending, torsion and axial loading.

Because of obvious reasons, the most frequently considered type of the loading conditions for pipes is pressure (internal and external). In [3, 4] three-dimensional elasticity method was applied to compute the stress distribution in composite pipes under pressure. Tsai-Hill failure criterion was used for failure prediction. The detailed analysis of failure coefficients along radial direction for different fibre orientations was presented in [4].

Some experiments on filament wound composite pipes were performed in [5, 6]. In the papers the fatigue failure was analysed for the pipes subjected to internal pressure. The whitening, leakage and final pipes' failure were clearly presented. Pipes with surface cracks under inner pressure were considered in [7, 8], where the fatigue was investigated.

The studies of layered composite pipes subjected to inner pressure and thermomechanical loading were published in [9, 10], where the developed three-dimensional anisotropic elasticity method was applied to compute the thermal stress and deformations. With the improvement of computational technics, the thermoplastic composite pipes under pressure and thermomechanical load were studied by Computer-Aided Engineering simulation software [11, 12]. Finite Element Analysis (FEA) results were validated by the available analytical solutions, the detailed through-thickness temperature distributions, and failure coefficients for different stacking sequences were presented and analysed.

IOP Conf. Series: Materials Science and Engineering **936** (2020) 012046 doi:10.1088/1757-899X/936/1/012046

The composite pipe subjected to the bending load were investigated in [13–15]. Xia, Takayanagi and Kemmochi [13] presented elasticity solution using the equilibrium equations and the boundary and interface conditions, and analysed the appropriate stresses and strains. Their approach was further developed and utilised by Menshykova and Guz [14], who focused on the distribution of failure coefficients for multi-layered pipes with specific lay-ups. The effects of winding angle on the radial, hoop and axial stress distributions through the wall thickness were also presented. Finally, the computation model of stress distribution and failure coefficients of pipes subjected to combined loading was presented in [15].

Considering the emerging applications of composite cylindrical structures (rods and pipes) in the energy generation and automotive industries (e.g., wind turbine towers and driveshafts), and for the sake of presentation of the current study, here we will present the analytical solution and 3D FEA results for composite pipes under torsion load only. The Tsai-Hill failure criterion will be applied to compute the failure coefficients, predict the pipe failure and formulate the design recommendations.

2. Problem statement

2.1. *Stress analysis*

The distribution of stresses and strains for each layer of the composite pipe (Figure 1) are given by the following matrix system [3, 4, 16]:

$$\begin{Bmatrix} \sigma_z \\ \sigma_\theta \\ \sigma_r \\ \tau_{\theta r} \\ \tau_{zr} \\ \tau_{z\theta} \end{Bmatrix}^{(l)} = \bar{C} \begin{Bmatrix} \varepsilon_z \\ \varepsilon_\theta \\ \varepsilon_r \\ \gamma_{\theta r} \\ \gamma_{zr} \\ \gamma_{z\theta} \end{Bmatrix}^{(l)} \tag{1}$$

where $\bar{C}$ is the matrix of off-axis stiffness constants.

The radial, hoop, and axial displacements depend on spatial coordinates.

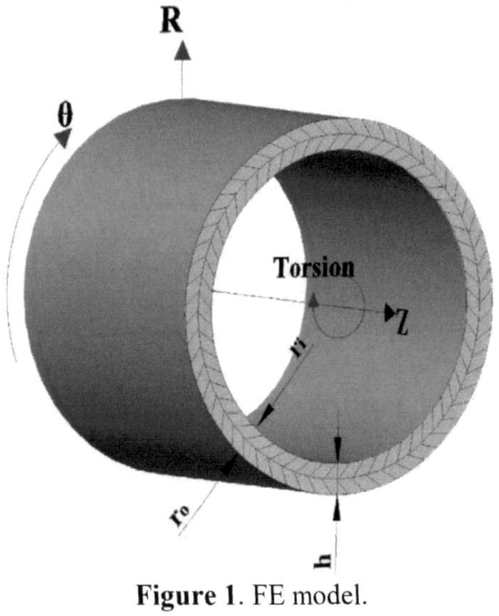

Figure 1. FE model.

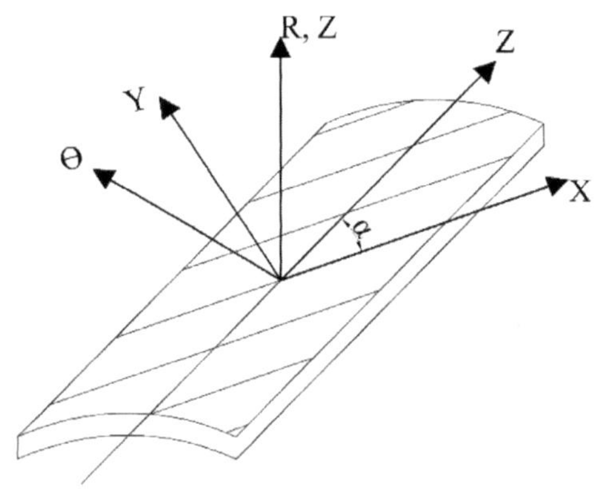

Figure 2. Coordinate systems.

The strains can be calculated from the appropriate displacements as [3, 4]:

$$\varepsilon_r^{(l)} = \frac{du_r^{(l)}}{dr},\ \varepsilon_\theta^{(l)} = \frac{u_r^{(l)}}{r},\ \varepsilon_z^{(l)} = \varepsilon_0, \tag{2}$$

$$\gamma_{zr}^{(l)} = 0,\ \gamma_{\theta r}^{(l)} = 0,\ \gamma_{z\theta}^{(l)} = \gamma_0 r, \tag{3}$$

here $\gamma_0$is the twist of pipe per unit length and ε_0 is the total axial strain.

The equilibrium equations for axisymmetric composite pipe are presented as in [3, 16]:

$$\frac{d\sigma_r^{(l)}}{dr} + \frac{\sigma_r^{(l)} - \sigma_\theta^{(l)}}{r} = 0,$$
$$\frac{d\tau_{\theta r}^{(l)}}{dr} + \frac{2\tau_{\theta r}^{(l)}}{r} = 0, \tag{4}$$
$$\frac{d\tau_{zr}^{(l)}}{dr} + \frac{\tau_{zr}^{(l)}}{r} = 0.$$

The boundary conditions under pure torsion load are expressed as

$$\sigma_r^{(1)}(r_i) = 0,$$
$$\sigma_r^{(N)}(r_o) = 0,$$
$$\tau_{\theta r}^{(1)}(r_i) = \tau_{zr}^{(1)}(r_i) = 0, \tag{5}$$
$$\tau_{\theta r}^{(N)}(r_o) = \tau_{zr}^{(N)}(r_o) = 0,$$

where r_i is the inner radius and r_o is the outer radius of the pipe consisting of N layers.

For the perfectly bonded layers (as we assumed in the current study), the continuity conditions for stresses and displacements must be satisfied. The radial displacements as well as radial and shear stresses must coincide at the interfaces between layers.

$$\sigma_r^{(l)}(r_l) = \sigma_r^{(l+1)}(r_l),$$
$$u_r^{(l)}(r_l) = u_r^{(l+1)}(r_l), \tag{6}$$
$$\tau_{\theta r}^{(l)}(r_l) = \tau_{\theta r}^{(l+1)}(r_l).$$

Furthermore, by integrating the moment of the shear stress $\tau_{z\theta}$ over the cross-sectional area one may derive the following integral equation [3, 16]:

$$2\pi \sum_{l=1}^{N} \int_{r_{l-1}}^{r_l} \tau_{z\theta}^{(l)}(r) r^2 dr = T, \tag{7}$$

where T is the torque.

The integral equation for axial force is the following [3, 16]:

$$2\pi \sum_{l=1}^{N} \int_{r_{l-1}}^{r_l} \sigma_z^{(l)}(r)\, r dr = 0. \tag{8}$$

From equation (1)–(4) the radial displacement can be written as [3]:

$$u_r^{(l)} = A^{(l)} r^{\beta(l)} + B^{(l)} r^{-\beta(l)} + \alpha_1^{(l)} \varepsilon_0 r + \alpha_2^{(l)} \gamma_0 r^2, \tag{9}$$

where $A^{(l)}$and $B^{(l)}$are unknown integration constants and

$$\beta^{(l)} = \left(\frac{\overline{C}_{22}^{(l)}}{\overline{C}_{33}^{(l)}}\right)^{\frac{1}{2}}, \alpha_1^{(l)} = \frac{\overline{C}_{12}^{(l)} - \overline{C}_{13}^{(l)}}{\overline{C}_{33}^{(l)} - \overline{C}_{22}^{(l)}}, \alpha_2^{(l)} = \frac{\overline{C}_{26}^{(l)} - 2\overline{C}_{36}^{(l)}}{4\overline{C}_{33}^{(l)} - \overline{C}_{22}^{(l)}}. \tag{10}$$

The transformed strains can be calculated as [3]:

$$\{\varepsilon\} = \begin{Bmatrix} \varepsilon_1 \\ \varepsilon_2 \\ \varepsilon_3 \\ \gamma_{23} \\ \gamma_{13} \\ \gamma_{12} \end{Bmatrix} = \begin{bmatrix} cos\alpha^2 & sin\alpha^2 & 0 & 0 & 0 & cos\alpha sin\alpha \\ sin\alpha^2 & cos\alpha^2 & 0 & 0 & 0 & -cos\alpha sin\alpha \\ 0 & 0 & 1 & 0 & 0 & 0 \\ 0 & 0 & 0 & cos\alpha & -sin\alpha & 0 \\ 0 & 0 & 0 & sin\alpha & cos\alpha & 0 \\ -2cos\alpha sin\alpha & 2cos\alpha sin\alpha & 0 & 0 & 0 & cos\alpha^2 - sin\alpha^2 \end{bmatrix} \begin{Bmatrix} \varepsilon_z \\ \varepsilon_\theta \\ \varepsilon_r \\ \gamma_{\theta r} \\ \gamma_{zr} \\ \gamma_{z\theta} \end{Bmatrix}, \tag{11}$$

$$\{\varepsilon\} = [T_\varepsilon]\{\overline{\varepsilon}\}$$

where α is the winding angle, $\overline{\varepsilon}$ is the strain vector in the off-axis coordinate system (Figure 2).

The relation between off-axis and on-axis stiffness constants is the follow:

$$\left[\overline{C}_{ij}^{(k)}\right] = [T_\varepsilon]^T \left[C_{ij}^{(k)}\right][T_\varepsilon] \tag{12}$$

Consequently, $\left\{\overline{C}_{ij}^{(k)}\right\} = [A]\left\{C_{ij}^{(k)}\right\}$, where $[A]$ is the transformation matrix of sin and cos [3].

Substituting expressions for displacements and stresses into boundary and continuity conditions and integral equations for axial force (8) and torque (7) one may obtain the system of algebraic equations and consequently find the unknown constants and stresses in the pipe.

2.2. *Tsai-Hill failure criterion*

The failure coefficients are computed using the following stresses in principal material directions

$$\begin{bmatrix}\sigma_x\\ \sigma_y\\ \tau\end{bmatrix} = \begin{bmatrix} cos\alpha^2 & sin\alpha^2 & 2cos\alpha sin\alpha \\ sin\alpha^2 & cos\alpha^2 & -2cos\alpha sin\alpha \\ -cos\alpha sin\alpha & cos\alpha sin\alpha & cos\alpha^2 - sin\alpha^2 \end{bmatrix}\begin{bmatrix}\sigma_z\\ \sigma_\theta\\ \tau_{z\theta}\end{bmatrix} \tag{13}$$

The Tsai-Hill failure coefficient is [4]

$$Failure\ Coefficient = \frac{\sigma_x^2}{X_T^2} - \frac{\sigma_x\sigma_y}{X_T^2} + \frac{\sigma_y^2}{Y_T^2} + \frac{\tau^2}{S^2}, \tag{14}$$

where X_T and Y_T are longitudinal and transversal tensile strength values, and S is the shear strength. The failure happens when the failure coefficient becomes greater than 1.

3. Case studies

3.1. *3D Finite Element modelling*

Abaqus/CAE 2019 was used to develop a numerical model in order to calculate stresses in a section of thick-walled composite pipe under torsion load. All layers are assumed to be perfectly bonded. Carbon/epoxy fibre composite (T300/LY5052) was considered in this study, and T300/5208 composite was used for validation. The material properties are shown in Table 1.

Table 1. Material properties [4, 16].

Properties	T300/LY5052	T300/5208
$E_1(GPa)$	135	132
$E_2(GPa)$	8	10.8
$G_{12}(GPa)$	3.8	5.65
ν_{12}	0.27	0.24
ν_{23}	0.49	0.59
$X_T(MPa)$	1860	1513
$Y_T(MPa)$	76	43.4
$S(MPa)$	98	86.87

The torque was applied and the rigid body constraints were used. The model was meshed using Abaqus C3D20R elements (quadratic brick elements with reduced integration).

3.2. *Validation*

To validate the results, obtained by developed analytical and numerical models, we compared them with the one presented in Herakovich [16]. Two-layer composite pipe with the ply sequence $[\pm 45°]$ was modelled, and the distributions of the stresses in the pipe wall were computed and normalised for comparison.

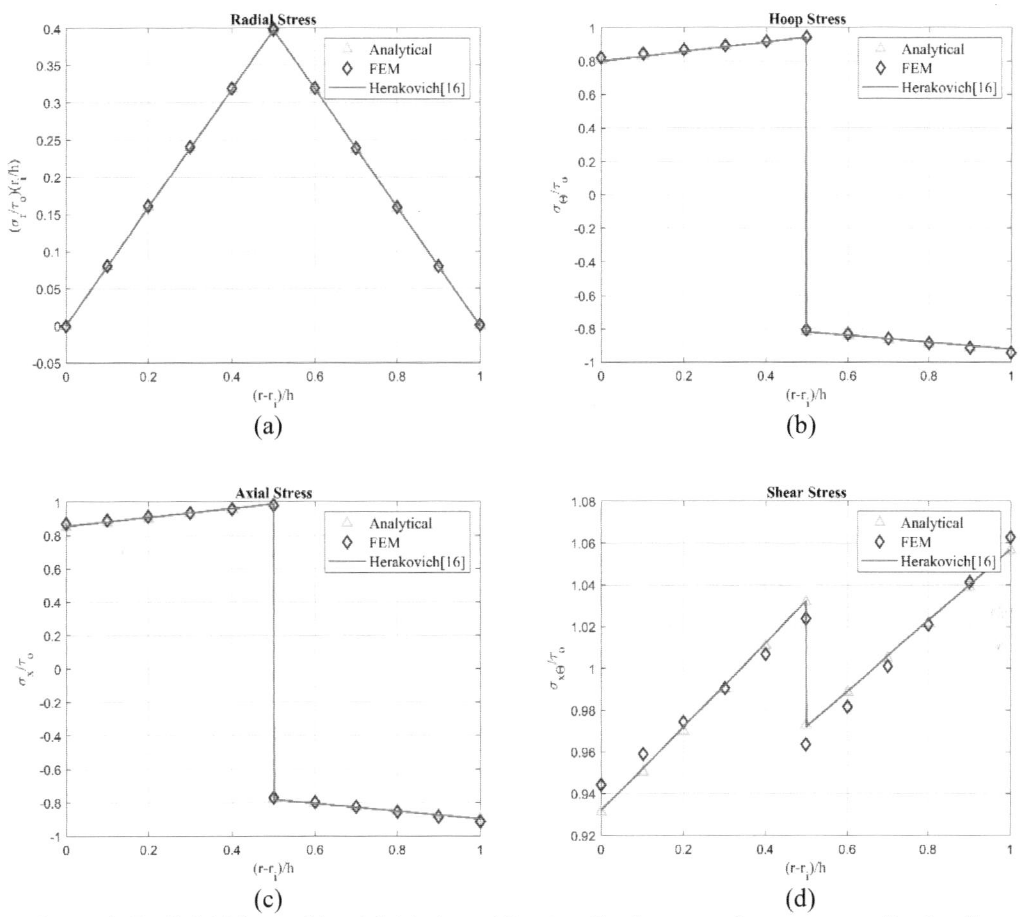

Figure 3. Radial (a) hoop (b) axial (c) shear (d) normalised stress values vs. normalised radius.

The average axial stress σ_o is

$$\sigma_o = \frac{P_x}{\pi(r_o^2 - r_i^2)}, \tag{15}$$

and the average shear stress τ_o is

$$\tau_o = \frac{3T_x}{2\pi(r_o^3 - r_i^3)}. \tag{16}$$

In this study $r_i/h = 5$ (where h is the total thickness of the composite structure) and the magnitude of the torque is $1kNm$.

Figure 3 shows that the analytical and numerical results coincide with the model ones [16].

3.3. Numerical results

In this section, we computed the failure coefficients for the four-layer composite pipe subjected to torques of different magnitude.

Inner and outer radii were $30mm$ and $36mm$, respectively, and the layer thickness was $1.5mm$.

The failure coefficients for the composite pipe $[\pm 55°/\pm 60°]$ for a range of torques from 5 to $30kNm$ are presented in Figure 4.

Failure coefficient clearly rises with the rise of the torque magnitude and the failure happens when the torque exceeds $20kNm$.

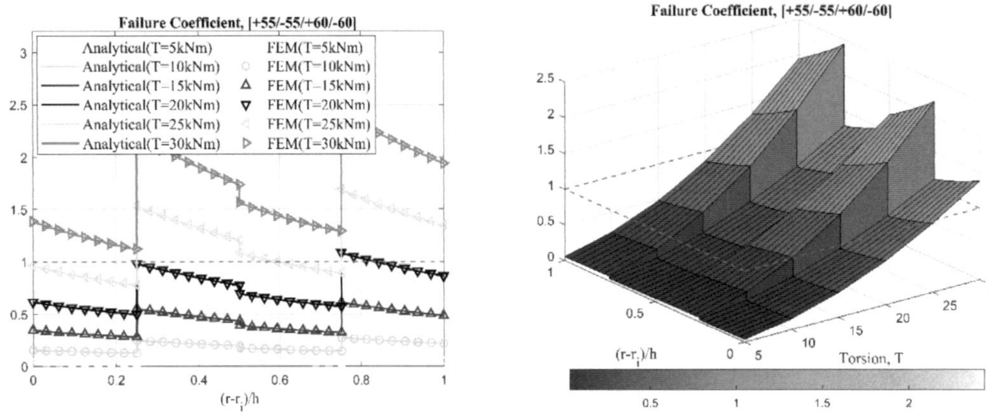

Figure 4. The failure coefficients for composite pipes subjected to torque.

4. Conclusions

Multi-layered long fibre reinforced composite pipes under torque load were investigated. Numerical and analytical results were validated by comparing with the model one presented in Herakovich [16]. Comparison of numerical results with the analytical ones was carried out for the pipes of different lay-ups subjected to the range of torsional load. That would be beneficial for the future analysis to investigate the interlaminar crack behaviour in thick walled composite pipes under different types of load [18].

5. References

[1] Yu K, Morozov E V, Ashraf M A and Shankar K 2017 A review of the design and analysis of reinforced thermoplastic pipes for offshore applications *J. Reinf. Plast. Compos.* **36** 1514–30

[2] Rafiee R 2016 On the mechanical performance of glass-fibre-reinforced thermosetting-resin pipes: A review *Compos. Struct.* **143** 151–64

[3] Xia M, Takayanagi H and Kemmochi K 2001 Analysis of multi-layered filament-wound composite pipes under internal pressure *Compos. Struct.* **53** 483–91

[4] Guz I A, Menshykova M and Paik J K 2017 Thick-walled composite tubes for offshore applications: an example of stress and failure analysis for filament-wound multi-layered pipes *Ships Offshore Struct.* **12** 304–22

[5] Tarakçioğlu N, Gemi L and Yapici A 2005 Fatigue failure behavior of glass/epoxy ±55 filament wound pipes under internal pressure *Compos. Sci. Technol.* **65** 703–8

[6] Gemi L, Tarakçioğlu N, Akdemir A and Şahin Ö S 2009 Progressive fatigue failure behavior of glass/epoxy (±75)2 filament-wound pipes under pure internal pressure *Mater. Des.* **30** 4293–8

[7] Tarakcioglu N, Samanci A, Arikan H and Akdemir A 2007 The fatigue behavior of (±55°)3 filament wound GRP pipes with a surface crack under internal pressure *Compos. Struct.* **80** 207–11

[8] Arikan H 2010 Failure analysis of (±55°)3 filament wound composite pipes with an inclined surface crack under static internal pressure *Compos. Struct.* **92** 182–7

[9] Bakaiyan H, Hosseini H and Ameri E 2009 Analysis of multi-layered filament-wound composite pipes under combined internal pressure and thermomechanical loading with thermal variations *Compos. Struct.* **88** 532–41

[10] Xia M, Kemmochi K and Takayanagi H 2001 Analysis of filament-wound fiber-reinforced sandwich pipe under combined internal pressure and thermomechanical loading *Compos. Struct.* **51** 273–83

[11] Hastie J C, Kashtalyan M and Guz I A 2019 Failure analysis of thermoplastic composite pipe (TCP) under combined pressure, tension and thermal gradient for an offshore riser application *Int. J. Press. Vessel. Pip.* **178** 103998

[12] Hastie J C, Guz I A and Kashtalyan M 2019 Effects of thermal gradient on failure of a thermoplastic composite pipe (TCP) riser leg *Int. J. Press. Vessel. Pip.* **172** 90–9

[13] Xia M, Takayanagi H and Kemmochi K 2002 Bending behavior of filament-wound fiber-reinforced sandwich pipes *Compos. Struct.* **56** 201–10
[14] Menshykova M and Guz I A 2014 Stress analysis of layered thick-walled composite pipes subjected to bending loading *Int. J. Mech. Sci.* **88** 289–99
[15] Cox K, Menshykova M, Menshykov O and Guz I A 2019 Analysis of flexible composites for coiled tubing applications *Compos. Struct.* **225** 111118
[16] Herakovich C T 1998 *Mechanics of Fibrous Composites* (New York: John Wiley)
[17] Cagdas I U 2017 Optimal design of variable stiffness laminated composite truncated cones under lateral external pressure *Ocean Eng.* **145** 268–76
[18] Menshykova M V, Menshykov O V, Guz I A 2009 Linear interface crack under plane shear wave *CMES – Comp. Model. Eng.* **48(2)** 107–20

IOP Conf. Series: Materials Science and Engineering **936** (2020) 012047 doi:10.1088/1757-899X/936/1/012047

Comparative Study of Beam Theories on the Effect of Span-Depth Ratio for Symmetric and Un-symmetric Loadings

Akash Gaur and Priyanka Dhurvey*

Department of Civil Engineering, Maulana Azad National Institute of Technology, Bhopal, India

*pdhurvey@gmail.com

Abstract. In the present work, a comparative study is done between Shear-indeformable (Euler-Bernoulli) model and Shear-deformable (Timoshenko) model for two different types of beams (simply supported and fixed beam) under uniformly distributed loading condition and analysed for different Span-Depth ratio (L/D). For un-symmetric loading, a simply supported beam subjected to external point moment is also analysed for different L/D ratios to check the behaviour of beam. Transverse Displacement is taken as a yardstick. Shape functions for Euler-Bernoulli and Timoshenko beam models are developed in MATLAB software by using the Finite Element Method for calculating deflection at different points on the beam. It was observed that for large L/D ratio, the deflections are same for both the beam theory models but as L/D ratio decreases, the deflections for both the beam theories vary from each other. For small L/D ratio, Timoshenko beam model gives more accurate results, since the Timoshenko beam theory is a higher order beam theory than the Euler-Bernoulli beam theory, it is known to be superior in predicting the response of the deep beam.

1. Introduction

The static and dynamic characteristics of beam element in a structure are evaluated by using the classical or refined beam theories. Euler-Bernoulli beam theory (EBT) is generally used for the analysis of the distribution of stresses and displacements in the beam element with a small thickness. It has been observed that in case of deep beams, the beam section warp under loaded conditions. This is because of shear deformations are significant in the deep sections and cannot be neglected.

As per IS Code – 456 (2000), a beam is considered as a deep beam when the ratio of effective span (L) to overall depth (D) is:-

1. L/D < 2 for simply supported beam
2. L/D < 2.5 for continuous beam

Timoshenko beam theory is the extension of Euler-Bernoulli beam theory. The basic difference in the assumptions of EBT and TBT is that in EBT the plane section remains plane and perpendicular to neutral axis before and after bending, but in TBT plane section before and after bending remains plane but is not perpendicular to the neutral axis. Mosavi et al. [9] investigated beam equation with track motion dynamic stability using fourth order partial differential of the Euler-Bernoulli and Timoshenko model on an Elastic Foundation. Labuschagne et al. [3] considered three different linear theories: Euler-Bernoulli, Timoshenko and two-dimensional elasticity models for a cantilever beam analysis. It is shown that the Timoshenko model gives remarkably accurate results compared to other two models. Edem B. [7] used a new model with two-node Timoshenko beam element for representing beam rotation in a shear deformable beam. Sayyad [5] compared various shear deformation theories for the free vibration analysis of thick isotropic beams. Wang C. [17] presented single span Timoshenko and Euler-Bernoulli beams subjected to transverse loading including the effect of transverse shearing strain for deflection and stresses without flexural-shear-deformation analysis which is more complicated. Patel R. et al [15] studied the effect of depth of an RCC beam using method of initial functions (MIF). Three different cases were discussed with respect to span-depth ratios which were compared different theories with the results from ANSYS software.

In the present study, MATLAB program is developed to compare various beam theories for the analyses of two different types of beams i.e. simply supported and fixed for different span-depth ratios for symmetric and un-symmetric loadings especially for deep beam section.

2. Mathematical formulation

2.1. *Euler Bernoulli beam theory (EBT)*

Euler–Bernoulli beam theory (also known as classical beam theory) is a simplified form of the linear theory of elasticity. It covers the case for small deflections of a beam that are subjected to lateral loads only. It is thus a special case of Timoshenko beam theory.

Governing equations (EBT) are as follows

$$\frac{M}{EI} = \frac{d^2y}{dx^2} \tag{1}$$

$$\frac{dM}{dx} = \frac{d^3y}{dx^3} \tag{2}$$

$$\frac{d^4y}{dx^4} = \frac{q}{EI} \tag{3}$$

where, M = Bending Moment; E = Modulus of Elasticity; I = Second Moment Inertia

2.2. *Timoshenko beam theory (TBT)*

Unlike the Euler-Bernoulli beam, Timoshenko beam theory accounts for shear deformation and rotational inertia effects. Therefore, the Timoshenko beam can be used for the analysis of deep beams. The stiffness of the Timoshenko beam is lower than the Euler-Bernoulli beam, which results in larger deflections under static loading and buckling. As the effect of shear is also considered in TBT, the plane section no longer remains normal to the neutral axis of the beam. The rotation of the plane section is the summation of bending rotation (θ) and rotation due to shear (γ). So the total rotation of the plane can be written as –

$$\frac{dw}{dx} = \theta - \gamma \tag{4}$$

$$\frac{d\left(EI\,\frac{d\theta}{dx}\right)}{dx} = GA\left(\theta - \frac{dw}{dx}\right) \tag{5}$$

where, E= Elastic Modulus; I= Second Moment Inertia; G= Shear Modulus; A=Area of cross-section

2.3. *Stiffness matrix*

The finite element stiffness matrix for Euler Bernoulli and Timoshenko beam models are-

For Euler Bernoulli beam –

$$K_E = \frac{EI}{L^3}\begin{bmatrix} 12 & 6L & -12 & 6L \\ 6L & 4L^2 & -6L & 2L^2 \\ -12 & -6L & 12 & -6L \\ 6L & 2L^2 & -6L & 4L^2 \end{bmatrix} \tag{6}$$

For Timoshenko beam –

$$K_E = \frac{EI}{(1+\emptyset)L^3}\begin{bmatrix} 12 & 6L & -12 & 6L \\ 6L & (4+\emptyset)L^2 & -6L & (2-\emptyset)L^2 \\ -12 & -6L & 12 & -6L \\ 6L & (2-\emptyset)L^2 & -6L & (4+\emptyset)L^2 \end{bmatrix} \tag{7}$$

$$\text{where, } \emptyset = \frac{12EI}{L^2GA} \tag{8}$$

Shape Function for EBT

$$[N] = [\,N_1 \;\; N_2 \; N_3 \; N_4\,] \tag{9}$$

where,

$$\left.\begin{aligned} N_1 &= 1-\frac{3x^2}{L^2}+\frac{2x^3}{L^3};\\ N_2 &= x-\frac{2x^2}{L}+\frac{x^3}{L^2};\\ N_3 &= \frac{3x^2}{L^2}-\frac{2x^2}{L^3};\\ N_4 &= \frac{x^3}{L^2}-\frac{x^2}{L} \end{aligned}\right\} \tag{11}$$

Shape Function for TBT

$$[N] = [\,N_1 \;\; N_2 \; N_3 \; N_4\,] \tag{10}$$

where,

$$\left.\begin{aligned} N_1 &= 1-\frac{x}{L};\\ N_2 &= \frac{-x^2}{2L}+\frac{x}{2};\\ N_3 &= \frac{x}{L};\\ N_4 &= \frac{x^2}{2L}-\frac{x}{2} \end{aligned}\right\} \tag{12}$$

Accuracy of the formulated/generated shape function of TBT is examined with the previously available investigations. The geometric, loading and the boundary conditions are same as stated in the literature, Patel R. et al. [8].Table 1 illustrates the values of transverse deflection of the simply supported beam with L/D ratio of 7.5. The present model is in good agreement with the transverse deflection obtained by Patel R. et al [14].

Table 1. Validation of programming

L/D ratio	Deflection Patel et al. [14]	Deflection Present model	Percentage difference $\frac{\delta 1-\delta 2}{\delta 1}\times 100$
7.5	177.58 (mm)	182.42 (mm)	2.73 %

3. Beam under consideration

The analysis is carried out for a beam (Figure 1)made up of isotropic material subjected to uniformly distributed load (UDL), q = 20kN/m with dimensions as length, L = 3 m, width, B = unity and depth, D = 0.375, 1, 1.2, and 1.5m. The following material properties are taken, Elastic Modulus, E = 22,360×10^{3}kN/m^2, Shear Modulus, G = 10,164×10^3 kN/m^2, δ_E= maximum deflection of Euler Bernoulli beam element, δ_T = maximum deflection of Timoshenko beam element.

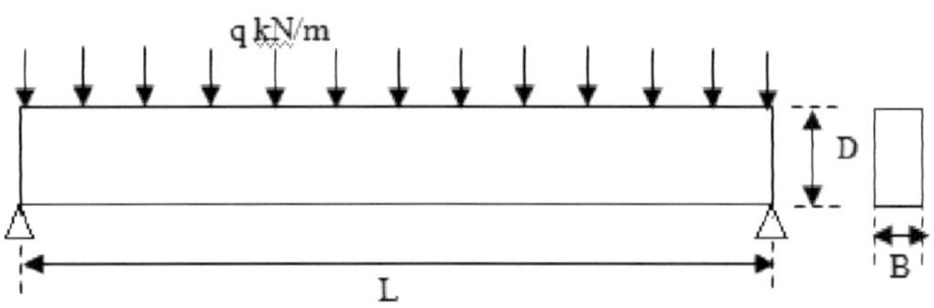

Figure 1. Schematic diagram of beam under consideration

4. Results and discussions

In present work, the analyses carried out were as follows-

Case 1. Beam under consideration was examined for symmetrical loading subjected to uniformly distributed load for simply supported and fixed beam.

Case 2. For unsymmetrical loading the beam was subjected to moment at the left end and examined for deflection using the formulated EBT and TBT for simply supported beam.

The variation of deflection (mm) with distance along the beam is discussed in figure 2 to figure 4 and in table 2 to table 4.

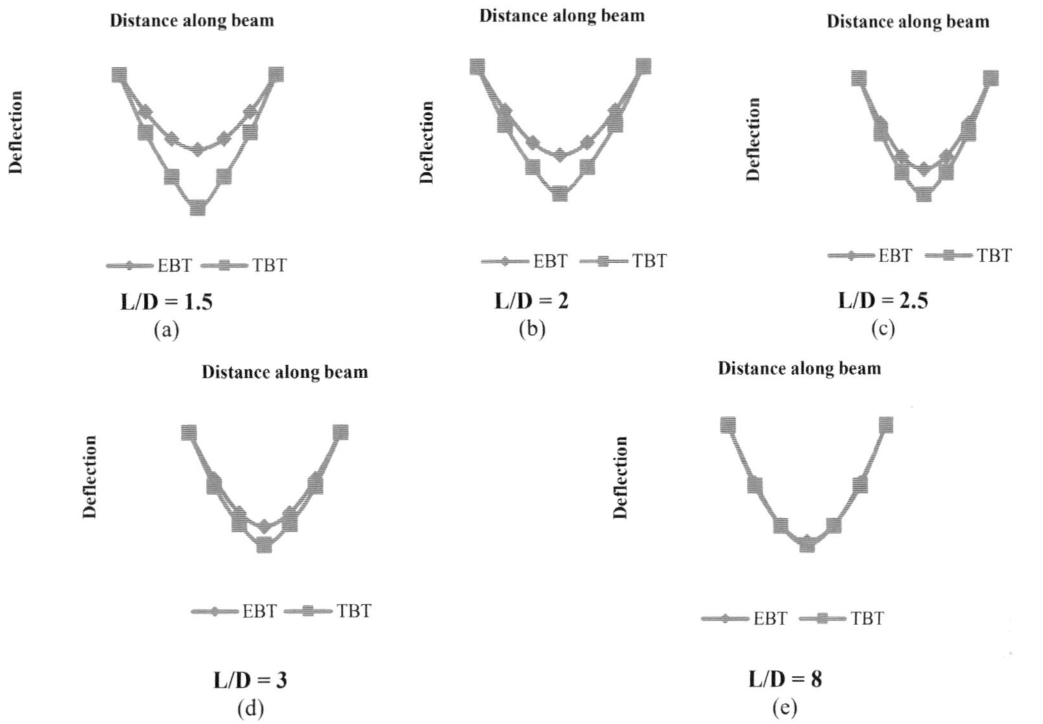

Figure 2. Variation in deflection (mm) of simply supported beams with different L/D ratio calculated using Euler and Timoshenko theories.

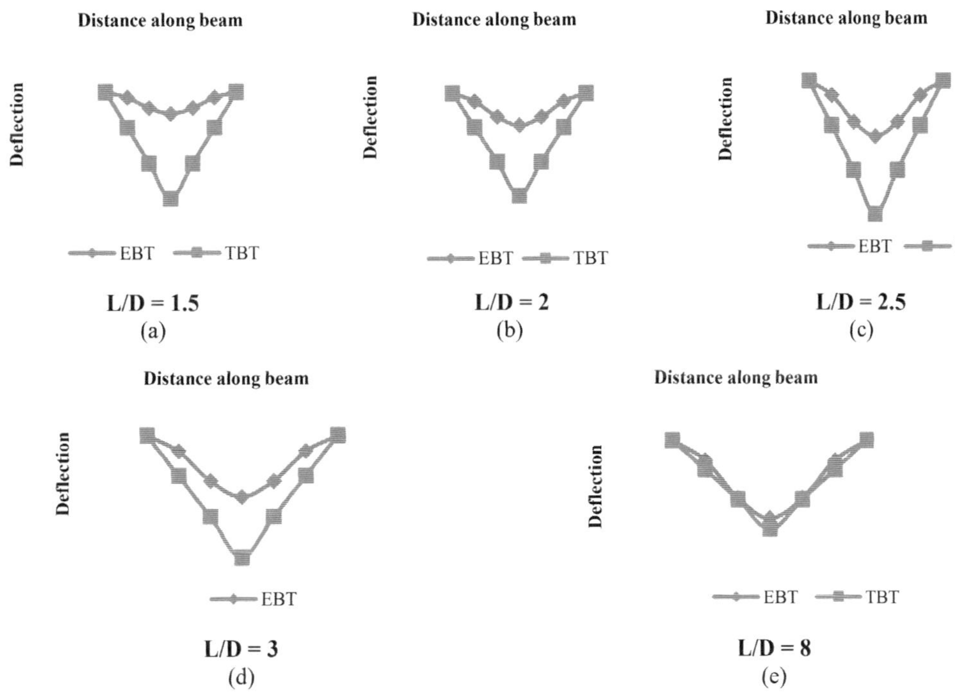

Figure 3. Variation in deflection (mm) of fixed beams with different L/D ratio calculated using Euler and Timoshenko theories.

As observed in figure 2 to figure 3 and table 2 and table 3 for different beams i.e. simply supported and fixed with the increase of L/D ratio for symmetric loading, the deflections calculated using the Timoshenko theory and Euler theory are so close and beams with lower L/D ratio variation in deflection was large because of the shear effect in Timoshenko theory which is not included in Euler theory.

Table 2. Variation in deflection of simply supported beams calculated using Euler and Timoshenko theories.

Sl. No.	L/D ratio	δ_E(mm)	δ_T(mm)	$\frac{\delta_T - \delta_E}{\delta_T} \times 100$ %
1	1.5	-1.42×10^{-3}	-2.52×10^{-3}	36.89
2	2	-3.35×10^{-3}	-4.83×10^{-3}	30.64
3	2.5	-6.55×10^{-3}	-8.40×10^{-3}	22.02
4	3	-1.13×10^{-2}	-1.35×10^{-2}	16.29
5	8	-2.15×10^{-1}	-2.21×10^{-1}	2.71

Table 3. Variation in Deflection of fixed beam calculated using Euler and Timoshenko theories.

Sl. No.	L/D ratio	δ_E(mm)	δ_T(mm)	$\frac{\delta_T - \delta_E}{\delta_T} \times 100$ %
1	1.5	-0.00028301	-0.0013899	79.64
2	2	-0.00067084	-0.0021466	68.75
3	2.5	-0.0013102	-0.003155	58.47
4	3	-0.0022641	-0.0044778	49.44
5	8	-0.0429338	-0.048837	12.09

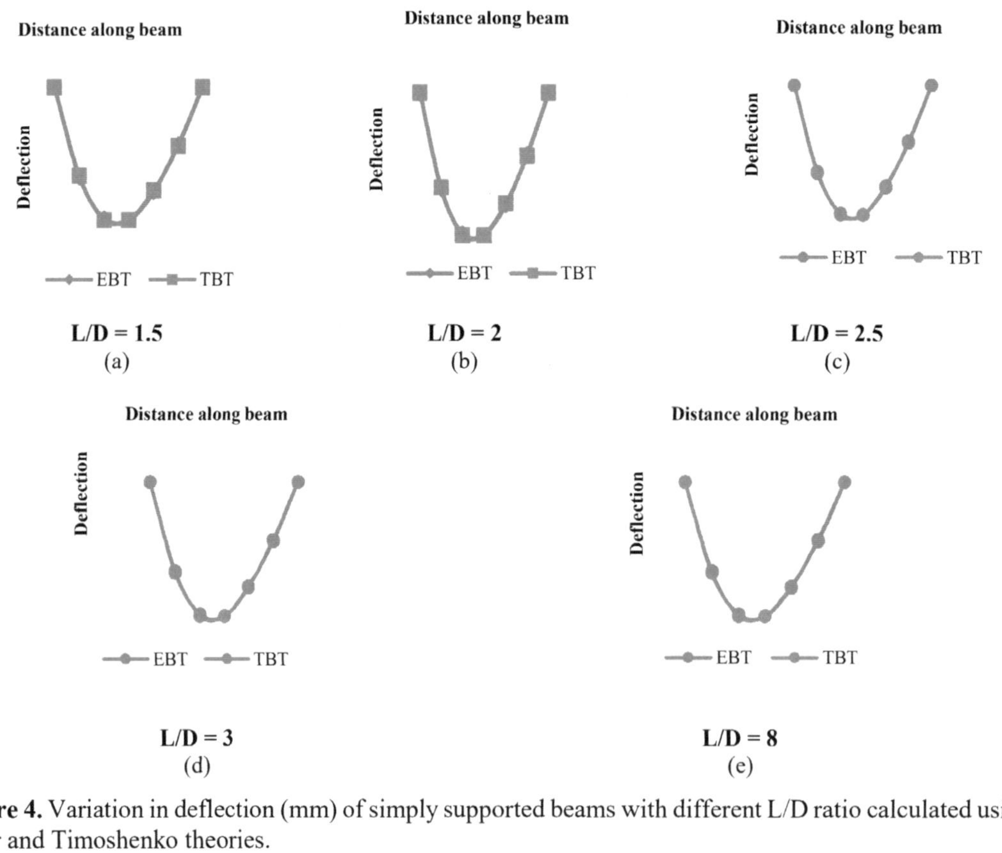

Figure 4. Variation in deflection (mm) of simply supported beams with different L/D ratio calculated using Euler and Timoshenko theories.

Similarly, when simply supported beam was subjected to moment at left end for un-symmetric loading i.e. Case-2, the results show that the Timoshenko beam model is remarkably accurate. As observed from figure 4, the TBT and EBT gave same values of deflection because the shear effect is negligible for this loading condition as only moment is applied in beam section.

Table 4. Variation in deflection of simply supported beams calculated using Euler and Timoshenko theories.

S.No.	L/D ratio	δ_E(mm)	δ_T(mm)	$\frac{\delta_T - \delta_E}{\delta_T} \times 100\ \%$
1.	1.5	-1.162×10^{-3}	-1.180×10^{-3}	1.525
2.	2	-2.753×10^{-3}	-2.795×10^{-3}	1.502
3.	2.5	-5.378×10^{-3}	-5.458×10^{-3}	1.465
4.	3	-9.295×10^{-3}	-9.432×10^{-3}	1.442
5.	8	-176.27×10^{-3}	-178.84×10^{-3}	1.437

5. Conclusions

In the present work, a MATLAB program is developed for the comparative study of beam theories to study the effect of span-depth ratio for different beam sections subjected to symmetrical UDL and un-symmetrical loading. The main objective was to formulate shape function for TBT and a program which can be used for the analysis of deep beam sections. Based on the obtained results, the following conclusions can be drawn –

1. The developed MATLAB programme is capable of analysing different deep beam sections and yields good agreement with literature.
2. The MATLAB programme can be considered for different design parameters like depth for deep beam, boundary conditions, and loading conditions.

6. References

[1] Beck A et al. 2010 Timoshenko versus Euler beam theory: Pitfalls of a deterministic approach *Structural Safety* **33(1)** 19–25

[2] Bazoune A et al. 2002 Shape functions of three dimensional Timoshenko beam element *Journal of Sound and Vibration* **259(2)** 473–80

[3] Labuschagne et al. 2009 Comparison of linear beam theories *Mathematical and Computer Modeling* **49(1-2)** 20–30

[4] Do K D 2017 Stabilisation of exact nonlinear Timoshenko beams in space by boundary feedback *Journal of Sound and Vibration* **422** 278–99

[5] Sayyad A 2011 Comparison of various refined beam theories for the bending and free vibration analysis of thick beams *Applied and Computational Mechanics journal* **5** 217–30

[6] Ding Hu et al. 2017 Natural frequencies of a supercritical transporting Timoshenko beam *European Journal of Mechanics A/Solid* **66** 79–93

[7] Edem B 2007 The exact two-node Timoshenko beam finite element using analytical bending and shear rotation interdependent shape functions *International Journal for Computational Methods in Engineering Science and Mechanics* 429–31

[8] Faghidian S 2017 Unified formulations of the shear coefficients in Timoshenko beam theory *ASCE* **143(9)**

[9] Mosavi A et al. 2017 Comparison of Euler-Bernoulli and Timoshenko beam equations for railway system dynamics *Recent Advances in Technology Research and Education* 32–40

[10] Janecka A et al. 2016 Euler–Bernoulli type beam theory for elastic bodies with nonlinear response in the small strain range *Archives of Mechanics* **68(1)** 3–25

[11] Krysko A V et al. 2017 Stability of curvilinear Euler-Bernoulli beams in temperature fields *International Journal of Non-Linear Mechanics* **94** 207–15

[12] Lignola G et al. 2017 Exact stiffness–matrix of two nodes Timoshenko beam on elastic medium An analogy with Eringen model of nonlocal Euler–Bernoulli nanobeams *Computers and Structures* **182** 556–72

[13] Muhammad I 2018 Laminated Timoshenko beams with viscoelastic damping *Journal of Mathematical Analysis and Applications* **466(1)** 619–41

[14] Patel R et al. 2014 Effect of depth span ratio on the behaviour of beams *International Journal of Advanced Structural Engineering (IJASE)* **6(2)** 1–7
[15] Patel R et al. 2014 Analysis of Isotropic Beams Using Method of Initial Functions (MIF) *Electronic Journal of Structural Engineering* **14** 1–6
[16] Radenkovic G et al. 2018 Linear static isogeometric analysis of an arbitrarily curved spatial Bernoulli–Euler beam *Computer Methods in Applied Mechanics and Engineering* **341** 360–96
[17] Wang C 1995 Timoshenko Beam-Bending Solutions in Terms of Euler-Bernoulli Solutions *Journal of Engineering Mechanic* **121(6)** 763–9

IOP Conf. Series: Materials Science and Engineering **936** (2020) 012048 doi:10.1088/1757-899X/936/1/012048

Effects of the Variational Consistency and Variational Inconsistency on Free Flexural Vibration Frequencies of Simply-supported Rectangular Isotropic Shear Deformable Beams

Kedar S Pakhare*, **P J Guruprasad and Rameshchandra P Shimpi**

Department of Aerospace Engineering, IIT Bombay, Mumbai 400076 India

*kedar188200@gmail.com

Abstract. In order to derive governing differential equations of a displacement-based shear deformation beam theory, one can either utilise the variationally consistent approach (*VCA*) or the variationally inconsistent approach (*VIA*). In the *VCA*, variational principles are utilised to derive beam governing equations. Whereas in the *VIA*, beam gross / integrated two-dimensional equilibrium equations are utilised to derive beam governing equations. Shimpi et al. [2] have proposed the single variable shear deformation beam theory (*SVSDBT*) which is based on the *VIA*. Whereas, Shimpi et al. [1] have proposed the two variable shear deformation beam theory (*TVSDBT*) which is based on the *VCA*. In the *SVSDBT* and *TVSDBT*, the beam axial displacement and beam transverse displacement consist of bending and shearing components. In both theories, bending components do not take part in the cross-sectional shearing force, and shearing components do not take part in the cross-sectional bending moment. Both theories utilise linear strain-displacement relations. Displacement functions of both theories give rise to the beam transverse shear strain (and hence to the beam transverse shear stress) which varies quadratically through the beam thickness and maintains transverse shear stress-free beam surface conditions. For the case of dynamics, the *SVSDBT* has only one governing equation, and the *TVSDBT* has two governing equations which are inertially coupled. The aim of this paper is to present the comparison study on free flexural vibration frequencies of simply-supported rectangular isotropic shear deformable beams obtained by utilising the *SVSDBT* and *TVSDBT*. These frequencies are obtained for various values of the beam thickness-to-length ratio and beam vibration mode parameter. This comparison shows that for both theories, the *VCA* and *VIA* have a minimal effect on resulting beam free flexural vibration frequencies.

1. Introduction

It is well established that transverse shear deformation effects for microscopic as well as macroscopic shear deformable beams have a significant effect on their flexural vibration frequencies. Because of the effects of shear in the beam deformation, non-dimensional vibration frequencies of shear deformable beams decrease in comparison to corresponding non-dimensional frequencies of slender beams (Shimpi et al. [1]).

Due to features associated with the assumed displacement field of the Bernoulli-Euler beam theory (*BEBT*), it is unable to capture effects of shear on the beam deformation behaviour. Various displacement-based first-order (*FOBT*) and higher-order beam theories (*HOBT*) have been reported in the literature in order to account for beam transverse shear deformation effects. *FOBT* result in the constant transverse shear strain and *HOBT* result in more realistic non-linear transverse shear strain throughout the thickness of the beam.

In order to derive governing equations of *FOBT* or *HOBT*, the variationally consistent approach (*VCA*) or the variationally inconsistent approach (*VIA*) can be used. In the *VCA*, variational principles are utilised to minimise the beam potential energy (for the case of statics) or to minimise the Lagrangian (for the case of dynamics), to derive beam governing equations. Whereas in the *VIA*, beam gross/integrated two-dimensional equilibrium equations are utilised to derive beam governing equations. It should be noted that as the number of primary unknowns associated with the beam theory increases, it becomes increasingly challenging to derive beam governing equations using the *VIA*. Moreover, beam boundary conditions must be specified explicitly based on the physical understanding of the beam deformation behaviour in case of the *VIA*. Aforementioned challenges are not encountered when the *VCA* is utilised for deriving beam governing equations. Recently Shimpi et al. [1] have proposed the two variable shear deformation beam theory (*TVSDBT*) and Shimpi et al. [2] have proposed the single variable shear deformation beam theory (*SVSDBT*). These theories are applicable for beams undergoing small deformations. In both theories, the beam axial and transverse displacements have been split into bending components and shearing components. This splitting is such that bending components do not take part in the cross-sectional shearing force, and shearing components do not take part in the cross-sectional bending moment. Both theories result in the parabolic distribution of the beam transverse shear strain (and hence the beam transverse shear stress) through the beam thickness, maintaining transverse shear stress-free beam surface conditions, and therefore do not require a shear correction factor. Governing equations of *TVSDBT* have been obtained based on the *VCA*. Whereas, governing equation of *SVSDBT* has been obtained based on the *VIA*. For the case of dynamics, the *TVSDBT* has two governing equations which are inertially coupled, and the *SVSDBT* has only one governing equation.

In this paper, free flexural vibration frequencies of simply-supported rectangular isotropic shear deformable beams are presented which are obtained using the *SVSDBT* and *TVSDBT* for various values of the beam thickness-to-length ratio and beam vibration mode parameter. This comparison study aims to show the very small to none effect that the *VCA* and *VIA* have for a given shear deformation theory on the beam free flexural vibration frequencies. Corresponding results obtained from the *BEBT* and the theory of Touratier [3] adapted for beams by Sayyad [4] are also presented for the comparison purpose.

2. Theoretical formulation details of the *SVSDBT* and *TVSDBT*

In this section, brief details about the theoretical formulation of the *SVSDBT* and *TVSDBT* are presented. The beam geometry, co-ordinate system considered and symbols appearing in subsequent equations are the same as those of Shimpi et al. [1].

2.1. Expressions for the beam axial and transverse displacements

For the *SVSDBT* and *TVSDBT*, the displacement field was defined as follows

$$u = u_b + u_s = -z\frac{\partial w_b}{\partial x} + h\left[-\frac{5}{3}\left(\frac{z}{h}\right)^3 + \frac{1}{4}\left(\frac{z}{h}\right)\right]\frac{\partial w_s}{\partial x} \tag{1}$$

$$w = w_b + w_s \tag{2}$$

2.2. Expressions for strains

As the *SVSDBT* and *TVSDBT* are valid for beams undergoing small deformations, expressions for the beam axial and transverse shear strain by utilising linear strain displacement relations were as follows

$$\epsilon_x = -z\frac{\partial^2 w_b}{\partial x^2} + h\left[-\frac{5}{3}\left(\frac{z}{h}\right)^3 + \frac{1}{4}\left(\frac{z}{h}\right)\right]\frac{\partial^2 w_s}{\partial x^2} \tag{3}$$

$$\gamma_{zx} = \left[-5\left(\frac{z}{h}\right)^2 + \frac{5}{4}\right]\frac{\partial w_s}{\partial x} \tag{4}$$

2.3. Expressions for stresses

For the *SVSDBT* and *TVSDBT*, expressions for the beam axial and transverse shear stress were as follows

IOP Conf. Series: Materials Science and Engineering **936** (2020) 012048 doi:10.1088/1757-899X/936/1/012048

$$\sigma_x = -E\,z\,\frac{\partial^2 w_b}{\partial x^2} + E\,h\left[-\frac{5}{3}\left(\frac{z}{h}\right)^3 + \frac{1}{4}\left(\frac{z}{h}\right)\right]\frac{\partial^2 w_s}{\partial x^2} \tag{5}$$

$$\tau_{zx} = \frac{E}{2\,(1+\mu)}\left[-5\left(\frac{z}{h}\right)^2 + \frac{5}{4}\right]\frac{\partial w_s}{\partial x} \tag{6}$$

2.4. Expressions for governing equations

Governing equations of the *TVSDBT* obtained using the *VCA* were as follows

$$E\,I\,\frac{\partial^4 w_b}{\partial x^4} - \rho\,I\left[\frac{\partial^2}{\partial t^2}\left(\frac{\partial^2 w_b}{\partial x^2}\right)\right] + \rho\,A\,\frac{\partial^2 w_b}{\partial t^2} + \rho\,A\,\frac{\partial^2 w_s}{\partial t^2} = q \tag{7}$$

$$\frac{E\,I}{84}\,\frac{\partial^4 w_s}{\partial x^4} - \frac{\rho\,I}{84}\left[\frac{\partial^2}{\partial t^2}\left(\frac{\partial^2 w_s}{\partial x^2}\right)\right] - \frac{5\,E\,A}{12\,(1+\mu)}\,\frac{\partial^2 w_s}{\partial x^2} + \rho\,A\,\frac{\partial^2 w_b}{\partial t^2} + \rho\,A\,\frac{\partial^2 w_s}{\partial t^2} = q \tag{8}$$

Governing equation of the *SVSDBT* obtained using the *VIA* was as follows

$$E\,I\,\frac{\partial^4 w_b}{\partial x^4} - \rho\,I\left[1+\frac{12\,(1+\mu)}{5}\right]\left[\frac{\partial^2}{\partial t^2}\left(\frac{\partial^2 w_b}{\partial x^2}\right)\right] + \rho\,A\,\frac{\partial^2 w_b}{\partial t^2} + \frac{\rho^2 I}{E}\left[\frac{12\,(1+\mu)}{5}\right]\frac{\partial^4 w_b}{\partial t^4} = q \tag{9}$$

3. Free flexural vibrations of simply-supported isotropic shear deformable beams

In this section, the *SVSDBT* and *TVSDBT* are utilised to obtain respective free vibration characteristic equations of the simply-supported isotropic shear deformable beams.

3.1. Details with regard to SVSDBT

For the *SVSDBT*, boundary conditions for simply-supported ends were prescribed as follows:

$$\left[w_b - \frac{h^2(1+\mu)}{5}\,\frac{\partial^2 w_b}{\partial x^2}\right]_{x\,=\,0,\,L} = 0 \tag{10}$$

Following solution for w_b was assumed which satisfies boundary conditions (10):

$$w_b = W_{bn}\sin(\omega_n\,t)\,\sin\left(\frac{n\,\pi\,x}{L}\right) \tag{11}$$

Substituting this assumed expression for w_b in governing equation (9), following characteristic equation was obtained:

$$\left(\omega_n^{\,2}L^4\,\frac{\rho\,A}{E\,I}\right)^2\left(\frac{1+\mu}{60}\right)\left(\frac{h}{L}\right)^4 - \left(\omega_n^{\,2}L^4\,\frac{\rho\,A}{E\,I}\right)\left[1+\left(\frac{n\,\pi\,h}{L}\right)^2\left(\frac{1}{12}+\frac{1+\mu}{5}\right)\right] + (n\,\pi)^4 = 0 \tag{12}$$

3.2. Details with regard to TVSDBT

For the *TVSDBT*, boundary conditions for simply-supported ends were prescribed as follows:

$$\left[w_b\right]_{x\,=\,0,\,L} = 0 \quad \text{and} \quad \left[E\,I\,\frac{\partial^2 w_b}{\partial x^2}\right]_{x\,=\,0,\,L} = 0 \tag{13}$$

$$\left[w_s\right]_{x\,=\,0,\,L} = 0 \quad \text{and} \quad \left[\frac{E\,I}{84}\,\frac{\partial^2 w_s}{\partial x^2}\right]_{x\,=\,0,\,L} = 0 \tag{14}$$

IOP Conf. Series: Materials Science and Engineering **936** (2020) 012048 doi:10.1088/1757-899X/936/1/012048

Following solutions for w_b and w_s were assumed which satisfy boundary conditions (13) and (14):

$$\begin{Bmatrix} w_b \\ w_s \end{Bmatrix} = \begin{Bmatrix} W_{bn} \\ W_{sn} \end{Bmatrix} \sin(\omega_n t) \sin\left(\frac{n \pi x}{L}\right) \tag{15}$$

Substituting these assumed expressions for w_b and w_s in governing equations (7) and (8), following equations were obtained which are of the standard eigenvalue form:

$$\left[\begin{bmatrix} k_{11n} & k_{12n} \\ k_{21n} & k_{22n} \end{bmatrix} - (\omega_n)^2 \begin{bmatrix} m_{11n} & m_{12n} \\ m_{21n} & m_{22n} \end{bmatrix}\right] \begin{Bmatrix} W_{bn} \\ W_{sn} \end{Bmatrix} = \begin{Bmatrix} 0 \\ 0 \end{Bmatrix} \tag{16}$$

with

$$k_{11n} = \frac{n^4 \pi^4 E I}{L^4} \qquad m_{11n} = \frac{n^2 \pi^2 \rho I}{L^2} + \rho A$$

$$k_{12n} = k_{21n} = 0 \qquad m_{12n} = m_{21n} = \rho A$$

$$k_{22n} = \frac{n^4 \pi^4 E I}{84 L^4} + \frac{5 n^2 \pi^2 E A}{12 (1 + \mu) L^2} \qquad m_{22n} = \frac{n^2 \pi^2 \rho I}{84 L^2} + \rho A$$

4. Numerical results

In this section, numerical results pertaining to free flexural vibration frequencies of simply-supported rectangular isotropic shear deformable beams obtained by utilising the *SVSDBT*, *TVSDBT*, *BEBT* and theory of Touratier adapted for beams are presented in tabular form in tables 1 through 4 for various values of the beam thickness-to-length ratio h / L and beam vibration mode parameter n. It is then followed by the relevant discussion.

The non-dimensional natural frequency of the beam vibration $\bar{\omega}_n$ was defined as follows:

$$\bar{\omega}_n = \sqrt{\frac{\rho A L^4}{E I}}\, \omega_n \tag{17}$$

Following points were considered while obtaining results which are presented in tables 1 through 4:

- The Poisson's ratio μ was assumed to be 0.3.
- The beam vibration mode parameter n was assumed to have values 1, 2, 3 and 4.
- The beam thickness-to-length ratio h / L was assumed to have values 0.01, 0.05, 0.10, 0.15 and 0.20.
- Numerical results obtained by utilising the higher-order trigonometric theory of Touratier and *BEBT* have been calculated by the present authors.
  i. In case of the higher-order trigonometric theory of Touratier, eqs. (14), (30) through (36) of Sayyad [4] were utilised.
  ii. In case of the *BEBT*, eq. (12) of Timoshenko [5] (wherein effects of rotary inertia are taken into consideration) was utilised.
- In tables 1 through 4, the percentage difference quoted in parentheses against results reported by the particular theory (either by the *SVSDBT*, *TVSDBT* or *BEBT*, as the case may be) was calculated as follows:

$$\%\ \text{difference} = \left(\frac{\text{The result reported by the particular theory}}{\text{The corresponding result of the theory of Touratier}} - 1\right) \times 100 \tag{18}$$

IOP Conf. Series: Materials Science and Engineering **936** (2020) 012048 doi:10.1088/1757-899X/936/1/012048

Table 1. Free flexural vibration frequencies of simply-supported rectangular isotropic shear deformable beams in the non-dimensional form, $n = 1$ and $\mu = 0.3$.

Theory	Non-dimensional natural frequencies of the beam vibration $\bar{\omega}_n = \sqrt{\frac{\rho A L^4}{E I}}\,\omega_n$ (Values in parentheses indicate the percentage difference)[$]				
	$h/L =$ 0.01	$h/L =$ 0.05	$h/L =$ 0.10	$h/L =$ 0.15	$h/L =$ 0.20
SVSDBT	9.8679	9.8281	9.7075	9.5180	9.2740
	(0.00 %)	(0.00 %)	(0.00 %)	(-0.01 %)	(-0.01 %)
TVSDBT	9.8679	9.8281	9.7075	9.5182	9.2745
	(0.0 %)	(0.00 %)	(0.00 %)	(0.00 %)	(-0.01 %)
BEBT [5]	9.8692	9.8595	9.8293	9.7795	9.7112
	(0.01 %)	(0.32 %)	(1.25 %)	(2.74 %)	(4.70 %)
Touratier [4]	9.8679	9.8282	9.7077	9.5186	9.2752

$ % difference is calculated by utilising eq. (18).

Table 2. Free flexural vibration frequencies of simply-supported rectangular isotropic shear deformable beams in the non-dimensional form, $n = 2$ and $\mu = 0.3$.

Theory	Non-dimensional natural frequencies of the beam vibration $\bar{\omega}_n = \sqrt{\frac{\rho A L^4}{E I}}\,\omega_n$ (Values in parentheses indicate the percentage difference)[$]				
	$h/L =$ 0.01	$h/L =$ 0.05	$h/L =$ 0.10	$h/L =$ 0.15	$h/L =$ 0.20
SVSDBT	39.4517	38.8299	37.0962	34.7354	32.1665
	(0.00 %)	(0.00 %)	(-0.01 %)	(-0.04 %)	(-0.09 %)
TVSDBT	39.4517	38.8301	37.0981	34.7431	32.1847
	(0.00 %)	(0.00 %)	(-0.01 %)	(-0.02 %)	(-0.03 %)
BEBT [5]	39.4719	39.3171	38.8446	38.0937	37.1120
	(0.05 %)	(1.25 %)	(4.70 %)	(9.63 %)	(15.27 %)
Touratier [4]	39.4517	38.8308	37.1009	34.7491	32.1948

$ % difference is calculated by utilising eq. (18).

5. Discussion on numerical results

Free flexural vibration frequencies of simply-supported rectangular isotropic shear deformable beams in the non-dimensional form (Tables 1 through 4) calculated using the *SVSDBT* and *TVSDBT* are in quantitative agreement with each other and with corresponding values calculated using the higher-order trigonometric theory of Touratier [4] adapted for beams for all values of the beam thickness-to-length ratio and beam vibration mode parameters considered. E.g., Percentage difference in calculating $\bar{\omega}_n$ for $h/L = 0.20$ and $n = 4$ is −0.54 in case of the *SVSDBT* and −0.13 in case of the *TVSDBT*. Hence for both theories, the *VCA* and *VIA* have a minimal effect on resulting beam free flexural vibration frequencies.

As it is well-known, the *BEBT* does not account for the beam transverse shear deformation effects, and hence it overestimates beam free flexural vibration frequencies. E.g., Percentage difference in calculating $\bar{\omega}_n$ for $h/L = 0.20$ and $n = 4$ in case of the *BEBT* is 36.32 %.

Table 3. Free flexural vibration frequencies of simply-supported rectangular isotropic shear deformable beams in the non-dimensional form, $n = 3$ and $\mu = 0.3$.

Theory	Non-dimensional natural frequencies of the beam vibration $\overline{\omega}_n = \sqrt{\frac{\rho A L^4}{E I}}\,\omega_n$ (Values in parentheses indicate the percentage difference)$				
	$h/L=$ 0.01	$h/L=$ 0.05	$h/L=$ 0.10	$h/L=$ 0.15	$h/L=$ 0.20
SVSDBT	88.6914 (0.00 %)	85.6619 (-0.01 %)	78.1547 (-0.04 %)	69.5062 (-0.12 %)	61.4581 (-0.26 %)
TVSDBT	88.6914 (0.0 %)	85.6634 (0.00 %)	78.1719 (-0.02 %)	69.5629 (-0.04 %)	61.5746 (-0.07 %)
BEBT [5]	88.7936 (0.12 %)	88.0158 (2.74 %)	85.7108 (9.63 %)	82.2414 (18.18 %)	78.0234 (26.62 %)
Touratier [4]	88.6915	85.6671	78.1855	69.5908	61.6192

$ % difference is calculated by utilising eq. (18).

Table 4. Free flexural vibration frequencies of simply-supported rectangular isotropic shear deformable beams in the non-dimensional form, $n = 4$ and $\mu = 0.3$.

Theory	Non-dimensional natural frequencies of the beam vibration $\overline{\omega}_n = \sqrt{\frac{\rho A L^4}{E I}}\,\omega_n$ (Values in parentheses indicate the percentage difference)$				
	$h/L=$ 0.01	$h/L=$ 0.05	$h/L=$ 0.10	$h/L=$ 0.15	$h/L=$ 0.20
SVSDBT	157.4877 (0.00 %)	148.3846 (-0.01 %)	128.6660 (-0.09 %)	109.2588 (-0.26 %)	93.2594 (-0.54 %)
TVSDBT	157.4878 (0.00 %)	148.3924 (-0.01 %)	128.7389 (-0.03 %)	109.4660 (-0.07 %)	93.6436 (-0.13 %)
BEBT [5]	157.8099 (0.20 %)	155.3785 (4.70 %)	148.4480 (15.27 %)	138.7083 (26.62 %)	127.8170 (36.32 %)
Touratier [4]	157.4882	148.4036	128.7792	109.5453	93.7660

$ % difference is calculated by utilising eq. (18).

6. Concluding remarks

In this paper, details regarding the variationally consistent two variable refined beam theory by Shimpi et al. [1] and variationally inconsistent single variable refined beam theory by Shimpi et al. [2] are presented. The main feature of these theories is that both of them have identical displacement fields. However, these theories differ in the approach utilised for deriving their respective governing equations. These two theories were used for calculating free flexural vibration frequencies of simply-supported rectangular isotropic shear deformable beams for various values of the beam thickness-to-length ratio and beam vibration mode parameters. It is observed that for these two theories, variationally consistent and variationally inconsistent approach utilised for deriving their respective governing equations has minimal effect on resulting beam free flexural vibration frequencies.

7. References

[1] Shimpi R, Guruprasad P and Pakhare K 2020 *J. Appl. Comput. Mech.* **6** 394–415
[2] Shimpi R, Shetty R and Guha A 2017 *P. I. Mech. Eng. C-J. Mech.* **231** 4576–91
[3] Touratier M 1991 *Int. J. Eng. Sci.* **29** 901–16
[4] Sayyad A 2011 *Appl. Comput. Mech.* **5** 217–30
[5] Timoshenko S 1921 *Philos. Mag.* **41** 744–46

IOP Conf. Series: Materials Science and Engineering **936** (2020) 012049 doi:10.1088/1757-899X/936/1/012049

Optimisation of Trapezoidal Corrugated Plate Girder

Sarang Dhakate[1] and A S Balu[2*]

[1] PG student, Department of Civil Engineering, National Institute of Technology Karnataka, Surathkal, Mangalore – 575 025, India

[2] Associate Professor, Department of Civil Engineering, National Institute of Technology Karnataka, Surathkal, Mangalore – 575 025, India

*asbalu@nitk.edu.in

Abstract. Trapezoidal Corrugated web girders is a newly developed structural design. The major advantage of such design of web is that the corrugated webs enhance the stability of beam against buckling, which results in a very economical design by the reduction in the use of web stiffeners. The flanges are made up of flat plates and welded to the trapezoidal web sheet with the modern manufacturing process and modern advance welding technology. The flanges are mainly used to provide flexural strength to the beam and web are used to increase the shear capacity of the beam. Main reason for the failure of the web is the steel yielding or web buckling. Other possible failure reasons are lateral torsional buckling of the girder and local flange buckling, separately or in combination. This paper presents a new technological solution of such a system, composed by web made of trapezoidal shape. The buckling strength of the girder is studied under different geometrical modifications performing a nonlinear finite element analysis in ANSYS.

1. Introduction

The main advantage of using corrugated web girder is to enhance the stability of beam against web buckling which may result in reduction in the use of stiffeners and which will directly make the corrugated web girder an economical system of construction. With the help of new welding technology like automatic fabrication it is possible to use corrugated web thickness up to 6mm. With the reduction in the thickness of the web the cost of the material is saved up to 30% in comparison with conventional girder and more than 30 - 35% when compared with standard hot-rolled Girder. This paper is all about the design and modelling of I section and trapezoidal corrugated plate girder and to study the effect of corrugation on strength of beam. The conventional I-section plate girder are widely used in construction of bridges, trusses, industrial structures, high-rise buildings, transmission towers, etc. It is commonly used for general construction purpose particularly suitable for structure subjected to dynamic loads and impacts such as bridge decking, girders and crane girders. With the help of modern manufacturing and highly precise welding process flanges are welded to the trapezoidal corrugated web. The flanges mainly provide flexural strength to the girder and corrugated web contribute to the shear capacity. The possible failures are web buckling or steel yielding and local flange buckling and lateral torsional buckling of girder and or in intersection [1–3].

1.1. Scope of work

An attempt has been made to study the following points.

- Study the behaviour of trapezoid corrugated web girder with the variation in web thickness and geometry.
- Numerical investigation to study the buckling behaviour.
- Compare the buckling strength and weight with the conventional plane web I-section girder.

1.2. Purpose of corrugations

The resistance of the girder against plane bending increases if we keep the web at some distance from the neutral axis. If we decrease the thickness of the web and increase the depth of the section, then there are chances that the girder may fail due to web buckling and to avoid the web buckling failure, we use the stiffeners to provide stability against elastic buckling. But the stiffeners can now be replaced with the corrugations provided on the web.

2. Numerical modelling of plate girder

2.1. Geometry

Figures 1(a) and 1(b) show the dimensions of the sections considered for the study with a span of 10 m subjected to a mid-span concentrated load of 100 kN, and Table 1 presents the variation in the geometry of corrugation.

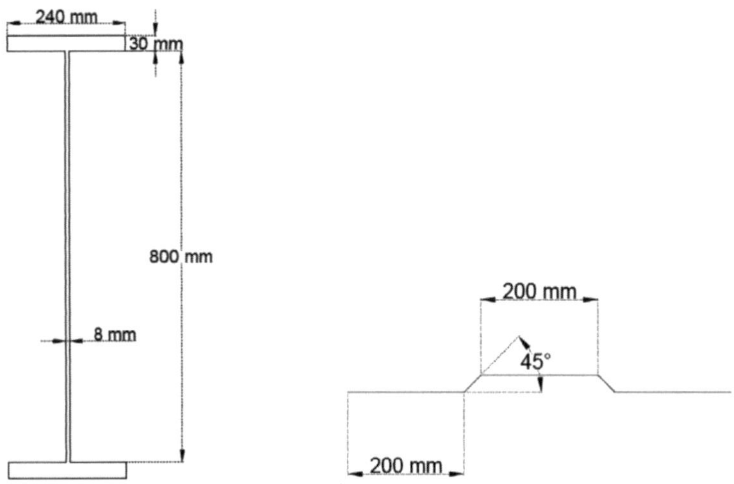

Figure 1. (a) Dimension of Conventional I Section Plain Web Plate Girder, (b) Trapezoidal Beam Interior Design Parameters.

Section classification as per IS 800 – 2007:

$$\frac{b}{t_f} = 8 < 8.4\varepsilon \text{ (Flange is plastic)}$$

$$\frac{d}{t_{w1}} = 133.33 > 126\varepsilon \text{ (Web is slender)}$$

$$\frac{d}{t_{w2}} = 100 > 84\varepsilon \text{ (Web is plastic)}$$

Table 1. Variation in the Geometry of Corrugation.

Corrugation length, C_L (mm)	Corrugation width, C_w (mm)	Thickness of web, t_w (mm)	Angle of corrugation		
200	20	6	30°	45°	60°
		8	30°	45°	60°
	40	6	30°	45°	60°
		8	30°	45°	60°
400	20	6	30°	45°	60°
		8	30°	45°	60°
	40	6	30°	45°	60°
		8	30°	45°	60°

2.2. Finite element analysis using ANSYS
A flow chart showing analysis in ANSYS is shown in figure 2.

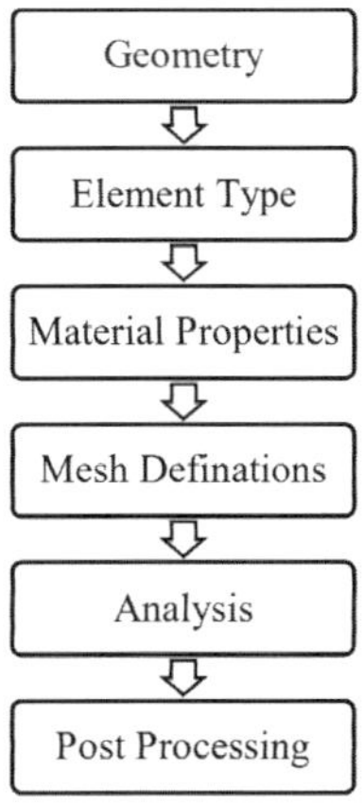

Figure 2. Steps used in the process of ANSYS.

2.2.1. Material specification.
- Mass density: 7850 kg/m^3, Poisson's Ratio: 0.3
- Modulus of elasticity: 2×10^5 MPa, Shear modulus: 75×10^3 MPa
- Yield strength: 250 MPa, Ultimate tensile strength: 450 MPa

2.2.2. Boundary conditions. After giving material properties proper meshing is to be done for both flange and web with the proper mesh size. After meshing it is required to give proper boundary conditions. For all the models prepared in ANSYS, both the ends are fixed and the web is restrained in U_x direction along the length. After proper meshing and boundary conditions loading is assigned as shown in Figures 3 and 4.

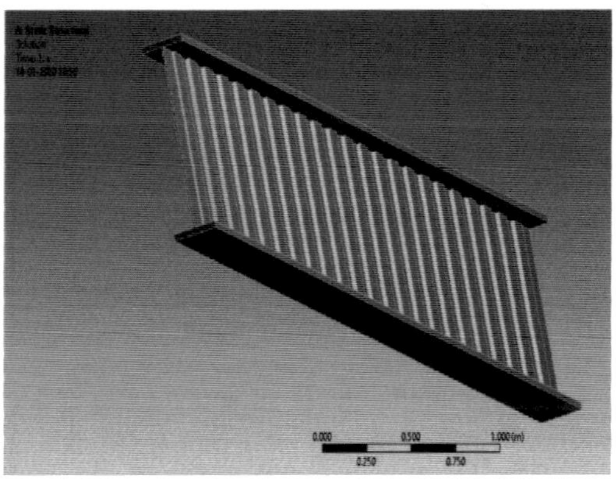

Figure 3. Trapezoidal Corrugated Girder.

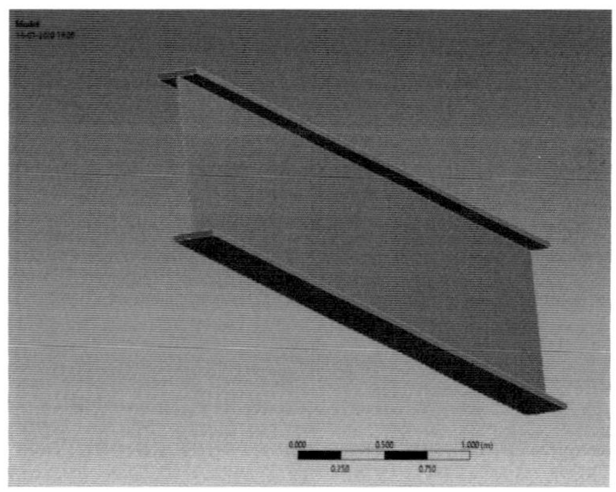

Figure 4. Conventional I Girder.

3. Analysis and weight comparison

3.1. Analysis using ANSYS

3.1.1. Structural analysis. The strength of components is the important requirement in gaining knowledge about product's performance, lifecycle and possible modes of failure. Mechanical loading, thermal stress, bolt tension, pressure conditions and rotational acceleration are number of individuals that will show strength requirements for determine which beam is strong for same loading condition. It helps us to determine the internal forces, stresses and deformation of structures under various loading conditions.

3.2. Dead weight comparison

a) For Conventional I-girder as shown in Figure 4.
Web plane thickness = 6 mm, Total span = 10m, Depth of web = 800mm, Buckling resistance = 308.76 kN, Self-weight = 1507.20 kg.

b) For corrugated web plate girder as shown in Figure 3.
Web plane thickness = 6mm, Total span = 10m, Depth of web = 800mm, Angle of corrugation = 30°, Corrugation width = 20mm, Corrugation length = 400mm, Buckling resistance = 1029.5 kN, Self-weight of web = 1511.4 kg.

4. Results and discussions

4.1. Comparison of buckling load

The variation of buckling load with the geometry is presented in Table 2.

Table 2. Buckling load of corrugated web beam with variation in geometry.

Corrugation length, (mm)	Corrugation width, (mm)	Thickness of web, (mm)	Buckling load (kN) 30°	45°	60°
Conventional I Girder		6		632.73	
Conventional I Girder		8		308.76	
200	20	6	1039.4	1046.7	1049.5
		8	1062.9	1092.7	1067.5
	40	6	1094.9	1159.5	1106.9
		8	1110.4	1177.8	1121.0
400	20	6	1029.5	1036.7	1029.6
		8	1045.3	1080.0	1050.5
	40	6	1080.0	1108.6	1082.7
		8	1096.4	1128.7	1101.3

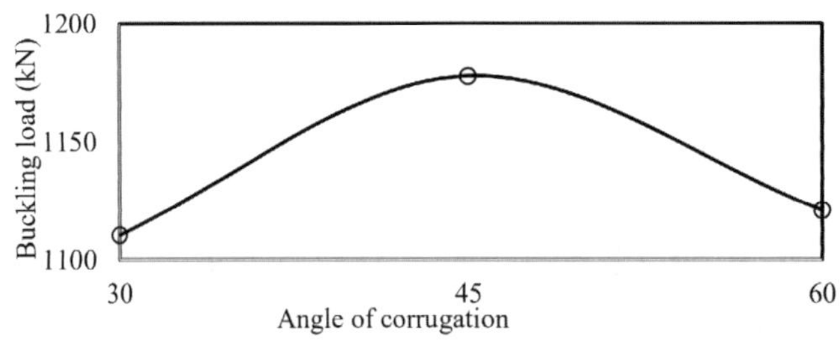

Figure 5. Variation of buckling load due to corrugation angle.

Figure 5 shows the graph of the buckling load vs variation in the angle of corrugation. The Comparison is made by varying the angle of corrugation with the constant length of corrugation plate of 200mm and constant web width of 200mm and thickness of 8mm. From Figure 5, it is seen that maximum buckling load is found at 45°. From Table 2, it is found that the other girders also show maximum buckling strength at the corrugation angle of 45°.

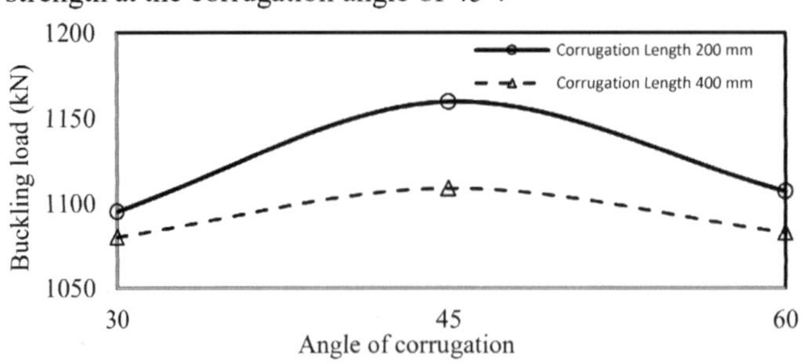

Figure 6. Variation of buckling load due to corrugation length.

Figure 6 shows the graph of buckling load vs variation of corrugation plate length. The comparison is made with the beam having constant web width of 40 mm, constant web thickness of 6mm and varying corrugation length of 200 mm and 400 mm. It can be seen that the maximum buckling load is for the corrugation plate length of 200 mm having a corrugation angle 45°. From Table 2, it is found that other girders also show maximum buckling strength with corrugation length of 200 mm with the other parameters being same.

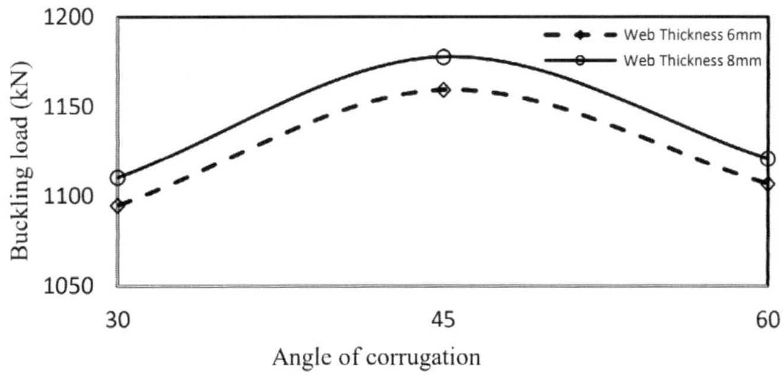

Figure 7. Variation of buckling load due to web thickness

Figure 7 shows the graph of buckling load vs variation of corrugation web thickness. The comparison is made for the beam having constant corrugation length of 200 mm and constant web width of 40mm and by varying the web thickness of 6 mm and 8 mm. From Figure 8, it can be seen that with the increase in the web thickness the buckling load also increases with the maximum buckling load at the corrugation angle of 45°. From Table 2 it is found that the other beams also show maximum buckling strength.

4.2. Comparison of weight

Table 3 shows the dead weight of the modelled beam for the span of 10 m and the comparison is made between the conventional I -girder and trapezoidal corrugated web beam.

Table 3. Total weight of corrugated web beam with variation in geometry.

Corrugation length, (mm)	Corrugation width (mm)	Thickness of web (mm)	Total weight (kg)		
			30°	45°	60°
Conventional I Girder		6		1507.2	
Conventional I Girder		8		1632.8	
200	20	6	1515.5	1520.7	1526.8
		8	1643.8	1650.8	1658.9
	40	6	1521.3	1531.5	1544.5
		8	1651.6	1665.3	1682.6
400	20	6	1511.4	1514.4	1517.2
		8	1638.5	1642.4	1646.1
	40	6	1514.9	1520.3	1525.5
		8	1643	1650.3	1657.2

5. Conclusions

Buckling strength is more for the trapezoidal corrugated web girder compared to conventional I-girder. The buckling strength is maximum for the corrugation angle of 45°. Buckling strength increases significantly for the corrugation width of 200 mm as compared to 400 mm. Buckling strength is more for the web of thickness 8mm as compared to 6mm thickness. Weight of corrugated in higher as compared to plane web. If we compared the buckling strength and weight for 8mm thick plane web and 8mm thick corrugated web with 30° angle and 400mm corrugation length and 20mm corrugation width the buckling strength increases by 65% with the 8% increase in weight.

IOP Conf. Series: Materials Science and Engineering **936** (2020) 012049 doi:10.1088/1757-899X/936/1/012049

6. References

[1] Karthikeyan K and Senthil P 2018 A study on flexure capacity of steel beams with corrugated web *International Journal of Civil Engineering and Technology* **9(4)** 679–89

[2] Limaye A A and Alandkar P M 2013 Strength of welded plate girder with corrugated web plate *International Journal of Engineering Research and Application* **3** 1925–30

[3] Sachin K G, Sowjanya G V and Muralidhar N 2014 Behaviour of plate girder with flat web and corrugated web *International Journal of Civil and Structural Engineering Research* **2(1)** 130–6

[4] Duggal S K 2010 *Limit State Design of Steel Structures* (Tata McGraw Hill Education Private Limited: New Delhi)

[5] IS 800 2007 *Code of Practice for General Construction in Steel* (Bureau of Indian Standards: New Delhi)

IOP Conf. Series: Materials Science and Engineering **936** (2020) 012050 doi:10.1088/1757-899X/936/1/012050

Analysis of Geometrically Nonlinear Euler-Bernoulli Beam using EFGM

D Paavani, M Aswathy*, **C O Arun and I R Praveen Krishna**
Department of Aerospace Engineering, Indian Institute of Space science and Technology, Trivandrum
*aswathymekkanakkil@gmail.com

Abstract. The current paper presents a meshless formulation for the analysis of geometrically nonlinear one-dimensional Euler-Bernoulli beams. Element-free Galerkin method (EFGM) is used as the meshless numerical tool and Newton-Raphson method as the iterative scheme. Numerical examples of beams with four different boundary conditions are solved, and the results are compared with that of the finite element method (FEM). EFGM results are found in good agreement with that of FEM. Also, it is found that EFGM eliminates the effect of membrane locking in thin beams.

1. Introduction

Finite element method (FEM) is the most widely used conventional numerical technique to solve complex problems in structural analysis. Generation of mesh in FEM is a complex task, and mesh dependency creates issues while dealing problems with large deformation, problems with moving discontinuities like crack propagation, stress concentration problems etc. In this respect, meshless method can be regarded as the most viable alternative to FEM. Unlike FEM, which requires predefined elemental connectivity, meshless methods need only nodes arbitrarily distributed in the problem domain as well as in the boundary of the domain. Among different types of meshless techniques [1–7], Element-free Galerkin method (EFGM) becomes very popular due to its simplicity and comparability of the formulations with FEM. It is first proposed by Belytschko et al. [2, 8]. Moving least-squares (MLS) interpolants [9] are applied for the construction of shape functions.

Valencia et al. [10] carried out a study on one-dimensional axially loaded bar, and Valencia et al. [11] analysed the beam bending problems to understand the influence of the characteristic parameters on the EFGM solution.

Though EFGM is very efficient, there exists issues in applying boundary conditions and point loads. Due to the loss of Kronecker-delta property, exact imposition of the essential boundary conditions is not possible. Several methods like Lagrange multipliers approach [2, 12], penalty factor [13], coupled FEM [14], transformation method [15–17] etc. are used to overcome this drawback.

In 2D and 3D EFGM formulations, the application of point loads is somewhat complicated because of the character of MLS interpolants. A method is suggested by Pang [18] in which concentrated forces are transformed into distributed forces by Dirac function. After such a transformation, the concentrated load can be made analogous to distributed load and thus, increasing the scope of EFGM. Thin-walled beams may undergo large deflection with small strains and moderate rotations, and the analysis of such structural component requires the incorporation of *von Kármán strains* [19]. Analysis of geometrically nonlinear 1D Euler-Bernoulli beam, incorporating *von Kármán strains*, using EFGM is explored in the current study.

2. von Kármán strains

Thin-walled beams are always prone to large deformations with small strains and moderate rotations leading to non-linear kinematic relationships. Figure 1 depicts the displacement kinematics of 1D Euler-Bernoulli beam in

X-Z plane. Let it be subjected to an arbitrary loading $q(x)$ in the positive Z direction and a moment M is applied as shown.

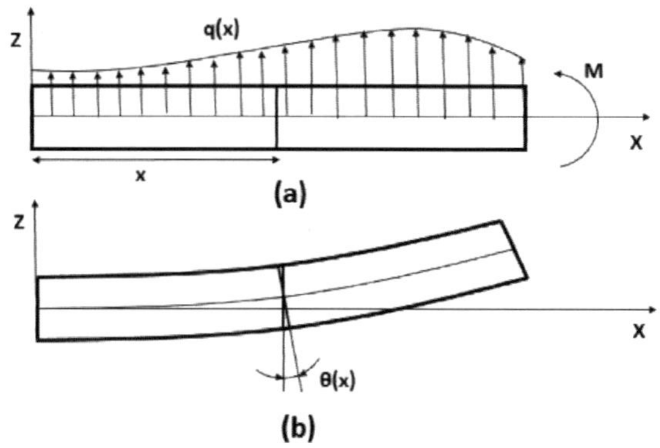

Figure 1. (a) Applied loading, (b) Response of the beam to applied loading.

Let U, V and W be the displacements in X, Y and Z directions respectively, and which are functions of x and z, given by

$$\begin{aligned} U(x,z) &= u(x) + z\theta(x) \\ V(x,z) &= 0 \\ W(x,z) &= w(x) \end{aligned} \tag{1}$$

where, u and w are the center-line displacements along x and z directions respectively, which are functions of 'x' alone. From the Green-Lagrange strain definitions, the nonlinear strain (ε) - displacement (U, V and W) relation can be written as

$$\varepsilon_{ij} = \frac{1}{2}\left(\frac{\partial u_i}{\partial x_j} + \frac{\partial u_j}{\partial x_i}\right) + \frac{1}{2}\left(\frac{\partial u_m}{\partial x_i}\frac{\partial u_m}{\partial x_j}\right) \tag{2}$$

Note that $u_1 = U(x,z)$, $u_2 = V(x,z)$, $u_3 = W(x,z)$, $x_1 = x$, $x_2 = y$ and $x_3 = z$. Substituting the displacements, the strain-displacement relations are obtained as

$$\begin{Bmatrix} \varepsilon^0_{xx} \\ \varepsilon^1_{xx} \\ \gamma_{xz} \end{Bmatrix} = \underbrace{\begin{Bmatrix} \frac{du}{dx} \\ \frac{d\theta}{dx} \\ \theta + \frac{dw}{dx} \end{Bmatrix}}_{\varepsilon_L} + \underbrace{\begin{Bmatrix} \frac{1}{2}\left(\frac{du}{dx}\right)^2 + \frac{1}{2}\left(\frac{dw}{dx}\right)^2 \\ \frac{du}{dx}\frac{d\theta}{dx} + \frac{1}{2}z\left(\frac{d\theta}{dx}\right)^2 \\ \frac{du}{dx}\theta \end{Bmatrix}}_{\varepsilon_{NL}} \tag{3}$$

where, normal strain in the x direction is given by, $\varepsilon_{xx} = \varepsilon^0_{xx} + z\varepsilon^1_{xx}$. ε^0_{xx} is the axial strain and $z\varepsilon^1_{xx}$ at any section is the bending strain. ε_L and ε_{NL} denote the linear and nonlinear parts of strain. In ε_{NL}, $\frac{1}{2}\left(\frac{dw}{dx}\right)^2$ is the most dominant term and so, all the other terms can be neglected. Also from the thin beam assumption, the shear strain can be neglected i.e., $\gamma_{xz} = 0$, and it implies $\theta = -\frac{dw}{dx}$. Thus, the only strain experienced by the structure is the normal strain which can be written as

$$\varepsilon_{xx} = \frac{du}{dx} - z\frac{d^2w}{dx^2} + \frac{1}{2}\left(\frac{dw}{dx}\right)^2 \tag{4}$$

$$\varepsilon^0_{xx} = \frac{du}{dx} + \frac{1}{2}\left(\frac{dw}{dx}\right)^2, \quad \varepsilon^1_{xx} = -z\frac{d^2w}{dx^2} \tag{5}$$

These strains are known as *von Kármán strains.*

3. EFGM formulation

The Galerkin weak form of principle of virtual displacement is used here to formulate the discretized set of equations.

3.1. *Principle of virtual displacement*

The principle of virtual displacement states that if a body is in equilibrium, the total virtual work done by both the internal and external forces in moving the body through the virtual displacement is zero [19].

$$\delta W^e \equiv \delta W_I^e - \delta W_E^e = 0 \tag{6}$$

where δW_I^e is the virtual strain energy stored in the system and δW_E^e is the work done by externally applied forces.

$$\delta W_I^e = \int_V \delta\varepsilon_{ij}\sigma_{ij} dV \tag{7a}$$

$$\delta W_E^e = \int_l q\delta w dx + \int_l f\delta u dx + \sum_{i=1}^{3n} Q_i^e \delta\Delta_i^e \tag{7b}$$

Here, V is the volume of the body, l is the length of the beam, q is the distributed transverse load per unit length, f is the distributed axial load per unit length and Q_i are the nodal forces corresponding to the nodal displacements Δ_i. For a beam discretized to 'n' nodes, $\Delta = [u_1\ u_2\ ...\ u_n\ w_1\ \theta_1\ w_2\ \theta_2\ ...\ w_n\ \theta_n]^T$ where, u_i, w_i and θ_i are the respective axial displacement, transverse displacement and slope at the node i. $\delta\Delta_i^e$ are the generalized virtual displacements.

The virtual work statements for a Euler-Bernoulli beam incorporating *von Kármán strains* can be written in two sets of equations as,

$$\int_l EA\frac{\mathrm{d}\delta u}{\mathrm{d}x}\left[\frac{\mathrm{d}u}{\mathrm{d}x} + \frac{1}{2}\left(\frac{\mathrm{d}w}{\mathrm{d}x}\right)^2\right] dx - \int_l \delta u f(x) dx - \sum_{i=1}^{n} Q_i\delta\Delta_i = 0 \tag{8a}$$

$$\int_l \left\{EA\frac{\mathrm{d}w}{\mathrm{d}x}\frac{\mathrm{d}\delta w}{\mathrm{d}x}\left[\frac{\mathrm{d}u}{\mathrm{d}x} + \frac{1}{2}\left(\frac{\mathrm{d}w}{\mathrm{d}x}\right)^2\right] + EI\frac{\mathrm{d}^2\delta w}{\mathrm{d}x^2}\frac{\mathrm{d}^2 w}{\mathrm{d}x^2}\right\} dx - \int_l \delta w q(x) dx - \sum_{i=n+1}^{3n} Q_i\delta\Delta_i = 0 \tag{8b}$$

where, A is the cross-sectional area of the beam, I is the second moment about y-axis and E is the modulus of elasticity of the material.

3.2. *MLS shape functions for axial and transverse displacements*

MLS method is employed in EFGM to construct shape functions for the axial as well as transverse displacements.

3.2.1. Axial Displacement. The MLS approximate of $u(\mathbf{x})$ is $u^h(\mathbf{x})$ defined by

$$u^h(\boldsymbol{x}) = \boldsymbol{p}^T(\boldsymbol{x})a(\boldsymbol{x}) = \sum_{i=1}^{n} p_i(\boldsymbol{x})a_i(\boldsymbol{x}) \tag{9}$$

where, $\boldsymbol{p}^T(\boldsymbol{x}) = (p_0(\boldsymbol{x}), p_1(\boldsymbol{x}), ..., p_m(\boldsymbol{x})) = (1, x, x^2, ..., x^m)$ is the basis vector of order m, $\boldsymbol{a}^T(x) = \{a_0(x), a_1(x)...a_m(x)\}$ is the coefficient vector, which is also a function of x. From the basic definition of moving least squares method, the error norm is defined as

$$J(x) = \sum_{I=1}^{n} wt_I \left[\boldsymbol{p}^T(x_I)\boldsymbol{a}(x) - \hat{\boldsymbol{u}}_{\boldsymbol{I}}\right]^2 \tag{10}$$

where, x_I is the coordinate of the node, $\hat{\boldsymbol{u}}_{\boldsymbol{I}}$ is nodal displacement parameters given by $u(x_I)$, $wt_I(x)$ is the weight at point x corresponding to node I. In matrix notation, equation (10) can be written as

$$J(\boldsymbol{x}) = [\boldsymbol{P}\boldsymbol{a}(\boldsymbol{x}) - \hat{\boldsymbol{u}}]^T \boldsymbol{W} [\boldsymbol{P}\boldsymbol{a}(\boldsymbol{x}) - \hat{\boldsymbol{u}}] \tag{11}$$

where, $\hat{\boldsymbol{u}} = \{\hat{u_1}, \hat{u_2}, ...\hat{u_n}\}$, $W = diag[wt_1(\boldsymbol{x}), wt_2(\boldsymbol{x}), ..., wt_N(\boldsymbol{x})]$ and $\boldsymbol{P} = \left[\boldsymbol{p}^T(x_1),\ \boldsymbol{p}^T(x_2),\ ...,\boldsymbol{p}^T(x_n)\right]$. The error norm can be minimized by setting the partial derivatives $\nabla J = 0$, where $\nabla = \left[\frac{\partial}{\partial a_1}, \frac{\partial}{\partial a_2} \cdots \frac{\partial}{\partial a_{m-1}}\right]^T$. This results in equation of the form

$$\boldsymbol{A}(x)\boldsymbol{a}(x) = \boldsymbol{C}(x)\boldsymbol{d} \tag{12}$$

where, $\boldsymbol{A}(x) = \boldsymbol{P}^T \boldsymbol{W} \boldsymbol{P}$, $\boldsymbol{C}(x) = \left[\boldsymbol{P}^T \boldsymbol{W}\right]$ and, $\boldsymbol{d} = [\hat{\boldsymbol{u}}]^T$.

In terms of shape functions, the approximate $u^h(\boldsymbol{x})$ is written as

$$u^h(\boldsymbol{x}) = \sum_{I=1}^{n} \psi_{uI}(x)\hat{u}_I \tag{13}$$

where,

$$\psi_u^T(x) = \boldsymbol{p}(x)\boldsymbol{A}^{-1}(x)\boldsymbol{P}^T\boldsymbol{W} \tag{14}$$

3.2.2. Transverse Displacement. This formulation is adopted from [16]. The MLS approximate of $w(\mathbf{x})$ is $w^h(\boldsymbol{x})$ defined by

$$w^h(\boldsymbol{x}) = \boldsymbol{p}^T(\boldsymbol{x})a(\boldsymbol{x}) + \boldsymbol{p}_x^T(\boldsymbol{x})a(\boldsymbol{x}) = \sum_{i=1}^{n} p_i(\boldsymbol{x})a_i(\boldsymbol{x}) + \frac{\partial p_i(\boldsymbol{x})}{\partial x}a_i(\boldsymbol{x}) \tag{15}$$

where, $\boldsymbol{p}_x^T(\boldsymbol{x}) = (\frac{\partial p_0(\boldsymbol{x})}{\partial x}, \frac{\partial p_1(\boldsymbol{x})}{\partial x}, ..., \frac{\partial p_m(\boldsymbol{x})}{\partial x}) = (0, 1, 2x, ..., (m)x^{m-1})$ is the derivate of basis with respect to x. In matrix notation, the error norm can be written as

$$J(\boldsymbol{x}) = [\boldsymbol{P}\boldsymbol{a}(\boldsymbol{x}) - \hat{\boldsymbol{w}}]^T \boldsymbol{W} [\boldsymbol{P}\boldsymbol{a}(\boldsymbol{x}) - \hat{\boldsymbol{w}}] + \left[\boldsymbol{P}_{\boldsymbol{x}}\boldsymbol{a}(\boldsymbol{x}) - \hat{\boldsymbol{\theta}}\right]^T \boldsymbol{W} \left[\boldsymbol{P}_{\boldsymbol{x}}\boldsymbol{a}(\boldsymbol{x}) - \hat{\boldsymbol{\theta}}\right] \tag{16}$$

$$= \left[\begin{Bmatrix} \boldsymbol{P} \\ \boldsymbol{P}_{\boldsymbol{x}} \end{Bmatrix} \boldsymbol{a}(\boldsymbol{x}) - \begin{Bmatrix} \hat{\boldsymbol{w}} \\ \hat{\boldsymbol{\theta}} \end{Bmatrix}\right]^T \begin{bmatrix} \boldsymbol{W} & 0 \\ 0 & \boldsymbol{W} \end{bmatrix} \left[\begin{Bmatrix} \boldsymbol{P} \\ \boldsymbol{P}_{\boldsymbol{x}} \end{Bmatrix} \boldsymbol{a}(\boldsymbol{x}) - \begin{Bmatrix} \hat{\boldsymbol{w}} \\ \hat{\boldsymbol{\theta}} \end{Bmatrix}\right] \tag{17}$$

where, $W = diag[wt_1(\boldsymbol{x}), wt_2(\boldsymbol{x}), ..., wt_N(\boldsymbol{x})], \hat{\boldsymbol{w}} = \{\hat{w}_1, \hat{w}_2, ..\hat{w}_n\}$ and $\hat{\boldsymbol{\theta}} = \{\hat{\theta}_1, \hat{\theta}_2, ..\hat{\theta}_n\}$; $\boldsymbol{P} = \left[\boldsymbol{p}^T(x_1), \boldsymbol{p}^T(x_2), \ldots, \boldsymbol{p}^T(x_n)\right]$ and $\boldsymbol{P}_{\boldsymbol{x}} = \left[\boldsymbol{p}_{\boldsymbol{x}}^T(x_1), \boldsymbol{p}_{\boldsymbol{x}}^T(x_2), \ldots, \boldsymbol{p}_{\boldsymbol{x}}^T(x_n)\right]$.

After minimization of error norm, the approximation for $w^h(\boldsymbol{x})$ is written as

$$w^h(\boldsymbol{x}) = \sum_{I=1}^{n} \phi_{wI}(x)\hat{w}_I + \phi_{\theta I}(x)\hat{\theta}_I \tag{18}$$

where,

$$\phi_w^T(x) = \boldsymbol{p}(x)\boldsymbol{A}^{-1}(x)\boldsymbol{P}^T\boldsymbol{W} = \boldsymbol{p}(x)\boldsymbol{A}^{-1}(x)\boldsymbol{C}_{\boldsymbol{u}}(x) \tag{19}$$

$$\phi_\theta^T(x) = \boldsymbol{p}(x)\boldsymbol{A}^{-1}(x)\boldsymbol{P}_{\boldsymbol{x}}^T\boldsymbol{W} = \boldsymbol{p}(x)\boldsymbol{A}^{-1}(x)\boldsymbol{C}_{\boldsymbol{\theta}}(x) \tag{20}$$

Substituting the MLS approximations in equation (8), we obtain

$$\sum_{j=1}^{n} \left[K_{ij}^{11}\hat{\boldsymbol{\Delta}}_j\right] + \sum_{J=n+1}^{3n} \left[K_{iJ}^{12}\hat{\boldsymbol{\Delta}}_J\right] - F_i^1 = 0 \quad (i = 1, 2...n) \tag{21a}$$

$$\sum_{j=1}^{n} \left[K_{Ij}^{11}\hat{\boldsymbol{\Delta}}_j\right] + \sum_{J=n+1}^{3n} \left[K_{IJ}^{12}\hat{\boldsymbol{\Delta}}_J\right] - F_I^2 = 0 \quad (I = n+1, n+2...3n) \tag{21b}$$

where, $\hat{\boldsymbol{\Delta}}$ is the nodal parameter vector given by $\hat{\boldsymbol{\Delta}} = [\hat{u}_1\ \hat{u}_2\ ...\ \hat{u}_n\ \hat{w}_1\ \hat{\theta}_1\ \hat{w}_2\ \hat{\theta}_2\ ...\ \hat{w}_n\ \hat{\theta}_n]^T$ and

$$\begin{aligned}
K_{ij}^{11} &= EA\int_l \frac{\mathrm{d}\psi_i}{\mathrm{d}x}\frac{\mathrm{d}\psi_j}{\mathrm{d}x}dx, \\
K_{iJ}^{12} &= 0.5EA\int_l \left(\frac{\mathrm{d}w}{\mathrm{d}x}\right)\frac{\mathrm{d}\psi_i}{\mathrm{d}x}\frac{\mathrm{d}\phi_J}{\mathrm{d}x}dx, \\
K_{Ij}^{21} &= EA\int_l \left(\frac{\mathrm{d}w}{\mathrm{d}x}\right)\frac{\mathrm{d}\phi_I}{\mathrm{d}x}\frac{\mathrm{d}\psi_j}{\mathrm{d}x}dx, \\
K_{IJ}^{22} &= EI\int_l \frac{\mathrm{d}^2\phi_I}{\mathrm{d}x}\frac{\mathrm{d}^2\phi_J}{\mathrm{d}x}dx + 0.5EA\int_l \left[\left(\frac{\mathrm{d}w}{\mathrm{d}x}\right)^2\right]\frac{\mathrm{d}\phi_I}{\mathrm{d}x}\frac{\mathrm{d}\phi_J}{\mathrm{d}x}dx \\
F_i^1 &= \int_l f\psi_i dx + Q_i \\
F_I^2 &= \int_l q\phi_I dx + Q_I
\end{aligned} \tag{22}$$

The stiffness and force are arranged in the matrix form as,

$$K = \begin{bmatrix} K^{11} & K^{12} \\ K^{21} & K^{22} \end{bmatrix}, \; F = \begin{Bmatrix} F^1 \\ F^2 \end{Bmatrix} \tag{23}$$

In the nonlinear case, the axial and transverse displacements are coupled and this coupling appears in the K^{12} and K^{21} terms.

The enforcement of essential boundary conditions is achieved here by the transformation method [17]. Thus, the force balance can be expressed as

$$\hat{K}(\{\hat{\mathbf{\Delta}}\})\{\hat{\mathbf{\Delta}}\} = \hat{F} \tag{24}$$

in which, $\hat{K} = \Lambda^{-T} K \Lambda^{-1}$ and $\hat{F} = \Lambda^{-T} F$, where Λ is transformation matrix.

In equation (24), stiffness matrix is a function of displacement, containing terms like $\frac{dw}{dx}$. This is to be solved iteratively by substituting the displacement and its derivatives from the previous iteration. In this study, Newton-Raphson method is used owing to its fast convergence [19].

4. Numerical examples

Beams with different boundary conditions are considered which are already solved using FEM by Reddy [19]. A beam of length $0.1m$ having area of cross-section $1cm^2$ with Young's modulus $E = 30 \times 10^8 N/m^2$, subjected to a uniformly distributed load q_0 N/m is considered. The beam is discretized using 20 uniformly distributed nodes. EFGM scaling parameter is taken as $d_{max} = 3$, a quadratic polynomial basis, an exponential weight function and 4 point Gauss-quadrature scheme for numerical integration are utilized. Beams with four different boundary conditions are studied. For each case, the load q_0 is varied from 0 *to* $2kN/m$ in 10 load steps and variation of deflection with load is plotted. For all the cases, the results obtained are validated with those obtained from FEM [19].

4.1. Pinned-Pinned beam

The geometric boundary conditions are

$$u(0) = w(0) = u(\frac{L}{2}) = \frac{dw}{dx}|_{x=\frac{L}{2}} = 0 \tag{25}$$

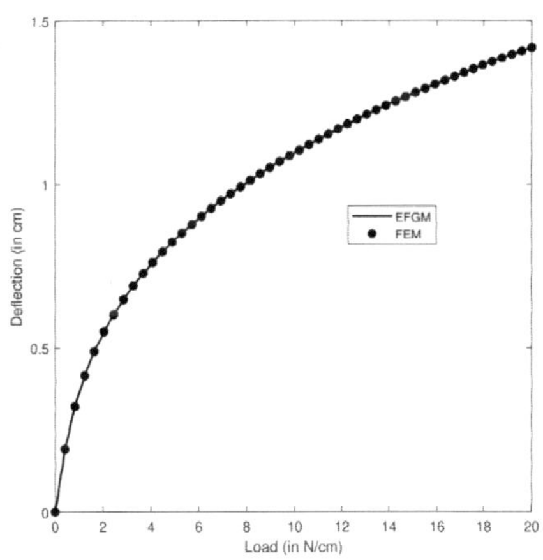

Figure 2. Load Vs Mid-point deflection of a pinned-pinned beam.

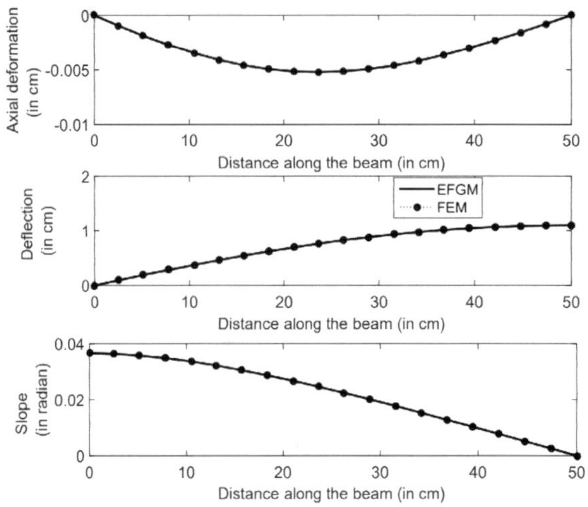

Figure 3. Symmetric deflection plot for a beam with pinned ends.

The nonlinear effect can be clearly witnessed in the Load-Deflection plot shown in figure 2. It is evidently a strain stiffening problem. Figure 3 shows the plot of deflection , axial deformation and slope along the length of the beam for a pinned-pinned beam. Due to symmetry, only half of the span is considered. The number of iterations required for each load step to converge is less than 5, since Newton-Raphson has fast convergence. The FEM results are almost coincident with the EFGM results.

4.2. Clamped-Clamped beam

The geometric boundary conditions are

$$u(0) = w(0) = \frac{\mathrm{d}w}{\mathrm{d}x}|_{x=0} = u(\frac{L}{2}) = \frac{\mathrm{d}w}{\mathrm{d}x}|_{x=\frac{L}{2}} = 0 \tag{26}$$

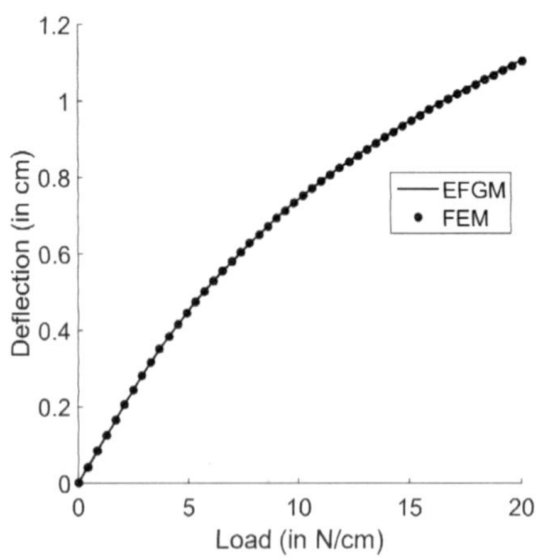

Figure 4. Load Vs Mid-point deflection of a clamped-clamped beam.

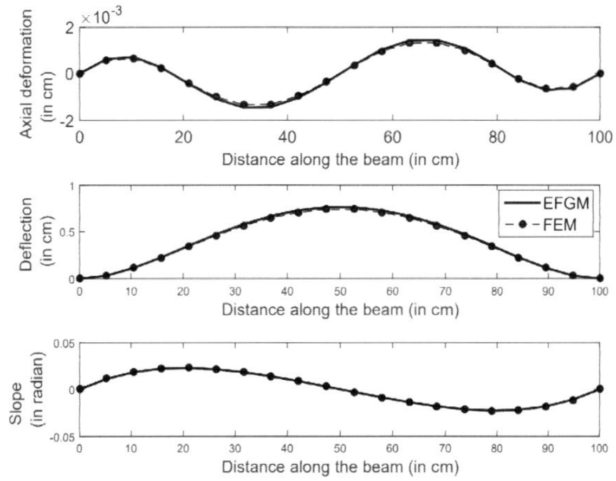

Figure 5. Deflection plot for a beam with clamped ends.

This is also a strain stiffening problem. The number of iterations required for each load step to converge is less than 10. Evidently, the deflection in this case is less than that of pinned-pinned case. The FEM results are quite close to EFGM results. Though no axial force is applied, due to the constrained boundaries, axial displacement and strains exist.

4.3. *Simply supported beam*

The geometric boundary conditions are

$$w(0) = u(\frac{L}{2}) = \frac{\mathrm{d}w}{\mathrm{d}x}|_{x=\frac{L}{2}} = 0 \tag{27}$$

Figure 6 shows the load-deflection behaviour and figure 7 shows axial deformation, transverse deflection and rotation along the length of a beam with roller constraints on both the ends using EFGM, FEM without reduced integration,and FEM with reduced integration.

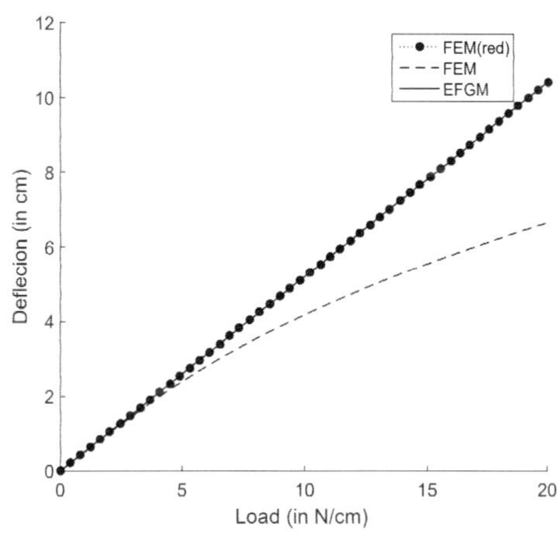

Figure 6. Load Vs Mid-point deflection of a simply supported beam.

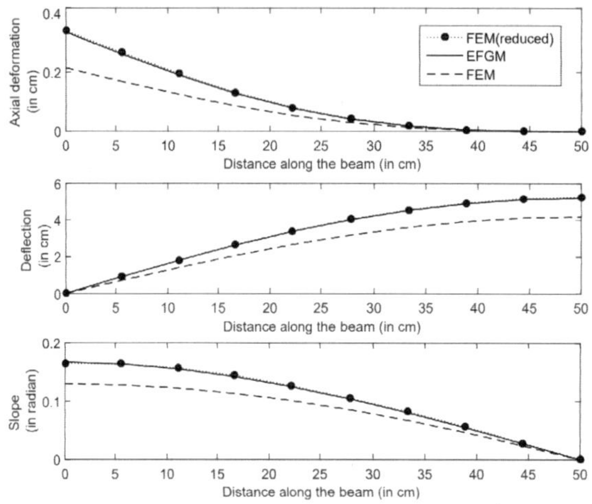

Figure 7. Symmetric Deflection plot for a simply supported beam.

A simply supported beam does not have any end conditions on u. Only symmetric boundary is applied at the center. Therefore, the beam does not experience any axial strain. $\Rightarrow \varepsilon_{xx}^0 = \frac{du}{dx} + \frac{1}{2}\left(\frac{dw}{dx}\right)^2 = 0$. In the case of FEM, for example, if u is interpolated linearly and w is interpolated using hermite functions, the order of $\frac{du}{dx}$ and $\left(\frac{dw}{dx}\right)^2$ is quite different. This introduces spurious constraints in the system. As a result, the element stiffness matrix becomes more stiff. This is known as *membrane locking*. In-order to satisfy this constraint, a reduced integration is used. However, in the case of EFGM, the order of the polynomial is certainly higher. From figures 6 and 7 it is clear that the results with reduced integration in FEM matches EFGM results and which should match with linear analytical solution. So, it can be concluded that the issue of membrane locking does not exist in EFGM.

4.4. Cantilever beam

The geometric boundary conditions are

$$u(0) = w(0) = \frac{dw}{dx}|_{x=0} = 0 \tag{28}$$

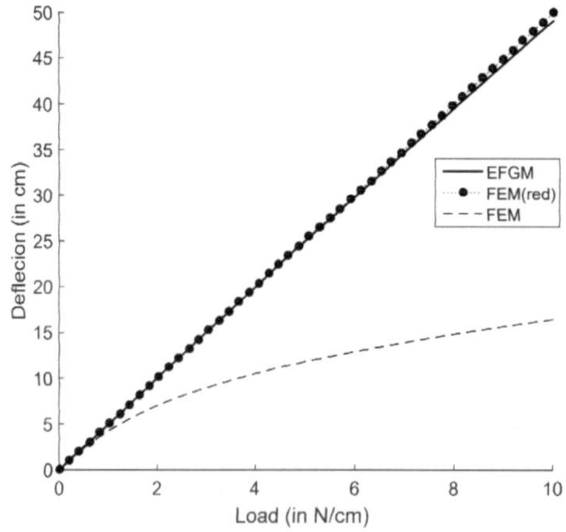

Figure 8. Load Vs Tip deflection of a cantilever beam.

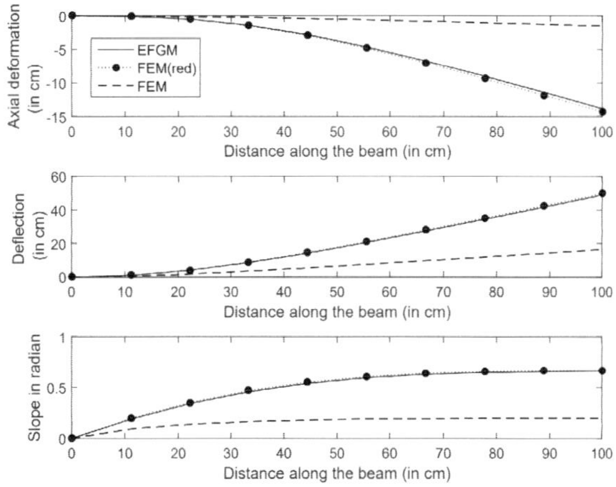

Figure 9. Deflection plot for a Cantilever beam.

In the case of cantilever beam also, the axial strain is zero, since only one end is constrained. Figures 8 and 9 again show that EFGM formulations are free from membrane locking. Figures 6 and 8 depict that *von Kármán strains* do not induce nonlinearity if the center line axial strain is zero.

5. Conclusions

In the present study, EFGM formulation for a geometrically nonlinear 1D Euler-Bernoulli beam is explained. Nonlinearity is introduced in the form of *von Kármán strains*. Numerical examples of Euler-Bernoulli beams with different boundary conditions are solved. The results are compared with those obtained from the converged FEM results. It is also found that the EFGM formulations are free of membrane locking. The reason for the same can be attributed to the fact that field approximation in EFGM is governed by the continuity of the weight function.

6. References

[1] Gingold R A and Monaghan J J 1977 *Monthly Notices of the Royal Astronomical Society* **181** 375–389
[2] Belytschko T, Lu Y Y and Gu L 1994 *International Journal for Numerical Methods in Engineering* **37** 229–256
[3] Atluri S N and Zhu T 1998 *Computational Mechanics* **22** 117–127
[4] Liu W K, Jun S, Li S, Adee J and Belytschko T 1995 *International Journal for Numerical Methods in Engineering* **38** 1655–1679
[5] Oñate E, Idelsohn S, Zienkiewicz O and Taylor R 1996 *International Journal for Numerical Methods in Engineering* **39** 3839–3866
[6] Liu G R and Karamanlidis D 2003 *Applied Mechanics Reviews* **56** B17
[7] Liu G R and Gu Y T 2005 *An introduction to meshfree methods and their programming* (Springer Science & Business Media)
[8] Dolbow J and Belytschko T 1998 *Archives of Computational Methods in Engineering* **5** 207–241
[9] Lancaster P and Salkauskas K 1981 *Mathematics of Computation* **37** 141–158
[10] Valencia O, Gómez-Escalonilla F and Díez J L 2008 *Proceedings of the Institution of Mechanical Engineers, Part C: Journal of Mechanical Engineering Science* **222** 1621–1633
[11] Valencia O, Gómez-Escalonilla F and López-Díez J 2009 *Proceedings of the Institution of Mechanical Engineers, Part C: Journal of Mechanical Engineering Science* **223** 1579–1590
[12] Lu Y Y, Belytschko T and Gu L 1994 *Computer Methods in Applied Mechanics and Engineering* **113** 397–414
[13] Liu G R and Yang K Y 1998 *Proc. 3rd HPC Asia* vol 98 pp 715–721
[14] Krongauz Y and Belytschko T 1996 *Computer Methods in Applied Mechanics and Engineering* **131** 133–145
[15] Rao B N and Rahman S 2000 *Computational Mechanics* **26** 398–408
[16] Dodagoudar G R, Rao B N and Sunitha N V 2015 *International Journal of Geotechnical Engineering* **9** 298–306
[17] Arun C O, Rao B N and Sivakumar S M 2009 *J. Struct. Eng* 182–194
[18] Pang Z H 2000 *Communications in Numerical Methods in Engineering* **16** 335–341
[19] Reddy J N 2014 *An Introduction to Nonlinear Finite Element Analysis: with applications to heat transfer, fluid mechanics, and solid mechanics* (OUP Oxford)

International Conference on Materials, Mechanics and Structures 2020 (ICMMS2020) IOP Publishing
IOP Conf. Series: Materials Science and Engineering **936** (2020) 012051 doi:10.1088/1757-899X/936/1/012051

A Review on Strut and Tie Methods

Kannan C Bhanu[1*], N Ganesan[2] and P V Indira[3]

[1] Research Scholar, Department of Civil Engineering, National Institute of Technology Calicut, Kerala, India

[2] Professor (Retd.), Department of Civil Engineering, National Institute of Technology Calicut, Kerala, India

[3] Professor Department of Civil Engineering, National Institute of Technology Calicut, Kerala, India

[*]kannanchandrabhanu23@gmail.com

Abstract. Strut and tie method (STM) provides simplified solutions to complicated design problems. It was an evolutionary discovery which lead to an improved understanding of shear and flexural deformations of the structures. STM is applicable to the regions of the structure where beam theory is not valid. Numerous studies were conducted in this field, and various modifications are being attached to the conventional STM. Indeterminate models replaced determinate models, and nonlinear analysis replaced linear elastic analysis. New ideas like automatic generation of models based on optimisation techniques, simplified inelastic analysis using secant stiffness method and 3-dimensional analysis were also introduced. This paper summarises some of these studies and discusses their merits and limitations. Some suggestions to simplify these complicated methods are also included.

1. Introduction

Strut and tie method is a consistent tool used for the analysis and design of structural concrete for several decades. Truss analogy suggested by Ritter and Mörsch [1,2] was the initial step of the evolution of the STM. Later, so many researchers have studied the possibilities of STM and developed it into a consistent tool. Determinate strut and tie models for several structural elements were developed. Linear elastic analysis was used for the analysis of these models. Hence, the models were able to predict only the elastic behaviour of the structures. One of the limitations of this method was the uncertainty in developing the models for complex loading and geometrical cases. New ideas like nonlinear STM, Topology optimised STM, Grid STM, Secant stiffness method, Compatibility STM were developed to overcome these limitations.

This paper includes a summary of the evolution of STM. Several new STMs are discussed, and their merits and limitations are highlighted.

2. Evolution of STM

The idea of considering cracked reinforced concrete (RC) beam as truss consisting of inclined concrete struts and steel ties was first presented by Ritter [1] in 1899. This truss analogy was based on the observation that cracks develop at an inclination of approximately 45° to the longitudinal axis of the reinforced concrete beams under shear stress. Later Mörsch [2] suggested that a continuous diagonal compression stress field reflects the behaviour of a beam better than that by prismatic strut elements. Mörsch used principal stress trajectories for proportioning the tensile reinforcement and also explained the ability of truss models to duplicate the behaviour of beams with bent up longitudinal bars.

The behaviour of reinforced concrete under shear remained a taxing issue for the researchers and engineers for several years. Since the application of Bernoulli hypothesis is limited to B regions that are away from the discontinuities, the disturbed regions (D regions) were designed using thumb rules or based on the previous experiences instead of sectional analysis. Figure 1 shows some examples of B and D regions. Later it was noticed that the loads at these disturbed regions are transferred to support or adjacent B region by some mechanism. This observation pointed out that STM can be effectively used for the analysis of D regions. These studies initiated the growth of strut and tie method. The evolution of STM is explained in the following sections.

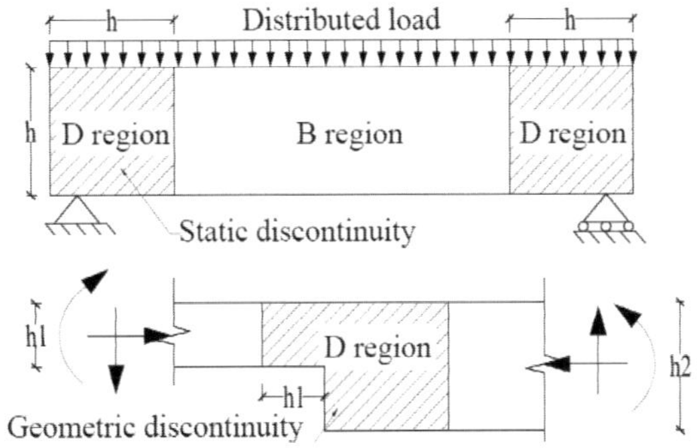

Figure 1. Examples of B and D regions.

2.1. Conventional strut and tie method

During '90s, STM was evolved and established as a consistent tool for the analysis and design of structures with both Bernoulli regions and disturbed regions and was recognised in many of the standards around the world [3-6]. Some of the classical works in STM was reported during this period [7-10]. These works explained the analysis and design procedure of several RC members with loading and geometrical complications like beam-column joints, corbels, deep beams and shear walls. It was recommended to use the elastic stress distribution or load path methods to develop determinate models in STM. In all these works, the capacity of the model was assessed based on the plasticity theorem which states that the stress field satisfying force equilibrium and boundary conditions and does not cause yielding at any point gives a lower bound solution for the capacity of elastic-perfectly plastic material. Since concrete is not such a material, it can endure only limited plastic deformations. Hence, one cannot choose any arbitrary strut and tie model for a structure but have to select the best model which has maximum stiffness and minimum weight. After the formulation of the proper strut and tie model for a structure or structural element, linear elastic analysis was conducted for obtaining the internal forces of each member. These member forces were checked with the corresponding member strength calculated [3-7]. Intersection points of these members termed as nodes were also checked with its strength. A conventional STM model for RC beam is shown in figure 1 (a). The required steel for carrying the tensile forces in tie elements of the model was designed and detailed.

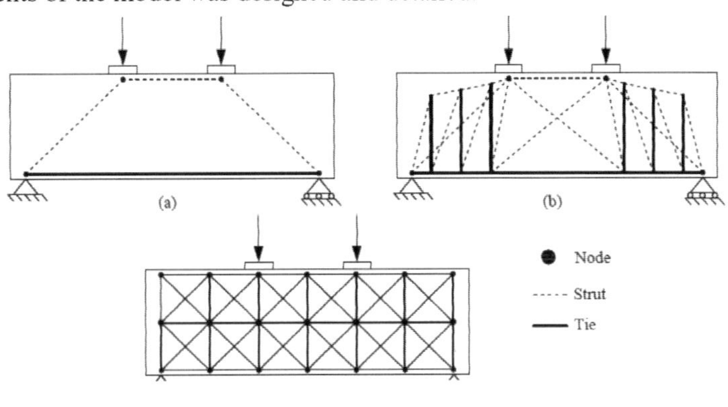

Figure 2. Different strut and tie models (a) Conventional (b) Model by Yun [13] (c) GSTM model[20].

2.2. Nonlinear STM

Basically, conventional STM uses force-based failure criteria and does not consider the inelastic deformations. The ability of determinate models in reflecting the exact load transfer mechanism of structures in certain complex cases was questionable. These limitations lead to the development of nonlinear STM and the use of indeterminate strut and tie models [11-16]. Yun [13] used nonlinear techniques in different stages of the proposed design algorithm. The model for an RC beam developed using the method is shown in figure 2 (b). Steel struts and concrete ties were positioned additionally at the locations of concrete struts and steel ties. The method was able to predict the load-deformation in the elastic and inelastic range of the structure precisely and properly design the complicated RC structures. Another important work in nonlinear STM was conducted by To et al. [16]. A specialised computer program was developed for the sectional analysis in this method. Nonlinear properties of concrete and steel were considered in the finite element analysis used in this method. To et al. [16] expanded the study from monotonic analysis to cyclic and dynamic analysis of complicated RC members. These nonlinear STMs were able to predict the behaviour of complex structures with increased precision. But the complexities and prolonged processing time made these methods impracticable for the generous use among the engineers.

2.3. Topology optimised STM

Another limitation of the conventional STM was the uncertainty in the selection of the model for each structural element. Various models were recommended in the classical works and codes for some of the important structural elements. But those were suitable only for certain geometrical and loading conditions which belonged to the domain studied up to then. Complexities like openings and different loading cases lacked proper strut and tie models and design recommendations. To overcome this constraint, topology optimisation methods [17-19] were adopted. Ali et al. [17] considered a ground structure as the model of the structure initially. Ground structure consisted of numerous truss elements which would be then optimised to the appropriate model (figure 3). Optimisation was based on minimum volume, and two factors (practicality factor and stress redistribution factor) were introduced to control the iterative procedure. Complex structural elements like deep beam with web opening, dapped end beam, beam-column joints and shear wall with openings were analysed using the method.

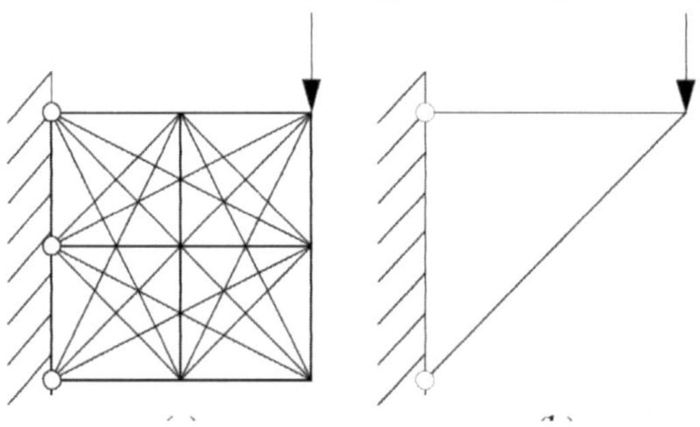

Figure 3. Topology optimisation [17] (a) Ground structure (b) Final model.

Another major work in this field was by Liang et al. [18] which was a performance-based optimisation procedure. It involved complicated iterative procedures comprising strain energy calculations and optimisation of weight of the model. Later the work was extended from the analysis of complex RC structural elements to prestressed concrete elements [19]. Optimisation methods were accurate and applicable to complex structures but were complicated and time-consuming. The whole iterative procedure must be repeated for each load cases to obtain the precise model for each load case.

2.4. Grid STM

A novel idea for obtaining a strut and tie model which can be used for different load cases of a structure was presented by Yun et al. [20]. It was termed as 2-Dimensional Grid Strut and Tie Method (GSTM). The structure was divided into several grids to construct the model (figure 2 (c)). Size of grid elements was governed by geometric compatibility. Strength of the members was optimised in the method, and a computer program was developed for all the iterative procedures. The method appeared to be a valid and improved analysis and design tool except for the need of the special program for carrying out all the graphical decision, optimisation and numerical calculations.

2.5. Secant stiffness based STM

As a simplified method for the inelastic analysis of strut and tie model, Park et al. [21] proposed secant stiffness method which was originally used by Park et al. [22] for the earthquake design of RC structures. Portions of the structure which would undergo inelastic deformations are identified first. Secant stiffness is allocated for the element at these regions and elastic stiffness for the elements where only elastic deformation is anticipated. Hence basically, the method depends on the design strategy adopted by the engineer. The method was able to predict the load-deformation behaviour of structures using the linear analysis of the strut and tie model. The concept of replacing nonlinear analysis with linear analysis using secant stiffness is explained in figure 4. This simplified method was not applicable in complex cases having buckling and fracture failure modes, because of the difficulties in attaining the accurate models.

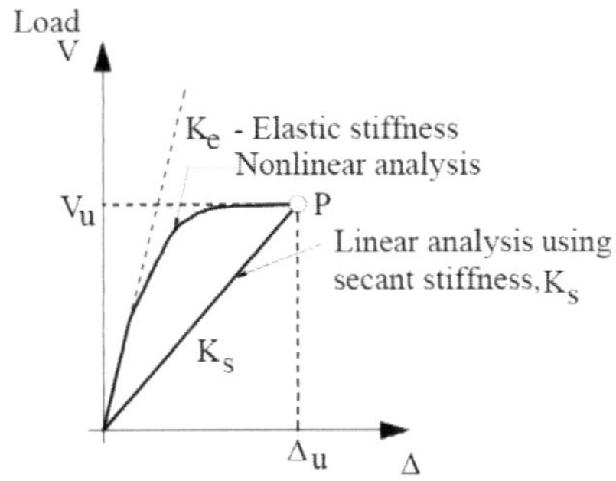

Figure 4. Comparison of nonlinear analysis and linear analysis using secant stiffness [21].

2.6. Compatibility STM

Yet another way for using strut and tie model was proposed by Scott et al. [23] namely compatibility STM. It was a nonlinear analysis method which considered both the reinforcement and concrete in each element of the model. Concrete and steel areas were constrained together as one element and assigned separate stiffness for the same. Truss action and arch action were combined together to form the compatibility strut and tie model (figure 5). Inelastic response of each element was attained using the nonlinear constitutive relationship of cracked concrete. This method was not capable of handling complex problems due to the complications in selecting suitable strut and tie models.

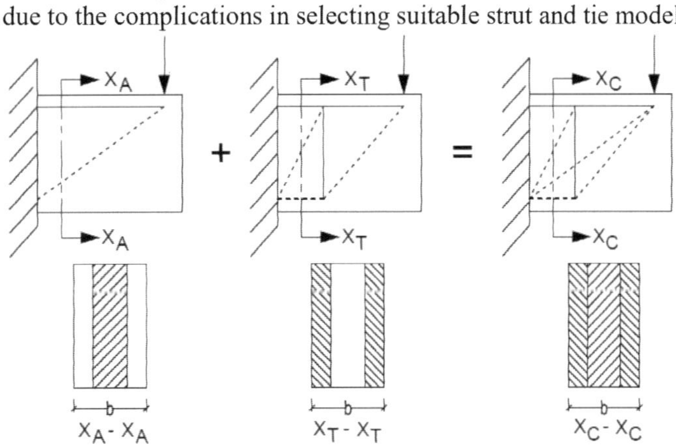

Figure 5. Compatibility STM modelling [23] (a) Arch action (b) Truss action (c) Compatibility model.

IOP Conf. Series: Materials Science and Engineering **936** (2020) 012051 doi:10.1088/1757-899X/936/1/012051

2.7. Evaluation and modification of existing codal recommendations
There were studies conducted to check the sensitivity of the STM recommendations in various standards. Tuchscherer et al. [24] compared the recommendations in ACI 318-11, AASHTO LRFD and CEB-FIP [25-27]. The work concluded that CEB-FIB recommendation gives more realistic results and the other two standards are overly conservative in the deep beam design. Some modifications regarding the recommendations on the calculation of the strut strength and nodal strength were also suggested.

2.8. Latest works in STM
As the latest works in this field, two more studies are discussed in this paper. The extended work of Yun et al. [28] in the field of GSTM is the first one. The work proposed a 3-dimensional GSTM for the design of concrete structures. The method was assuring more accuracy and understanding of structural behaviour. The complexities and processing time was also increased for this extension work. Generous use of this method among engineers remains impractical. Another new trend was presented by Meléndez et al. [29]. The work proposed a refined 3-dimensional strut and tie model for four-pile caps. The major innovative modification was the usage of optimisation technique to arrive at the precise inclination of the strut elements. For simplicity, the determinate model was used, and shear and flexural failure criteria were considered simultaneously, unlike existing strut and tie models of pile caps. The work was purely limited to four-pile cap without shear reinforcement.

3. Discussions
All of the STMs explained in this paper have improved accuracy than the conventional STM, but are complicated and inappropriate for general use among engineers. Some of the modifications that may be used to simplify the methods are suggested below.

i. Geometrical compatibility of the GSTM can be achieved in a simplified way if grids are developed by considering all the main reinforcement in the structure along with a portion of concrete bonded to it as the elements of grids.
ii. Consider the diagonal concrete portion inside each grid as the strut elements of the model.
iii. Using linear elastic analysis of the model created using the above modifications may be enough for a satisfactory prediction of the load path and failure pattern.
iv. For the design purpose, the initial model may be developed from the minimum reinforcement throughout the structure.

4. Conclusions
Strut and tie method has evolved from a simple truss analogy to a complicated nonlinear analysis tool. Several modifications are being introduced to the conventional STM. Improvements like the generation of strut and tie models using optimisation techniques and the ability to predict the nonlinear load-deformation behaviour were achieved by the numerous works that are discussed in this paper. But since now, none of these methods has reached to the practising engineers as a generous analysis and design tool. This is due to the complications of the iterative procedures included, tedious calculations and unavailability of specialised computer programs. Some modifications that can be applied to simplify these methods are also discussed.

5. References

[1] Ritter W 1899 Die bauweise hennebique (construction techniques of hennebique) *Schweizerische Bauzeitung* **33(7)** 59–61
[2] Mörsch E 1909 *Concrete-Steel Construction* (New York : McGraw-Hill)
[3] Canadian Standards Association (CSA) 1984 Design of concrete structures for buildings *CAN3-A23.3-M84* (Rexdale, Canada)
[4] Comite Euro-International du Beton (CEB) 1990 *CEB-FIP Model Code 1990* (Telford, London)
[5] AASHTO 1998 *AASHTO LRFD Bridge Design Specifications-2nd Ed.* (Washington D C)
[6] American Concrete Institute (ACI) 2005 Building code requirements for structural concrete *ACI 318-0 and Commentary ACI 318R-05* (Farmington Hills, MI :American Concrete Institute)
[7] Schlaich J, Schaefer K and Jennewein M 1987 Toward a Consistent Design of Structural Concrete *PCI Journal* **32(3)** 74–150

[8] MacGregor J G 1992 *Reinforced Concrete Mechanics and Design*, second edition (Prentice Hall, Englewood Cliffs, NJ)
[9] Collins M P 1978 Towards a rational theory for rc members in shear *J. Struct. Div., ASCE* **104(4)** 649–66
[10] Park R and Paulay T 1975 *Reinforced Concrete Structures* (New York : John Wiley & Sons, Inc.)
[11] Alshegeir A and Ramirez J 1992 Computer graphics in detailing strut-tie models *J. Comput. Civil Eng.* **6(2)** 220–32
[12] Abdalla K M, Alshegeir A and Chen W F 1996 Analysis and design of mushroom slabs with a strut-tie model *Comput. Struct.* **58(2)** 429–34
[13] Yun Y M 2000 Nonlinear strut-tie model approach for structural concrete *ACI Struct. J.* **97(4)** 581–90
[14] Tjhin T N and Kuchma D A 2002 *Alternative Design for the Non-Slender Beam- Examples for the Design of Structural Concrete with Strut-and-Tie Models* ed. K. H. Reineck (Farmington Hills, Mich : ACI)
[15] Park H G, Kim Y G and Uhm T S 2005 Direct inelastic strut-tie model using secant stiffness *J Korean Conc. Inst.* **17(2)** 201–212
[16] To N H T, Ingham J M and Sritharan S 2001 Monotonic nonlinear analysis of reinforced concrete knee joint using strut and tie computer models *Bulletin of the New Zealand society for Earthquake Engineering* **32(3)** 169–190
[17] Ali M A and White R N 2001 Automatic generation of truss model for optimal design of reinforced concrete structures *ACI Struct. J.* **98(4)** 431–42
[18] Liang Q Q, Xie Y M and Steven G P 2000 Topology optimisation of strut-and-tie models in reinforced concrete structures using an evolutionary procedure *ACI Struct. J.* **97(2)** 322–30
[19] Liang Q Q, Xie Y M and Steven G P 2001 Generating optimal strut-and-tie models in prestressed concrete beams by performance-based optimisation *ACI Struct. J.* **98(2)** 226–32
[20] Yun Y M and Kim B H 2008 Two-dimensional grid strut-tie model approach for structural concrete *J. Struct. Eng.* **134(7)** 1199–214
[21] Park H and Eom T 2007 Truss model for nonlinear analysis of RC members subject to cyclic loading *J. Struct. Eng.* **133(10)** 1351–63
[22] Park H and Eom T 2005 Direct inelastic earthquake design using secant stiffness *J. Struct. Eng.* **131(9)** 1355–62
[23] Scott R M, Mander J B and Bracci J M 2012 Compatibility strut-and-tie modeling: Part I-Formulation *ACI Struct. J.* **109(5)** 635–44
[24] Tuchscherer R G, Birrcher D B, Williams C S, Deschenes D J and Bayrak O 2014 Evaluation of existing strut-and-tie methods and recommended improvements *ACI Struct. J.* **111(6)** 1451–60
[25] ACI Committee 318 2011 *Building Code Requirements for Structural Concrete (ACI 318-11) and Commentary* (Farmington Hills, MI :American Concrete Institute)
[26] AASHTO 2010 LRFD *Bridge Design Specifications, fifth edition with 2010 Interim Revisions, American Association of State Highway and Transportation Officials* (Washington DC)
[27] FIP Report 1999 *Practical Design of Structural Concrete* Federation Internationale de la Precontrainte, CEB-FIP (fib), (Lausanne, Switzerland : Federation Internationale du Beton)
[28] Yun Y M, Kim B and Ramirez J A 2018 Three-dimensional grid strut-and-tie model approach in structural concrete design *ACI Struct. J.* **115(1)** 15–26
[29] Meléndez C, Sagaseta J, Miguel Sosa,P F and Pallarés Rubio L 2019 Refined three-dimensional strut-and-tie model for analysis and design of four-pile caps *ACI Struct. J.* **116(4)** 15–29

IOP Conf. Series: Materials Science and Engineering **936** (2020) 012052 doi:10.1088/1757-899X/936/1/012052

Analysis of Beams and Frames using Applied Element Method

D Lincy Christy*, T M Madhavan Pillai and Praveen Nagarajan

Department of Civil Engineering, National Institute of Technology Calicut, Kerala, India

*lincychristyd@gmail.com

Abstract. Applied Element Method (AEM) is an efficient tool for analysing structures numerically. It has a few benefits when conventional methods are concerned. In conventional numerical methods, node-to-node connection is important. Hence, when elements of different sizes are to be connected, transition elements will have to be used. In AEM, rigid elements are connected to each other with the help of springs. Therefore, node-to-node connection is not required. In this paper, some plane stress problems are analysed using AEM. For this, two-dimensional (2D) element is made use of. Continuous beam and 2D frame are analysed using AEM. The results show that AEM is able to analyse beams and frames accurately. Modal analysis of beam clamped on one end and hinged at the other end, gave accurate first natural frequency and mode shape. It was also possible to perform nonlinear analysis of cantilever beam accurately.

1. Introduction

Several numerical methods for the analysis of structures are readily available today. Applied Element Method (AEM) is one of them. AEM is a rigid body method in which springs are introduced between rigid elements to transfer forces. Kimiro Meguro and Hatem Tagel-Din [1-5] are behind the development of the method. Thereafter, the method was used for the analysis of different materials and structures [6-7]. Collapse behaviour of structures also can be predicted using AEM.

AEM is used for the analysis of continuous beam and two-dimensional (2D) frame in this paper. The results were compared with that obtained by the analysis using STAAD. Modal analysis of a beam clamped on one end and hinged at the other end is included in this paper. Nonlinear analysis of cantilever beam with tip load carried out by both force-based and displacement-based approaches is also included.

2. Applied Element Method

Conventionally, AEM has only two elements – 2D element and three-dimensional (3D) element. There are two translational degrees of freedom (DOFs) and one rotational DOF in a 2D element, as shown in figure 1. Two springs – one for transferring normal force and the other to carry shear force make a pair of spring for two-dimensional analysis. In the present study, 2D element was used for the analysis. The connection between the elements in AEM are shown in figure 1. The DOFs are marked at the centre of the elements.

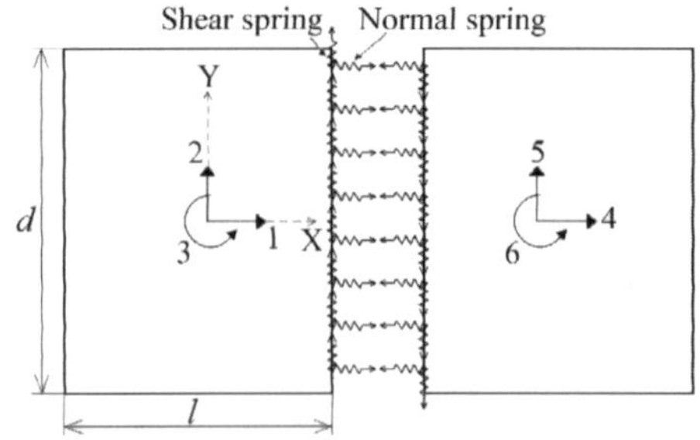

Figure 1. Element connection in 2D AEM.

Every pair of springs has a stiffness matrix, which depends on its position and stiffness as given by equation (1) [8].

$$K = \begin{pmatrix} k_n & 0 & -k_n y & -k_n & 0 & k_n y \\ 0 & k_s & k_s x & 0 & -k_s & k_s x \\ -k_n y & k_s x & k_n y^2 + k_s x^2 & k_n y & -k_s x & -k_n y^2 + k_s x^2 \\ -k_n & 0 & k_n y & k_n & 0 & -k_n y \\ 0 & -k_s & -k_s x & 0 & k_s & -k_s x \\ k_n y & k_s x & -k_n y^2 + k_s x^2 & -k_n y & -k_s x & k_n y^2 + k_s x^2 \end{pmatrix} \quad (1)$$

$$k_n = \frac{E d_s b_s}{l_s}; \qquad k_s = \frac{G d_s b_s}{l_s}$$

where, l_s, d_s and b_s are the length, depth and width of the portion represented by a spring; (x, y) is the coordinate of the location of the spring with respect to the centre of the first element; E and G are the Young's modulus and modulus of rigidity respectively.

3. Analysis of continuous beam

A two-span continuous beam shown in figure 2 was considered for analysis. The cross-section of both the members were 300 mm × 500 mm. The width of element was50 mm. 100 springs were used along all the faces. The Young's modulus and Poisson's ratio were25000 N/mm^2 and 0.2, respectively.

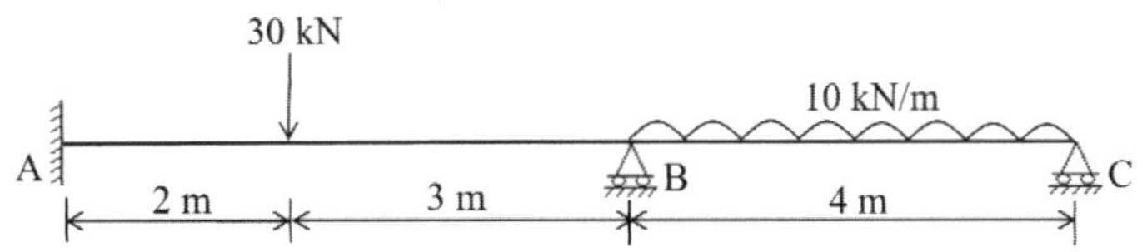

Figure 2. Continuous beam considered for analysis.

The stiffness matrix of all the members were determined. The stiffness matrix and force vector were condensed to retain the end DOFs. The stiffness matrix and force vector of both the members were assembled by considering the relevant DOFs alone. Thus, static condensation approach was made use of to carry out the analysis. Figure 3 shows the discretisation of the beam.

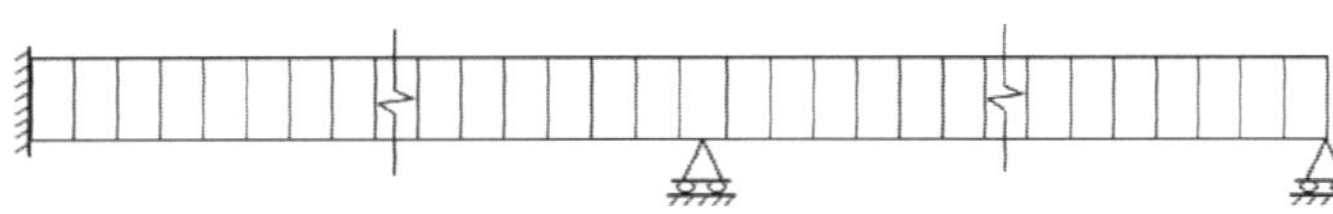

Figure 3. Discretisation of continuous beam.

The deflected shape and reactions obtained using AEM and by the analysis using STAAD are shown in figures 4 and 5, respectively.

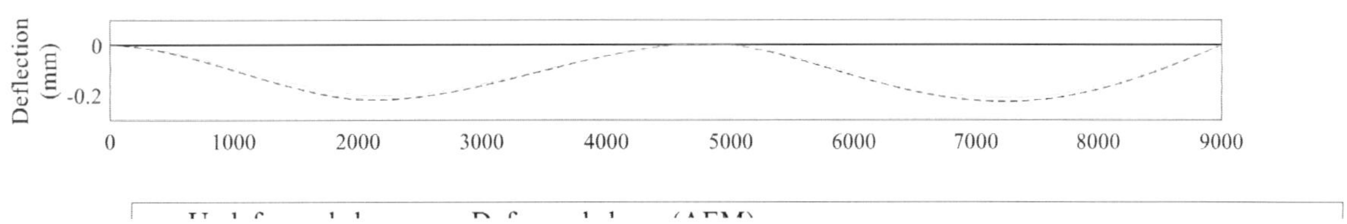

Figure 4. Deflected shape of continuous beam.

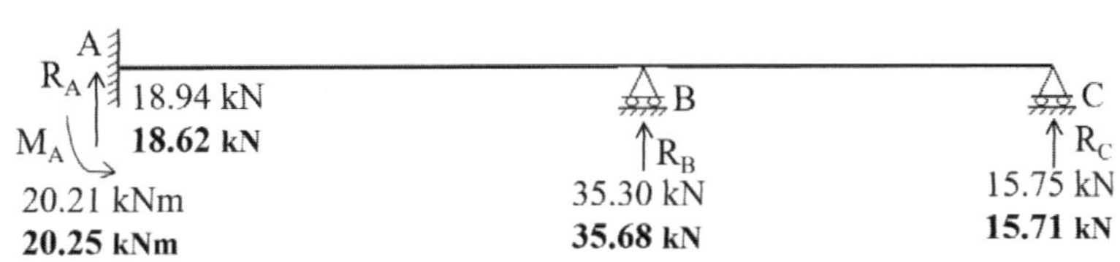

Figure 5. Support reactions by AEM and STAAD (boldface letters indicate results from STAAD).

The percentage difference in the reactions obtained are presented in table 1.

Table 1. Reactions in continuous beam.

Reaction	AEM	STAAD	Percentage difference (%)
R_A	18.944	18.616	1.76
M_A	20.211	20.247	-0.18
R_B	35.304	35.675	-1.04
R_C	15.752	15.709	0.27

* Forces are in kN and moment is in kNm

Figures 4 and 5 show that AEM is able to predict the deformation and reaction of continuous beam accurately. From table 1, it is seen that AEM predicts the reaction with less than 2% difference.

4. Analysis of 2D frame

The 2D frame shown in figure 6was considered for the analysis. The cross-section of all the members were 200 mm × 400 mm.The width of the element, number of pairs of springs and material properties were same as that of section 3.

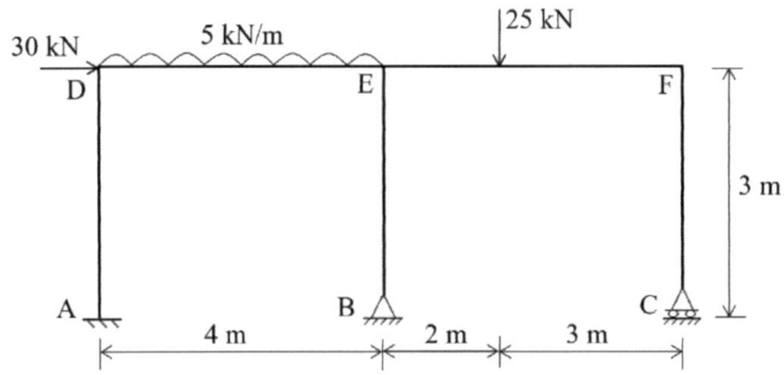

Figure 6. 2D frame.

The discretisation of the frame is shown in figure 7.

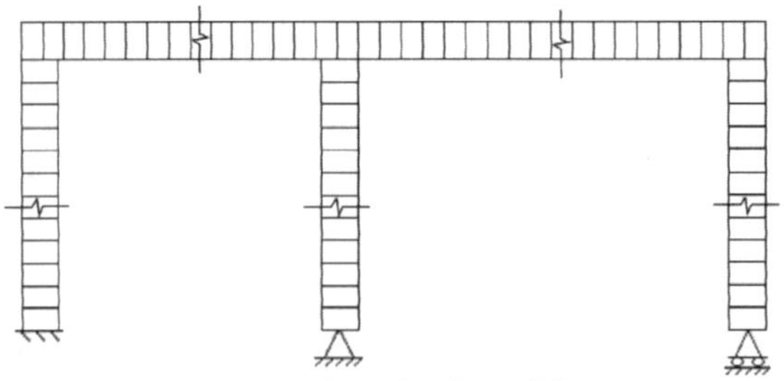

Figure 7. Discretisation of frame.

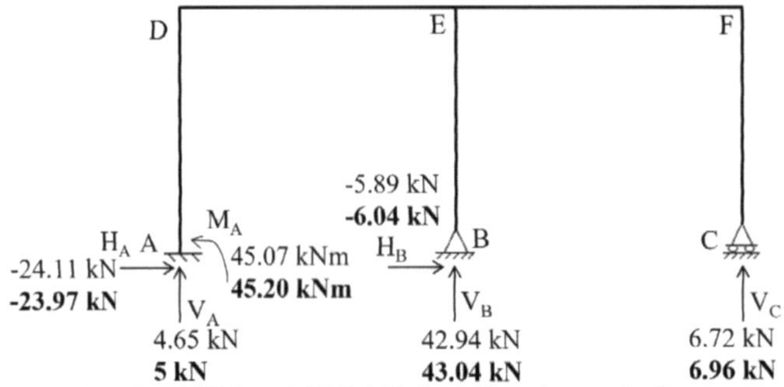

Figure 8. Support reactions by AEM and STAAD (boldface letters indicate results from STAAD).

The support reactions and deflected shape of the frame obtained by applied element analysis are shown in figures 8 and 9, respectively. The deflected shape obtained by the analysis using STAAD is also shown in figure 9.

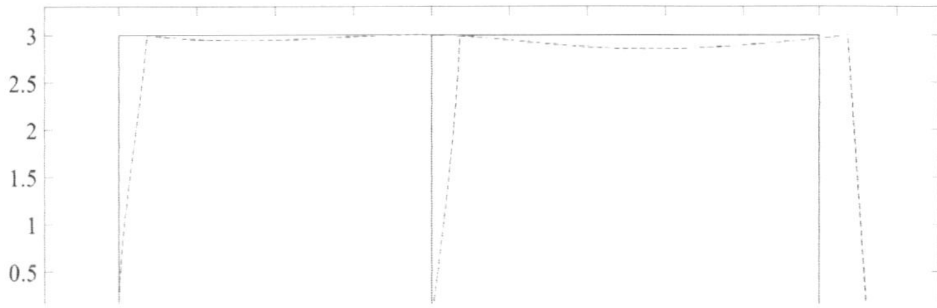

Figure 9. Deflected shape by AEM and STAAD
(* The displacement is magnified by a factor of 100).

Figures 8 and 9 show that AEM predicts reactions and deformations with reasonable accuracy. The displacement vector and reactions by AEM and by the analysis using STAAD are given in table 2 and 3.

Table 2. Displacement vector.

DOF	AEM	STAAD	Percentage difference (%)
u_A	0	0	0
v_A	0	0	0
θ_A	0	0	0
u_B	0	0	0
v_B	0	0	0
θ_B	-0.002	-0.002	0
u_C	6.1054	6.2268	-1.95
v_C	0	0	0
θ_C	0.001	0.001	0
u_D	3.6208	3.6854	-1.75
v_D	0.0070	0.0076	-7.89
θ_D	- 0.001	-0.001	0
u_E	3.6091	3.6734	-1.75
v_E	-0.0644	-0.0645	-0.16
θ_E	-0.001	-0.001	0
u_F	3.6091	3.6734	-1.75
v_F	-0.0101	-0.0105	-3.81
θ_F	0.001	0.001	0

* The translations are in 'mm' and rotations are in 'rad'. 3 significant digits are considered to find the percentage difference

Table 3. Reactions.

Reaction	AEM	STAAD	Percentage difference (%)
H_A	- 24.111	- 23.965	0.61
V_A	4.652	4.998	- 6.92
M_A	45.065	45.202	- 0.30
H_B	- 5.889	- 6.035	- 2.42
V_B	42.937	43.037	- 0.23
V_C	6.715	6.961	- 3.53

* Forces are in kN and moment is in kNm

Table 2 and 3 shows that AEM is capable of analysing 2D frames accurately.

5. Modal analysis of beam

A 2 m long beam with one end clamped and the other end hinged as shown in figure 10 was considered for the modal analysis. The cross-section of the beam was 200 mm × 200 mm. The beam was divided into 100 elements and they were connected by 100 springs at each face. The density, Young's modulus and Poisson's ratio of the material of the beam were 7850 kg/m^3, 25000 N/mm^2 and 0.2 respectively.

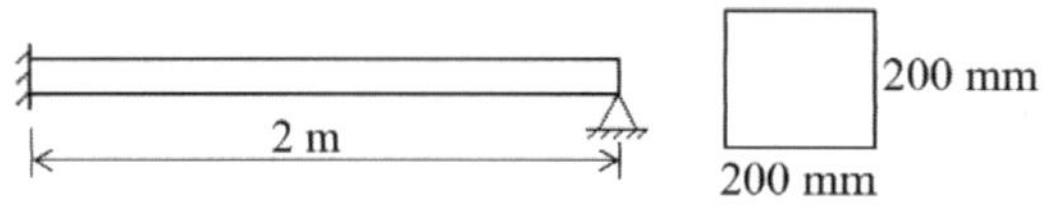

Figure 10. Beam considered for modal analysis.

The first natural frequency (f_1) of a beam considered is given by equation (2) [9].

$$f_1 = \frac{1}{2\pi}\frac{\lambda_1}{l^2}\sqrt{\frac{EI}{m}} \quad (2)$$

Where, λ_1, *l, E, m and I* are the frequency factor, length of the beam, Young's modulus, moment of inertia and mass per unit length respectively.

Frequency factor for first mode was 15.418 [9].

The first natural frequency of the beam obtained by AEM and analytically were 61.52 Hz and 63.21 Hz respectively, which are close enough.

The shape of first mode of vibration of the beam is shown in figure 11.

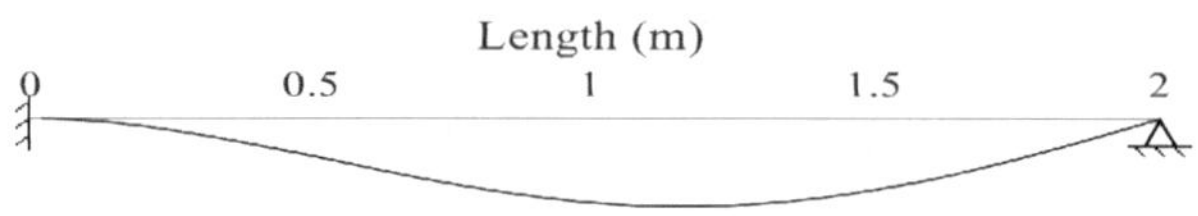

Figure 11. First mode shape of the beam.

Figure 11 shows that AEM can predict the mode shape of the beam with reasonable accuracy.

6. Nonlinear analysis

A nonlinear analysis done by Moideen and Dewangan [10] using ANSYS was considered for the study. The problem taken was a steel cantilever beam of span 0.5 m and cross-section 70 mm × 70 mm with point load of 60 kN at free end as shown in figure 12. The initial modulus of elasticity, Poisson's ratio and yield strength were 2.1×10^5 N/mm^2, 0.3 and 250 N/mm^2 respectively.

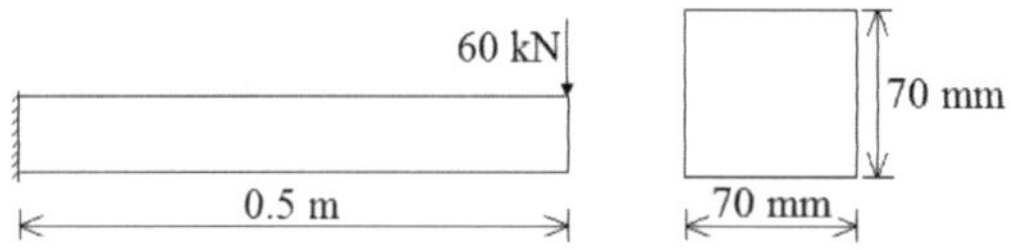

Figure 12. Cantilever beam for nonlinear analysis.

The bilinear stress-strain graphs as shown in figures 13(a) and (b) were adopted. These are regarded as case 1 and case 2 respectively.

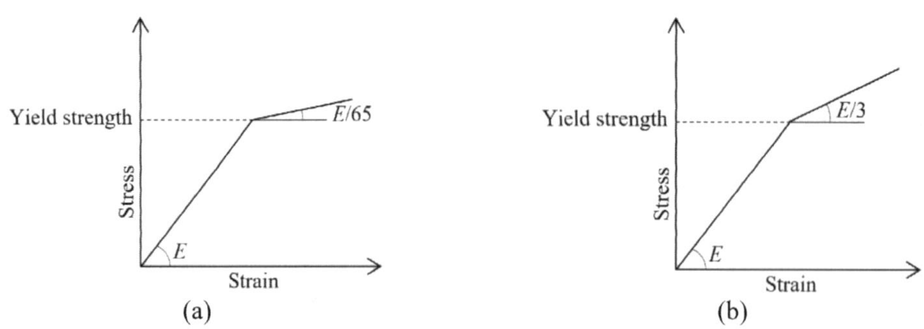

Figure 13. Stress-strain graph for (a) case 1 (b) case 2.

IOP Conf. Series: Materials Science and Engineering **936** (2020) 012052 doi:10.1088/1757-899X/936/1/012052

The nonlinear analysis was carried out using AEM by two approaches – force based and displacement based. An increment of 500 N and 0.05 mm were adopted for the force based and displacement based approaches, respectively. The beam was discretised into 100 elements along length and 10 elements along depth. 10 number of springs were provided at all element faces.

Table 4. Results of force based incremental method for case 1.

Load (kN)	Deflection (mm)		Percentage difference (%)	Maximum stress (N/mm^2)		Percentage difference (%)
	AEM	ANSYS		AEM	ANSYS	
1	0.10	0.10	0.88	9.08	8.31	-8.48
5	0.50	0.50	0.88	45.40	41.55	-8.48
10	0.99	1.01	1.89	90.80	83.10	-8.48
20	1.98	2.02	1.89	181.60	166.18	-8.49
30	2.98	3.03	1.83	250.48	249.27	-0.48
40	4.29	4.60	7.13	255.67	332.36	29.99
50	12.46	15.52	24.57	299.41	415.45	38.75
60	42.36	47.32	11.70	384.15	498.54	29.78

Table 5. Results of force based incremental method for case 2.

Load (kN)	Deflection (mm)		Percentage difference (%)	Maximum stress (N/mm^2)		Percentage difference (%)
	AEM	ANSYS		AEM	ANSYS	
1	0.10	0.10	0.88	9.08	8.31	-8.48
5	0.50	0.50	0.88	45.40	41.55	-8.48
10	0.99	1.01	1.89	90.80	83.10	-8.48
20	1.98	2.02	1.89	181.60	166.18	-8.49
30	2.97	3.03	1.86	258.88	249.27	-3.71
40	4.09	4.12	0.73	306.86	297.38	-3.09
50	5.62	5.89	4.80	373.46	318.48	-14.72
60	7.59	8.06	6.17	452.47	376.27	-16.84

Table 6. Results of displacement based incremental method for case 1.

Deflection (mm)	Load (kN)		Percentage difference (%)	Maximum stress (N/mm^2)		Percentage difference (%)
	AEM	ANSYS		AEM	ANSYS	
0.10	1.01	1.00	-0.82	9.16	8.31	-9.23
0.50	5.04	5.00	-0.82	45.78	41.55	-9.23
1.00	10.08	10.00	-0.82	91.55	83.10	-9.23
2.00	20.17	20.00	-0.82	183.11	166.18	-9.24
3.05	30.72	30.00	-2.34	250.69	249.27	-0.57
4.60	41.28	40.00	-3.10	257.15	332.36	29.25
15.50	50.84	50.00	-1.64	310.94	415.45	33.61
47.30	60.58	60.00	-0.95	393.80	498.54	26.60

The deflection under the load and the maximum stress in beam at various loads, using force based approach, for case 1 and case 2 are given in table 4 and 5 respectively. The load and corresponding maximum stress in the beam at various deflections at free end for Case-1 and Case-2 are given in table 6 and 7 respectively. The analysis has been carried out by displacement-based approach.

Table 7. Results of displacement based incremental method for case 2.

Deflection (mm)	Load (kN)		Percentage difference (%)	Maximum stress (N/mm^2)		Percentage difference (%)
	AEM	ANSYS		AEM	ANSYS	
0.10	1.01	1.00	-0.82	9.16	8.31	-9.23
0.50	5.04	5.00	-0.82	45.78	41.55	-9.23
1.00	10.08	10.00	-0.82	91.55	83.10	-9.23
2.00	20.17	20.00	-0.82	183.11	166.18	-9.24
3.05	30.73	30.00	-2.39	261.98	249.27	-4.85
4.10	40.06	40.00	-0.15	307.40	297.38	-3.26
5.90	51.51	50.00	-2.93	385.06	318.48	-17.29
8.05	62.03	60.00	-3.28	469.69	376.27	-19.89

Ta show that the load-deflection values obtained by three approaches – load based incremental method, displacement based incremental method and ANSYS; are similar. Although the maximum stress value obtained from both incremental approaches were similar, they differed from ANSYS results at higher loads.

The load-deflection curve for case 1 and case 2 are shown in figures 14 and 15, respectively.

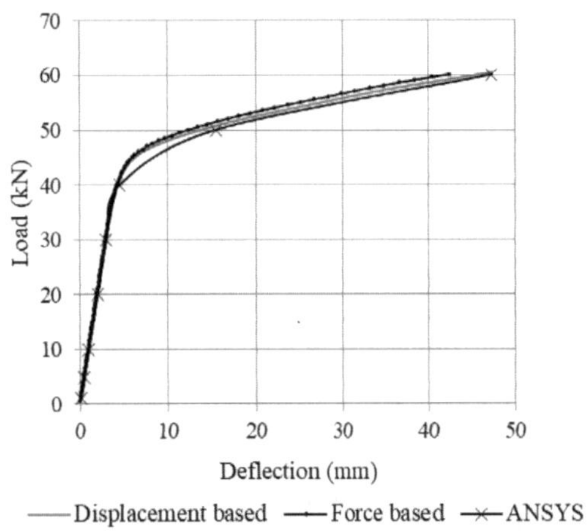

Figure 14. Load-deflection curve (case 1).

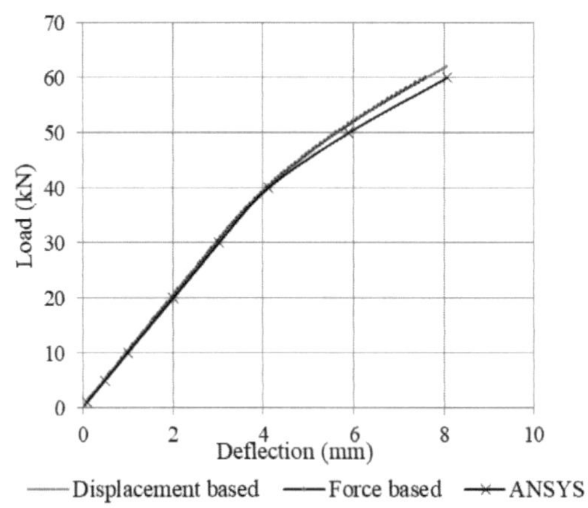

Figure 15. Load-deflection curve (case 2).

Figures 14 and 15 show that the load-deflection curve obtained by displacement based incremental method and force based incremental method approximately matched the load-deflection curve obtained from ANSYS.

7. Conclusions

In this paper, a continuous beam and a 2D frame are analysed using AEM. Nonlinear analysis considering material nonlinearity is also done. From the results, it is seen that AEM could determine the deformed shape and reactions of continuous beam and 2D frame accurately. Also, the natural frequency and mode shape of beam are also predicted with reasonable accuracy. The load-deflection curve of cantilever beam with tip load obtained from AEM agrees well with the results of ANSYS.

8. References

[1] Meguro K and Tagel-Din H 1997 A new efficient technique for fracture analysis of structures *Bulletin of Earthquake Resistant Structure Research Center, Institute of Industrial Science, The University of Tokyo* **30** 103–116

[2] Meguro K and Tagel-Din H 1998 A new simplified and efficient technique for fracture behavior analysis of concrete structures *Proceedings of the Third International Conference on Fracture Mechanics of Concrete and Concrete Structures (FRAMCOS-3)* vol 2 pp 911–920

[3] Tagel-Din H and Meguro K 1999 Applied element simulation for collapse analysis of structures *Bulletin of Earthquake Resistant Structures* 113–123

[4] Tagel-Din H and Meguro K 2000 Nonlinear simulation of RC structures using applied element method *Structural Engineering/Earthquake Engineering* **17(2)** 137–148

[5] Meguro K and Tagel-Din H S 2002 Applied element method used for large displacement structural analysis *Journal of Natural Disaster Science* **24(1)** 25–34

[6] Mayorca P and Meguro K 2003 Modeling masonry structures using the applied element method *Institute of Industrial Science, The University of Tokyo* **55(6)** 581–584

[7] Elkholy S and Meguro K 2004 Numerical simulation of high-rise steel buildings using improved applied element method *13th World Conference on Earthquake Engineering. Vancouver, BC, Canada*

[8] Christy D L, Pillai T M M and Nagarajan P 2018 Analysis of concrete beams using applied element method *IOP Conference Series: Materials Science and Engineering* **330(1)** 1–7

[9] Chen S S and Wambsganss M W 1974 Design guide for calculating natural frequencies of straight and curved beams on multiple supports *Components Technology Division, Argonne National Laboratory, Argonne, Illinois*

[10] Moideen K R and Dewangan U K 2016 Effect of material nonlinearity on deflection of beams and frames *Indian Journal of Science and Technology* **9(28)** 1–6

IOP Conf. Series: Materials Science and Engineering **936** (2020) 012053 doi:10.1088/1757-899X/936/1/012053

Optimal Rate of Printing of 3D Printed Concrete Columns and Walls to Avoid Buckling

K P Phadnis*, **M N Shariff and D Menon**

Indian Institute of Technology Madras, Chennai, Tamil Nadu, India

*kedarphadnis10@gmail.com

Abstract. Extrusion based construction of concrete structures has been identified as one of the alternative construction technologies which reduces the construction costs, delays in construction and the material consumption. The most attractive feature of 3D printed constructions is the ease of developing intricate forms and shapes (which otherwise require skilled manpower). Since this is a relatively new technology, not much research has been done with regard to its structural performance and stability. One of the major concerns is the possibility of buckling (under its self-weight) of the column or wall during the extrusion process. This sets a limitation of the acceptable height, thickness and rate of extrusion process. In this study simple ways (second order analysis using matrix methods) have been explored to determine the critical height and the optimal rate of extrusion. Matrix methods have inherent limitations due to the simplification adopted by ignoring the higher-order terms in Taylor's series. The study uses the evolution of modulus of elasticity of concrete as proposed in literature. The analysis method has been validated using pertinent buckling tests reported in literature. Parameters such as modulus of elasticity and length are considered to determine the critical height to thickness of the wall and the optimal rate of extrusion. The problem is solved using the MATLAB software. This method requires minimal input from the user such as material properties, geometry and printing speed. The analysis method used in this study provides a useful design tool to estimate the limiting parameters of the structural walls and columns manufactured through the process of 3D printing.

1. Introduction

Additive manufacturing is used in various disciplines such as medical devices, aerospace, automotive, bio-printing and construction. The research going in the field of 3D printing has increased drastically due to its benefits to the community [1]. The additive manufacturing of concrete structures is now gaining momentum and acceptance for the construction of full scale structures [2]. This method of construction can be especially useful for low cost construction by reducing the material consumption and manpower requirement. One of the major challenges faced during the extrusion process is the possibility of buckling of walls or columns. This happens because the concrete is still fresh and has not gained adequate strength. Tests have shown that when more layers are added to the member, at a relatively fast rate, the whole wall can buckle [3]. Hence it is of interest to know the critical height of the element, for a given thickness and printing speed to ensure buckling can be prevented.

Additive manufacturing of concrete is a process in which desired structure is built in layers of suitable concrete paste which is forced through a tube. This tube is guided by a machine that takes the design details from the user. Various studies have been done to get the structural properties of the 3D printed concrete using finite element analysis and differential equations, which are computationally expensive [3]. In this study, the use of stiffness matrix methods to estimate the buckling height of the 3D printed elements is explored using the procedure described in the literature [4].

In the extrusion process, as every new layer is laid, the hydration process initiates and the concrete starts gaining strength. The strength gain is generally dependent on the quantity of accelerator added in the concrete paste. With strength, the modulus of elasticity (E), also increases. The evolution of E with time can be defined using various functions: linear, exponential or quadratic. In the present study, where the buckling problem is solved, the modulus of elasticity is considered as the input. It is reasonable to assume that there is no variation of the stiffness properties along the length of the wall (z-direction) (see figure 1). This implies that the major variation of stiffness properties occurs along the vertical direction (y-direction), i.e. in the direction of extrusion. Therefore, the early age modulus of elasticity of concrete varies for each layer. This needs to be accounted for in the buckling calculations.

The presence of axial compression reduces the flexural stiffness of the element, and conversely, the presence of axial tension enhances the flexural stiffness. A critical condition of buckling instability is likely to occur when the flexural stiffness of the column or wall reduces to zero under axial compression [4]. The axial compression load is referred to as the critical buckling load, P_{cr}. This instability, in literature, is referred to as elastic instability. To estimate the critical buckling load of the member, a primary stiffness matrix in conjunction with a geometric stiffness matrix is employed. The total stiffness matrix for each element is shown in equation (1). The terms in the primary stiffness matrix can be derived using the slope-deflection method, while the terms in the geometric stiffness matrix can be derived using the stability functions. This is represented in figure 2.

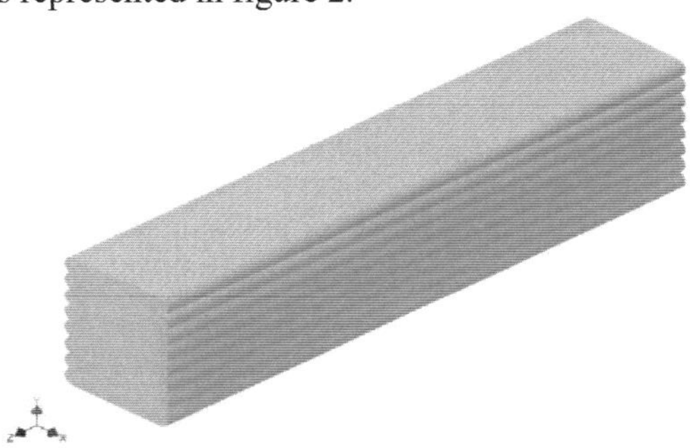

Figure 1. Schematic representation of a 3D printed wall.

To make the problem convenient to solve it is assumed that the elements are adequately braced against out of plane and torsional modes of buckling. The bottom of the wall is assumed to be fixed (rigid support). The results of the study have been compared with the experiments published in the literature [1].

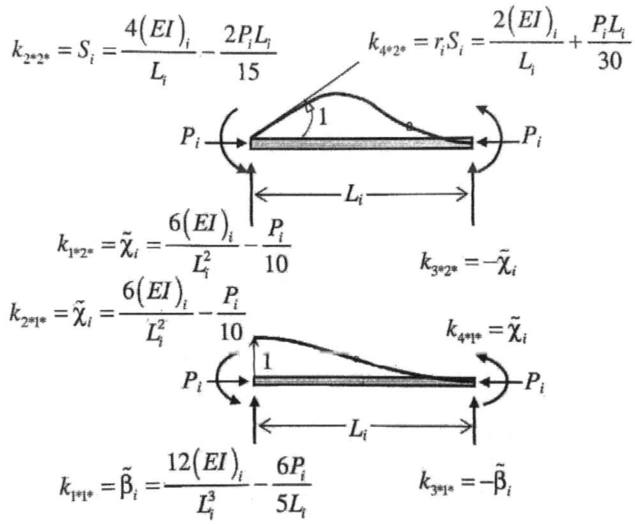

Figure 2. Simplified stiffness coefficients (for small (μL); axial compression).

$$k(i)=\frac{E(t)I}{h}\begin{pmatrix}\frac{12}{h^2} & \frac{6}{h} & -\frac{12}{h^2} & \frac{6}{h}\\ \frac{6}{h} & 4 & -\frac{6}{h} & 2\\ -\frac{12}{h^2} & -\frac{6}{h} & \frac{12}{h^2} & -\frac{6}{h}\\ \frac{6}{h} & 2 & -\frac{6}{h} & 4\end{pmatrix}-(p+iw)\begin{pmatrix}\frac{1.2}{h} & 0.1 & -\frac{1.2}{h} & 0.1\\ 0.1 & \frac{2h}{15} & -0.1 & -\frac{h}{30}\\ -\frac{1.2}{h} & -0.1 & \frac{1.2}{h} & -0.1\\ 0.1 & -\frac{h}{30} & -0.1 & \frac{2h}{15}\end{pmatrix} \quad (1)$$

where, $k(i)$ is the total stiffness matrix of the ith element, $E(t,i)$ is the modulus of elasticity of the ith element at time t, $I(i)$ is the area moment of inertia of the ith element, $h(i)$ is the height of ith element, p is the critical buckling load of the entire member and w is the self-weight of each layer.

2. Methodology

An algorithm used to determine the critical buckling height of the wall is shown in figure 3. First, the geometric parameters such as thickness of wall (t), length of wall (L) and height of each layer (h) are decided. The material parameters for the concrete used in the 3D printing process such as the variation of modulus of elasticity of concrete with time $E(t)$, Poisson's ratio (υ) and density of fresh concrete (ρ) are chosen. Further, the rate of printing speed (v) is also decided based on concrete rheology and printer capability. With this as the input, the analysis is carried out for the ith layer and the critical buckling load ($P_{cr,i}$) is evaluated. This is checked against the self-weight of the element (P_{sw}) and verified whether $P_{cr,i} < P_{sw}$. If the condition fails, the program is executed for the next layer, by suitably modifying the parameters based on the elastic analysis. Once again, for the two layers, the E is evaluated and used in the elemental stiffness matrix to determine the $P_{cr,i}$. This exercise is carried out till the critical buckling height is reached. Figure 4 shows a schematic representation of the layer creating in the numerical model.

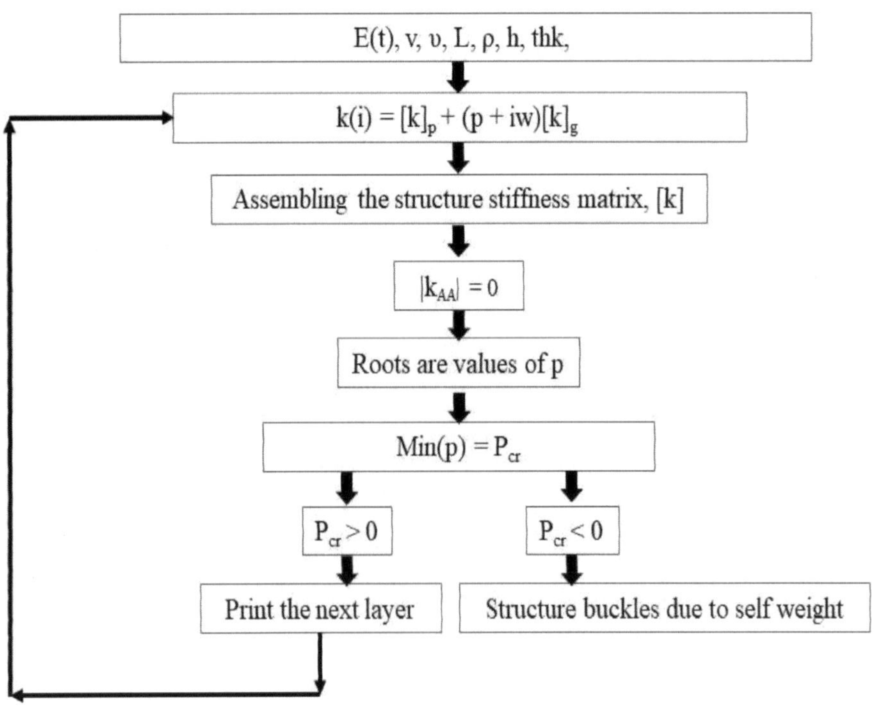

Figure 3. Flow chart of the program.

The steps involved in assessing the critical height of the member are:

a. Assigning input values of the code such as length of the wall (L), thickness (t), height (h), density of the concrete (or any material) (ρ), initial modulus of elasticity (E), curing rate of concrete (υ), velocity of the printer head (v).
b. Weight of each layer is calculated using the given dimensions of the section and density of the material.
c. Based on time taken to print a layer (from printer head velocity and length of the wall) variation of modulus of elasticity with time is determined with the help of specified curing function and curing rate constant.
d. Deformation of each layer due to the self-weight of concrete layers above the layer is determined by including the Poisson's effect.
e. After computing all the deformed dimensions, the stiffness matrix of every layer is calculated and combined to get a global stiffness matrix.
f. The determinant of the global stiffness matrix is equated to zero and if the minimum value of the load carried by the wall is greater than zero we add/print next layer and run the loop again and if it is less than zero the wall buckles by its self-weight and previous layer is assumed to be the critical buckling height for the given initial conditions.

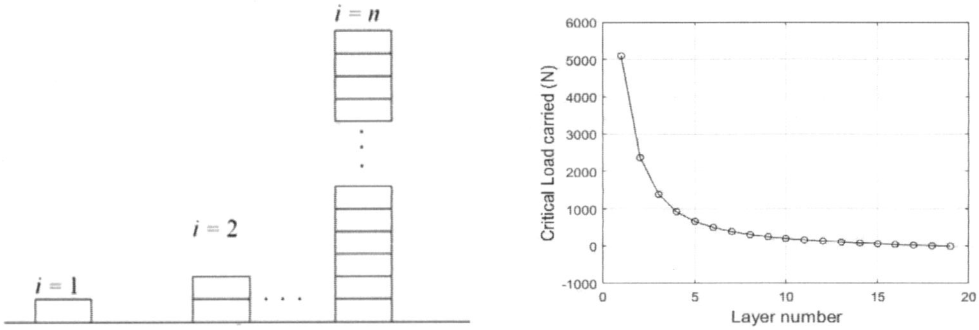

(a) (b)

Figure 4. (a) Schematic representation of the critical height evaluation for each layer, (b) P_{cr}-P_{sw} values for each layer.

Figure 4(a) shows the creating of each layer in the numerical program to evaluate the buckling load of the element. In figure 4(b) it is assumed that the self-weight of each layer is acting at the bottom of every layer of the member, therefore the top layer also carries just compressive load and the bottom-most layer carries the self-weight of the member and the additional compressive load. When the critical load carried by the wall is less than or equal to the self-weight of the wall then it is assumed that the member has failed due to self-buckling.

The limitations of the numerical program are:

a. The viscous effects of the concrete paste are not considered
b. The modulus of elasticity of concrete is assumed to be a constant for the entire length of the layer which is clearly not the case for long walls
c. The slip between layers is ignored
d. Poisson deformation is assumed to be uniform on all the sides
e. The curing function of concrete needs to suitably simulated to get numerical values close to the experimental data.

3. Validation using test reported in the literature

Wolf and Suiker (2019) reported tests carried out on 3D printed walls which failed during the extrusion process with a buckling mode of failure. In this paper, the proposed numerical approach is adopted to evaluate the buckling height of the tested walls. Figure 5 shows the comparison of the numerical results with the experimental results of 1m, 5m and 10.4m long walls. It is seen that for all the three layers the numerical results under-estimate the buckling height. This discrepancy could be attributable to the approximation made in the modulus of elasticity and Poisson's ratio of fresh concrete. Further, the viscous effects of fresh concrete are ignored. The deviation is further pronounced in the 10.4m length wall and this could be attributable to the fact that the numerical program is capable of handling only the instability due to height and not due to length. The 10.4m length wall buckled due to the length instability. Further, the numerical program assumes a constant value of modulus of elasticity of concrete for the entire layer at a given time instant. However, for longer walls, this is a highly simplified approach.

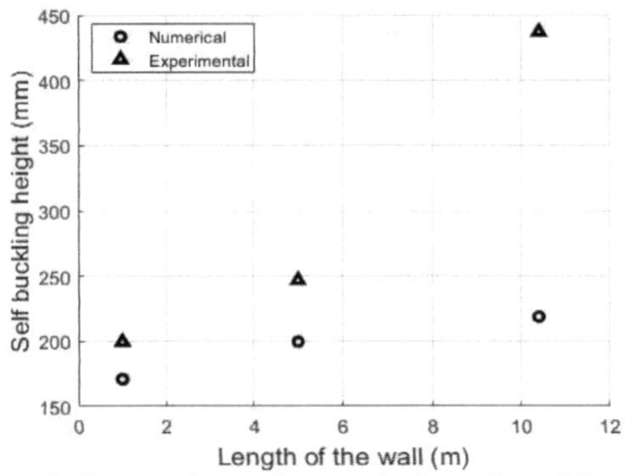

Figure 5. Comparison of numerical results with test data.

4. Parametric study

In order to establish the critical buckling height of the wall for a given thickness and rate of printing, a parametric study is carried out by adopting three different rates of printing (10mm/s, 20mm/s and 100mm/s), for two different lengths: 1m and 5m. The height of each layer is assumed to be 9.5mm. The evolution of the modulus of elasticity is assumed using a function described in equation (2).

$$E(t) = E_0(1 + \chi t) \tag{2}$$

where E(t) is the modulus of elasticity at any time t in seconds, E_0 is the initial modulus value and χ is the rate function (here assumed to be 7.2×10^{-4}).

In the present study, a simple rectangular section is considered, and the moment of inertia is calculated for each layer. However, in practice, more complex/optimized cross-sections are possible. The proposed numerical method is capable of solving the buckling problem for any cross-section geometry.

Figure 6 and 7 shows the critical buckling height versus thickness plots for three different rates of printing for walls having lengths of 1m and 5m respectively. As expected, it is seen from both figures 6 and 7, that as the thickness of the wall increases, the critical self-buckling height also increases. The effect of the slow speed of printing versus the fast rate of printing is evident. It is seen that for a given thickness, slow rates of printing, result in higher critical self-buckling height.

A clear distinction between the three rates of printing is evident in the 5m long wall (as seen in figure 7). This is due to the fact that time taken to print each layer increases with the length of the wall. This means that there is a significant difference in the modulus of elasticity of the subsequent layers in the member.

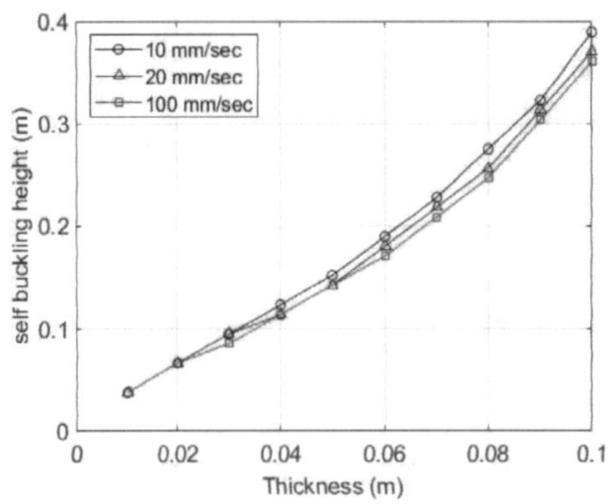

Figure 6. Self-buckling height of 1m long wall versus thickness for different rates of printing.

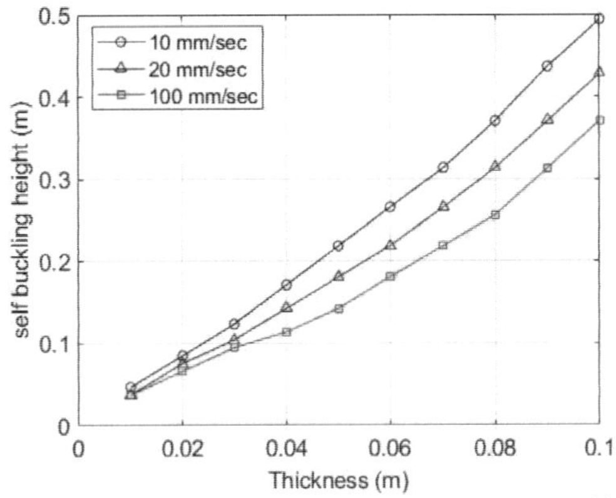

Figure 7. Self-buckling height of 5m long wall versus thickness for different rates of printing.

4.1. Optimal rate of printing

A critical choice in the 3D printing process is the optimal rate of printing. Slower printing enables us to build taller members but causes time delays. Hence, based on the practical parameters governing the concrete mixture properties and the printer capability, the optimal rate of printing can be determined. Based on the parametric study carried out, for the given variation of modulus of elasticity with time, it is seen that the rate of printing has negligible influence for a 1m long wall. However, for 5m long wall, there is a significant difference.

4.2. Optimal thickness of the wall

As indicated above, the geometrical parameters (length and thickness of the wall) and the rate of printing dictate the self-buckling height of the wall during the printing process. If the rate of printing and the length of the wall are predetermined the only choice left to the designer is to alter the thickness of the wall. It is interesting to know what is the optimal thickness which needs to be used for the wall to avoid self-buckling. The present study shows the procedure using which this can be determined.

5. Conclusions

The following conclusions can be drawn from this study:

- A numerical procedure to evaluate the critical buckling height of a wall during its 3D printing process is developed using matrix methods of analysis.
- The proposed analytical procedure is validated using tests reported in the literature.
- It is seen that for the choice of the evolution of modulus of elasticity a function used in the present study, there is no significant difference in the critical buckling heights for different rates of printing for 1m long wall. However, there is a significant difference in the 5m length wall.
- Using the present study, an optimal rate of printing or an optimal thickness of wall can be determined to avoid buckling due to its self-weight during the extrusion process.
- This procedure can be extended to any cross-section of the wall or column element.

More studies are needed to ascertain the correct behaviour of the wall, by including the effects of viscosity of fresh concrete and realistic modulus of elasticity of fresh concrete.

6. References

[1] Suiker A S J 2018 Mechanical performance of wall structures in 3D printing processes: Theory, design tools and experiments *Int. J. Mech. Sci.* **137** 145–70
[2] Buswell R A, Leal de Silva W R, Jones S Z and Dirrenberger J 2018 3D printing using concrete extrusion: A roadmap for research *Cem. Concr. Res.* **112** 37–49
[3] Wolfs R J M and Suiker A S J 2019 Structural failure during extrusion-based 3D printing processes *Int. J. Adv. Manuf. Technol.* **104** 565–84
[4] Menon D 2009 *Advanced Structural Analysis* (Alpha Science)

IOP Conf. Series: Materials Science and Engineering **936** (2020) 012054 doi:10.1088/1757-899X/936/1/012054

A Comparative Study of Aerodynamic Coefficients on Tall Structures using Experimental Studies with International Codes and Standards

Shalini R Nair[1*], Jessy Rooby[2] and R John Branesh[3]

[1] Assistant Professor, Hindustan Institute of Technology and Science, Chennai, India

[2] Professor, Hindustan Institute of Technology and Science, Chennai, India

[3] P.G. Student, Hindustan Institute of Technology and Science, Chennai, India

[*]shalinirn@hindustanuniv.ac.in

Abstract. This paper illustrates the comparison of aerodynamic forces and coefficients on rectangular tall buildings. The forces and coefficients were obtained using Wind Tunnel Experiment and various international code provisions and standards. Rectangular tall building models selected for the study were made of acrylic sheets with various geometric scales. The models were tested using low-speed wind tunnel for various angles of incidents of wind varying from 0° to 90°. Drag coefficients obtained from experiments were compared with values of mean drag coefficients obtained from codes and practices, (a) IS: 875(Part3): Wind Loads on Buildings and Structures (India, 2015), (b) AS/NZS1170.2:2011, the combined Australia/New Zealand Standard on Wind Actions (Standards Australia,2011); (c) ASCE7–10 (ASCE, 2010); and (d) HKCoP-2019, the Code of Practice on Wind Effects in Hong Kong, (Buildings Department HongKong, 2019). Buildings with the same plan aspect ratio show the same aerodynamic behaviour. Wind tunnel study over-predicts the coefficient values compared to standards and practices. The average coefficient of drag values from codes is in good agreement with wind tunnel data with a deviation of less than 15%. The study also includes an examination of parameters such as pitching moment and its coefficient.

1. Introduction

The era of intelligent buildings has begun with the advancement of smart city concepts worldwide. Tall buildings serve as a solution for population issues and lack of land in most of the major cities. As this scenario increases, the structural concerns related to tall buildings have become more prominent in the areas of research. When the heights of the building were limited, the effect of aerodynamics on them could be approximated. As the height increases, the analysis on aerodynamic parameters needed a change from approximate to accurate predictions. In the initial development stages of analysis, the along wind and across wind responses of skyscrapers were made based on the random process and vibration theory[3]. The codes and standards followed the same during its initial formation years. With the use of wind tunnel experiments for studying the responses of tall structures, there has been a drastic change in the method of evaluation of response towards wind effects [1]. These developments brought revisions in the existing codes and practices. This paper discusses the wind-induced aerodynamic parameters for five selected models from wind tunnel study and compares the along-wind response from the wind tunnel study with various standards and practices. The study involves an investigation of lift, drag and torsional behaviour of selected models in the experimental study.

2. Experimental Study

2.1. Wind Tunnel

The experimental study was conducted in a low-speed wind tunnel at Hindustan Institute of Technology and Science Chennai. The wind tunnel has a test section of 600mm x 600mm and a length of 2m. The parts of the tunnel are: Bell mouthed section, honeycomb, settling chamber, contraction cone, test section, transition, diffuser, fan duct, motor and stand, 2 screens, 8 meshes and 16 mesh stainless steel (Figure1a). The test section has a proper window for viewing the specimen test section. The fabrication of the tunnel is made by using Teak Wood and waterproofing plywood.

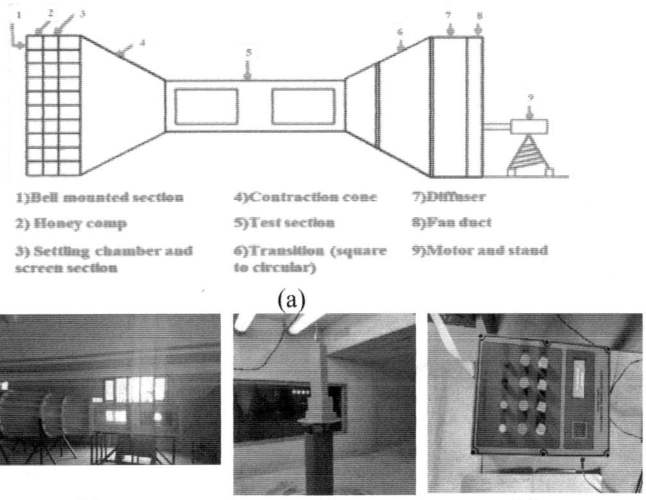

(a)

(b) (c) (d)

Figure 1. (a) Components in the wind tunnel, (b) View of wind tunnel in the laboratory, (c) Model mounted over dynamometer in the test section, (d) 3- component instrumentation for reading values.

The model to be tested is mounted over dynamometer (Figure 1c and Figure 4). Wind speed in the tunnel is controlled by varying the RPM. The lift, drag and moment values are displayed in 3-component instrumentation which is connected to a dynamometer.

2.2. Simulation of wind

The accuracy of wind tunnel experiments depends on the characteristics of wind flow inside the tunnel[2]. Here the wind is scaled suitably. The rectangular building models are subjected to simulated flow conditions to a reduced scale with respect to the model. Before the experimental study on five selected models, an experiment is conducted on the scaled down model of Empire State Building (Figure 2) as a validation. Coefficient of drag for Empire State Building being a known value, it was compared with results from the experiment to verify the accuracy of test. The value of the drag coefficient from the experimental study was compared with the actual drag coefficient of structure and found to be the same.

(a) (b) (c)

Figure 2. (a) Model of Empire State Building for calibration of test setup, (b) 3D drawing of model, (c) Model inside test section.

IOP Conf. Series: Materials Science and Engineering **936** (2020) 012054 doi:10.1088/1757-899X/936/1/012054

2.3. Models for Experiment
All five models used for the experimental study are made of acrylic sheets of 6mm thick (Figure 3). All models were subjected to wind flow at angles of attack of 0°,20°,40°,75°,90°. Details of models are given in table 1.

Table 1. Details of five models.

	Dimensions (mm)			Aspect Ratio	
	L	B	H	Plan Aspect Ratio $\frac{B}{D}$	Height Aspect Ratio $\frac{H}{D}$
Model 1	50	50	300	1.0	6.0
Model 2	33	33	300	1.0	9.1
Model 3	100	50	150	2.0	3.0
Model 4	50	50	150	1.0	3.0
Model 5	34.5	69	103.5	0.5	1.5

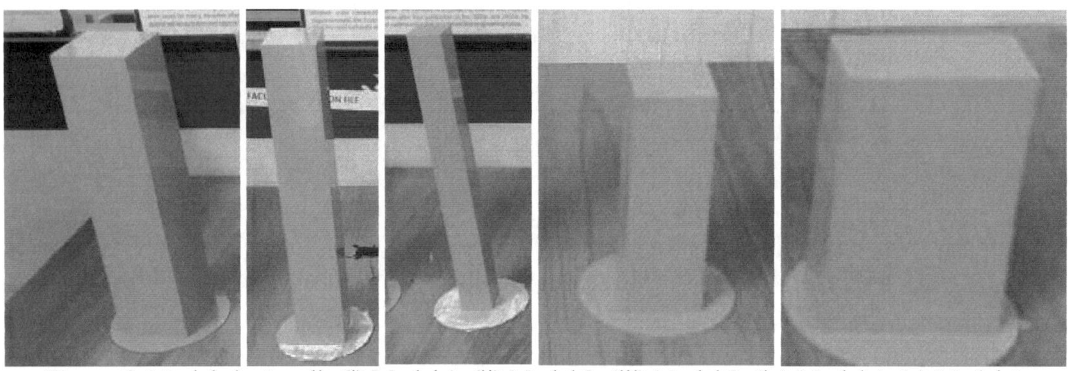

Figure 3. Models in Acrylic (i) Model 1, (ii) Model 2, (iii) Model 3, (iv) Model 4, (v) Model 5.

Figure 4. Models mounted on a dynamometer in the test section.

3. Data Analysis
The measured data from wind tunnel such as lift, drag forces and pitching moment are processed to obtain the coefficients of along wind, across wind and torsional behaviour of each models.

3.1. *Along wind response*

The measured drag force from the instrument is processed using the following relation to obtain the drag coefficient of the building[2].

$$C_d = \frac{D}{q_{\propto} B'} \tag{1}$$

where,

C_d : Coefficient of Drag
D : Measured Drag Force
q_α : Dynamic Pressure ($\frac{1}{2}\ \rho\ \vartheta^2$)
B' : Projected width normal to wind direction

3.2. *Across wind response*

The measured lift force from the instrument is processed using the following relation to obtain the lift coefficient of the building[2].

$$C_L = \frac{L}{q_{\propto} D'} \tag{2}$$

where,

C_L : Coefficient of Lift
L : Measured Lift Force
q_α : Dynamic Pressure ($\frac{1}{2}\ \rho\ \vartheta^2$)
D' : Projected width parallel to wind direction

3.3. *Torsional behavior*

The measured pitching moment from the instrument is processed using the following relation to obtain the pitching moment coefficient of the model[2].

$$C_M = \frac{M}{q_{\propto}\ B' D'} \tag{3}$$

where,

C_M : Coefficient of Pitching Moment
M : Measured Pitching Moment
q_α : Dynamic Pressure ($\frac{1}{2}\ \rho\ \vartheta^2$)
B' : Projected width normal to wind direction
D' : Projected width parallel to wind direction

3.4. *Comparison with codes*

Four standard design codes were used for the comparison of along wind response calculations of the wind-tunnel data, (a) IS: 875(Part3): Wind Loads on Buildings and Structures (India, 2015), (b) AS/NZS1170.2:2011, the combined Australia/New Zealand Standard on Wind Actions (StandardsAustralia, 2011); (c) ASCE7–10 (ASCE, 2010); and (d) HKCoP-2019, the Code of Practice on Wind Effects in Hong Kong,(Buildings Department HongKong, 2019).

4. Results and Discussions

4.1. *Aerodynamic forces and force coefficients*

Aerodynamic forces such as lift and drag obtained from the instrumentation of wind tunnel were analysed and the lift and drag coefficients are computed using equations (1) and (2). The Lift and Drag forces are plotted against the angle of attack as shown in figure 5. It can be noted that the along wind forces are more compared to across wind forces. The along wind component tends to increase with the increase in angle of incident wind, whereas the across wind component shows an increase between 0° to 20°. In all five models, the results exhibit that critical angle of attack is between 20° to 40°. From 40° to 90° it has a tendency to reduce. It can also be noted that the L/D ratio has a specific pattern with respect to the angle of attack as the aspect ratio changes (Figure 6). From the graph, it is evident that the buildings with the same plan aspect ratio, i.e. Model 1, 2 and 4 shows same pattern of behaviour. Thus, analysing this will be helpful to optimise the shape of a building in the preliminary design stage.

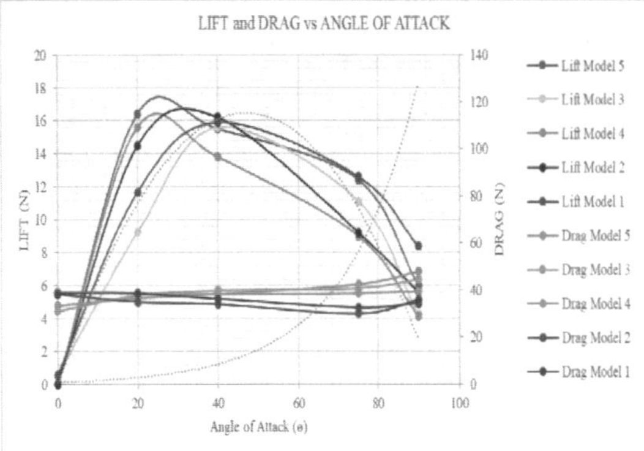

Figure 5. Variation of Lift and Drag with Angle of attack – all five models.

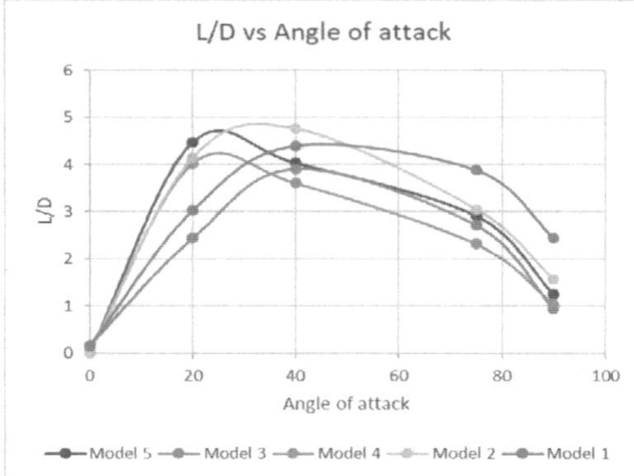

Figure 6. Variation of L/D ratio with angle of attack.

4.2. Torsional Behavior

From the obtained pitching moment through experiment, the pitching moment coefficient is calculated using equation (3)[2]. The behaviour of moment with respect to angle of incident wind was plotted (Figure 7). Here it can be observed that the coefficient of moment has tendency to increase in the areas where there is a notable increase in lift force and reduces at the areas where lift forces decrease. It can be said that the torsional moment will affect the structure when there is a combined action of along wind and across wind forces.

The coefficient of pitching moment and variation of pitching moment for model 2 has a mismatch with the other models. This is due to the higher aspect ratio (9.09), i.e. the aspect ratio is more than the preferable aspect ratio of 6.

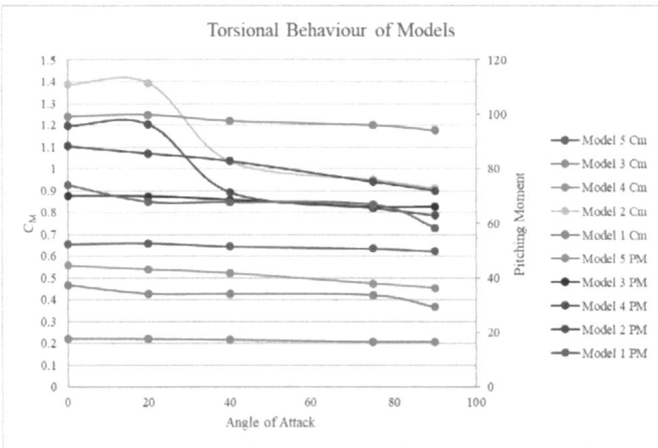

Figure 7. Variation of Pitching Moment and coefficient of pitching moment with Angle of attack

4.3. Comparison of along wind Response with standards and codes

On comparison of the coefficient of drag from wind tunnel experiments with the four codes and practices namely, (a) IS: 875(Part3): Wind Loads on Buildings and Structures (India, 2015), (b) AS/NZS1170.2 :2011, the combined Australia/New Zealand Standard on Wind Actions (StandardsAustralia, 2011); (c) ASCE7–10 (ASCE, 2010); and (d) HKCoP-2019, the Code of Practice on Wind Effects in Hong Kong, (Buildings Department HongKong, 2019), it is clear that the wind tunnel study over predicts the coefficient values compared to standards and practices (shown in table 2, figure 8). Codes and Standards do not consider Ekman effect while analysing solid air interaction whereas in wind tunnel experiments the dynamic nature of wind including all its features are taken into account. Due to this, it is observed that there is a general nature of the variation of coefficient of drag from experimental study compared to the one obtained from Codes and Standards.

Table 2 Comparison of along wind Response with standards and codes.

Building Model	Plan Aspect Ratio (B/D)	Height Aspect Ratio (H/D)	Effective Drag coefficient (Cd)				
			Wind Tunnel Experiment	IS 875: 2015	AS/NZS/ ASCE7	Hong Kong Code of Practice 2019	Average of Codes
Model 1	1.0	6.0	1.325	1.40	1.425	1.325	1.4
Model 2	1.0	9.1	1.544	1.50	1.601	1.410	1.5
Model 3	2.0	3.0	1.425	1.20	1.600	1.270	1.4
Model 4	1.0	3.0	1.601	1.30	1.700	1.210	1.4
Model 5	0.5	1.5	1.380	1.15	1.410	1.200	1.3

From table 2 it is evident that the average of standard predictions is close to wind tunnel data for drag and the average of coefficient of drag values from codes are in good agreement with wind tunnel data with a deviation less than 15%.

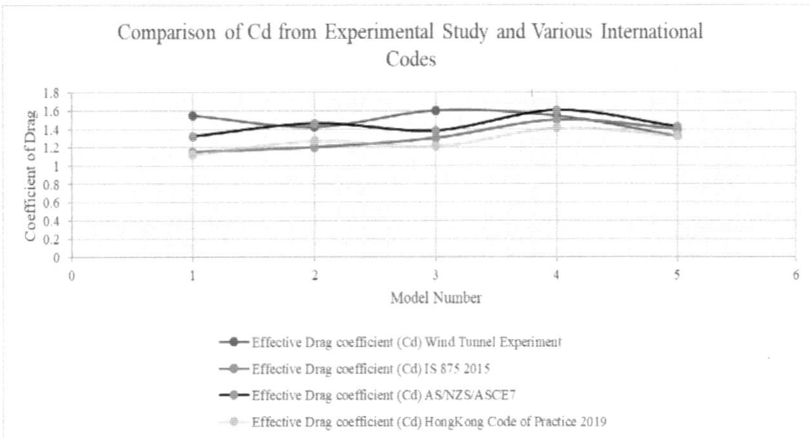

Figure 8. Comparison of Coefficient of Drag from Experimental Study and Various International Codes and practices

5. Conclusions

The study was carried out on five selected models and comparison of results from wind tunnel tests and various codes and standards were made. From the study, the following conclusions have been derived.

- The along wind component has a tendency to increase with the increase in angle of incident wind.
- The across wind component shows an increase between 0° to 20°. In all five models, the results exhibit that the critical angle of attack is between 20° to 40°.
- Buildings with the same plan aspect ratio show the same aerodynamic behavior. Analysing the aspect ratio will be helpful to optimise the shape of a building in the preliminary design stage.
- The model with a higher aspect ratio of 9.1 exhibited variation in behaviour of pitching moment and its coefficient compared to those models which has a lesser aspect ratio. Hence optimising aspect ratio for a project is important during initial stages in design of tall structure.
- On comparison of the coefficient of drag from wind tunnel experiments with codes and practices, it is clear that the wind tunnel study over predicts the coefficient values compared to standards and practices. The average of coefficient of drag values from codes are in good agreement with wind tunnel data with a deviation of less than 15%.
- Codes and standards do not consider Ekman effect while analysing solid air interaction whereas in wind tunnel experiment the dynamic nature of wind including all its features are taken into account. Due to this, it is observed that there is a general nature of the variation of coefficient of drag from experimental study compared to the one obtained from Codes and Standards.

IOP Conf. Series: Materials Science and Engineering **936** (2020) 012054 doi:10.1088/1757-899X/936/1/012054

6. References

[1] Shalini R Nair, Daniel Nayagam J and Jessy Rooby 2018 A solution to enhance the efficiency of CFD analysis results for real time usage *International Journal of Civil Engineering and Technology (IJCIET)* **9(4)** 731–6

[2] H Sarath Kumar, S Selvi Rajan, A Joseph Andrew and G Ramesh Babu 2016 Aerodynamic coefficients for a rectangular tall building under sub-urban terrain using wind tunnel *Asian Journal Of Civil Engineering (Bhrc)* **17(3)** 325–33

[3] John D Holmes 2014 Along- and cross-wind response of a generic tall building: comparison of wind-tunnel data with codes and standards *Journal of Wind Engineering and Industrial Aerodynamics* 136–41

[4] Yin Zhou, Trac Kijewski, S M and Ahsan Kareem 2003 Aerodynamic loads on tall buildings: interactive database *International Journal of Structural Engineering* **129(3)** 394–404

[5] Shalini R Nair and Shyam R Nair 2014 Study and comparison of aerodynamic forces on an aircraft and suspension bridge *IOSR Journal of Engineering* **4(4)** 43–6

[6] B Samali, K C S Kwok, G S Wood and J N Yang 2005 Wind tunnel test for wind-excited benchmark building *Journal of Engineering Mechanics* **130(4)** 447–50

[7] Ahmed Elshaer 2017 Enhancing wind performance of tall buildings using corner aerodynamic optimaisation *Engineering Structures* **136** 133–48

[8] Jun Yi and Q S Li 2015 Wind tunnel and full scale study of wind effects on a super-tall building *Journal of Fluid and Structures* 236–53

[9] Ahmed Aslam 2016 Effects of across-wind response on tall slender towers of circular cross section *ASCE Structures Congress* 1–9

[10] Bharat S Chauhan and Ashok K Ahuja 2018 Modifications of wind power distribution on tall buildings due to variation in height of interfering building *ASCE India Conference* 424–34

IOP Conf. Series: Materials Science and Engineering **936** (2020) 012055 doi:10.1088/1757-899X/936/1/012055

Seismic Performance of High Rise Buildings with Different Types of Shear Wall

Vineeth Vijayan[1*], M Helen Santhi[2] and Romy Mohan[3]

[1] PG Student, Division of Structural Engineering, School of Civil Engineering, Vellore Institute of Technology, Chennai 600127, India

[2] Professor, Division of Structural Engineering, School of Civil Engineering, Vellore Institute of Technology, Chennai 600127,India

[3] Structural Engineer,WVA Consulting Engineers Pvt.Ltd.,Cochin 682311,India

*vineethvijayan094@gmail.com

Abstract. In today's construction world, aesthetic tall structures play a major role. The design of aesthetic high - rise buildings need to be checked for its safety against extreme load conditions such as earthquake, wind, etc. The provision of composite shear wall is one of the efficient methods to make high rise buildings seismic resistant. In this paper different types of shear walls, namely, concrete shear wall, silica fume concrete shear wall, steel plate shear wall and steel-silica fume concrete composite shear wall are considered for lift wall in 22 and 52 storeyed high-rise buildings. The seismic performance of these buildings is analysed using response spectrum method by ETABS. The factors such as storey displacement, storey drift and storey shear are studied and found that there is a significant reduction when compared to the conventional shear wall.

1. Introduction

At present, the construction world moves towards the execution of tall buildings with vertical and horizontal irregularities. Generally, regular and uniform buildings are preferred as they resist earthquakes better than irregular structures. For the construction of lift wall in the medium and tall structures, RCC is being used worldwide. In the recent past, RCC lift walls are replaced with composite materials which are having better strength and stiffness properties [1 - 6]. While going for the structural design of high-rise buildings for seismic loading it must be checked for seismic resistance to avoid loss of life and economic loss due to earthquake. In this project, an attempt has been made to investigate analytically the behavior of composite shear wall as a lateral-load resisting system in comparison to other types of normal shear walls. Different types of shear walls such as concrete shear wall, silica fume concrete shear wall, steel plate shear wall and steel-silica fume concrete composite shear wall are considered. The composite shear wall is made with steel plate inside and with reinforced silica fume concrete outside. Using ETABS software the shear walls are modeled in 22 and 52 storeyed building and their behavior are compared using response spectrum analysis as specified in IS 1893:2016 [7]. The factors such as storey displacement, storey drift and storey shear of each building are plotted and are compared.

2. Objective

- To analyse the high rise buildings with different types of shear wall using response spectrum method.
- To compare the seismic performance of high rise building with shear wall under study.

3. Structural modeling and analysis

For the analysis, a G+21, G+ 51 and G +20 storeyed RCC framed buildings irregular in plan are modeled in ETABS software. The plan of both G +21, G + 51 is similar; but the number of storey is different (Figure 1a). The plan of the third building, G + 20 storey is different from the other two buildings as shown in figure 1b. The details of the building are given below.

The floor height is taken as 3 m; thickness of the shear wall is 250 mm; the size of rectangular columns is 0.3m x 0.8m,0.3m x 0.6m; the size of circular columns is 0.6 m dia; the size of the beam is 0.25 m x 0.45m, 0.25m x 0.30m; thickness of slab is 120 mm; materials used are M25 concrete, Fe 415 &Fe 500 HYSD bars, silica fume concrete, Fe 250 steel plate; soil type is type 2; seismic zone is IV. For modeling steel plate-silica fume concrete composite shear wall, the stress-strain properties of silica fume concrete are taken from the experimental work done by Vindhya and Suresh Babu (2016). Response spectrum analysis as specified in [7] is carried out to study the seismic performance of the buildings.

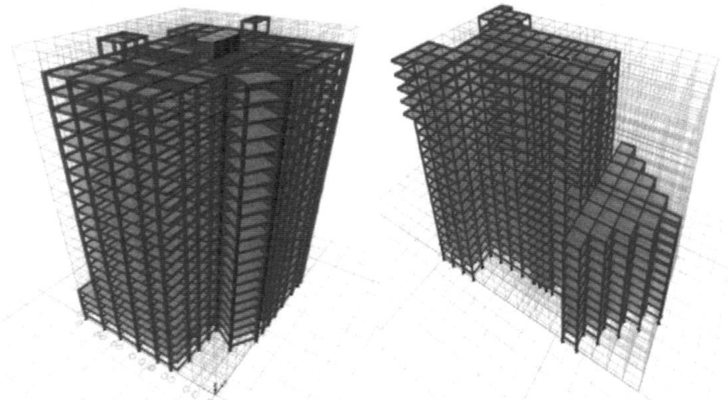

Figure 1. 3D view of buildings under study.

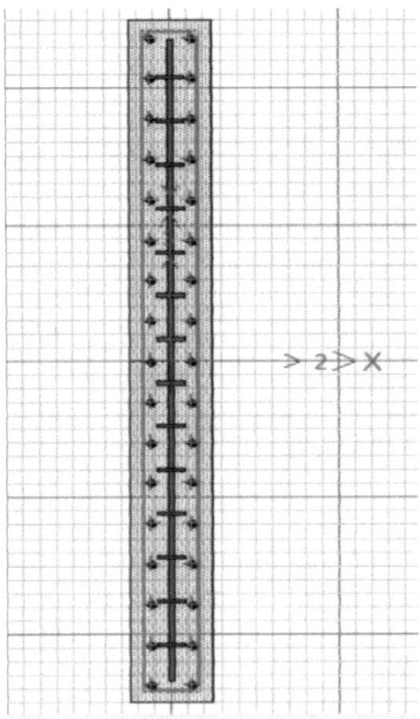

Figure 2. Modeling of steel plate-silica fume concrete composite shear wall.

4. Results and Discussion

4.1. Response of the G +21 storeyed building

Figure 3 shows the storey drift in X and Y direction due to seismic loads obtained from response spectrum analysis of 22 storeyed building with different types of shear walls. In both RESP-X and RESP-Y there is a large reduction in storey drift using composite shear wall over other types of shear walls considered in this project. The storey drift is slightly more in Y-direction than X-direction; however, the storey drift is within the limit in both directions.

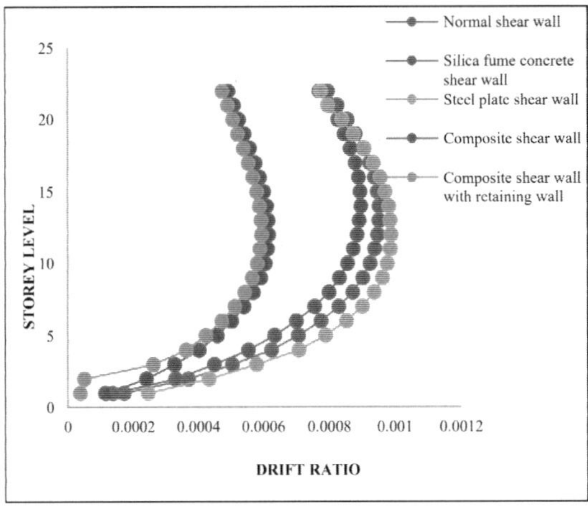

Storey drift (RS-X)

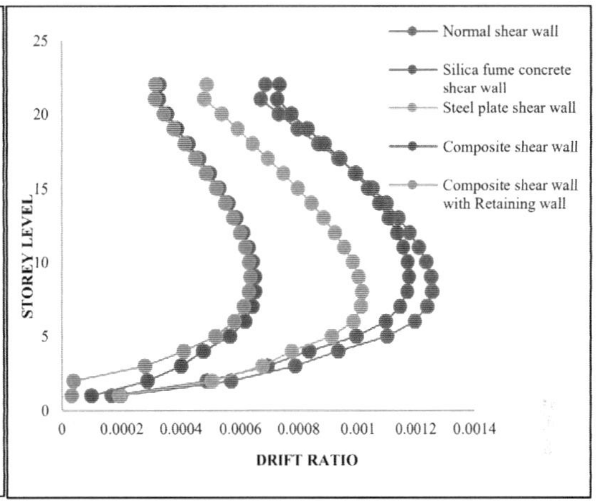

Storey drift (RS-Y)

Figure 3. Storey drift of G +21 storeyed building.

Figure 4 shows the reduction in displacement value by providing composite shear wall when compared with normal RCC shear wall, silica fume concrete shear wall and steel plate shear wall. Around 33 % reduction in displacement by providing steel-silica fume concrete composite shear wall is observed.

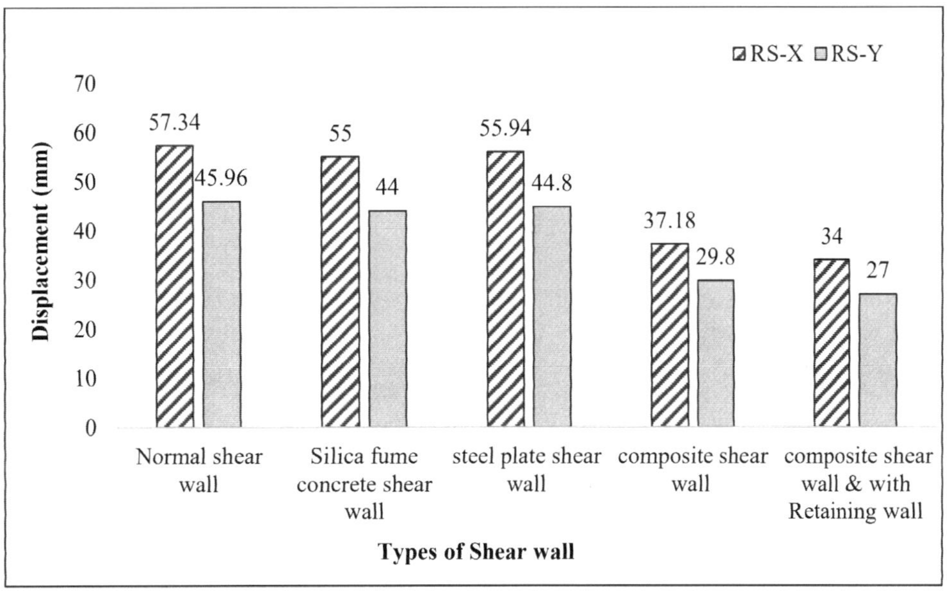

Figure 4. Max storey displacement of G + 21 storeyed building.

In figure 5, storey shear is plotted for different types of buildings in X & Y directions. It is seen that the storey shear is reduced in composite type of buildings when compared to the conventional and other shear wall type buildings. Almost 50% reduction in storey shear is noticed using composite shear wall and this will in turn reduces the demand for bending and shear in the structural elements.

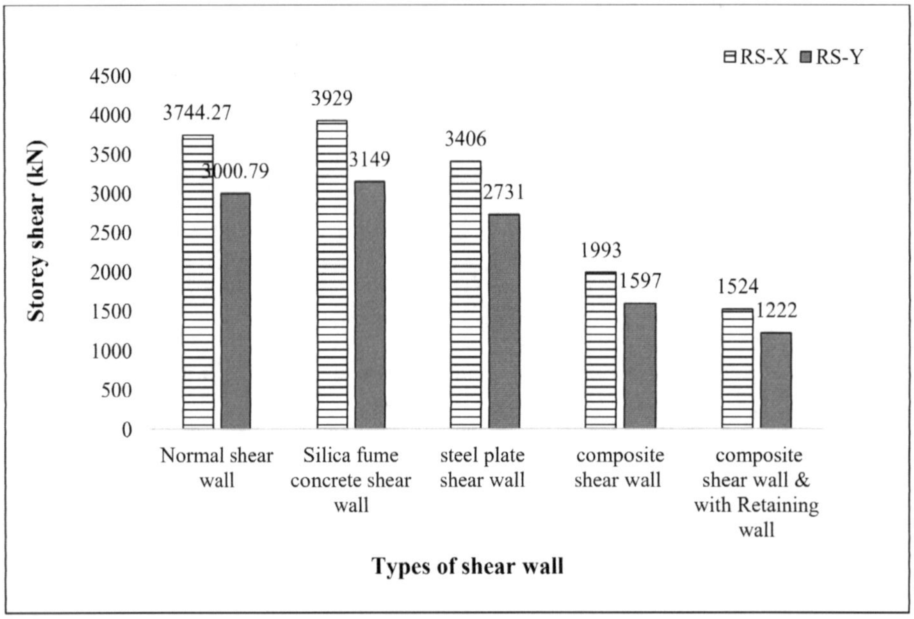

Figure 5. Max storey shear of G +21 storeyed building.

4.2. Response of G + 51storeyed building.

Figure 6 represents the storey drift RESP-X & RESP-Y of the same building when the storey number is increased into 52. Hence the storey number is increased. Also there is a large reduction in storey drift by providing composite shear wall. From the graph it is evident that composite shear wall can play a major role in the reduction of storey drift during earthquake. It will help to reduce the seismic effect on tall buildings.

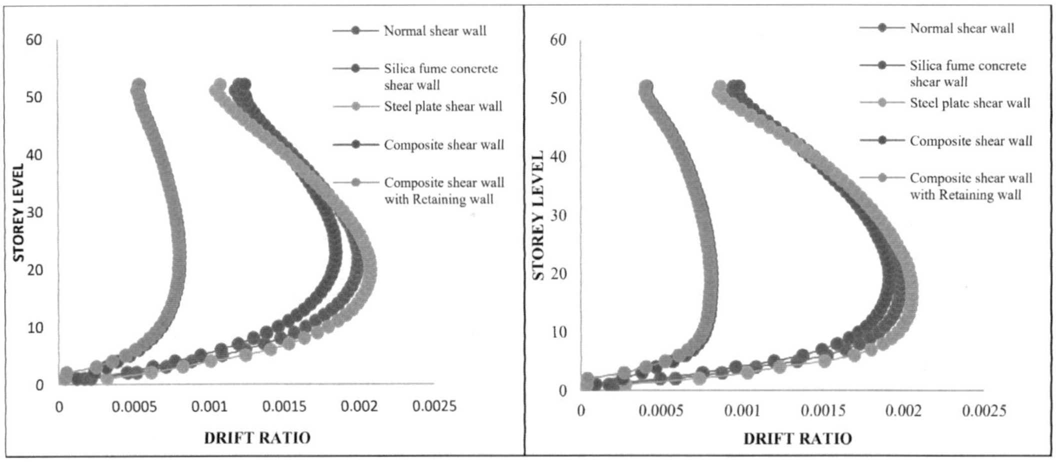

Figure 6. Storey drift of G + 51 storeyed building.

There is a large reduction in displacement value by providing composite shear wall in 52 storeyed building. From the graph plotted in figure 7, it is clear that almost 60% reduction in displacement occurs while providing composite shear wall instead of normal shear wall.

In figure 8, it is evident that there is a large reduction in storey shears by providing composite shear wall (A reduction of storey shear from 3806 kN to 1353 kN).

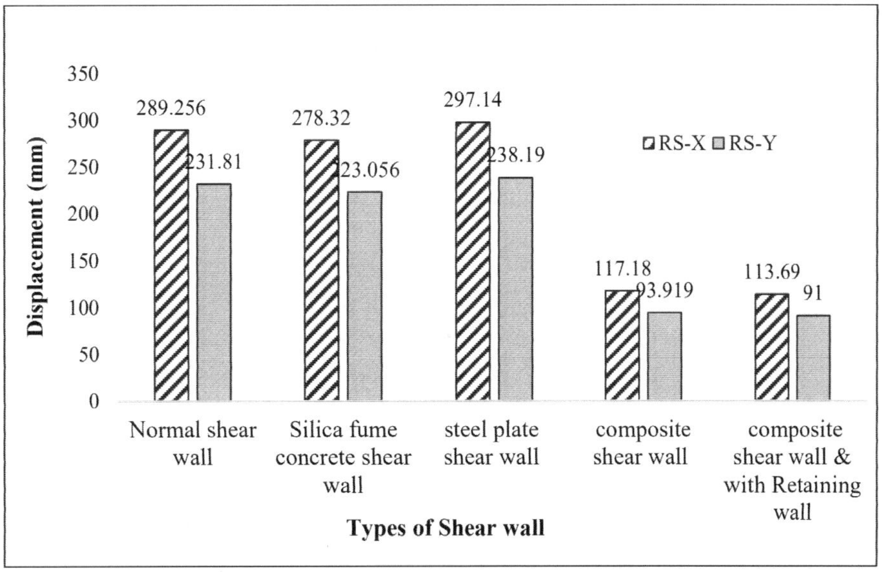

Figure 7. Max storey displacement of G + 51 storeyed building.

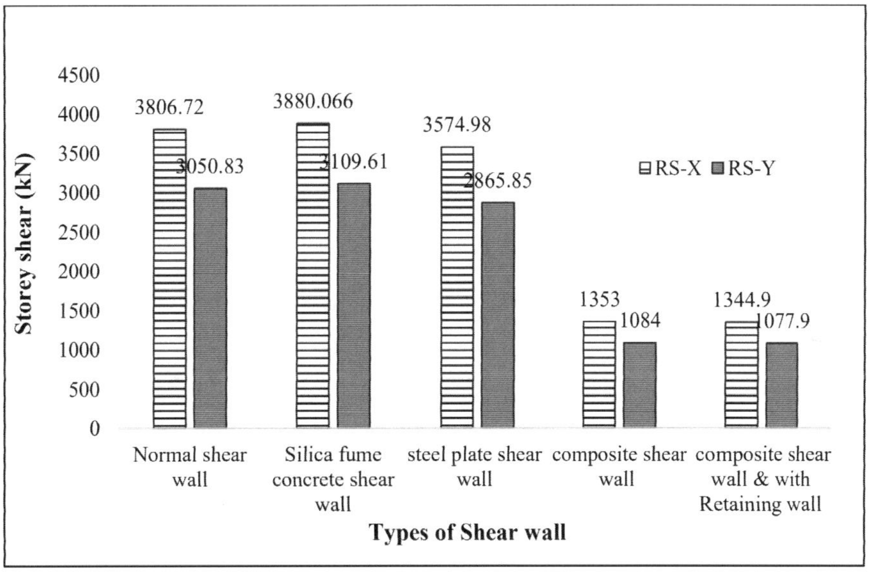

Figure 8. Max storey shear of G + 51 storeyed building.

4.3. Response of G+20 storeyed building.

In this case, building consists of more number of cantilever portions and is irregular in both plan and elevation. Figure 9,10 and 11 represent the response of building subjected to seismic load in both X and Y directions. In figure 9, silica fume concrete shear wall has less storey drift than steel plate and normal shear wall. While comparing with normal shear wall, composite shear wall reduces the drift in each floor to a large extent. Hence while going for aesthetic appearance, composite shear wall would be the best choice to reduce the seismic effect on tall structures.

Figure 10 shows the displacement of the building when different types of shear walls are provided. Here steel plate and normal shear wall have almost the same displacement. By providing silica fume concrete shear wall, displacement is reduced from 83.7 mm to 78.6 mm in RESP-Y and 42 to 37.8 mm in RESP-X. Hence silica fume concrete is effective in reducing displacement due to seismic load. While comparing with normal shear wall, it is evident that composite shear wall can reduce the displacement due to seismic load by 60%.

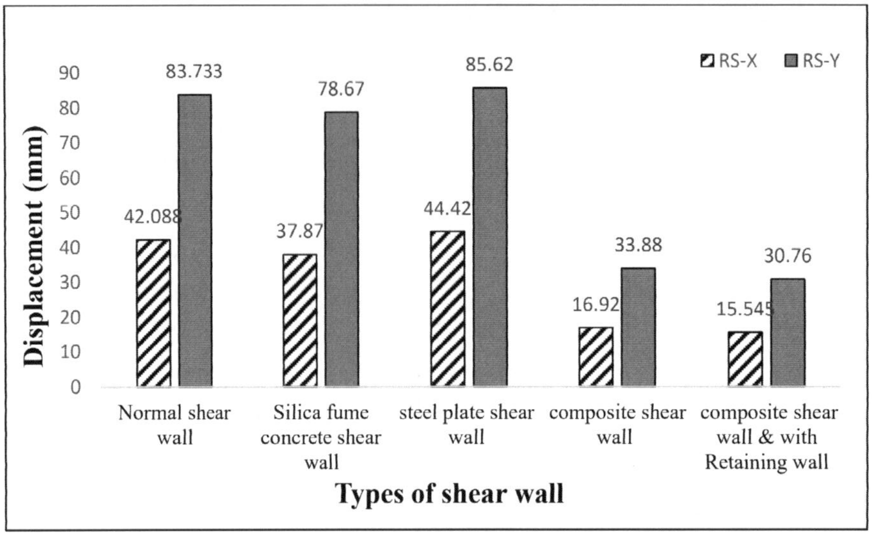

Figure 10. Max storey displacement of G+20 storeyed building.

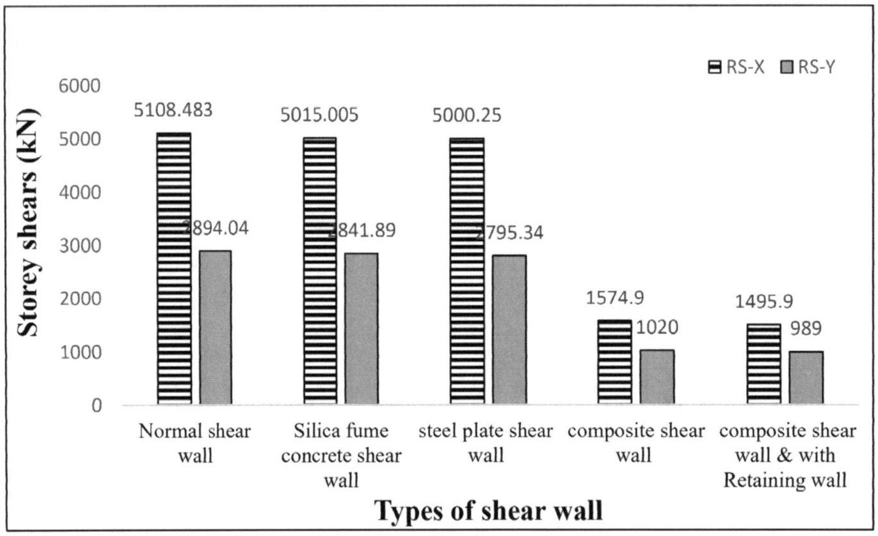

Figure 11. Max storey shear of G+20 storeyed building.

From figure 11, it is clear that by providing composite shear wall in aesthetic tall building, storey shear will also reduce considerably (here it reduces from 5000 kN to 1574 kN).

5. Conclusions

Providing steel-silica fume concrete composite shear wall instead of normal RCC shear wall, in aesthetic tall buildings, the building becomes more seismic-resistant since it provides attractive structural properties in regions. The advantages provided to the structure due to the provision of composite shear wall are.

1. A significant reduction in the storey drift of the aesthetic tall building due to composite shear wall compared to the normal RCC shear wall. The storey drift is also reduced when number of storey increased in building with composite shear wall.
2. While comparing with normal shear wall it is evident that composite shear wall can reduce the seismic effect to a larger extent because nearly 60% reduction in displacement occurs by providing composite shear wall.
3. The study reveals that the composite shear wall has a major role in the reduction of storey shear.

6. References

[1] Reslan N and Masri A 2018 Composite shear walls an efficient seismic resistant system for multi-story buildings *J. Steel Struct. Constr.* **4(140)** 2472–0437

[2] M Vidhya and T Suresh Babu 2016 Utilisation of silica fume & steel slag in concrete as a partial replacement of cement and coarse aggregate *International Journal of Applied Research* **29** 566–78

[3] Zhenya Huang and J Y Richard Liew 2016 Structural behavior of steel-concrete-steel sandwich composite wall subjected to compression and end moment *Thin-Walled Structures* **98** 592–606.

[4] Nam H Nguyen and Andrew S Whittaker 2017 Numerical Modelling of steel-plate concrete composite shear walls *Engineering structures* **150** 1–11

[5] Wei Wang and Yingzi Ren 2019 Experimental study of the hysteretic behaviour of corrugated steel plate shear walls and steel plate reinforced concrete composite shear walls *Journal of Constructional Steel Research* **160** 136–52

[6] Wenyuan Zhang and Ke Wang 2019 Experimental study of the seismic behaviour of composite shear walls with stiffened steel plates and infilled concrete *Thin-walled Structure* **144** 106279

[7] IS:1893(Part1)2016 Indian Standard *Criteria for earthquake resistant design of structures* (NewDelhi)

AUTHOR INDEX